Sensors & Their Applications XVII 2013

AA008649

Journal of Physics: Conference Series Volume 450

Dubrovnik, Croatia
16-18 September 2013

ISBN: 978-1-62748-678-1
ISSN: 1742-6588

Printed from e-media with permission by:

Curran Associates, Inc.
57 Morehouse Lane
Red Hook, NY 12571

Printed by Curran Associates, Inc. (2013)

Institute of Physics
Dirac House, Temple Back
Bristol BS1 6BE UK

Phone: 44 1 17 929 7481
Fax: 44 1 17 920 0979

techtracking@iop.org

Additional copies of this publication are available from:

Curran Associates, Inc.
57 Morehouse Lane
Red Hook, NY 12571 USA
Phone: 845-758-0400
Fax: 845-758-2634
Email: curran@proceedings.com
Web: www.proceedings.com

TABLE OF CONTENTS

INVITED CONTRIBUTIONS

OPTICAL SENSORS

CHEMICAL AND GAS SENSORS

SMART SENSORS AND INTERFACE ELECTRONICS

NANOTECHNOLOGY FOR SENSORS AND ACTUATORS

SENSOR MODELLING

SENSORS IN BIOLOGY AND MEDICINE

THICK AND THIN FILM SENSORS

IMAGING AND INTEGRATED ACTUATORS

GENERAL

PREFACE

This volume records the Proceedings of the seventeenth conference in the biennial Sensors and Their Applications series that took place at Rixos Libertas, Dubrovnik, Croatia from 16–18 September 2013. The conference is organised by the Instrument Science and Technology Group of the Institute of Physics. The conference was the first organised by the Institute of Physics to be held outside of the UK and Ireland, thus continuing the collaborative and adventurous nature of the meeting.

The conference proceedings record the continuing health, diversity and activity of the sensors community worldwide, bringing together contributions from academics and industrial researchers to provide excellent networking opportunities. It is interesting to note some continuing themes such as Optical Sensors and Electromagnetic Sensors, as well as trends in Environmental Sensing and Glacial Monitoring that reflect our changing world, and Sensors in Biology and Medicine that have a growing importance with an ageing population. The conference also accounts for research specialisms and unique strengths from the local community in Croatia, including demining and metal detector sensing.

We should like to thank all of our colleagues and friends in the sensor community who have supported this event by contributing manuscripts. Our thanks go also to members of the Technical Programme Committee for their support, and in particular for refereeing the submitted manuscripts. We are also pleased to express our thanks to the Conference Department of the Institute of Physics for their invaluable support in organising this event. We are especially grateful to Dawn Stewart for her responsive and day-to-day handling of this conference, as well as Claire Garland for help in planning and managing this international event.

We hope that the conference authors, participants and a wider audience will find these proceedings to be of interest and to serve as a useful reference text.

V Bilas, G McConnell and P Kyriacou

Organising Committee

Conference Organising Committee

V Bilas, conference chair, University of Zagreb, Croatia
G McConnell, conference chair, University of Strathclyde, UK
P Kyriacou, conference chair, City University London, UK
D Stewart, conference co-ordinator, Institute of Physics, UK

Technical Programme Committee

L Benini, University of Bologna, Italy
M Butta, Czech Technical University in Prague, Czech Republic
M Cifrek, University of Zagreb, Croatia
G Collier, Kingston University London, UK
J Deur, University of Zagreb, Croatia

H Dzapo, University of Zagreb, Croatia
M Gasulla, Universitat Politenica de Catalunya, BarcelonaTech, Spain
S Hadjiloucas, University of Reading, UK
P Kyriacou, City University of London, UK
I Lackovic, University of Zagreb, Croatia
R Magjarevic, University of Zagreb, Croatia
G McConnell, University of Strathclyde, UK
A O'Riordan, Tyndall National Institute, UK
K Ozanyan, University of Manchester, UK
A Peyton, University of Manchester, UK
S Reilly, National Physical Laboratory, UK
T Sun, City University London, UK
A Tickle, Coventry University, UK
D Vasic, University of Zagreb, Croatia
S Welch, ESPKTN, UK
Y Yan, University of Kent, UK
H Zangl, Technical University of Graz, Austria

Peer review statement

All papers published in this volume of *Journal of Physics: Conference Series* have been peer reviewed through processes administered by the proceedings Editors. Reviews were conducted by expert referees to the professional and scientific standards expected of a

Security applications of magnetic sensors

Pavel Ripka

Czech Technical University, Faculty of Electrical Engineering,
Technicka 2, 166 27 Praha 6, Czech Republic
ripka@fel.cvut.cz

Abstract. Magnetic sensors are often used for security and military applications such as detection, discrimination and localization of ferromagnetic and conducting objects, navigation, position tracking and antitheft systems. We give only general overview, few remarks and some interesting references on these applications.

1. Introduction

Magnetic sensors are key elements in many security and military systems. Traditional sensors such as fluxgates, induction coils and resonance magnetometers are complemented by new sensor types such as AMR (Anisotropic MagnetoResistors), GMR (Giant Magneto-Resistance), SDT (Spin-Dependent Tunelling) and GMI (Giant Magneto-Impedance) sensors [1]. The aim of this paper is to shortly review the mentioned applications, show few examples and give references to interesting papers.

The main requirements for the sensors in critical applications are:

- reliability
- high sensitivity
- selectivity (low false alarm rate)
- resistance against overload and vandalism
- self-adaptability (e.g. for changing background signal)
- self-testing features
- environmental resistance including radiation [2]
- some applications require small size, weight and power consumption
- price is of concern mainly in multi-sensor systems

2. Detection of ferromagnetic objects

DC magnetic detectors usually use the distortion of the Earth's field caused by the permeability of the ferromagnetic object ("soft component" of magnetization). The ferromagnetic object may also have some permanent magnetization ("hard component"), but it is not predictable, unless the object is intentionally marked by permanent magnets.

This technique can be effectively used for objects in larger distances from the sensor. Either the sensor is moving over the buried object such as UXO (Unexploded ordnance) [3], or the stable sensor detects moving target such as vehicle [4].

If the detected object is small and close to the sensor, it is often more effective to use active field generator instead of the passive method which rely on the Earth's field.

2.1. Detectors using the Earth's field.
The Earth's field has a magnitude of 30 to 60 μT and direction from horizontal (in the equatorial region) to vertical (at the poles). The practical detection limit is 1 nT, which corresponds to signature of 1 kg iron object in 3 m distance or 1000 kg object in 40 m distance. The challenge is to detect such small field anomaly under the field variations (man-made or natural), which may reach hundreds of nT. This is usually solved by using gradient sensors or more complex sensor fields. Localization of the target and estimation of its magnetic moment under uncertainty is a complex task [5].

Another sources of errors arise when the measurement is made by 3-axial vector sensors from the moving platform: recalculation into geographical reference system is possible only when the sensors are very linear and orthogonalized. Resonance magnetometers are resistant to this type of error, but they provide neither directional field components nor their gradients which are important for target localization [6].

2.2. Detectors using artificial magnetic field.
This technique is used to detect small objects close to the sensor. In such case the artificially generated field can be much stronger than the Earth's field. The generated field is usually low-frequency bipolar squarewave, which allows suppressing the background Earth's field. Using low frequency decreases the influence of conducting objects in the vicinity.

Magnetic imaging systems are used to scan magnetic ink on banknotes and important documents. Some secure paper has embedded magnetic particles or pieces of magnetic wires [7].

3. Detection of conducting objects using eddy currents
Eddy current metal detectors utilize excitation (primary) field in the frequency range of 1 kHz to 100 kHz. The two main types are continuous-wave (CW) and pulse detectors. The magnetic field sensor measuring the secondary field caused by the eddy currents is traditionally an induction coil [8], recently also AMR sensors are employed [9]. The major task is to suppress the large excitation field in order to measure the small secondary field. Pulse detectors utilize time discrimination: the measurement period starts after the decay of the excitation field. CW detectors use gradient coil system or the detection coil just perpendicular to the excitation coil. Similar techniques are used both for the portable metal detectors and detection frames for access control.

3.1. Time-domain detectors
In the early years the pulse detectors with time-domain signal processing were popular due to their natural ability to discriminate spurious response of "difficult" soil caused by magnetic particles. The magnetic field from these particles decays much faster than the response from the circulating eddy currents. The disadvantage is that the time domain signal is susceptible to deformation by spurious capacitances which change with changing humidity and temperature. More sophisticated methods of signal processing are therefore very difficult to implement. Typical pulse duration is 400 μs with 200 Hz repetition rate. The eddy-current response is monitored between 10 μs and 1 ms after the decay of the excitation pulse.

3.2. CW metal detectors
These devices allow processing the secondary field in the frequency domain using synchronous detectors. Modern CW detectors have multi-tone excitation in order to acquire more information about the measured object. The amplitude and phase at each frequency depends on the conductivity, permeability and geometry of the target and complex information allows discriminating false targets such as metal clutter and magnetic rocks. The signal from soils containing superparamagnetic particles with magnetic viscosity can be suppressed by measurement of the frequency characteristics [10].

3.3. Eddy current imaging systems

Imaging systems allow even more effective target discrimination. Line plots or 2-D or even 3-D magnetic maps contain images which can be recognized by comparing them to the catalogue of standard images of known targets such as landmines [11]. Creating the map needs precise position information which is non-trivial task. Many imaging systems use sensor fields in order to reduce the necessary scanning.

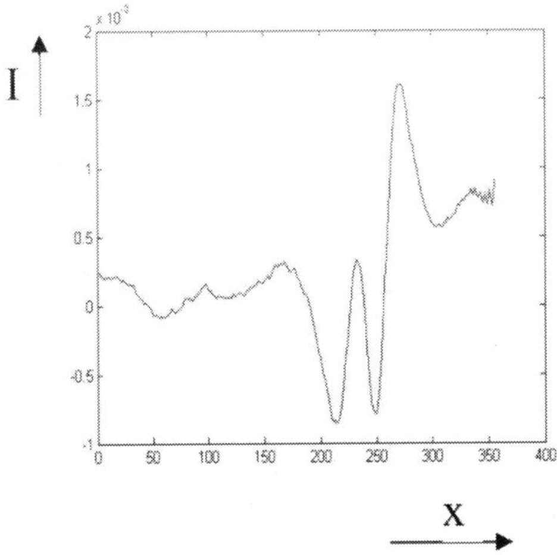

Fig. 1: Example of intensity line plot of landmine [12]

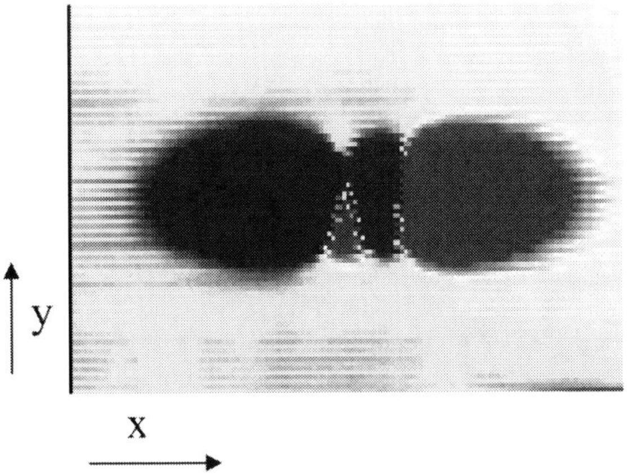

Fig. 2: Example of 2-D intensity map of landmine [12]

4. Navigation using the Earth's field

Modern magnetic compass is electronically gimbaled: it uses 3-axial magnetic sensor and two inclinometers to calculate azimuth in any position. The precision achievable with fluxgate sensors is well below 0.1° and it depends mainly on the precision of the inclinometer. Compasses using AMR or GMI sensors usually have 1° precision [13].

Some advanced military navigation systems utilize also local anomalies of the Earth's field.

5. Position tracking and local navigation using artificial field

These devices are used to precisely monitor the local position and orientation of important objects such as helmet of the pilot, or for underwater or indoor navigation [14]. The system usually consists of stationary field source in the form of coils or long current conductors. The moving object carries either set of magnetic sensors or passive markers.

In the first case calculation of the position and rotation angles requires 3-axial magnetic sensor and multiple independent field sources which can be distinguished either by time or frequency multiplex. This can be reduced to a single field source if the information from other sensor types such as inclinometers or gyrocompass is used.

Passive resonant targets are monitored by stationary magnetic sensors. Using sensor field allows increasing the detection range while keeping high spatial resolution. This technology is also used by antitheft systems using magnetic markers. The main types of magnetic markers are

- LC tanks utilizing electrical resonance [15]
- Magnetoelastic tapes and wires utilizing mechanical resonance [16]
- Wiegand wires utilizing large domain jump

All these effects are easily detectable by remote coil.

Some older antitheft and tracking systems utilize permanent magnet and DC magnetic sensors such as AMRs or fluxgate [17], [18]. 1cm^3 /5g NdFeB magnet has magnetic moment of about 5Am^2, which gives 500 nT field in 1 m distance [19]. The range and accuracy of such system is therefore strongly limited.

6. Conclusions

Magnetic sensors are essential elements for many security applications as magnetic field is invisible and penetrates through most of materials. The Earth's field is omnipresent and cannot be switched off, so it gives backup to all navigation systems.

References

[1] Ripka P, Janosek M: Advances in Magnetic Field Sensors, IEEE Sens. J. 10 (2010) Issue: 6 , 1108-1116
[2] O' Brien H., P. Brown, T. Beek, C. Carr, E. Cupido, T. Oddy: A radiation tolerant fluxgate magnetometer, Meas. Sci. Tech. 18 (2007), 3645-3650.
[3] Ripka P., A. M. Lewis, P. Kaspar, and J. Vcelak: DC Gradiometers for Bomb Detection: Scalar Versus Vectorial, Sensor Lett. 5, 271–275 (2007)
[4] Kang, MH Choi, BW Koh, KC Lee, JH Park, GT: Experimental study of a vehicle detector with an AMR sensor Sens. Actuator A-Phys. 118 (2005) pp. 278-284.
[5] Sheinker A, Lerner B, Salomonski N, et al.: Localization and magnetic moment estimation of a ferromagnetic target by simulated annealing, Meas. Sci. Technol. 18 (2007) 3451-3457.
[6] Nara Takaaki, Satoshi Suzuki, and Shigeru Ando: A Closed-Form Formula for Magnetic Dipole Localization by Measurement of Its Magnetic Field and Spatial Gradients, IEEE Trans. Magn. 42 (2006), 3291
[7] Vázquez M and A Hernando: A soft magnetic wire for sensor applications J. Phys. D: Appl. Phys. 29, Number 4, 1996, 939-951
[8] Tumanski S.: Induction coil sensors, Measurement Sci Technol 18, 2007, R31-R46
[9] Tumanski S, Thin film magnetoresistive sensors, IOP (2001)
[10] Ripka P, Lewis AM, Kubik J: Mine detection in magnetic soils, Sens. Lett. 5 (2007),15-18
[11] Nováček, P. - Roháč, J. - Šimánek, J. - Ripka, P.: Metal Detector Signal Imprints of Detected Objects, IEEE Trans. Magn. 49 (2013), 69-72
[12] P. Ripka: Position Unit for Metal Detection, Czech patent CZ-302406
[13] Vcelak J., P. Ripka, A. Platil, J. Kubík and P. Kašpar: Errors of AMR compass and methods of their compensation, Sens Act.A129 (2006), pp 53-57
[14] Tsujimura T. and Tetsuya M.: Electromagnetic system navigating tunneling robots, Robotics and Autonomous Systems, Volume 40, Issue 4, 30 September 2002, Pages 229-238.
[15] Kim Yong-Seok, Seong-Cho Yu, Jeong-Bong Lee and Heebok Lee, A new class of LC-resonator for micro-magnetic sensor application, Journal of Magnetism and Magnetic Materials, Volume 304, Issue 1, 2006, Pages 117-121.
[16] Herzer G.: Magnetoelastic Sensors for Electronic Article Surveillance Sensor Lett. 5, 259–262 (2007)
[17] Cordova-Fraga, T; Carneiro, AAO; de Araujo, DB; Oliveira, RB; Sosa, M; Baffa, O: Spatiotemporal evaluation of human colon motility using three-axis fluxgates and magnetic markers, MEDICAL & BIOLOGICAL ENG. & COMPUTING 43 (6): 712-715, 2006
[18] Park, HS Hwang, JS Choi, WY Shim, DS Na, KW Choi, SO: Development of micro-fluxgate sensors with electroplated magnetic cores for electronic compass Sens. Actuator A-Phys. 114 (2004) 224-229
[19] Grönefeld M.: Permanent Magnets: Sensor Applications, in J. Buschow (ed.) Encyclopedia of Materials: Science and Technology, Elsevier 2007

Journal of Physics: Conference Series **450** (2013) 012002 doi:10.1088/1742-6596/450/1/012002

New tricks with old sensors: Pervasive Technologies for Novel Applications

Geoff Merrett

ECS, Faculty of Physical and Applied Sciences, University of Southampton, Southampton, SO17 1BJ, United Kingdom
Email: gvm@ecs.soton.ac.uk

Abstract. Sensors are interleaved into society, instrumenting considerable aspects of our lives without our comprehension. Sensors such as the MEMS accelerometer have transitioned from their original domains to applications that they were never conceived for: from games controllers to contextually rotating the screen on your smartphone. Further advances in technologies such as pervasive computing and networked embedded sensing are enabling new applications and smart devices which utilise sensors in new ways. In this presentation I will highlight new trends, applications and research in these fields, and show how "simple" sensors are being used in larger connected systems – from assistive technologies to distributed monitoring.

From hemodynamic towards cardiomechanic sensors in implantable devices

Bozidar Ferek-Petric

Medtronic Adriatic Region, Folnegoviceva 1c, 10000 Zagreb, Croatia

E-mail: bozidar.ferek-petric@medtronic.com

Abstract. Sensor could significantly improve the cardiac electrotherapy. It has to provide long-term stabile signal not impeding the device longevity and lead reliability. It may not introduce special implantation and adjustment procedures. Hemodynamic sensors based on the blood flow velocity and cardiomechanic sensors based on the lead bending measurement are disclosed. These sensors have a broad clinical utility. Triboelectric and high-frequency lead bending sensors yield accurate and stable signals whereby functioning with every cardiac lead. Moreover, high frequency measurement avoids use of any kind of special hardware mounted on the cardiac lead.

1. Introduction

Cardiac electrotherapy is used either to maintain the heart rate, or terminate the arrhythmia, or even to improve the heart pump function. Pacemakers, implantable cardioverter-defibrillators (ICDs), and cardiac resynchronization therapy (CRT) devices currently deliver the therapy based upon sensing of the myocardial electrical depolarization events. A cardiac implantable electrotherapy system comprises a device can and one or more leads in contact with the myocardium in order to deliver pacing therapy or even a shock [1]. Finally it has to maintain physiologic myocardial contractions and consequent hemodynamics. Information based solely on the electrical events does not give the insight into the electromechanical effect of the therapy.

Instead of monitoring only the electrical activity, knowing the outcomes of the pacing stimulus, i.e. the contraction itself would be the real value for clinical practice. A sensor able to determine either the intracardiac pressure, or the blood flow or mechanical contraction, being integrated in cardiac electrotherapy systems would improve the therapy in the aspect of its automatic regulation and optimization. Apart from ability to measure mechanical action in the heart or circulation, the sensor should satisfy other requirements such as robustness, stability, ease of clinical use, longevity, and low cost. It is expected that custom leads comprising a sensor exhibit similar longevity and stability to standard bipolar leads usually implanted for the lifetime of a patient. The lead should have low failure rate for a long period of time whereby the sensor output has to be stable because it can only be tested and calibrated during pulse generator implantation or replacement procedures. The sensor stability is also required under various patient conditions such as posture variability, exercise, etc. Improvement of patient therapy may not increase the burden of physicians to implant the sensor in specific anatomic position, or to perform additional tests and adjustment procedures. Any additional hardware on the lead, protruding out of the lead body diameter may be the physical obstacle for endovenous implantation whereby additional lead wires would decrease the reliability of the system. Any future sensor definitively may not require redesign of the connector assembly for electrical connection between the lead and a device. Finally it may not increase the current consumption and cause the reduction of the device longevity. These are sophisticated tasks and many efforts have been put in the development of the sensors that require custom leads different from those standard and commercially available [2], [3] that comprise either a pressure transducer or an intracardiac accelerometer. This paper discloses results of sensor designs, utilizing hemodynamic or cardiomechanic parameters, not yet being implemented for commercial use.

2. Blood Flow Sensors

From the aspect of clinical application, the most convenient would be the standard implantation of a lead in its position of the tip within the right ventricular apex whereby the sensor would be mounted fixed proximally on the lead body in order to measure the flow through the tricuspid valve[4].

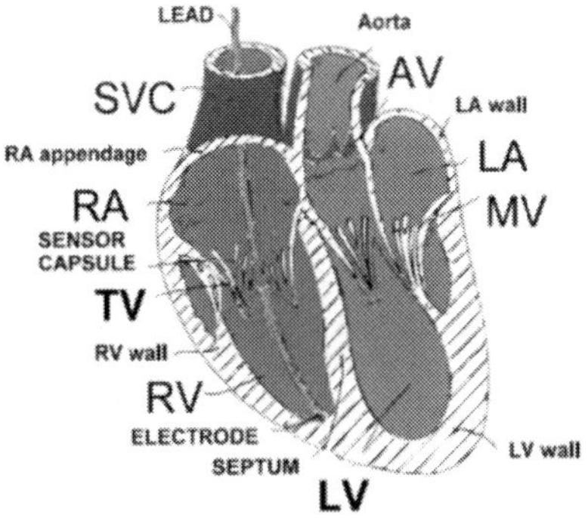

Figure 1. Cross-section of the heart with a flow measurement lead implanted within the right ventricle. Abbreviations: RV-right ventricle, RA-right atrium, LV-left ventricle, LA-left atrium, SVC-vena cava superior, TV-tricuspid valve, MV-mitral valve, AV-aortic valve.

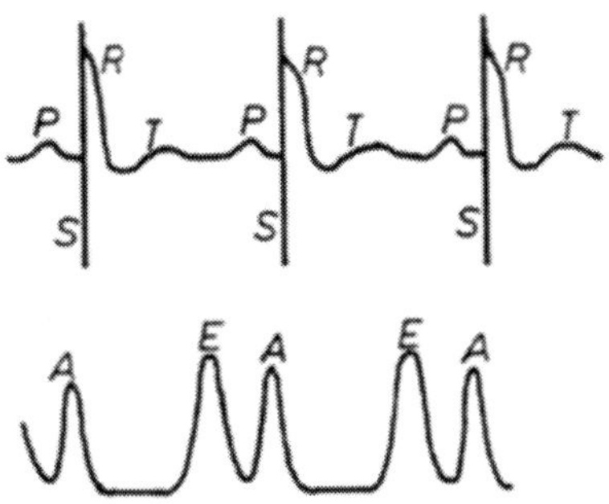

Figure 2. ECG waveform and corresponding tricuspid flow waveform in atrial-synchronized ventricular pacing with normal A-V delay. It looks the same in sinus rhythm

The flow velocity waveform through the tricuspid valve is disclosed in figure 2 in time correlation to the ECG waveform. It comprises the E-wave being the consequence of the ventricular relaxation and the A-wave being the consequence of the atrial contraction. The integral of the flow velocity waveform is equal to the blood volume [5]. This is true for flow within every valve within the heart. The ratio of peak velocities E/A is a hemodynamic parameter showing the cardiac muscle performance [6]. Atrial fibrillation detections is easy because it is detection of disappearance of the atrial filling wave. Circulatory catecholamine directly influences interval between the QRS complex and the corresponding following E wave as well as the rate of diastolic filling. The peak velocity of the E-wave, rapid diastolic filling period which is the E-wave duration and the first derivative of the E-wave onset being the diastolic filling acceleration (dE/dt) are hemodynamic parameters. In the case of exercise there is a significant increase in velocity time integrals and in peak flow velocities at the aortic, pulmonic, mitral, and tricuspid valves [7][8]. There is a broad scope of clinical utility of this sensor: tachycardia detection and classification according to hemodynamic stability, vasovagal syncope prediction, heart failure monitoring, CRT optimization, hypertension, pulmonary hypertension, aortic stenosis, and cardiomyopathy monitoring. It is also convenient for various pacing algorithms such as automatic thresholds measurements.

2.1 Ultrasonic Doppler Sensor

Experimental ultrasonic Doppler circuit has been developed being ready for implementation within implantable devices[9]. Utilizing the waveform sampling system[10] ultrasonic transducer power in implantable devices may be less than 1 μWatt for chronic long-term measurements and within the range of 10 μWatts for beat-to-beat measurements. Additional power would be consumed in functional system by signal processing circuits. Though technically possible, this sensor is already obsolete due to further more feasible developments.

2.2 Other Ultrasonic Flow Sensors
The implantable leads comprising either the autocorrelation ultrasonic sensor [11] or the transit time system [12] would be very complicated in design. Though never chronically investigated in-vivo, feasibility of these systems may be questionable. Autocorrelation method is attractive because the aortic flow could be measured by means of the sensor implanted within the vena cava superior.

2.3 Electrochemical Flow Sensors
2.3.1. Galvanic flow sensor is a sensor that does not consume any power. A cardiac lead at least two electrodes made of two different biocompatible materials mounted on the catheter. One of the electrodes is formed as a polarizable electrode mounted in the flow detecting position. Variation in an over-voltage or in a galvanic voltage is detected and is used as a blood flow velocity signal [13]. Our animal experiments revealed that in chronic setting, fibrous tissue overgrows the electrodes and significantly attenuates the signal. Therefore this system is not feasible for application in living organism and further evaluation was terminated.

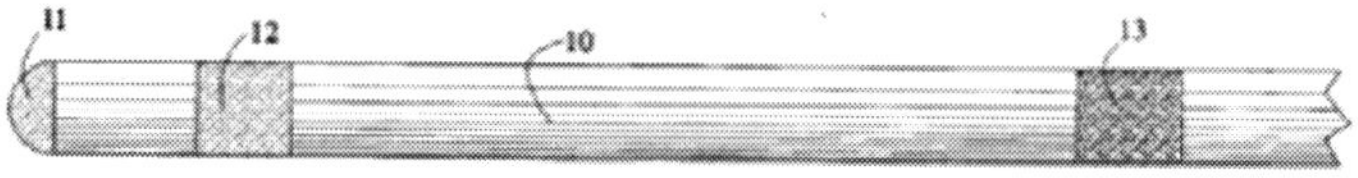

Figure 3 The lead body 10 comprises an active electrode 11 at its tip, an indifferent ring electrode 12 and a highly polarizable electrode 13 at the position within the tricuspid valve. Galvanic potential measured between electrodes 13 and 12 is modulated by the blood flow.

2.3.2 Polarographic flow sensor [14] was only demonstrated in acute setting because it exhibits the same problem of signal attenuation in chronic phase. Moreover, DC biasing potential stimulates the fibrous tissue overgrowth.

2.3.3. Differential oxygen diffusion sensor showed to be very reliable and stabile in acute and chronic animal study [9]. The current through an electrode is proportional to the number of molecules arriving

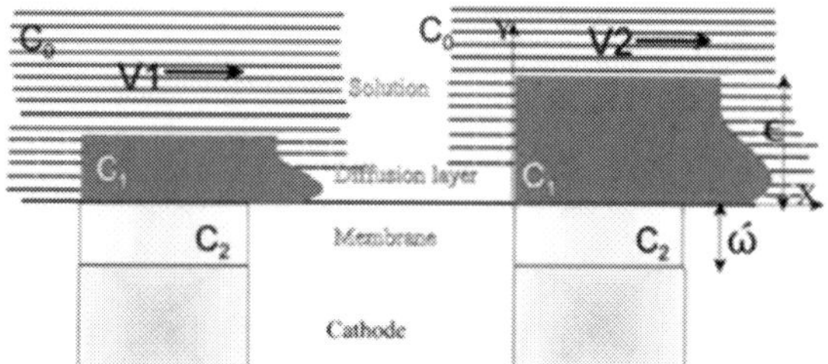

Figure 4 Diffusion layer thickness is a function of solvent velocity: greater velocity yields thinner layer.

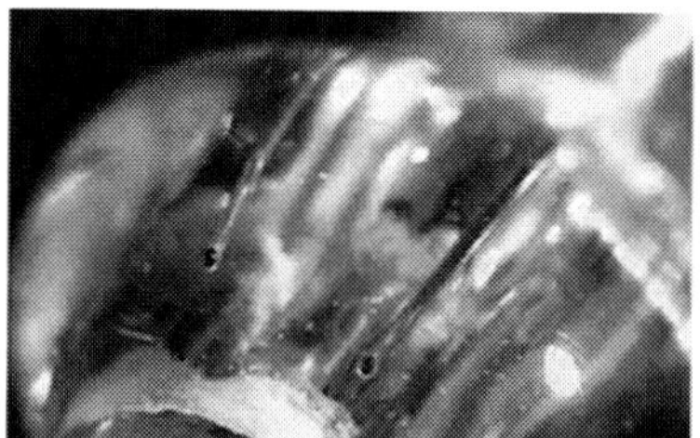

Figure 5 Flow sensor comprises two Platinum dots having 100 μm diameter in a glass capsule.

at the electrode surface per unit time. The molecules supply is dependent on diffusion process. When connecting the differential charge amplifier to the Platinum dots, the output represents the flow signal.

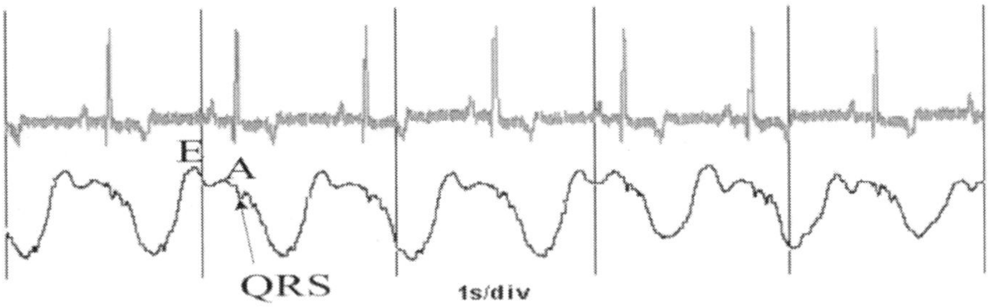

Figure 6 Tricuspid flow in acute porcine model that exhibits E and A waves and QRS artefact

2.4 Other Flow Sensors
2.4.1 Hydrodynamic flow sensor [15] requires two pressure sensors on one lead and therefore very complicated lead wires arrangement. It could become feasible if the entire sensor were designed as a wireless sensor capsule. Due to uniform fibrous tissue overgrowth over both pressure sensors, it could exhibit the long-term stability.

2.4.2 Thermal diffusion sensor [16] utilizes the temperature changes, caused by the flow, on a thermocouple being inserted into the tissue and heated at constant electrical power. This method is used to provide a qualitative indication of tissue blood flow changes.

3. Cardiac Contraction Sensors

The first proposed system comprises a tensiometric element disposed at a location on the lead body subject to bending due to cardiac contractions [17]. The tensiometric element may be made either of piezoelectric material or variable resistivity material. The lead body bending causes the element to produce a voltage or a resistivity variation respectively representing cardiac contractions. Research of this system has been terminated at the in-vitro experiments stage due to further attractive developments of this idea. Two different systems were developed not requiring any lead design change and functioning with every cardiac lead.

3.1 Triboelectric sensor
The triboelectric effect, more commonly known as electrostatic discharge, occurs when two materials slide against each other. The frictional force involved in this process causes electrons of one material to separate and reattach themselves to the second material, creating a charge imbalance between the two surfaces. The induced current in the wiring resulting from this charge imbalance is responsible for unwanted noise and interference in measuring-, controlling-, monitoring- or audio applications. These disturbances called mechanically induced noises or cable microphony are superimposed to the signals that the cable is carrying. Flexed or otherwise deformed cardiac pacing leads seen as coaxial or parallel cables generate electrical noise. Charge measurements have been done between one of the pacing lead conductors and a pacing lead stylet that is normally inserted in the internal lumen of the lead only during a regular implant procedure in order to provide additional stiffness and support for the lead placement. Instead of a classic stylet made of uninsulated steel wire, different kinds of insulated wires were tested made of various materials in the effort to obtain better signal [18]. Acute animal and acute human study was done [19]. Triboelectric sensor does not require a specially designed lead and it could theoretically be used with any existing lead at any anatomical site of implantation. The sensor could be easily replaced whenever an implanted device is replaced. The stylet may be retracted from the chronically implanted lead and the new one may be inserted.

Accordingly, its functional lifetime and its long-term stability could be equal to the implantable device longevity. The patient could receive a new sensor with the new device.

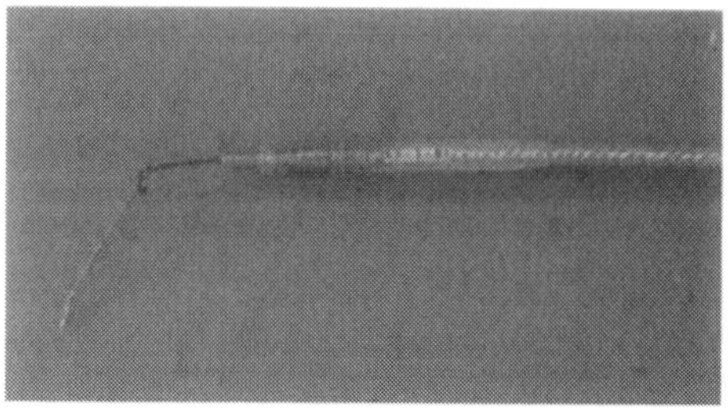

Figure 7 Cardiac lead connector having an insulated stylet inserted into the lead channel

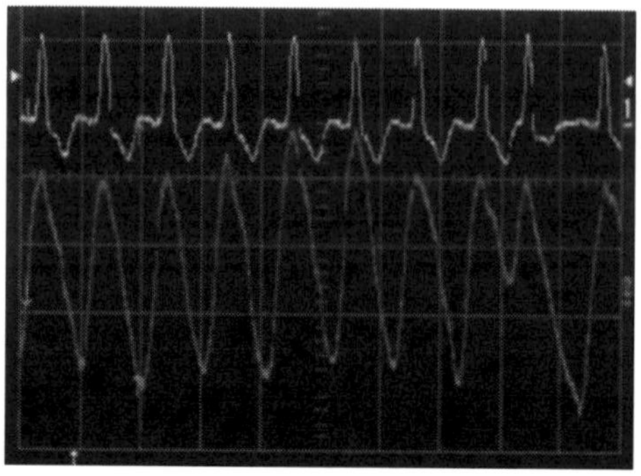

Figure 8 Yellow waveform is the ECG and red waveform is the triboelectric potential of a human patient in supraventricular tachycardia. The negative slope of the triboelectric signal was caused by the contraction and consequent bending of the lead, while the positive slope was caused by the relaxation and consequent elongation of the lead. The signal has magnitude of 30 pC_{pp} measured by differential charge amplifier. Shorter coupling interval causes the contraction signal's amplitude decrease due to the lower magnitude of the lead bending. Longer coupling interval causes the contraction signal's amplitude increase due to the higher magnitude of the lead bending.

3.2 High frequency (HF) sensor

The myocardial contraction causes the lead to bend. The flexion of the lead body changes its geometry and impedance. An accurate measurement of the HF impedance or any other characteristic dependent on HF impedance such as reflection or transmission coefficient, can precisely represent the cardiac contraction. In vitro experiments showed that such measurements can be easily performed on any cardiac lead having at least two conductors that are parallel, helicoidal (co-radial) or coaxial. The chronic stability of the HF characteristics of various leads has been proven [20]. After completing the animal study [21], the acute human study of patients undergoing implantations of ICD was initiated whereby reflection coefficient of pacing-sensing part of the lead is measured by network analyser [22].

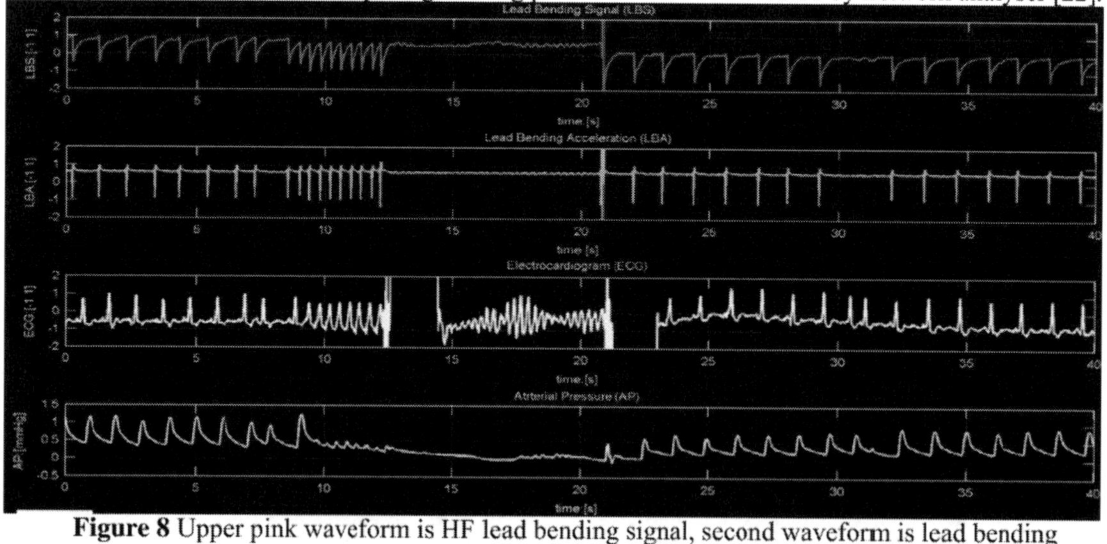

Figure 8 Upper pink waveform is HF lead bending signal, second waveform is lead bending acceleration, yellow waveform is ECG and lower blue waveform is arterial pressure. Following the rapid pacing train, shock on T provokes the ventricular fibrillation. After a detection period, rescue shock is delivered and sinus rhythm restored.

HF sensor does not require any additional hardware on the lead, but an additional HF circuit in a device that will not significantly impede the battery longevity.

References

[1] Chow AWC and Buxton AE., eds. 2006 *Implantable Cardiac Pacemakers and Defibrillators: All You Wanted to Know* Blackwell Publishing Ltd Oxford UK
[2] Daum DR, Jones BR and Lang DJ 2000 Hemodynamic sensors *Am J Cardiol.* **86** pp. 95K100K
[3] Lau CP and Tse HF 2008 Hemodynamic sensors in heart failure devices In *Devices for Cardiac Resynchronization* (S. S. Barold and P. Ritter, eds.), pp. 253-268, Springer US
[4] Ferek-Petric B and Breyer B 1993 Tricuspid flow synchronized cardiac electrotherapy system with blood flow measurement transducer and controlled pacing signals based on blood flow measurement. *U.S.Patent No.* 5,243,976
[5] Yellin EL et al. 1992 The Scientific Bases for the Relations Between Pulsed-Doppler Transmitral Velocity Patterns and Left Heart Chamber Properties *Echocardiography* **3** pp. 313-338.
[6] Zile MR 1992 Hemodynamic Determinanats of Echocardiographically-Derived Indices of Left Ventricular Filling *Echocardiography* **3** pp. 289-300.
[7] Maslow A 1996 Pulsed wave Doppler measurement of cardiac output from the right ventricular outflow tract *Anesth Analg* **83** 466-71.
[8] McEntee K et al. 1999 Doppler echocardiographic study of left and right ventricular function during dobutamine stress testing in conscious healthy dogs *Am J Vet Res* **60** 865-71.
[9] Ferek-Petric B 2005 Cardiac Pacing Based on Intracardiac Flow Measurement *PhD thesis* University of Zagreb, Faculty of Electrical Engineering and Computing, Zagreb
[10] Ferek-Petric B and Breyer B 2006 Ultrasound methods and implantable medical devices using same *U.S.Patent No.* 7,037,266
[11] Breyer B and Ferek-Petric B 1998 Blood flow measurement device. *U.S.Patent No.* 5,785,657
[12] Franklin DL, Baker DW and Rushmer RF 1962 Pulsed ultrasonic transit time flowmeter *IRE Trans. Bio Med. Electron.* **9** 44
[13] Ferek-Petric B and Breyer B 1998 Blood flow velocity measurement device *U.S.Patent No.* 5,799,350
[14] Williamson DE 1976 Intravascular liquid velocity sensing method using a polarographic electrode *U.S. Patent No.* 3,930,493
[15] Breyer B and Ferek-Petric B 1993 Hydrodynamic system for blood flow measurement. *U.S.Patent No.* 5,271,408
[16] Delhomme G et al. 1994 Thermal Diffusion Probe and Instrument System for Tissue Blood Flow Measurements: Validation in Phantoms and In Vivo Organs *IEEE Transactions on Biomedical Engineering*, **41** pp 656 – 662
[17] Ferek-Petric B 1993 Cardiac lead with tensiometric element for providing signals corresponding to heart contractions *U.S.Patent No.* 5,261,418
[18] Ferek-Petric B 2010 Cardiac therapy system including a triboelectric sensor *U.S. Patent* 7,780, 607
[19] Tomasic D, Ferek-Petric B, Brusich S et al. 2011 Acute animal and human study of tensiometric pacing lead sensor based on triboelectricity. *IEEE Trans Biomed* **8** 2290-2296
[20] Tomasic D, Ferek-Petric B, Brusich S et al. 2012 Acute and chronic high frequency properties of cardiac pacing and defibrillation leads. *Med Biol Eng Comput* **50** 827-837
[21] Brusich S, Tomasic D, Sovilj S et al. 2013 New cardiomechanic pacing lead sensor based on high frequency parameters: experimental studies in sheep *J Cardiovasc Electrophysiol* **24** 338-46
[22] Brusich S, Tomasic D, Sovilj S et al. 2012 VF detection utilizing the lead bending sensor. *Cardiostim abstract* n° 1812

Plastic optical fibre sensor for spine bending monitoring

M A Zawawi[1,2], S O'Keeffe[1] and E Lewis[1]
[1] Optical Fibre Sensors Research Centre, Department of Electronic and Computer Engineering, University of Limerick, Ireland
[2] Faculty of Electrical and Electronics Engineering, Universiti Malaysia Pahang, 26600 Pekan, Pahang, Malaysia

E-mail: mohd.zawawi@ul.ie

Abstract. This paper presents a study on the application of plastic optical fibre for spine bending monitoring based on an intensity modulation. The bending angle is measured as the angle between the emitting and receiving fibres is changed. The measured light attenuation is compared with a theoretical evaluation and the differences between these values are discussed. It was found that the light attenuation for the light intensity agreed well (margin of error < 15%) with the theoretical value for the range between 180° (representing no bend) and 200° and it was significantly increased for the bending angle beyond that value due to the effect of fibre gap increment which resulted in a less reliable experimental estimation.

1. Introduction

The importance of optical fibre in sensing applications has been highlighted by many researchers and these have been summarized in [1] according to the type of measurands. In general, the optical fibre type can be classified into glass optical fibre (silica) and plastic optical fibre (POF). These two types of fibre can be distinguished from several point of views such as fibre handling issues (connection, flexibility and technical precision level requirement), fibre optic parameters (signal loss, wavelength operating range, numerical aperture and bandwidth) and cost of handling (component cost, system cost and test equipment) which have been summarized in [2].

As compared to POF, glass fibre potentially offers a more sensitive and accurate measurement. However as the overall cost is more expensive, plastic optical fibre has been widely studied due to the overall use of low cost components such as LEDs, photodiode, multimode fibre and basic amplifier circuit for signal amplification. Furthermore, the strains encountered in this application are likely to be a few percent in magnitude in which glass optical fibres cannot endure and can easily break. Therefore the use of POF in this investigation represents a better solution. However, the intensity modulation technique employed in this investigation is dependent on many extraneous factors such as noise and drift of the LED and photodetector circuit, intensity fluctuation due to source temperature changes, external light source coupling and fibre bending effects. Such errors have been minimized using forms of referencing techniques such as balanced bridge, divided beams and two-wavelength referencing [3].

This paper presents the development of a simple and low cost optical fibre bending sensor based on intensity modulation for use in physiotherapy clinical environments. To define the minimum working range of the sensor, typical range of motions of human spine comprising of lumbar, cervical and thoracic are presented. The working range and the light attenuation characteristics of the sensor are presented and compared with the theoretical values. To understand the effects of fibre gap (longitudinal displacement) and fibre tilt angle (bending angle) with respect to the light attenuation, both results of light attenuation (in dB) versus fibre gap (in mm) and bending angle (in degree) are initially presented. A brief overview of currently available optical-based sensors for monitoring human spine bending is also presented in orders to provide an up to date comparison between of those techniques and the one of this investigation.

2. Human spine monitoring

Spine monitoring activities have widely been conducted as a measure to differentiate between healthy and back-pain patients, apart from interviews and discussions approaches which are considerably difficult to provide meaningful data for further analysis [4]. As for the back-pain sufferers, the spinal examination is an understandably difficult task for them to undergo as this process would normally require some exaggerated activities of back movements such as flexion, extension, lateral bending and axial rotation.

Moreover, for most measuring techniques, the devices are placed directly on the skin surface at the back of the body which adds more distress to the patient condition. Besides that, for certain patients' condition it is necessary to continuously monitor the spine movement up to several hours in order to identify the actual cause of the back problems [5]. To reduce this discomfort experience, several non-invasive techniques have been widely studied to provide alternatives to physiotherapists in handling these tasks and to increase more comfort to respective patients.

2.1. Human spine range of motion and sensor working range

As stated above, human spine monitoring involves several parameters which are flexion angle, extension angle, lateral bending and axial rotation. Besides that, several other parameters can also be measured in the same area along the human spine such as point to point position, angular velocity and acceleration as well as range of motion. Range of motion of human spine is defined as the angle between the maximum bending between flexion and extension.

There are many reference text that facilitate the understanding of the anatomy and basic terminology related to human spine, e.g. [6]. In order to define the sensor working range, it is necessary to find the bending angle limitation of the lumbar, cervical and thoracic spines in the intended directions of the spine movement. Typical range of motion of human spine can be found in [7] and presented in summary in table 1. However, the actual range of motion may depend on several other factors such as age, gender and spinal health condition.

Table 1. Typical range of motion of human spine [7].

Type of movement	**Cervical**	**Thoracic/ Lumbar**
Flexion	45°	75°
Extension	55°	30°
Lateral bending	40°	35°
Rotation	70°	30°

3. Review on current optical fibre sensor for spine bending monitoring

A wide range of sensors have previously been studied and tested on patients to monitor the spine bending movement with differing degrees of success. As all these research work have been conducted by different persons and under widely differing testing environments and subject groups, it is not possible to accurately compare each sensor's performance. The prime reason for providing this brief background is to identify the differences between each sensor type in terms of sensing mechanism, type of measurand and sensor installation requirement as compared to the sensor design presented here.

The first optical fibre sensor is designed using plastic optical fibre with abraded fibre side [8]. To measure the light attenuation at multiple points along the spine, the fibre was cut at several positions along the same side of the fibre using the technique proposed in [9] as shown in figure 1. This sensor was proven to be comfortably wearable and extensible in sensing length. Besides that, it is also possible to measure the direction of the bending throughout the measurement. If the abraded area is at the external side (convex) of the bending fibre, more light will escape the fibre, thus giving a higher attenuation and vice versa. On the other hand this sensor was sensitive to noise and fluctuation of the light source and detector as well as external light coupling effect at the abraded fibre areas.

The second sensor was realised using a positional optical fibre [10] to record the curvature of the spine during static flexion and extension position. It was designed using 8 paired optical fibre sensing loops periodically spaced at 6 cm along a flexible metal ribbon. Each sensor loop detected bend and twist along the substrate and converted the data into orthogonal orientation vectors. Then, the position of each sensor along the tape, relative to a base or reference point was tracked and recorded using the software developed by the ShapeTape^{TM} supplier company (Measurand Inc). More description on the sensing mechanism for this type of sensor is presented in [11] and illustrated in figure 2. This optical fibre sensor was shown to have the ability to resolve spinal shape change across time at multiple spinal regions. However, as the measurement was based on sensor orientation with respect to the base sensor, the attachment process was difficult and time-consuming. Furthermore, it was not possible to measure any lateral bending as metal ribbon is stiff in sideway directions.

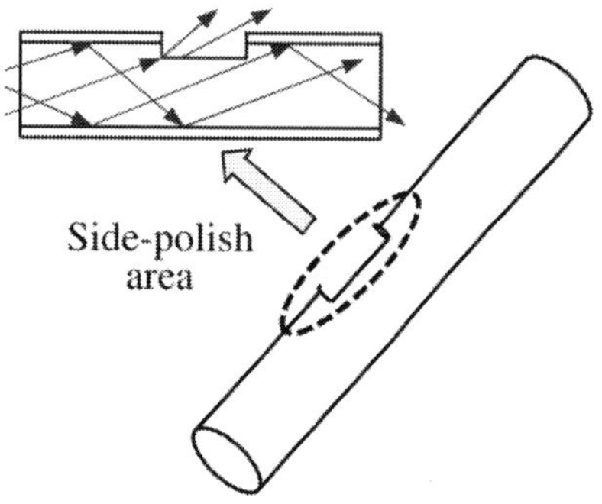

Figure 1. Side-polish optical fibre bending sensor

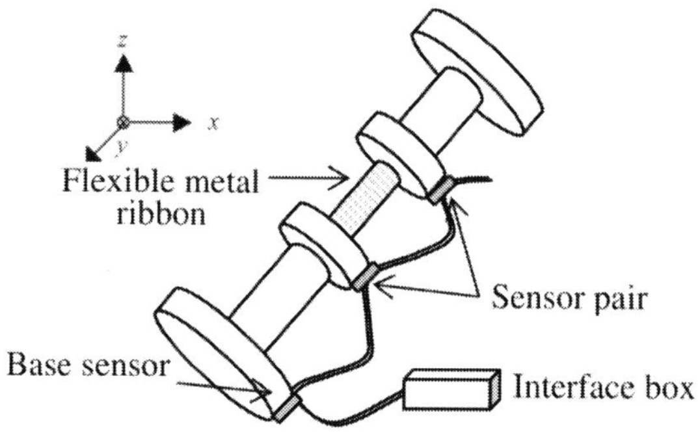

Figure 2. Positional optical fibre sensor

The next sensor was implemented using optical fibre goniometer (FOG) system [12]. The overall sensor comprised of a baseplate, a flexible rod, infrared LED source, a phototransistor and a modified 1 mm diameter optical fibre. The baseplate was placed on the sacrum level (S1) using adhesive tape. The fibre was inserted in a flexible rod and it was placed onto the lumbar spine by using two slider tubes. Similar to the previously considered positional optical fibre sensor, this FOG sensor was also not able to measure any movement in lateral directions.

4. Sensor configuration implementation

Several aspects have been considered before a suitable configuration is proposed for the development of spine bending sensor using plastic optical fibre such as the overall size of the sensor, interface medium between the sensor and the human body surface as well as approach to reduce the effect of external disturbances. For the initial experimental setup of this study, the type of fibre used was Mitsubishi Rayon GHV4001 (multimode plastic optical fibre (PMMA) with a core diameter of 1 mm, numerical aperture at 0.5, step index profile), a green LED source (IF-E93) with peak wavelength at 530 nm which was ideally mapping to the lowest attenuation window of PMMA fibre [13] and a stabilized RiFOCs detector. A few millimetres of fibre gap was allowed to give a smooth bending angle movement between the emitting fibre and receiving fibre.

In this section, two separate configurations of the OFS of this investigation are considered:-

4.1. Fibre longitudinal displacement loss

Both input and output fibres end were polished with a flat end tip. The emitting fibre was attached on a static stage while the other fibre was placed on the micrometer positioning stage (moveable part) using an adhesive tape. The receiving fibre was moved in a longitudinal direction and the change of the output intensity of the detector was measured at every 0.25 mm movement. The relationship between the light intensity loss and the distances between fibres is presented in [14] as shown below:-

$$\alpha = -10\log\left(1 - \frac{2 \cdot s \cdot NA}{3 \cdot d \cdot n_0}\right) \tag{1}$$

Where s is the distance between the fibre gap, NA is the numerical aperture of the fibre, d is the diameter of the fibre core and n_0 is the refractive index of the medium in between those fibres (in this case, air).

4.2. Fibre tilt angle loss

This type of configuration is based on the changes of the light intensity as the angle between the emitting fibre axis and the receiving fibre axis is increased. Both end tips of the emitting and receiving fibres were polished with flat surface and placed into V-pin crimp simplex connector (HFBR-4501Z). Clear silicon soft rubber tubing with inner diameter of 4 mm and outer diameter of 6 mm was used to firmly hold both fibres (with an initial gap of 1.5 mm) in their positions as shown in figure 3. The sensor was then placed on top of a custom-made bending rig to allow a better sensor movement. An 8 inch spinal goniometer was placed next to the bending rig for measuring the bending angle. The light intensity in the form of voltage output was measured for each 5° of bending angle between 130° and 230°. The maximum initial voltage value was measured at the output for a straight fibre (180°) and the light intensity was found to be linearly dependent on the fibre axis changes between 180° and 220° and between 180° and 140°.

In general, the light attenuation that is caused by the changes of the angle between the fibre axes can be presented in an equation [15]:

$$\alpha = -10\log\left(\cos\theta\left\{\frac{1}{2} - \frac{p}{\pi}\sqrt{1-p^2} - \frac{\arcsin p}{\pi} + q\left[\frac{r}{\pi}\sqrt{1-r^2} + \frac{\arcsin r}{\pi} + \frac{1}{2}\right]\right\}\right) \tag{2}$$

Where

$$p = \frac{\cos\theta_c(1-\cos\theta)}{\sin\theta_c \sin\theta},\quad q = \frac{\cos^3\theta_c}{(\cos^2\theta_c - \sin^2\theta)^{3/2}},\quad r = \frac{\cos^2\theta_c(1-\cos\theta) - \sin^2\theta}{\sin\theta_c \cos\theta_c \sin\theta}$$

Where θ is the angle between fibre axes and θ_c (θ_c = arcsin [NA/n_0]) is the acceptance angle for the fibre, which is depending on the fibre numerical aperture. For the type of fibre used in this measurement, the acceptance angle, θ_{max} is 30°, which limits the light attenuation calculation using the above equation up to an angle of fibre axes of 59°.

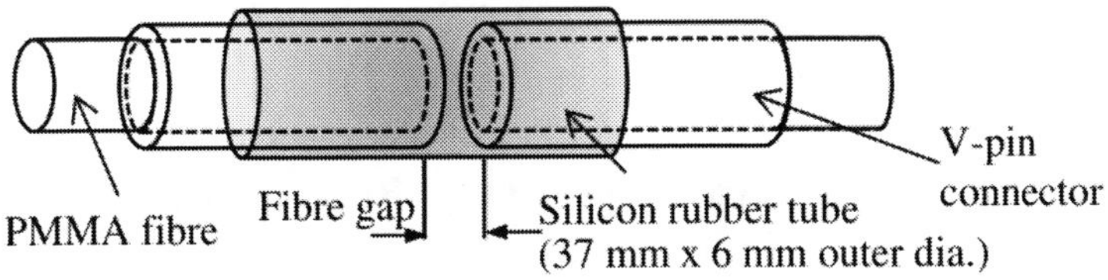

Figure 3. Proposed spine bending sensor

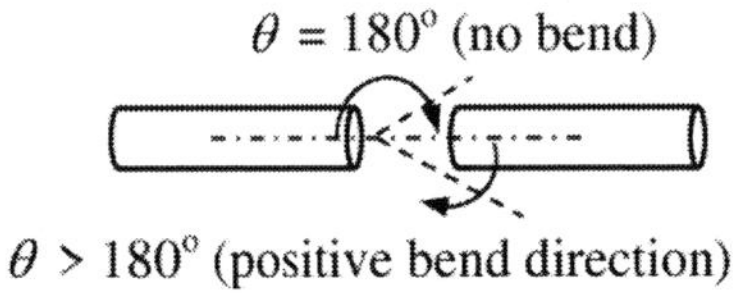

Figure 4. Fibre tilt angle

5. Results and discussion

To accurately define the light attenuation characteristic of the fibre bending angle, both fibre tilt angle and fibre gap (displacement) were measured separately. The measurements of light intensity versus fibre gap and fibre tilt angle are presented in figure 5 and figure 6, respectively.

From figure 5, the intensity of the light is in linear relationship with the fibre gap at three different slope degrees. The maximum slope ramp is between 0 mm and 2.75 mm, while the minimum slope is between 5.5 mm and 10 mm. For simplicity output intensity estimation, a suitable linear analysis can be found for each fibre-gaps sub-group separately using a regular linear equation approach.

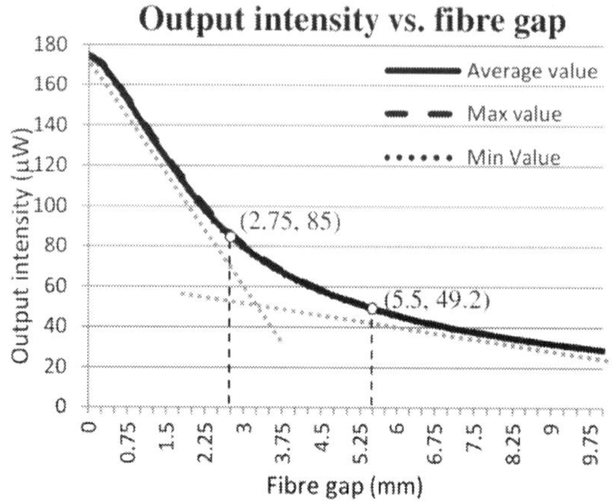

Figure 5. Light intensity (μW) vs. fibre gap (mm)

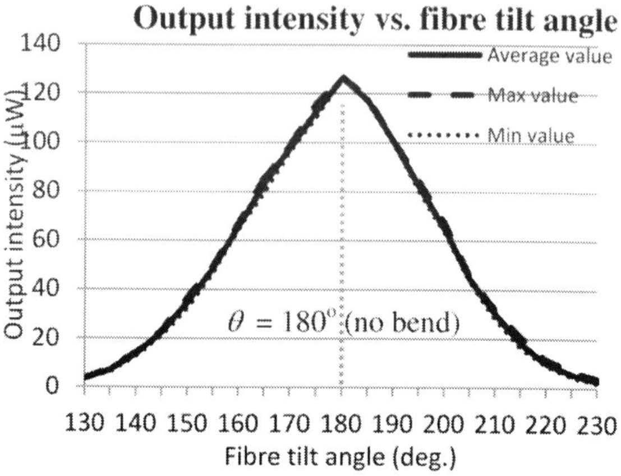

Figure 6. Light intensity (μW) vs. fibre tilt angle (degree)

It is also noted from figure 6 both positive and negative directions of the fibre axes angle are relatively linear between 180° and 215° and between 140° and 180°. When the bending angle is increased beyond that angles ($\theta < 140°$ and ($\theta > 215°$) less light loss is measured. Fibre optic position with respect to the bending angle is illustrated in figure 4 where 180° represents no bending. The difference between the theory and measurement values for the fibre gap in figure 7 is significantly increased as the distance is increased which is understandable because the equation was initially derived for splicing loss estimation of fibre gap of less than 1 mm assuming uniform mode distribution [14]. The light attenuation characteristic for the applied tilt angle is shown in figure 8 and summarized in table 2. The differences are significant for $\theta > 200°$ due to fibre gap increment caused by tilt angle.

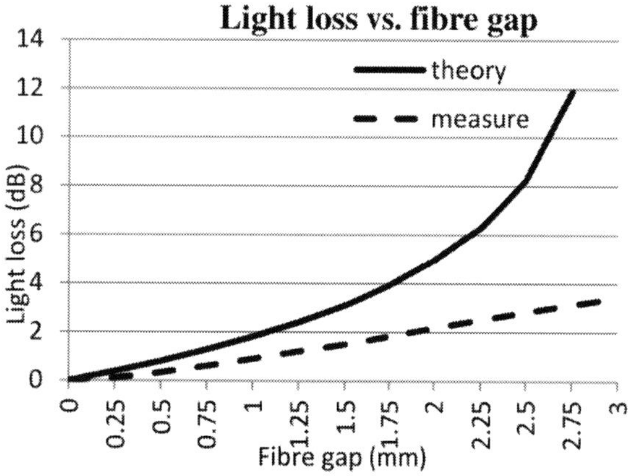

Figure 7. Light loss (dB) vs. fibre gap (mm)

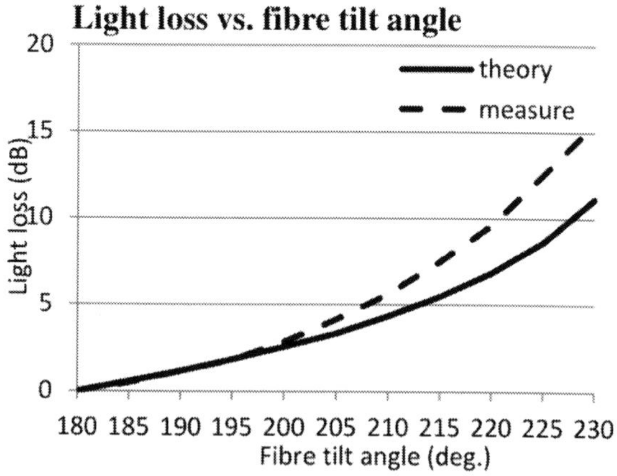

Figure 8. Light loss (dB) vs. fibre tilt angle (mm)

Table 2. Light loss comparison between theory and measurement for different tilt angle

Tilt Angle, θ (°)	**180**	**185**	**190**	**195**	**200**	**205**	**210**	**215**	**220**	**225**	**230**
Measurement (dB)	0.00	0.49	1.16	1.81	2.85	4.15	5.61	7.50	9.61	12.53	15.55
Theory (dB)	0.00	0.57	1.18	1.84	2.57	3.34	4.35	5.47	6.85	8.64	11.15
Difference	0.00	-0.08	-0.02	-0.03	0.28	0.81	1.26	2.03	2.76	3.89	4.40
Margin of error (%)	0.01	14.0	1.7	1.6	10.9	24.3	29.0	37.1	40.3	45.0	49.5

6. Conclusion

Plastic optical fibre sensor based on intensity modulation for measurement of bend angles in the spines of human physiotherapy patients has been presented and discussed. The measurements of light intensity with respect to fibre gap and fibre tilt angle were compared with theoretical values. The light intensity is decreases with the increase in the gap between emitting and receiving fibres as well as increases in bending angle between them. As the fibre gap increases sharply when the fibre was bend at a greater angle, it was found that the measurement values departed from the theoretical ones, but this was due to a higher degree of non-linearity in the measured intensity with the increasing angle. However a signal was still detected in this range.

References

[1] Lee B 2003 Review of the present status of optical fiber sensors *Optical Fiber Technology* **9(2)** 57-79

[2] Zawawi M A, O'Keeffe S and Lewis E 2013 Intensity-modulated fiber optic sensor for health monitoring applications: a comparative review *Sensor Review* **33(1)** 57-67

[3] Jones B E, Medlock R S and Spooncer R C 1989 Intensity and wavelength-based sensors and optical actuators *Optical Fiber Sensors: Systems and Applications* **vol. II** Culshaw B and Dakin J Artech House Inc 431-473.

[4] De Vet H C W, Heymans M W, Dunn K M, Pope D P, van der Beek A J, Macfarlane G J, Bouter L M, and Croft P R 2002 Episodes of Low Back Pain: A Proposal for Uniform Definitions to Be Used in Research *Spine*, **27(21)** 2409-16

[5] Spenkelink C D, Hutten M R, Hermens H J and Greitemann B OL 2002 Assessment of activities of daily living with an ambulatory monitoring system: a comparative study in patients with chronic low back pain and non-symptomatic controls *Clinical Rehabilitation*, **16(1)** 16-26

[6] Netter F H 2011 Back and spinal cord *Atlas of human anatomy* 5th ed. Saunders Elsevier 149-174 ISBN : 978-1-4160-5951-6

[7] McCullough P A 2003 Introduction to clinical exercise physiology *Clinial Exercise Physiology* Ehrman J K, Gordon P M, Visich P S and Keteyian S J, Human Kinetics Inc. 41-57 ISBN : 0-7360-0252-9

[8] Dunne L E, Walsh P, Hermann S, Smyth B, and Caulfield B 2008 Wearable Monitoring of Seated Spinal Posture *IEEE Transactions on Biomedical Circuits and Systems*, **2(2)** 97-105

[9] Kuang K S C, Cantwell W J and Scully P J 2002 An evaluation of a novel plastic optical fibre sensor for axial strain and bend measurements *Meas. Sci. and Tech.* **13** 1523-34

[10] Williams J M, Haq I and Lee R Y 2010 Dynamic measurement of lumbar curvature using fibre-optic sensors *Medical Engineering & Physics* **32** 1043-49

[11] Morin E and Reid S 2002 The validation of measurand shapetape for measuring joint angles *Proc. of 27th Canadian Medical and Biological Engineering Conference*, Ottawa

[12] Bell J A and Stigant M 2008 Validation of a fibre-optic goniometer system to investigate the relationship between sedentary work and low back pain *Int. J. of Industrial Ergonomics* **38** 934-941

[13] Ziemann O, Krauser J, Zamzow P E and Daum W 2008 Optical Fibers: Fundamentals of Optical Fibers *POF Handbook: Optical Short Range Transmission Systems*, **2nd ed.** Springer 47. ISBN 978-3-540-76628-5

[14] Ziemann O, Krauser J, Zamzow P E and Daum W 2008 Passive Components for Optical Fibers: Basis for Calculating Connector Losses *POF Handbook: Optical Short Range Transmission Systems*, **2nd ed.** Springer 259-268. ISBN 978-3-540-76628-5

[15] Thiel F L and Hawk R M 1976 Optical waveguide cable connection *Applied Optics* **15(11)** 2785-91

Integration of Leaky Waveguide Detection with Electrowetting on Dielectric Digital Microfluidic Devices

Ruchi Gupta[1] **and Nick Goddard**

School of Chemical Engineering and Analytical Science, University of Manchester, Oxford Road, Manchester M13 9PL, United Kingdom

Email: ruchi.gupta@manchester.ac.uk

Abstract. Typically, Electrowetting on dielectric (EWOD) digital microfluidic devices consist of an array of metal electrodes covered with a continuous hydrophobic dielectric layer. The monitoring of droplet position and detection in EWOD is usually achieved via microscopy, thereby resulting in increasing the size and complexity of the instrumentation associated with such devices. This work for the first time demonstrates that metal clad leaky waveguide (MCLW) is suitable for detection in EWOD devices. MCLW devices typically consist of a metal layer covered with a dielectric layer in which the leaky waveguide mode propagates. The two structures are fundamentally compatible provided the metal and dielectric layer thicknesses and refractive indices can be optimised to permit both electrowetting and waveguiding. In this work, it has been shown that titanium electrodes covered with a fluoropolymer layer can be used to perform MCLW detection of droplets on EWOD platforms.

1. Introduction

Electrowetting on dielectric (EWOD) devices are attractive for a range of applications as a result of their programmability and portability [1-5]. Optical detection, typically based on microscopy [5-6] and transmission [7], is commonly used to measure the output (e.g. antibody-antigen interactions) and monitor the position of droplets within EWOD devices. Such detection methods necessitate that the electrodes on at least one side of the EWOD device are made from a transparent material such as Indium Tin Oxide (ITO). The use of transmission based detection methods also requires a "clean" (e.g. non-turbid or non-absorbing) sample. Surface Plasmon Resonance imaging (SPRi) [8], whispering gallery mode (WGM) resonators [9] and integrated waveguides [10] have also been used to perform label-free sensing within EWOD devices. The use of SPRi for detection within EWOD devices, however, requires the formation of local detection zones that are free of hydrophobic dielectric [8]. The use of the WGM resonator and integrated waveguides also require additional fabrication steps.

A typical metal clad leaky waveguide (MCLW) device consists of a dielectric waveguide and metal layers on a glass substrate. In a MCLW, Fresnel reflection is used at one waveguide boundary to partially confine light and a peak (or dip) in reflectivity is observed around the resonance angle. The shifts in the position and intensity of the peak (or dip) has been used to perform refractive index sensing and absorption measurements respectively [11-12]. To the best of our knowledge, however, the suitability of MCLW for detection within EWOD devices has not been reported.

[1] To whom all correspondence should be addressed

In this work, the inherent similarity of MCLW and EWOD structures has, for the first time, been exploited to perform detection and actuation using the same structure (See Figure 1 (a)). This reduces the complexity of fabrication and the integration of the EWOD device with optical instrumentation. In addition, MCLW devices can provide greater sensitivity than microscopy-based detection and relax the requirements for non-turbid and non-absorbing samples.

2. Experimental

2.1. Chemicals and Materials

The chemicals used are as follows: hydrogen peroxide (30%), sulphuric acid (98%), acetone (>99.9%), glycerol (>99%), titanium (>99.99%, all: Sigma-Aldrich, Gillingham, UK), Norland optical adhesive (NOA) 13685 (Tech Optics, Kent, UK), Teflon® AF 1600 (DuPont, Geneva, Switzerland), Fluorinert™ FC-40 (3M, Bracknell, UK), SPR™ 220-7.0 and MF-26A (both: Dow, Middlesbrough, UK). 18.2 MΩ water (Elga Maxima Ultra Pure, Vivendi Water Systems, Buckinghamshire, UK) was used. 1.2 mm thick glass slides were purchased from VWR International (Leicestershire, UK).

2.2. Microfabrication

2.2.1. Fabrication of MCLW devices

The glass slides were cleaned in piranha solution (H_2O_2:H_2SO_4=1:3 (v:v)) and dried in an oven at 110 °C for 1 hour. The cleaned glass slides were cut into 25 mm×25 mm squares using a diamond scribe. These squares were loaded into an electron beam deposition system (Auto500, Edwards, West Sussex, UK) to deposit 9 nm thick titanium at a rate of 0.2 nm/s and under a pressure of 10^{-6} mbar. Film thickness was monitored by a quartz crystal oscillator (FTM7, Edwards, West Sussex, UK).

Other cleaned glass squares were spin coated with SPR™ 220-7.0 at 5000 rpm for 30 s, baked at 110°C for 90 s on a hot plate and exposed for 62 s using a maskless photolithography system (SF-100, Intelligent Micro Patterning, Florida, USA). The resist coated slides were developed in MF-26A, rinsed in water and baked at 110°C for 30 min. 9 nm thick titanium was then deposited by following the procedure described above. Subsequently, the resist was removed using acetone to obtain rectangles of different size of titanium on the glass slides. The size of the titanium rectangles investigated in this work was 0.36 mm×0.48 mm, 0.72 mm×0.96 mm and 1.44 mm×1.92mm.

Unless stated otherwise, NOA 13685 was spin coated on the slides with continuous and patterned titanium layers at a speed of 2000 rpm for 30 s. The substrates were UV-cured (2000-EC UV curing flood lamp, Dymax, Germany) under nitrogen dust for 10 min. Finally, a solution of 1% Teflon® AF 1600 in Fluorinert™ FC-40 was spin coated at 5000 rpm, 30 s and cured at 110°C for 30 min.

2.2.2. Fabrication of MCLW integrated EWOD devices

The lower plate of EWOD devices was fabricated by depositing patterned titanium electrodes using the lift-off photolithography process described above. The contact pads were protected using tape. NOA 13685 was then spin coated at 2000 rpm, 30 s and UV-cured under nitrogen for 10 min. Subsequently, Teflon® AF 1600 was spin coated at 5000 rpm, 30 s and cured at 110 °C for 30 min to form a hydrophobic layer. The tape was then removed from the contact pads and wires were bonded using silver-loaded epoxy (RS components, Northamptonshire, UK). The upper plate was constructed in a similar way on a continuous titanium layer, but without the deposition of NOA 13685. A double-sided adhesive with a thickness of 275 µm (3M 467MP, Viking Industrial Products Ltd, Keighley, UK) was used as a spacer between the upper and lower plates.

2.3. Instrumentation

The optical set-up used to probe MCLW devices is shown in Figure 1 (b). A prism made of N-BK7 (Qioptic Photonics, Denbighshire, UK) was used to couple light into the device. The angular position of the light source and the detector were controlled by mounting them on rails, which were connected to goniometers. A polystyrene flow cell was clamped firmly on top of the MCLW device, following which solutions were pumped using a peristaltic pump (Minipuls® 3, Gilson, Bedfordshire, UK) at a flow rate of 0.2 ml/min.

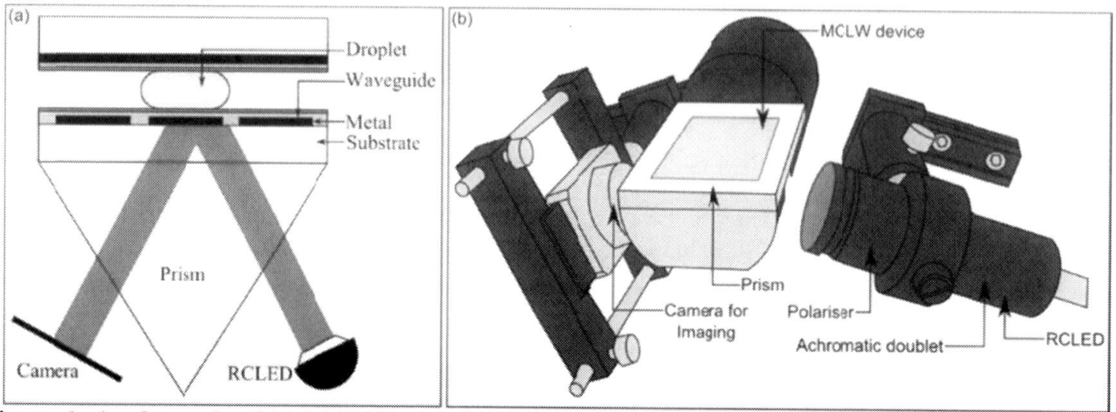

Figure 1: A schematic of (a) MCLW with EWOD (b) optical set-up (the flow cell is not shown)

Preliminary studies were performed using a TE-polarised laser (RS Components, Northamptonshire, UK) with a peak wavelength of 650 nm and a photodiode (Centronic, Surrey, UK).

In order to study the refractive index sensitivity of the MCLW device, a resonant cavity light emitting diode (RCLED, PR65-F1P0T, Roithner Lasertechnik, Vienna, Austria) was used. The light emitted from the RCLED was collimated and TE-polarised using an achromatic doublet (63 DQ 25) and polariser (32 CA 25, both: Comar Instruments, Cambridge, UK) respectively. A cylindrical lens was then placed in the path of light to obtain a beam with a wedge angle of ~34.7° in air and ~21.1° in N-BK7. The angle of the RCLED was fixed at 72° to obtain the peak at the centre of the image sensor. The output of the MCLW device *versus* time as different concentrations of glycerol solutions were introduced in the flow cell was captured on a camera (PL-B781, Pixelink, Ontario, Canada) using an integration time of 25 ms. The image was converted to a 1D profile by integrating gray values along x-axis and plotting them as a function of pixel values along y-axis (where x- and y-axis refers to the position across the width of the device and angle of incidence respectively). The data on shift in peak position, which was estimated using a centre of gravity algorithm and a threshold of 60% of the height of the peak, was used to determine the refractive index sensitivity of MCLW devices.

Two-dimensional (2D) imaging was performed using the collimated beam emitted from the TE-polarised RCLED and the camera. 2 µl drops of different concentrations of glycerol solutions were placed on the MCLW and an output image was captured at angles of incidence in the range between 60° and 80°. The camera integration time was 25 ms. A plot of average gray value as a function of different concentrations of glycerol solutions was used to determine the refractive index sensitivity of the MCLW at a chosen angle of incidence. The spatial resolution of the MCLW was determined by probing devices constructed by depositing NOA 13685 on titanium rectangles of different sizes.

EWOD actuation was performed by applying a DC potential of ~80 V between the electrodes on the lower and upper plates using a high voltage power supply (PS350, Stanford Research Systems, Hertfordshire, UK). The angle of incidence of the light source was fixed at 68° and the camera integration time was 25 ms. The movies were made using Windows movie maker and the frames were extracted via DVDVideoSoft.

3. Results and Discussion

3.1. Preliminary study

Figure 2 (a) shows the profile of the optical mode travelling in the MCLW structure for TE-polarised light, which was obtained using a transfer matrix modelling package (L-Pro version 4.0).

The effect of spin coating speed of the NOA 13685 on the output curve of the MCLW device is shown in Figure 2 (b). The film thickness obtained by using spin speeds between 4000 rpm and 6000 rpm was close to cut-off for supporting an optical mode. At 1500 rpm, on the other hand, the NOA 13685 layer was comparatively non-uniform. As a result, devices consisting of NOA 13685 waveguide layer deposited at 2000 rpm were selected to perform the rest of the work.

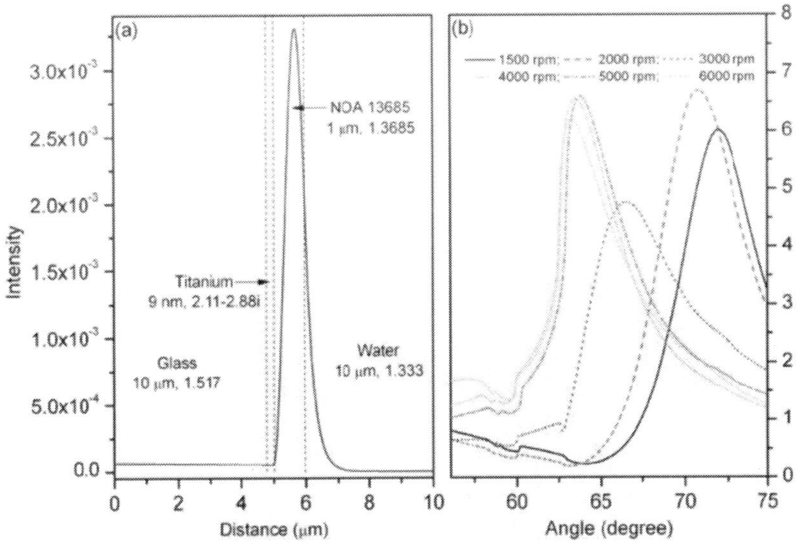

Figure 2: Profile of the (a) optical mode obtained via modelling and (b) output of MCLW devices consisting of NOA 13685 spun at different speeds

A typical image captured by the camera placed on the output side of the MCLW device and the shift in peak position *versus* time for different concentrations of glycerol solutions is shown in Figure 3. The relationship between the change in reflectivity peak position and the refractive index of glycerol solutions was linear and is given by pixel= -5644.9+4236.3×refractive index with r^2=0.983. Thus, the sensitivity of the RM device is 4236.3 pixels per refractive index unit (RIU).

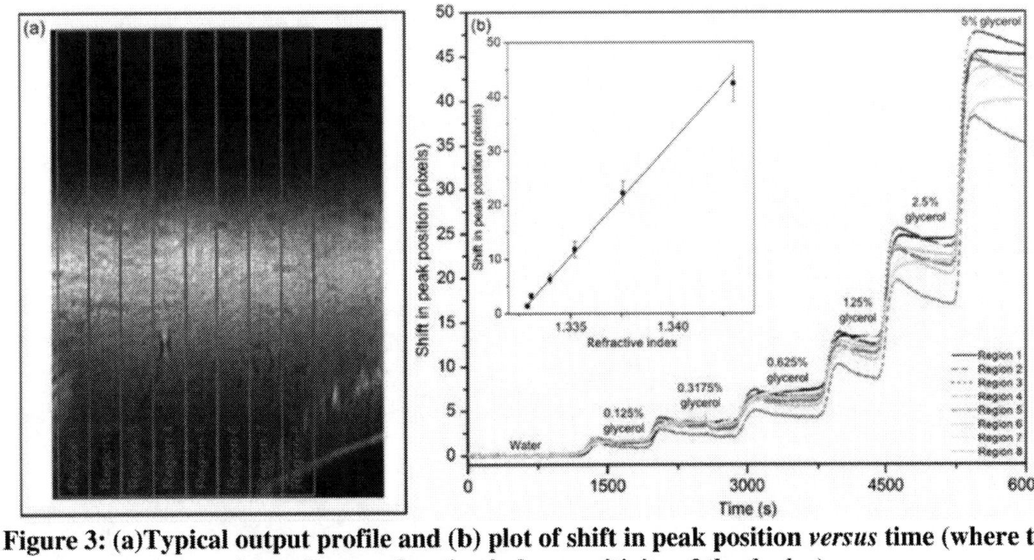

Figure 3: (a)Typical output profile and (b) plot of shift in peak position *versus* time (where the inset shows refractive index sensitivity of the device)

3.2. 2D imaging

The output of the MCLW with droplets of different concentrations of glycerol solutions at selected angles of incidence of the light source is shown in Figure 4 (a). As shown in Figure 4 (b), the normalised gray scale value decreases as the concentration (and hence the refractive index) of the droplet of glycerol solution increases at 68°. This is because the resonance angle of the MCLW increases as it is flushed with high concentrations of glycerol solutions. Similarly, the normalised gray scale value increases at the concentration of droplet of glycerol solution increases at 77°. Figure 7 (b) also shows that the sensitivity of detection of the MCLW device is highest at 68°. As a result, the angle of incidence of the light source was fixed at 68° in the rest of the work.

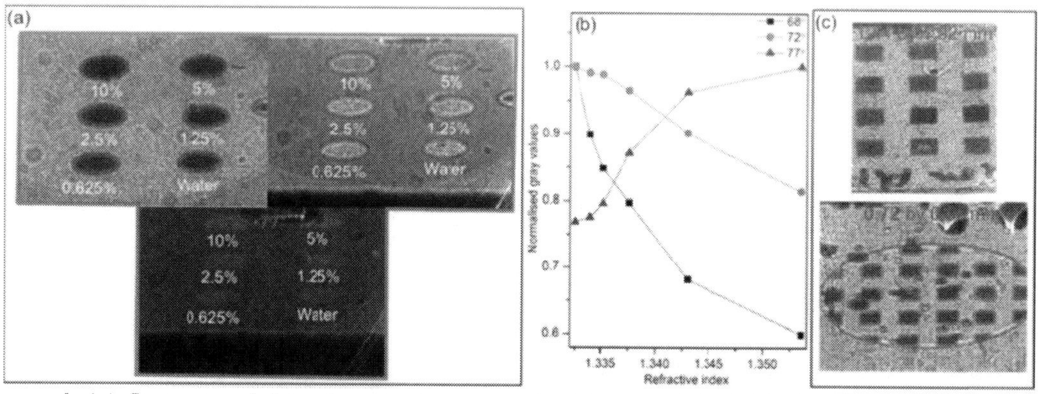

Figure 4: (a) Output and (b) refractive index sensitivity of the MCLW with continuous titanium layer and (c) output of the MCLW with patterned titanium layer

With the set-up currently used to probe the MCLW, devices consisting of titanium rectangles of 0.72 mm×0.96 mm and 1.44 mm×1.92 mm were successfully imaged (see Figure 4 (c)). The edges of titanium rectangles of size 0.36 mm×0.48 mm were, however, blurred and could not be resolved using the current set-up. The spatial resolution of the MCLW can be enhanced by using a highly collimated non-coherent (to avoid laser speckle) beam emitted by for example, superluminescent diodes or a supercontinuum source.

3.3. MCLW integrated with EWOD

MCLW detection was then used to monitor the position of water droplet as a function of time on applying a potential between a top and bottom pair of electrodes (see Figure 5). The droplet moves between 0.5 and 1 seconds in Figure 5. The droplets appear elliptical because of foreshortening caused by the high angle between the EWOD device and the camera.

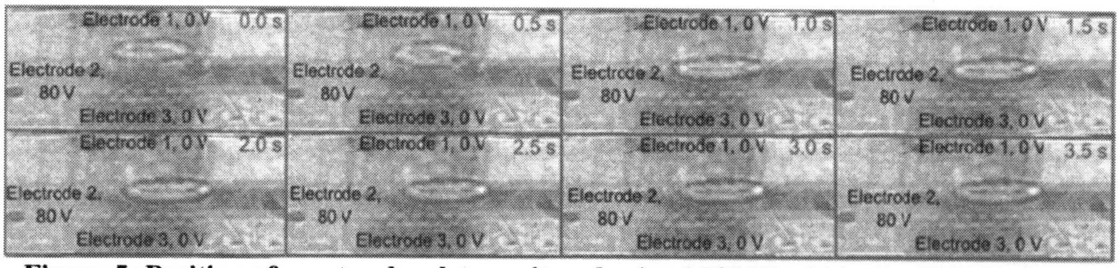

Figure 5: Position of a water droplet monitored using MCLW within the EWOD device

Subsequently, as shown in Figure 6, MCLW detection was shown to be suitable to monitor the mixing of two water droplets within an EWOD device. Future work will focus on increasing the spatial resolution of the MCLW detection by using highly collimated light sources. In addition, an optimised electrode design will be used to lower the required actuation voltage. Finally, the suitability of the MCLW integrated EWOD device to solve real-world problems will be demonstrated.

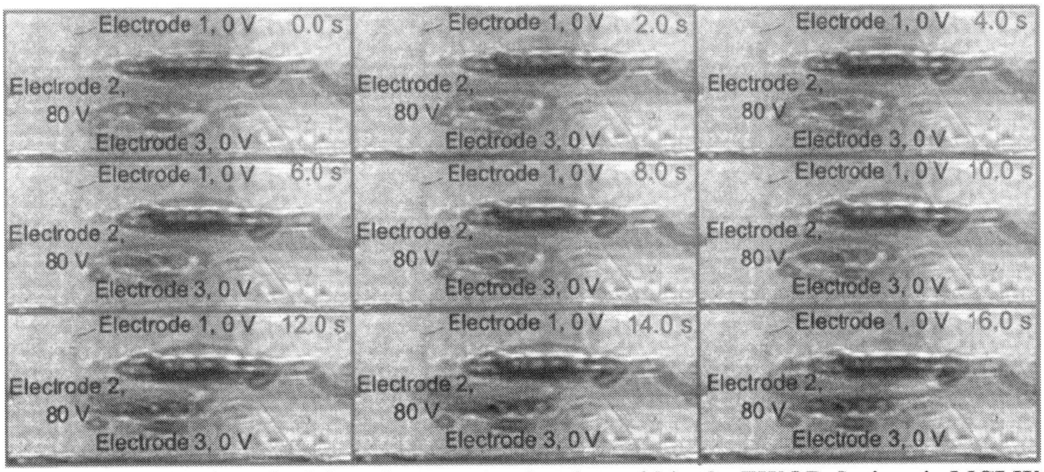

Figure 6: Monitoring of mixing of two water droplets within the EWOD device via MCLW

4. Conclusions

Currently used optical techniques for detection within EWOD devices require "clean" samples, additional fabrication steps and complex integration. In order to address these issues, this work demonstrates the suitability of metal clad leaky waveguide (MCLW) for detection within EWOD.

MCLW sensing has been shown to be directly compatible with EWOD devices as the basic structure of both is the same. Preliminary work has shown that high optical quality hydrophobic films can be deposited on both continuous and patterned titanium layers and that these give reproducible responses to refractive index change. Both peak position monitoring with wedge beams and intensity monitoring using whole chip imaging with collimated light have been shown to be capable of detecting small changes in refractive index. In addition, water droplets have been imaged directly and their position monitored in real time. This real time capability was exploited to follow the movement of droplets and droplet merging under EWOD actuation.

References

[1] I. Barbulovic-Nad, H. Yang, P.S. Park, A.R. Wheeler, Lab Chip 8 (2008) 519.
[2] Y.-H. Chang, G.-B. Lee, F.-C. Huang, Y.-Y. Chen, J.-L. Lin, Biomed. Microdevices 8 (2006) 215.
[3] M.J. Jebrail, A.H.C. Ng, V. Rai, R. Hili, A.K. Yudin, A.R. Wheeler, Angew. Chem. Int. Ed. 49 (2010) 8625.
[4] V.N. Luk, A.R. Wheeler, Anal. Chem. 81 (2009) 4524.
[5] R.S. Sista, A.E. Eckhardt, V. Srinivasan, M.G. Pollack, S. Palanki, V.K. Pamula, Lab Chip 8 (2008) 2188.
[6] E.M. Miller, A.R. Wheeler, Anal. Chem. 80 (2008) 1614.
[7] V. Srinivasan, V.K. Pamula, M.G. Pollack, R. Fair, in MEMS, IEEE 16th International, IEEE, Kyoto, Japan, 2003, p. 327.
[8] L. Malic, T. Veres, M. Tabrizian, Lab Chip 9 (2009) 473.
[9] C. Lerma Arce, D. Witters, R. Puers, J. Lammertyn, P. Bienstman, Anal. Bioanal. Chem. 404 (2012) 2887.
[10] F. Ceyssens, D. Witters, T. Van Grimbergen, K. Knez, J. Lammertyn, R. Puers, Sens. Act. B 181 (2013) 166.
[11] M. Zourob, N.J. Goddard, Biosens. Bioelectron. 20 (2005) 1718.
[12] R. Gupta, B. Bastani, N.J. Goddard, B. Grieve, Analyst 138 (2013) 307.

Optical detection of microcystin produced by cyanobacteria

R. Al-Ammar[1], A. Nabok[1], A. Hashim[1], T. Smith[2]
[1] Sheffield Hallam University, Material & Engineering Research Institute
[2] Sheffield Hallam University Biomedical Research centre
City campus, Howard Street, Sheffield, S1 1WB

E-mail: a9046817@my.shu.ac.uk

Abstract. Microcystin (MC-LR) produced by cyanobacteria (blue-green algae) was detected in direct immunoassay with specific monoclonal antibody MC10E7 using an optical method of Total Internal Reflection Ellipsometry (TIRE). The minimal detected concentration of MC-LR of 0.1 ng/ml is a remarkable achievement for direct immunoassay against such low molecular weight analyte molecule. The study of binding kinetics of MC-LR to MC10E7 antibody allowed the evaluation of the association constant KA of about 10^8 (l/Mol) typical for highly specific immune reactions. Concentration of MC-LR in aqueous solutions was reduced using an absorbent made of polyelectrolyte-coated microparticles functionalized with MC10E7 antibodies.

1. Introduction

Photosynthetic cyanobacteria (also known as blue-green algae) exist in any types of surface waters and depending on the environmental conditions may release a number of toxins microcystins [1, 2] which possess a serious threat to human, animals and generally to the environment. Microcystin-LR (MC-LR) is perhaps the most toxic from a large family of cyanotoxins; it is known to cause liver damage and also acts as carcinogen [3]. World Health Organization set the limit of 1μg/l for MC-LR in drinking water, and 0.04 μg/kg of body weight, based on liver pathology [4].
There are a number of analytical methods currently available for detection of microcystin-LR in water in a desired concentration range of 0.1-1μg/l, for example HLPC which requires expensive equipment and highly professional personal to carry out complex chromatography procedures and long time. The method of ELISA has been widely employed to monitor microcystin-LR at levels below 1μg/l [5] but it is also based on quite complex preparation procedures which require the use of expensive chemicals. Another method of protein phosphates inhibition assays (PPIA) [6] can provide the required sensitivity but may yield false positives if the enzyme is inhibited by other compounds present in the sample. The method of lateral flow dipstick (Microcystin Immuno Strip) based on colloidal gold nano-particles [7] was recently introduced for detection of MC-LR however its sensitivity is not sufficient. Surface plasmon resonance (SPR) has become a widely used analytical technique which lead to development of SPR based immunosensor for the determination of MC-LR [8], the required high sensitivity can be achieved using competitive immunoassay format. Furthermore, the Quartz Crystal Microbalance (QCM) method was developed in combination with sandwiched immunoassay amplified by Au nano-particles which reduces the detection limit for (MC-LR) down to 0.1 pg/l [9]. Another detection method using color-changeable polydiacetylene vesicle achieved the 1 pg/l detection level [10]. Relatively low molecular weight of MC-LR (995 m/mol) makes it difficult to detect using direct

immunoassya in combination with conventional QCM and SPR based analytical methods; the sensitivity enhancement can be acheieved using more complex and expensive assyas, i.e competitive or sandwich immunoassays.

Therefore, the development of highly sensitive, reliable, and (at the same time) inexpensive and easy-to-use methods of detection of microcystin is of very high importance nowadays. The method of Total Internal Reflection Ellipsometry (TIRE) [11] is particularly attractive for the above task considering its high sensitivity and particular suitability for detection of low molecular weight analytes in direct immoassay format [12]. This method has been used recently for detection of different toxins, i.e. mycotoxins (T2, aflatoxin, zearalenone) [13-15] alkylphenols (nonylphenol) [16] and herbicides (simazine and atrazin) [17] and achieved the detection of the above substances in sub-ppb range (0.1 ng/ml) in direct immunoassay format. In this work we tried to utilize the method of TIRE for detection microcystin LR.

2. Experimental details

2.1 Chemicals for immunoassay and samples preparation

Microcystin LR (MC-LR) was detected in direct immunoassay with specific monoclonal antibody MC10E7 raised against MC-LR in rabbit (the immune pair was acquired from Enzo Life Scientific). Other chemicals used were: systamine hydrochloride, (poly)allylamine hydrochloride (PAH) protein G, Tris-HCl and PBS buffers (all from Sigma Aldrich).

The antibodies were immobilized on the surface of gold coated microscopic glass slides which were prepared by consecutive thermal evaporation of layers of chromium (3 nm) and gold (25 nm) without breaking the vacuum of about 10^{-6} Torr in the Edwards E360A evaporation unit. The presence of thin Cr layer improves the adhesion of Au layer to glass. The surface of gold was charged negatively by immersing gold coated slides overnight into 100mM solution of mercapto-ethylsulfonate sodium salt ($SH\text{-}(CH_2)_2\text{-}SO_3^-$ Na^+ in methanol.

The following immobilization steps were performed by injection of required solutions into reaction cell. First, a layer of polycationic PAH was electrostatically adsorbed on the surface of gold from its 2 mg/ml aqueous solution in water. Secondly, protein G molecules, carrying net negatively charge at pH 8 in Tris-HCl buffer, were electrostatically adsorbed on the surface. Then, antibodies MC10E7 were attached to protein G, and therefore have their Fab-fragments oriented towards the solution.

Microcystin LR (MC-LR) is a hyrophobic molecule with molecular wieght of 995D. In order to reduce the probability of formation of micelles of microcystin in aqueous solutions, the 100 μg/ml stock solution of MC-LR was prepared in a mixture of 10% of methanol and 90% of deionised water (MQ Elga). The series of samples of different concentrations of MC-LR from 0.1 ng/ml up to 10 μg/ml were prepared by multiple dilution of the stock solution in deionised water.

2.2. TIRE measurements.

Total Internal Reflection Ellipsomerty (TIRE) measurements were performed using the experimental set-up described in detail earlier [12-17] which is based on J.A.Woollam spectroscopic ellipsometer M2000. As shown schematically in Fig. 1 (inset), the key element of the above set-up is a 68^0 glass prism in optical contact (via index matching fluid) with the gold coated glass slide. Such configuration allows coupling the light into the thin gold film at an angle close to the critical angle of total internal reflection at the glass/water interface, and therefore forcing the ellipsometer to operate in the surface plasmon resonance (SPR) mode. The main advantage of TIRE method, however, lies in the ability of recording spectra of two parameters Ψ and Δ representing, respectively, the ratio of amplitudes of and the phase shift between p- and s- components of polarized light, as compared to one parameter of the reflected light intensity in conventional SPR. The modelling showed that the parameter of Δ is about 10 fold more sensitive than Ψ to small changes in the refractive index and thickness of adsorbed layer on the surface of gold [12]. A PTFE reaction cell of 0.2ml in volume sealed against the gold layer has the inlet and outlet tubes enabling the injection of different solutions into the cell.

Two types of TIRE measurements were performed: (i) dynamic spectral measurements where a number of Ψ(λ) and Δ(λ) spectra were recorded during the adsorption or binding of all molecules (the incubation time of 15 min was used for all adsorption or binding stages); (ii) single spectroscopic scans which performed in the same PBS buffer solution (pH 7.5) after washing our non-specifically bound material by purging of up to 6 ml of water (or PBS) buffer through the cell.

2.3. TIRE data fitting

Figure 1 shows typical Ψ(λ) and Δ(λ) spectra recorded on bare gold coated glass slides. Ψ(λ) spectrum resembles traditional SPR curve while Δ(λ) spectrum exhibits a characteristic phase drop from 270^0 to -90^0. Because of higher sensitivity of Δ, all further analysis is based on Δ(λ) spectra. The thickness evaluation was performed by the fitting of the obtained spectra of Ψ(λ) and Δ(λ) using a least-square minimization software available in J.A. Woollam instruments [18]. Four-layer model containing (i) BK-7 glass, (ii) Cr/Au layer, (iii) adsorbed molecular layer, and iv) water was used for TIRE data fitting [12]. The effective values of the thickness (d) and effective dispersion functions of the refractive index n(λ) and extinction coefficient of k(λ) for a mixed Cr/Au layer were obtained by fitting the spectra recorded on the samples with bare surface of gold and then kept fixed at consecutive fittings on the same sample. All molecular layers adsorbed on the gold surface are considered to be transparent in the 370-1000nm spectral range of M2000 ellipsometer so that k = 0, and the dispersion function of the refractive index is described by Cauchy model [18]:

$$n = A + \frac{B}{\lambda^2} + \frac{C}{\lambda^4}, \tag{1}$$

Because the refractive indices of all amino acid-based bio-molecules are very close [19] we used the fixed values of A=1.39, B=0.01, and C=0 which yield n = 1.42 at 630 nm and therefore the effect of molecular adsorption is associated with the changes in the effective adsorption layer thickness only.

3. Experimental results and discussion.

3.1. TIRE study of the immune reaction between MC-LR and MC10E7

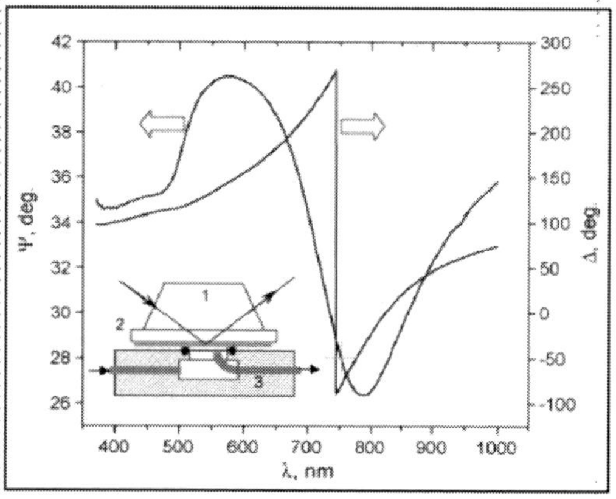

Figure 1. Ψ(λ) and Δ(λ) spectra recorded on gold coated glass slides; inset shows the prism-cell assembly.

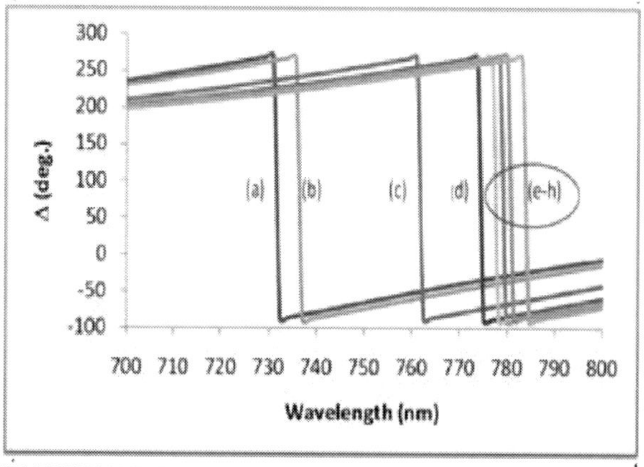

Figure 2. Typical set of Δ(λ) spectra of: bare Au (a), after deposition of PAH (b), protein A (c); antibodies (d), and after binding MC-LR on increasing concentrations (e-h)

A typical set of Δ spectra in Figure 2 shows consistent spectral shift to high wavelengths stared from bare gold treated with $SH-(CH_2)_2-SO_3^-\ Na^+$ (curve a), then after electrostatic deposition of PAH layer (curve b), after electrostatic deposition of protein G (c), and after deposition of MC10E7 antibodies (d). The following group of curves (e - h) corresponds to sequential binding of MC-LR of different concentrations (starting from the smallest 0.1 ng/ml up to 10μg/ml). All deposition and binding stages (e–h) were performed by injecting corresponding solutions into TIRE cell. Intermediate rinsing was performed between adsorption (binding) steps by purging 2-3 ml of water or buffer through the cell. The dependence of the of the thickness increment of the molecular layers the concentration of MC-LR in shown in Figure 3 was obtained by fitting TIRE data. The lower limit of detection for MC-LR is between 0.1 ng and 1ng/ml. The maximal thickness increment of about 1.2 nm is reasonable for non complete coverage by MC-LR molecules of about 3 nm in size. The obtained dependence in Fig. 3 can be used in future as a calibration curve for testing samples of water containing unknown concentration of MC-LR.

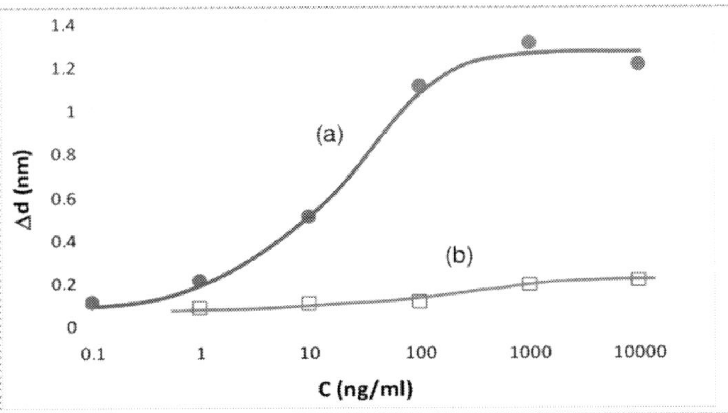

Figure 3. Dependences of the molecular layer thickness increase vs concentration of MC-LR: (a) for untreated MC-LR solution, (b) for MC-LR solution treated with microparticles functionalized with MC10E7 antibodies.

3.2. The study of MC-LR/MC10E7 binding kinetics.

Dynamic spectral measurements described earlier were performed in-situ during all adsorption and binding steps. Typical time dependencies of Ψ and Δ at particular selected wavelength (670 nm this time) during binding MC-LR to MC10E7 are shown in Figure 4a. As one can see the noise is much larger on Y curve which is quite obvious considering different angle spans 0.5^0 for Ψ and 2.5^0 for Δ. It illustrates once again much high sensitivity of the parameter Δ and justify the use of time dependences of Δ for further analysis. Figure 4a also shows that the binding reaction reaches saturation after 15 min of incubation.

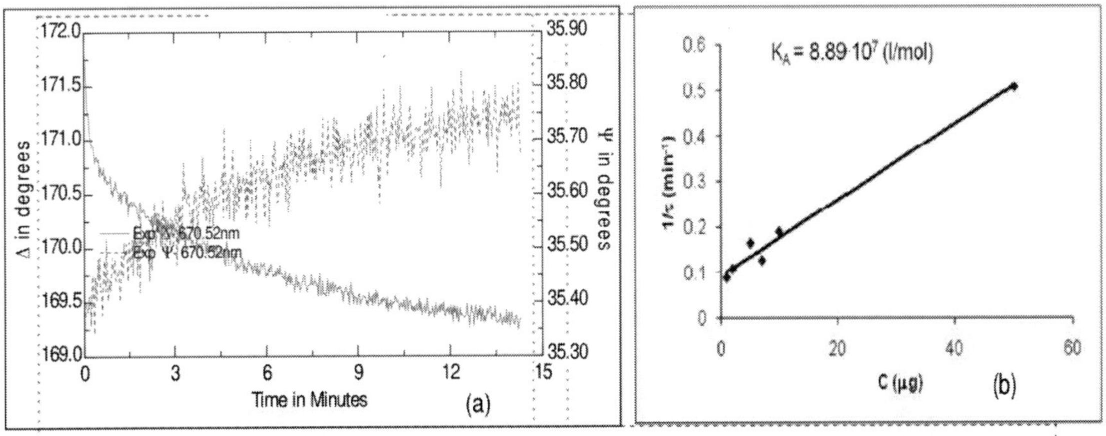

Figure 4. (a) Typical time dependencies of Ψ and Δ during binding MC-LR from its 100 ng/ml aqueous solution to MC10E7 antibodies; (b) The evaluation of the association constant K_A from the dependence 1/τ □□vs. concentration of MC-LR .

Both curves in Figure 4a can be described by exponential dependences, and the characteristic time constant (τ) can be found by curve fitting. Further analysis and the evaluation of the association or/and affinity constants is based on the solution of a first order differential equation for molecular adsorption on the surface [17]. It appeared that reciprocal of the time constant 1/τ is linearly dependent of the concentration of adsorbed molecules (C).

$$\frac{1}{\tau} = k_a C + k_d \tag{2}$$

The values ka and kd being the rates of adsorption and de-sorption can be evaluated from the linear graph 1/τ vs. C in Figure 4b as the gradient and intercept, respectively. Then both the association and affinity constants can be found as:

$$K_A = \frac{k_a}{k_d}, \quad K_D = \frac{1}{K_A} \tag{3}$$

The analysis of the data presented in Fig. 6b yield the values in the range of 10^7-10^8 l/mol ($8.89.10^7$ l/mol for the particular set of data presented) with the accuracy of 10-15%. The obtained values of K_A are typical or highly specific immune reaction between MC-LR and MC10E7 antibodies.

3.3. Formation of functionalized polyelectrolyte microcapsules for MC-LR cleansing

Polyelectrolyte microcapsules were prepared by coating inorganic $CaCO_3$ template particles with the layers of polyanions (sodium poly-stryrene sulfonate or PSS) and polycations (poly-allylamine hydrochloride or PAH). The obtained micro-particles were further functionalized with protein G and MC10E7 antibodies specific to MC-LR following the procedure described in [20]. Fluorescence microscopy image of the obtained micro-particles stained with SYTO-9 green fluorescent dye is shown in Figure 5 (a), SEM images of different magnifications are shown in Figure 5 (b, c).

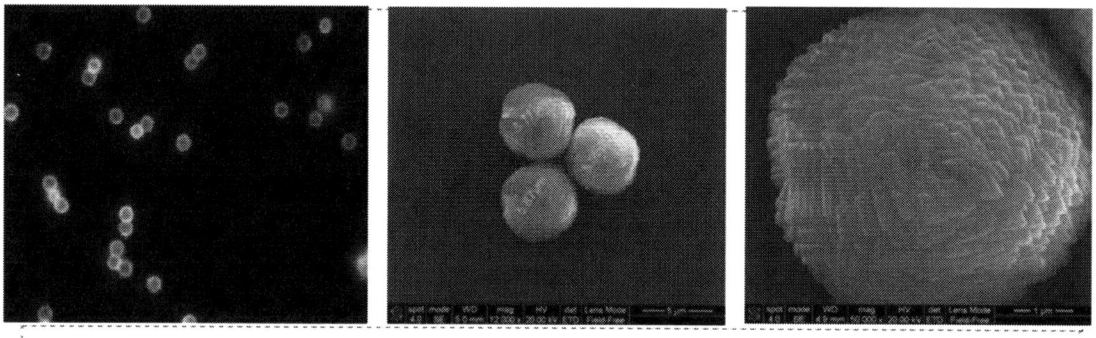

Figure 5. 1SEM image (a) and SEM images (b, c) of 6 µm $CaCO_3$ micro particles coated with PSS/PAH and functionalized with MC10E7 antibodies.

Following the procedure developed earlier [20] micro-particles functionalized with MC10E7 antibodies were utilized for purification of solutions containing MC-LR. The solutions of MC-LR of different concentrations were mixed 1:1 with suspension of functionalized microparticles and incubated for up to 1 hour. Heavy particles sediment on the bottom of sample tubes leaving purified solution above.

UV-vis spectral measurements of MC-LR samples in Figure 6 show the reduction in the intensity of the main absorption band of MC-LR at about 240 nm after treatment with microparticles functionalized with MC10E7 antibodies. The effect of purification was also confirmed with TIRE measurements taken on MC-LR samples after treatment with functionalised capsules. As one can see in Figure 2, the data points of the thickness increment for the samples treated with microparticles (curve b) lie well below the original calibration curve (a).

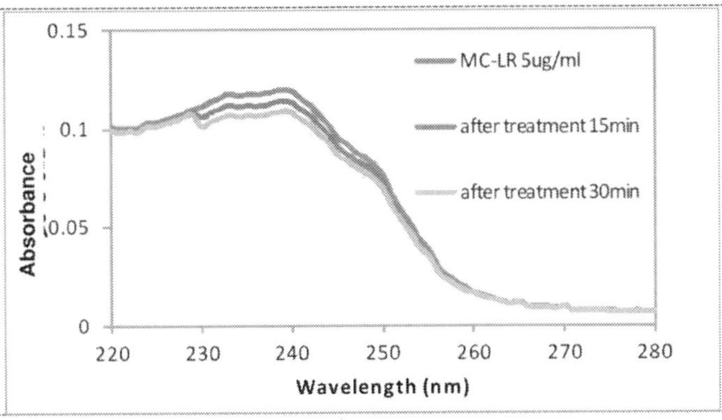

Figure 6. UV-vis absorption spectra of 5 μg/ml MC-LR aqueous solution sample: original (a), and after treatment with functionalized microparticles for 15 min (b) and 30 min (c).

4. Conclusions

The method of TIRE was proved to be capable of detection of Microcystin LR in low concentrations down to 0.1ng/ml in direct immunoassay with specific monoclonal antibodies MC10E7. The above immune reaction is highly specific according to the results of the binding kinetics which yielded the values of the association constant in the range of 10^7 - 10^8 (l/mol). $CaCO_3$-based microparticles functionalized with MC10E7 antibodies are acting as absorbent for MC-LR in water solutions, and could be used in future for purification of contaminated samples. Further work will focus on detection of microcystin produced by cyanobacteria in real water samples.

References

[1] Lindholm T, Eriksson J E, Meriluoto J A O 1989 *Water research* **23** 481-486
[2] Gregora J, Marsˇaleka B, Sˇıpkova H 2007 *Water Research* **41** 228–234
[3] Botes D P et al. 1985 *J Chem. Soc., Perkin transactions*, **I** 2747-2748
[4] Cyanobacterial toxins: Microcystin-LR in Drinking-water 1998 World Health Organization, Geneva.
[5] Mountfort D O, Holland P, Sprosen J 2005 *Toxicon* **45** 199-206
[6] Jayaraj R, Lakshmana Rao P V 2006 *Toxicon* **48**, 272-277
[7] Herranz S, Marciello M, Olea D, Velez M, Marazuela M D, Guisan J M , Moreno-Bondi M C 2012 Eurotrode XI, 1-4 April, Barcelona, O- B5.1
[8] Campas M, Marty J-L 2007 *Biosensors and Bioelectronics* **22** 1034-1040
[9] Xia Y, Zhang J, Jiang L 2011 *Colloids and Surfaces B: Biointerfaces* **86** 81 -86
[10] Herranz S, Bockova M, Marazuela M D, Homola J, Moreno-Bondi M C 2010 *Anal. Bioanal. Chem.* **398** 2625-2634
[11] Arwin H, Poksinski M, Johansen K 2004 *Appl. Optics* **43** 3028-3036
[12] Nabok A, Tsargorodskaya A 2008 *Thin Solid Films* **516** 8993-9001
[13] Nabok A, Tsargorodskaya A, Holloway A, Starodub N F, Gojster O 2007 *Biosensors & Bioelectron.* **22** 885-890
[14] Nabok A, Tsargorodskaya A, Mustafa M K, Szekacs I,.Starodub N F, Szekacs A 2011 *Sensors & Actuators B, Chemical* **154** 232-237
[15] Nabok A, Mustafa M K, Tsargorodskaya A, Starodub N F 2011 *BioNanoScience* **1** 38-45
[16] Nabok A, Tsargorodskaya A, Holloway A, Starodub N F, Demchenko A 2007 *Langmuir* **23** 8485-8490
[17] Nabok A, Tsargorodskaya A, Hassan A K, Starodub N F 2005 *Appl. Surf. Sci.*, **246,** 381-386
[18] Guide to using WVASE32 2001 J A Woollam Co. Inc, New York
[19] Szekacs A, Adanyi N, Szekacs I, Majer-Baranyi K, Szendro I 2009 *Appl. Opt.* **48** B151-B158
[20] Nabok A V, Erokhin V, Erokhina S, Szekacs A, Mustafa M K, Al-Ammar R, 2013 *BioNanoScience* **3** 79-84

A Novel Optical Sensor Platform Designed for Wireless Sensor Networks

Shuo Yang, Bochao Zhou, Tong Sun, Kenneth T.V. Grattan

School of Engineering and Mathematical Sciences and City Graduate School
City University London, London, EC1V 0HB, UK

Shuo.Yang.2@city.ac.uk, Bochao.Zhou.1@city.ac.uk, T.Sun@city.ac.uk, K.T.V.Grattan@city.ac.uk.

Abstract. This paper presents a novel design of an optical sensor platform, enabling effective integration of a number of optical fibre ('wired') sensors with wireless sensor networks (WSNs). In this work, a fibre Bragg grating-based temperature sensor with low power consumption is specially designed as a sensing module and integrated successfully into a WSN, making full use of the advantages arising from both the advanced optical sensor designs and the powerful network functionalities resident in WSNs. The platform is expected to make an important impact on many applications, where either the conventional optical sensor designs or WSNs alone cannot meet the requirements.

1. Introduction

Wireless sensor networks (WSN) have provided an effective means in acquiring information such as temperature, pressure, acceleration, vibration and specific chemicals [1]. WSNs have been reported to fulfil complex tasks, for example, for environmental monitoring, industrial sensing and diagnostics, battle tracking and security detection.

A WSN contains a large number of small scale nodes, from several to hundreds. Each node is equipped with functionalities for sensing, computing and communicating. The sensor modules in WSNs often play an important role in determining both the application areas and the performance of the WSNs. WSN nodes are small, lightweight and of low-power consumption, therefore have attracted a significant amount of interest from researchers and developers [2].

To date, most conventional WSN sensing modules, particularly those used in commercial products, are based on electrical or mechanical sensing principles [3,4]. Optical fibre sensors, however, offer advantages in applications where electrical sensors failed to perform well, for example, due to the electromagnetic interference and/or the contamination from harsh working conditions. Despite the advanced development of optical fibre sensors and extensive research into the wireless sensor network, there is limited report on the integration of optical sensors into the wireless sensor network platform. This paper thus aims to establish a novel 'hybrid' WSN platform to allow the seamless integration.

Temperature is one of the most importance parameters to be measured in industrial process control, in scientific activities and in daily life. In this paper, an optical fibre Bragg grating-based temperature sensor is designed specifically, with low power consumption, to serve as an optical sensor module, to replace conventional MEMS based sensor module in a WSN node.

2. Platform Design

Figure 1 shows the generic platform, designed specifically in this work and it comprises three main modules: optical sensing module, WSN module and power module.

Figure 2 shows a detailed design of the optical sensing module, which contains a SLED light source (1550nm, Dense Light), a circulator (4 ports, JDS Fitel), an InGaAs photodiode (Thorlabs) and two fibre Bragg gratings (FBGs) Centred at the same wavelength (1550nm). FBG2 acts as a temperature sensor as its Bragg wavelength shifts as a function of temperature variation [5] and FBG1, located at the other port of the circulator, is used as a filter or a reference for ease of the interrogation of the sensor (FBG2) as its surrounding temperature is fixed (usually at room temperature). This specific sensor design is aimed to minimize the power consumption, by using low power light source (SLED) and a photodetector (PD), with a potential to share the same power module with the WSN. The SLED drive circuit includes a 100-mA constant current source and a high precision transimpedance amplifier (OPA380, Texas Instruments,) with an amplification gain of 10^7 V/W. The sensor signal obtained from FBG2 with reference to the signal from FBG1 is captured by a photodiode before being converted into digital signals using an analog-to-digital converter (ADC). The overall size of the sensing module is 15cm×5.5 cm×4.5 cm, which has been carefully designed to be able to interface with the WSN module shown in Figure 1.

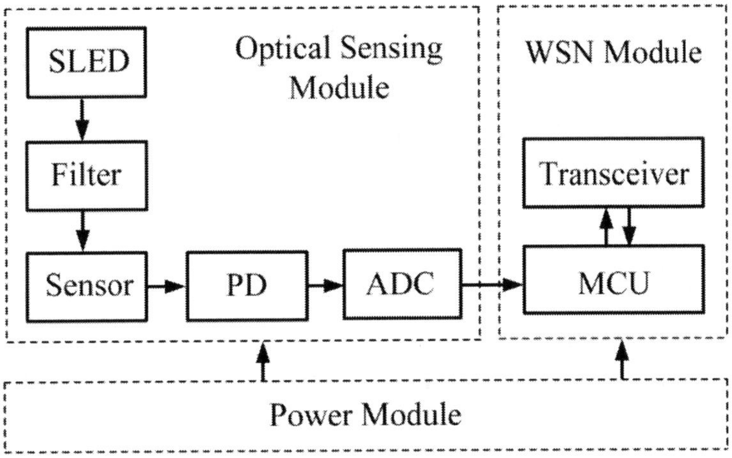

Fig.1. Architecture of the platform

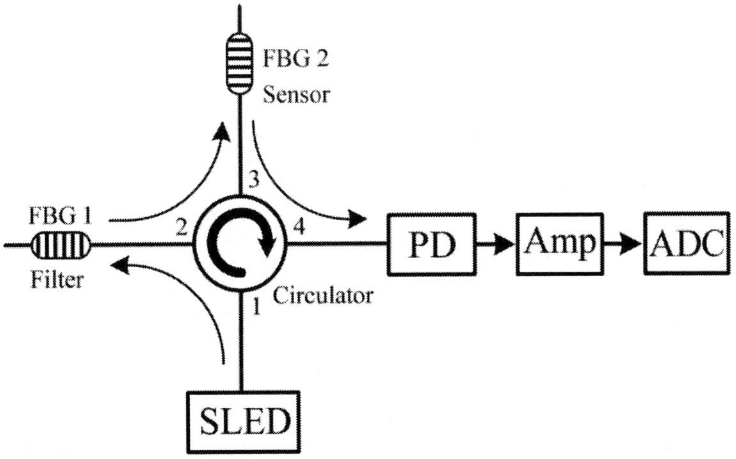

Fig.2. Schematic diagram of an optical sensing module

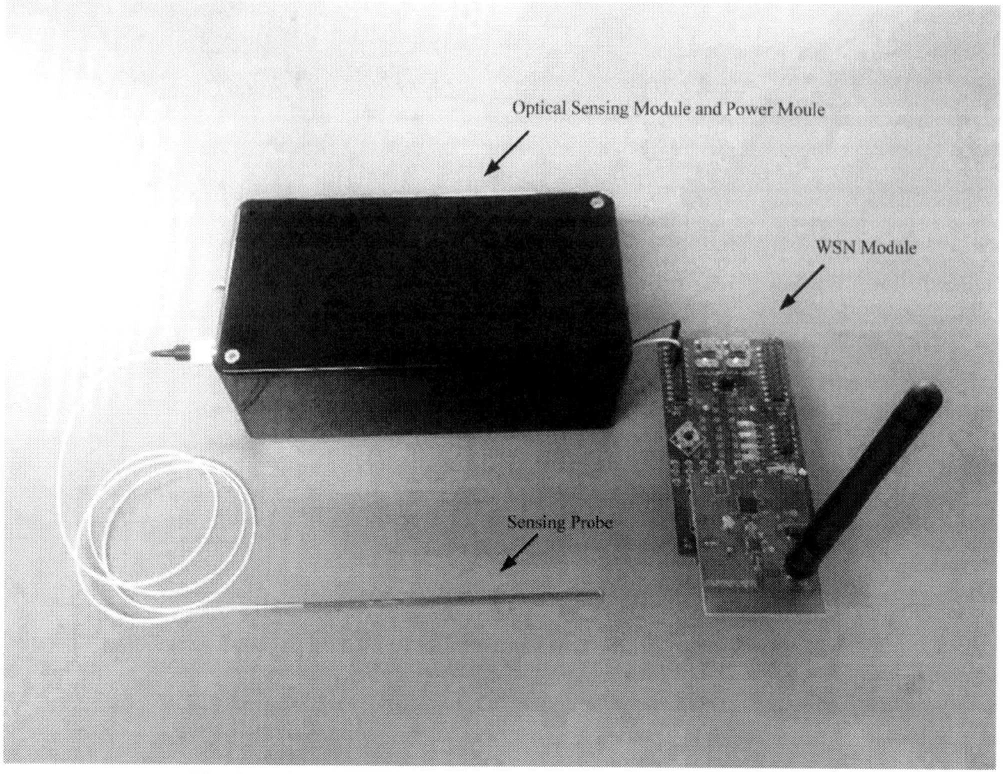

Fig.3. A fully integrated optical sensor platform for WSN

The WSN module is integrated with computing and group communication functionalities, allowing for both data transmission via multi-hop routing and data processing. For the WSN module, CC2530F256 (Texas Instruments Inc.) is chosen as the prototype module since it is a relatively mature platform that has enough extension sockets for further functionality development. The WSN module is composed of a transceiver (CC2431, Texas Instruments Inc.), an MCU (MSP430F5438, Texas Instruments Inc.) with a built-in analog-to-digital converter (ADC) and a power source for data processing and network communication.

The power module used in the platform includes two recycled AAA sized NiMH batteries (550mAh, Ansmann) and a voltage regulator (TPS61200, Texas Instruments Inc.) which fixes the output voltage at 3.3V.

Figure 3 is a photo of a complete optical sensor platform created for a WSN. The sensing probe, which contains FBG2 at the tip, is protected by an aluminium tube and this allows the following multiple experimental tests by subjecting the probe to various temperature conditions.

3. Implementation of the new platform created

Fig.4 shows the experimental data collected from the platform when the optical fibre sensor probe is placed into a temperature chamber with temperature varies from 20℃ to 80℃. The dynamic response of the sensor is obtained from the data-monitoring centre contained in the WSN module. Fig.5 shows the temperature measurement results obtained from the optical WSN node and it is noticeable that the signal intensity decreases with the increase of temperature when it varies.

The data obtained show that the remote sensor can be used effectively and conveniently in this configuration. Multiple tests have been undertaken, including long term tests. The stable output produced indicates clearly that the platform has been successfully created, which offers the seamless integration between the optical fibre sensor and the WSN.

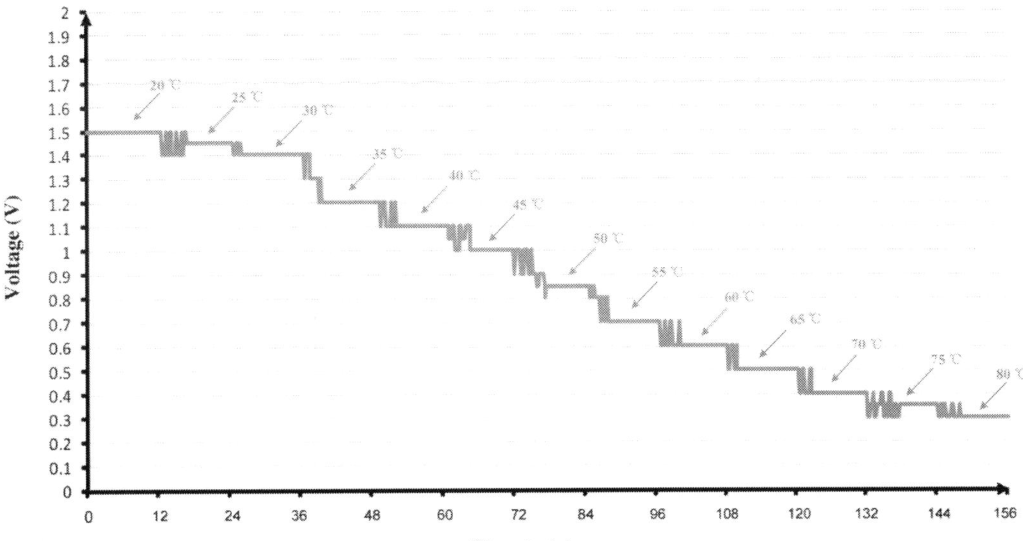

Fig.4 Dynamic response of a wireless sensor node, coupled with an optical fibre temperature sensor (FBG2) when it is subjected to different temperature conditions

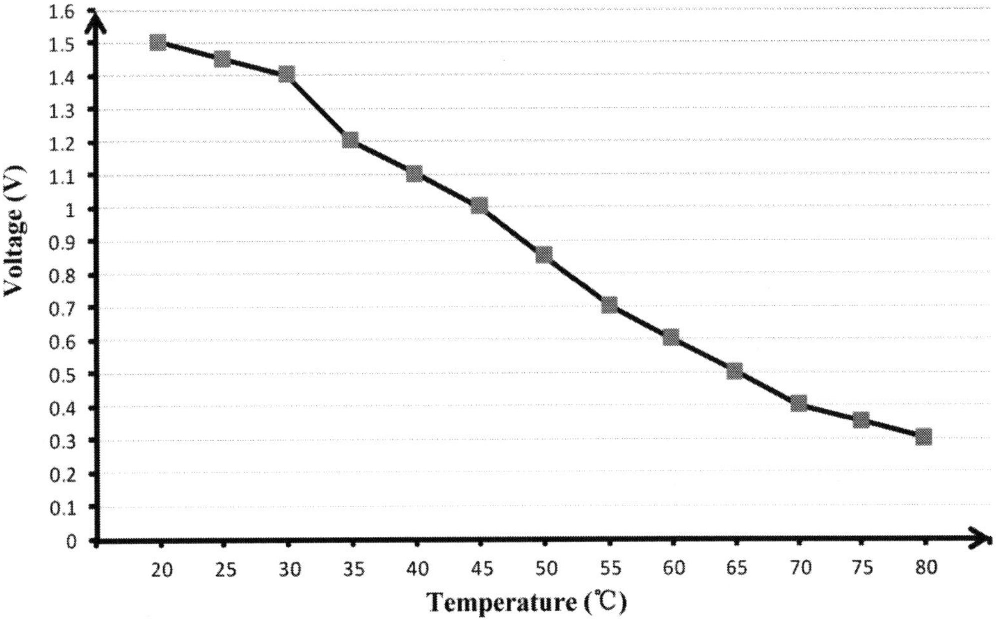

Fig.5 Temperature measurement results obtained from the optical WSN node

4. Conclusion and future work

In this paper, an optical fibre platform designed with a WSN node has been created and tested successfully in this work. The platform is established through hybrid integration of an optical sensor module (temperature sensor module), TI WSN module and a power module. The positive experimental data obtained confirms the compatibility of the optical sensor with WSN nodes, taking the advantage of both the flexible optical sensor design and the network capability of the WSN.

The designed optical platform for WSNs is generic and it can be easily modified to be integrated with an array of optical fibre sensors for various industrial applications where high sensitivities and real-time monitoring are needed. The research work is still on-going and future work consists of more practical field tests using other types of optical fibre sensors to improve further the overall performance of the system and implementation of a full hybrid WSN platform.

References

[1] Akyildiz I F, Su W, Sankarasubramaniam Y and Cayirci E 2002 Wireless sensor networks: a survey *Computer Networks* **38** 393–422

[2] Yick J, Mukherjee B and Ghosal D 2008 Wireless sensor network survey *Computer Networks* **52** 2292–330

[3] Li X, Cheng X, Yan K and Gong P 2010 A monitoring system for vegetable greenhouses based on a wireless sensor network. *Sensors (Basel, Switzerland)* **10** 8963–80

[4] Pakzad S N, Fenves G L, Kim S and Culler D E 2008 Design and Implementation of Scalable Wireless Sensor Network for Structural Monitoring *Journal of Infrastructure Systems* **14** 89–101

[5] Kersey a. D, Davis M a., Patrick H J, LeBlanc M, Koo K P, Askins C G, Putnam M a. and Friebele E J 1997 Fiber grating sensors *Journal of Lightwave Technology* **15** 1442–63

Journal of Physics: Conference Series **450** (2013) 012008 doi:10.1088/1742-6596/450/1/012008

The use of nanotechnology in the development of a distributed fibre-optic temperature sensor for subsea applications

G R McDowell[1*], A S Holmes-Smith[1], M Uttamlal[1], P A Wallace[1], D M Faichnie[2], A Graham[2] and D McStay[2]
[1]School of Engineering & Built Environment, Glasgow Caledonian University, Cowcaddens Road, Glasgow, G4 0BA, UK.
[2]Optoelectronics group, FMC Technologies, Finch Way, Strathclyde Business Park, Bellshill, ML4 3PE

E-mail[*]: gary.mcdowell@gcu.ac.uk

Abstract. Monitoring temperature in subsea pipelines and processing plants is paramount to ensuring the efficient extraction of oil and gas. Preliminary results from studies of an alternative optical temperature sensing methodology based on the fluorescence spectroscopy of quantum dots (QDs) immobilised in a sol-gel thin film are presented. A discussion of the potential for this approach to oil and gas industry applications is given.

1. Introduction

The increasing global demand for oil and gas combined with the decrease in the availability of "easy oil" (e.g. large reserves located on-shore) have forced the industry to seek to exploit more remote and difficult to access resources. A key component of this activity is the exploitation of offshore oil and gas reserves. In parallel the industry is investing in new technologies for increased oil recovery (IOR), improved production efficiency along with reduced costs and risk. Increasingly the industry is seeking to reconfigure offshore production such that more of the process is carried out subsea. This possesses a range of challenges and is driving new technological developments within the industry. The future subsea production system is illustrated in figure 1. Typically such a subsea production facility will be a complex array of specialised equipment that may include trees, subsea processing systems, pipelines and risers. This increase in complexity of the subsea system will require a great increase in the degree of instrumentation on the subsea equipment to allow effective monitoring and control. Optical fibre sensing and communications are one of the technologies that the industry is looking at to provide components of the increased monitoring. Optical fibres have a number of advantages over existing electronic systems for subsea oil and gas applications, which include being more reliable in hostile environments, flexible, immune from electromagnetic interference and the variety of sensing formats. These include the ability of a single optical fibre to be made into a distributed sensor or to support a multitude of discrete sensors.

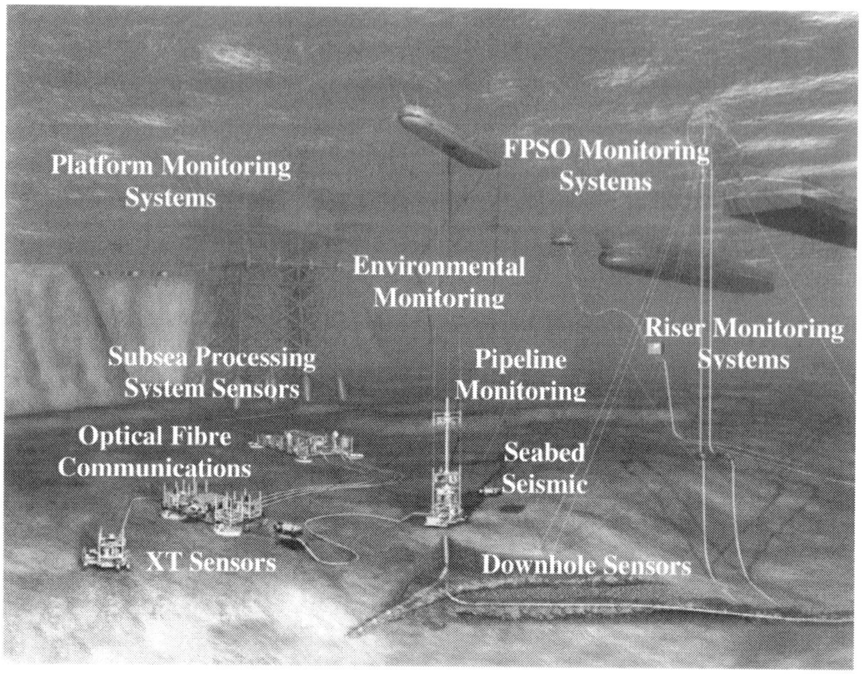

Figure 1. Subsea production system

Temperature measurement is a critical parameter for a range of oil and gas production processes as well as being an important tool in integrity monitoring. For example, knowledge of the temperature is of importance in oil and gas pipelines as the parameter can be used to monitor the formation of hydrates. Hydrates or more specifically gas (clathrate) hydrates are crystalline water based solids that resemble ice, which can have detrimental effects within an oil flow line, as the hydrates can restrict or completely block the flow. This occurs due to low temperatures in the subsea environment and high pressures [1]. There is thus considerable interest in new and improved temperature sensing schemes that will lend themselves to use in subsea oil and gas production systems.

In this paper we report results from preliminary studies into the potential of employing semiconductor Quantum Dot (QD) nanoparticles as the basis of a distributed array optical fibre temperature sensor scheme. Although still at a relatively early stage QD technology is an expanding area of interest, particularly in environmental and biomedical applications [2], with a number of the initial QD materials possessing properties that make them potentially suitable for temperature sensing applications. The QD materials are nanometer scale particles which follow the quantum confinement effect [3]. This means that the energy level of a QD is dependent on its size. Therefore, the smaller the QD the larger the band gap (the greater the difference in energy between the highest valence band and the lowest conduction band) resulting in more energy being required to excite a QD. This gives QDs unique optical properties which provide outstanding photostability, relatively high quantum yields (>50%) and with comparison to other luminescent dyes, have a narrow emission spectrum, are bright and can be detected through visible excitation.

Recently, it has been demonstrated that temperature has an effect on QDs luminescence, opening the field for thermometry applications. QDs have been shown to shift in wavelength with the increase of temperature, in studies ranging from room temperature to 100 °C and 200 °C respectively [4-5]. The highest reported temperature range using QDs was carried out by Pugh-Thomas et al [6] who carried out temperature studies of a QD encased in a sol-gel environment ranging from 22-252 °C resulting in a linear red wavelength shift and a linear decrease in luminescence intensity. A reversible wavelength shift is observed due to heat expanding the size of the QD, decreasing the band gap energy, hence less energy is required to excite the QD [6].

Journal of Physics: Conference Series **450** (2013) 012008 doi:10.1088/1742-6596/450/1/012008

2. Experimental

2.1 General

The QD (CdSe/ZnS) nanopowders were supplied by Ocean NanoTech (USA). The Cadnium selenide (CdSe) core shell QDs is coated with a passivating layer of zinc sulphide (ZnS) to yield stable and highly emissive QDs. The quantum yield (>50%) of the QDs was determined by the manufacturer using the integrating sphere All solvents were purchased from Sigma Aldrich UK (>99% purity) and were used as received. QD thin films were prepared on a glass substrate and the temperature response characterised using fluorescence spectroscopy.

2.2 Sensor preparation

A sol solution was prepared by mixing 3ml of tetraethyl orthosilicate (TEOS) with 3ml of ethanol, in which 1ml of deionised water was added dropwise. 150μl of 0.01M HCl was added to catalyse the polymerisation of the slightly acidified sol-gel precursor, avoiding too strong an acidic solution to prevent degradation of the QD nanoparticles. The sol precursor was mixed uncovered for 45 minutes using a magnetic stirrer and flee. After, the vial was sealed with parafilm and placed in an oven at 60°C for 4 hours. 2mg of each QD powder was mixed with 500μl of the sol-gel precursor using an ultrasonic bath for 2 minutes. The QD/sol mixture was then mixed for 30 minutes before being placed onto clean glass substrate. The QD films were then left to dry for 48 hours and then placed within an oven at 100°C, producing QD glass films. All films were kept in the dark until used.

2.3 Sensor characterisation

Absorption spectrometry of the QDs in solvent was performed using a Perkin Elmer Lambda 2 dual beam UV-Vis spectrometer. Luminescence spectrometry was carried out using a FluoroMax-3 Spectrometer (HORIBA Jobin Yvon Ltd). In order to characterise the emission of the QDs with temperature, the sensors were placed into a thermally controlled oven with an entry port, capable of supporting a high temperature resistive bifurcated fibre-optic bundle that was connected to a holder within the spectrometer. Excitation of the QD films was performed using a 405nm laser diode with an operating output power of 30mW. A 455nm cut-on long pass wavelength filter (Schott GG455) was placed in front of the emission monochromator, to exclude excitation light.

3. Experimental results

General characterisation of the QDs was performed by dissolving 1mg of each QD powder in dichloromethane (DCM). Absorbance and emission spectra were obtained for all QD nanoparticles.

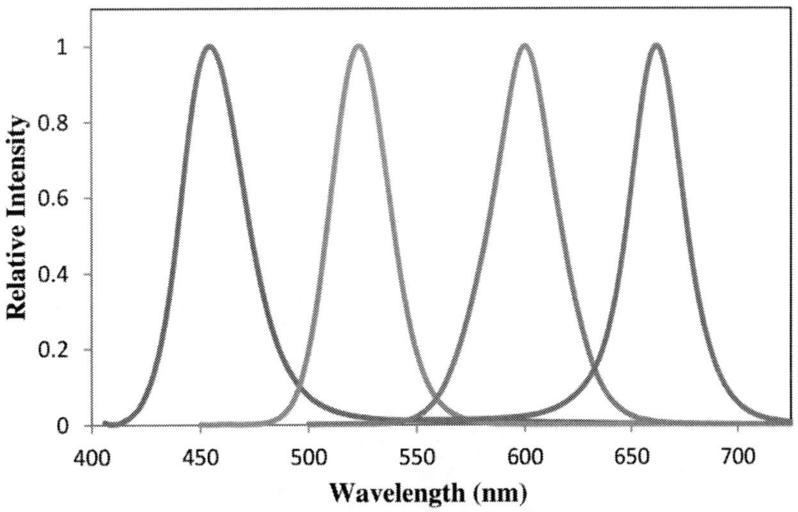

Figure 2. Luminescence emission spectra of the QD nanoparticles in DCM, $\lambda ex = 405$ nm

Figure 2 shows the luminescence spectrum of each of the QDs, highlighting that each signal can be easily distinguished from one another. Each QD has a specific absorption peak (table 1) overlaid on a broad absorption band allowing each QD to be excited using the same excitation wavelength (405nm). This is beneficial in the development of a distributed system, as the instrument development will be simplified by the use of one commercially available light source. Table 1 provides the corresponding data, with reference to the manufacturer's specifications.

Table 1. Specifications of the QD nanoparticle powders

	Manufacturer's Values		Measured Values	
QD powder (nm)	Absorbance (nm)	Emission (nm)	Absorbance (nm)	Emission (nm)
450	431	453	430	454
525	509	523	506	526
600	584	602	584	600
665	650	663	653	662

It is evident from table 1 that the optical properties meet the specifications outlined by the manufacturer, as the tolerance peak wavelength for each QD is ±5 nm. All Full Width Half Maximum (FWHM) were less than 35 nm.

Once prepared, the QD sol-gel films were then exposed to temperature measurements in which the QD luminescence peak wavelength and intensity were recorded. QD luminescence spectra of the green (526 nm) QDs and experimental data are seen in figure 2. Each QD showed a linear red wavelength shift as temperature increased (table 2). The green (526 nm) and red (662 nm) QDs showed a similar wavelength shift whereas that observed for the orange (600 nm) QDs was half. The QD luminescence intensity decreased linearly with increasing temperature and the FWHM increased by ~5 nm over the temperature range studied. The decrease in luminescence intensity can be attributed to temperature quenching of the luminescence, with the green QD showing the greatest decrease in peak luminescence intensity. Similar observations were reported by Biju et al [7] for QDs in solution.

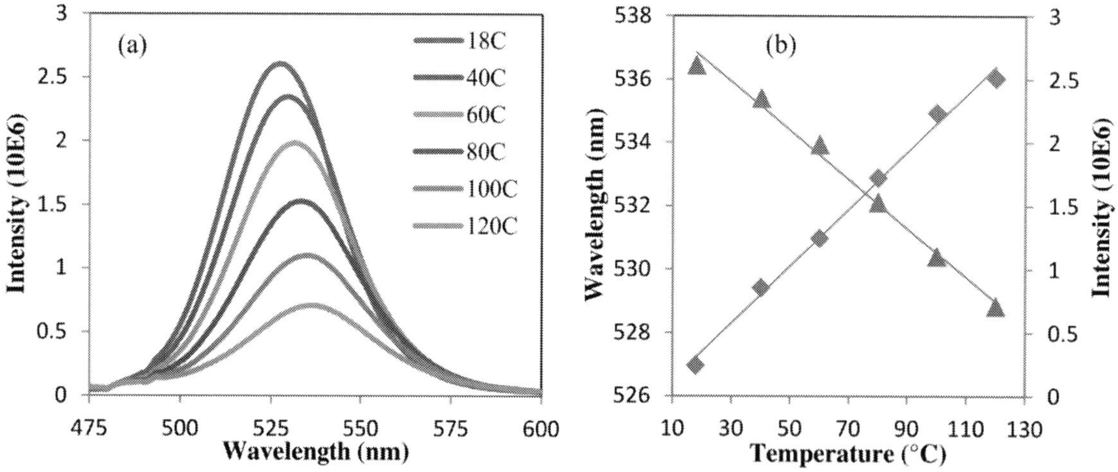

Figure 1. QD luminescence spectra of the 526 nm QDs with temperature (a) and corresponding temperature response (b) of luminescence wavelength (◊) and luminescence intensity (Δ). $\lambda ex = 405$ nm

 IOP Publishing
doi:10.1088/1742-6596/450/1/012008

Table 2. Temperature response from each of the QD sol-gel films

QD (nm)	Wavelength shift (nm/°C)	Coefficient of variation (R^2)
525	0.086 ± 0.006	0.996
600	0.041 ± 0.004	0.987
665	0.098 ± 0.008	0.994

The wavelength shift of the QDs observed with temperature introduces the possibility of implementing a distributed array sensor. One excitation source simplifies the build of such a system, in which, a large Stoke's Shift between excitation and luminescence signals, aids in the easy separation of the spectra.

Here the emission spectra of 3 QDs with temperature were measured. The manufacturer currently supplies 9 QD powders (450, 490, 525, 540, 575, 600, 630, 645 and 665 nm), therefore giving the possibility of 9 different QD sensing sites. Assuming that each QD shifts ~10 nm over a 100 °C range, then each QD emission spectrum can still be distinguished. This opens up the possibility of a multipoint temperature sensing system for use in subsea applications. Figure 3 shows a schematic representation of how this would be achieved. In this method the QD material would be located at the distal end of a spliced fibre. The excitation light (λ_{ex}) would be transmitted down the fibre, in which the 3 dB coupler would split the excitation light and a small percentage would be directed to the QD sensing tip, the rest of the light would continue down the fibre until it gets to the next coupler, and so on. The emission signals (λ_{em}) would be coupled back into the fibre and detected by a CCD spectrophotometer. The QD emission spectra can then be separated by their wavelength response.

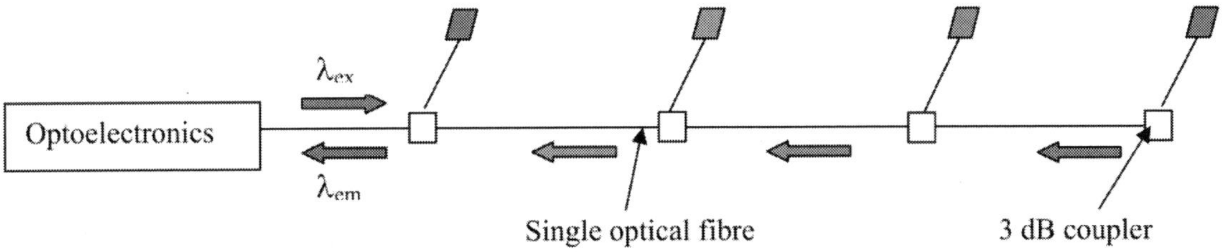

Figure 2. Schematic diagram of a distal end fibre-optic distributed system

4. Conclusions

The use of semiconductor QD nanoparticles in thermometry applications has been addressed. The technology described is beneficial for distributed temperature sensing and could prove useful in the oil and gas industry for subsea applications. Introducing a variety of luminescent QDs within the sensing system opens up the possibility of more temperature sensing points in, for example, an oil flow line. Preliminary results indicate a linear wavelength response from the QD films in going from room temperature to 120 °C. This temperature range would limit the sensors use i.e. it could not be used in down-hole environments, as temperatures here exceed 120 °C. Nevertheless, a luminescence based fibre-optic temperature sensor has been demonstrated with the use of semiconductor QD nanoparticles, which can be used in the application of a distributed sensing methodology.

References

[1] Gao S 2000 *Energy & Fuels.* **22** 3150-53
[2] Clapp A R, Medintz I L, Mauro M, Fisher B R, Bawendi M G and Mattoussi H 2004 *J. Am. Chem. Soc.* **126** 301-10
[3] Li-Gong Z, De-Zhen S, Xi-Wu F and Shao-Zhe L U 2002 *Chin. Phys. Lett.* **19** 578-80
[4] de Bastida G, Arregui F J, Goicoechea J and Matias I R 2006 *IEEE Sensors Journal.* **6** 1378-79
[5] Chen K J, Chen H C, Shih M H, Wang C H, Kuo M Y, Yang Y C, Lin C C and Kuo H C 2012 *J. of Lightwave. Tech.* **30** 2256-61
[6] Pugh-Thomas, D, Walsh B M and Gupta M C 2011 *Nanotechnology.* **22** 1-7
[7] Biju V, Makita Y, Sonoda A, Yokoyama H, Baba Y and Ishikawa M 2005 *J. Phys. Chem. B.* **109** 13899-905

Detection and classification of orange peel on polished steel surfaces by interferometric microscopy

M L Miranda-Medina, P Somkuti and B Steiger
AC^{2}T Research GmbH
Viktor Kaplan Strasse 2, Wiener Neustadt 2700, Austria

E-mail: medina@ac2t.at

Abstract. In this work, we provide a general description of the so-called orange peel defect produced on polished steel surfaces. By characterizing a prototype set of samples with various degrees orange peel, we attempt to create a simple model that allows the classification of additional samples through the study of surface parameters. On those surfaces, the orange peel structure has roughness amplitudes in the nanometer range. Detecting surface features on that range requires the implementation of a high-precision technique, such as phase shifting interferometry (PSI). Therefore, we can contribute to the improvement of the manufacturing of polished steel surfaces as well as to the quality control by using optical techniques.

1. Introduction

Recently, the manufacturing steel industry has increased the use of optical techniques in their respective quality monitoring schemes. Such optical systems are favourable due to the high speed in detection as well as the non-destructive measuring principle. In particular, highly polished metallic surfaces are easily damaged through contact with even very low load.

However, during polishing processes, different physical and chemical parameters are involved, and some of them can lead to specific artefacts on the surface. Unlike the orange peel texture in paints, which is usually produced on purpose, we focus on orange peel on metal that is an undesirable roughening on the surface. Consequently, the reflection of the light is not homogeneous on the surface, affecting the functionality of polished surfaces for particular applications such as the production of mirrors, where specular reflection is a quality criterion.

Commercial devices are currently available for the evaluation of orange peel in paint, plastic or even mirror-like metals [1], however the most of these devices measure orange peel structures with amplitudes in the micron range. Nevertheless, under specific conditions most chemical-mechanical polishing techniques produce an orange peel texture with amplitudes in the nanometer range.

Despite this, such types of orange peel are still detected by visual inspection performed by experts in the field. Thanks to the fact that the human eye is sensitive enough to perceive visual effects such orange peel, the appearance of polished surfaces is evaluated and classified according to levels of quality. Due to the subjective nature of this method, the support of additional techniques is needed for the standardization of the quality control process, independently of sample position, evaluator or lighting conditions. Hence, we propose the topography characterization of polished samples by means of interferometrical measurements.

Interferometry is a well-known technique that associates shapes of produced fringes (circular, elliptical, straight, irregular) to the shape of surface. Phase shifting interferometry (PSI) allows the measurement of the surface topography at high accuracy and is thus the method of choice to obtain topographies of polished steel surfaces. State of the art 3D interferometrical microscopes include a PSI mode, where sequence of images is obtained and processed by computational methods to extract information of the sample surface. Measurements done by PSI have a vertical resolution of less than 1 nm independent of the field size. The common feature of PSI instruments is that, unlike differential or phase contrast methods, surface heights are directly proportional to interference phase [2].

2. Detection and evaluation of orange peel

In collaboration with steel industry[1], we performed the evaluation of two sets of polished steel samples. The first set, composed of six samples, has been classified by visual inspection as prototypes of degrees of orange peel, and the second set, which contains ten samples, are unclassified specimens. In the set of prototypes, three degrees of orange peel are identified with the classifiers "strong", "intermediate" and "weak". In section 2.1 we provide detailed information about the methodology as well as the relevant results for each set. Afterwards, in section 2.2 we develop a simple model to classify the second set of samples according to their surface parameters.

2.1. Experimental details and results

A total of 16 samples were provided to us with the initial information summarized in table 1. The samples are made of stainless steel and have a lateral size of 10 cm x 10 cm.

Table 1. Samples information.

Degree of orange peel (visual)	Set 1 (prototypes)	Set 2
Weak	W1, W2	
Intermediate	I1, I2	
Strong	S1, S2	
Unclassified		Ui [a]

[a] For these samples i = 1, 2….10

The topographies of the samples have been determined by means of a commercial 3D interferometric microscope in the PSI operational mode, where the focused sample is scanned vertically in steps that are very accurate fractions of the wavelength. The profiling algorithms produce a phase map of the surface, which is converted to the corresponding height map via an unwrapping procedure [3]. The magnification of the selected Mirau objective is 10x, measuring a field of view (FOV) of 1270 µm × 950 µm, and uses a blue LED of 460 nm wavelength as light source. In order to increase the size of the evaluated area, we stitched 12 × 12 single images with 10 % of overlap by using a proper software [4] to achieve a total area of 11.8 mm × 8.7 mm. Afterwards, the full evaluated area of each sample was levelled via subtraction of a least-squares plane and subsequently the main shape was removed by subtracting a least-square second order polynomial. In figure 1, as an example of orange peel degrees, the topographies corresponding to samples W1, I1 and S1 are shown respectively.

[1] Berndorf Band GmbH, Leobersdorfer Strasse 26, A-2560 Berndorf, Austria.

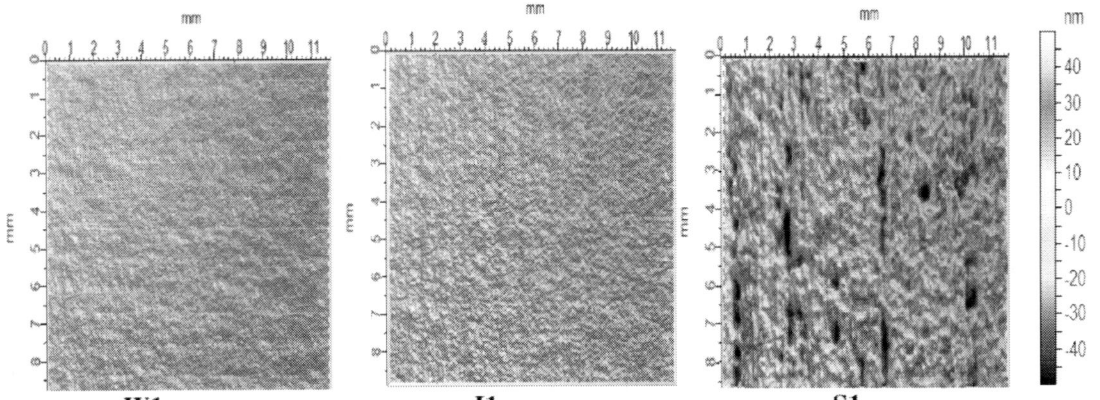

Figure 1. Three different degrees of orange peel: W1 (weak), I1 (intermediate) and S1 (strong).

In figure 1, the sample referred to as W1 exhibits the lowest grade of orange peel on the surface. On this area, the orange peel texture is manifested without any predominant orientation. In contrast, the samples I1 and S1 are perceived by the microscope as a pebble pattern with a specific orientation, which is created during the treatment of the base material. In addition, a peculiar characteristic that distinguishes degrees of orange peel is the amplitude of the surface (see color bar in figure 1) measured from a reference plane. For samples W1 it is in the range between 0 nm to 10 nm, while for S1 it can be between 20 nm to 50 nm. Therefore, in order to provide a quantitative description of orange peel topography, surface parameters are calculated from the measured data. These evaluate diverse characteristics of the surface, such as amplitude of peaks and valleys, spatial distribution of structures, texture orientation, slopes of on a surface, etc. Surface parameters are calculated according to ISO 25718 [5] and a full description of these can be found in reference [6]. To characterize the degrees of orange peel for the prototype samples, we have selected 20 surface parameters from the categories: areal height parameters (*Sq, Sa, Ssk, Sku*), areal spacing parameters (*Sal, Str*), areal hybrid parameters (*Sdq, Sdr*), functional parameters (*Sk, Svk, Spk, Vm, Vv, Vmp, Vmc, Vvc, Vvv*) and feature parameters (*S5p, S5v, S10z*). In figure 2, the corresponding normalized surface parameters of the two sets of samples are shown.

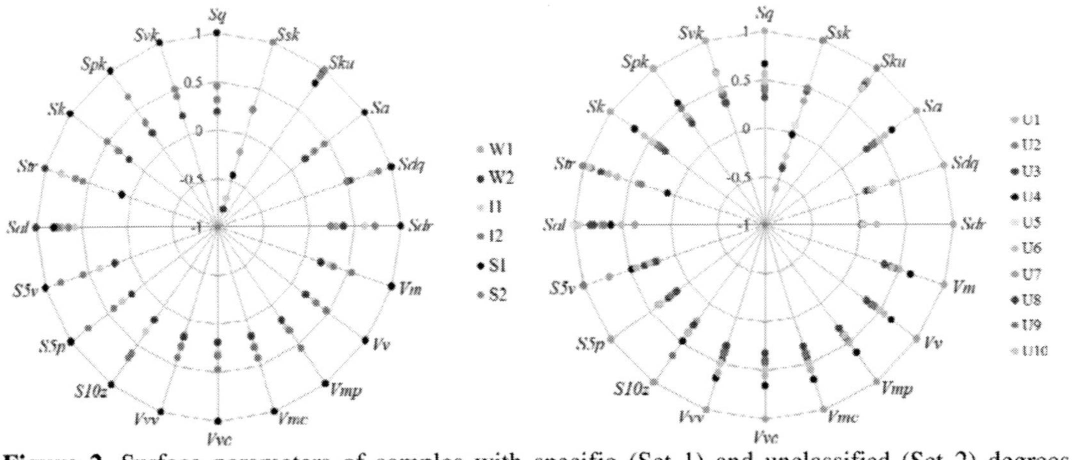

Figure 2. Surface parameters of samples with specific (Set 1) and unclassified (Set 2) degrees of orange peel.

For the calculation of feature parameters (*Sk, Svk, Spk*), a Gaussian filter with a cut-off wavelength of 1.5 mm was used, and the rest of the parameters were calculated with the standard variables defined in reference [6]. Additionally, in figure 2, degrees of orange peel in the samples of set 1 are well distinguished for most of the surface parameters, and this distinction agrees with the visual inspection criterion. On the other hand, parameters such as *Sal, Ssk, Sku* and *Str* cannot classify degrees of orange peel, since those parameters give us information regarding periodicity, skewness, kurtosis and orientation of the samples respectively. For example, the sample S1 exhibits the lowest *Str* parameter, which indicates that the texture of this sample is highly orientated as is confirmed from figure 1. Moreover, the sample U7 is the only one that can be distinguished from the rest of the samples according to the chosen surface parameters (figure 2, set 2), however we can not a priori assume that this sample has strong degree of orange peel, since the normalization of the parameters was done independently.

2.2. Classification of samples by feature recognition

Surface parameters corresponding to Set 1 have been adopted as an arrangement of descriptors represented by a prototype vector for each class or degree of orange peel. For example, weak degree of orange peel is denoted by $\vec{w} = (w_{Sq}, w_{Sa}, w_{Sdq}, w_{Sdr}, w_{Vm}, w_{Vv}, w_{Vmp}, w_{Vmc}, w_{Vvc}, w_{Vvv}, w_{S10z}, w_{S5p}, w_{S5v}, w_{Sk}, w_{Spk}, w_{Svk})^T$ and similar notation has been assigned for intermediate $\vec{i}$ and strong $\vec{s}$ orange peel. Therefore, an unclassified sample is assigned to the class to which the surface parameters or the corresponding vector is closest to a prototype vector. So, as a first step, we calculate the mean vector of the two samples assigned to each degree of orange peel, which is given by

$$\vec{m}_Z = \frac{1}{2}\sum \vec{z}_j \quad z = w,\ i,\ s \text{ and } \quad j = 1,\ 2. \tag{1}$$

Afterwards, we calculate the Euclidian distance between an unknown vector sample $\vec{x}$ and the mean vector $\vec{m}_z$ to each class of orange peel. From reference [7], we know that the smallest distance between an unclassified sample and a prototype sample is equivalent to calculate a decision function given by

$$d_z(\vec{x}) = \vec{x}^T\vec{m}_z - \frac{1}{2}\vec{m}_z^T\vec{m}_z\,. \tag{2}$$

In order to define the class membership of sample $\vec{x}$, one of the following conditions has to be satisfied:

1) $\vec{x} \in$ weak orange peel class if : $d_w(\vec{x}) > d_i(\vec{x})$ and $d_w(\vec{x}) > d_s(\vec{x})$,
2) $\vec{x} \in$ intermediate orange peel class if : $d_i(\vec{x}) > d_w(\vec{x})$ and $d_i(\vec{x}) > d_s(\vec{x})$,
3) $\vec{x} \in$ strong orange peel class if : $d_s(\vec{x}) > d_i(\vec{x})$ and $d_s(\vec{x}) > d_w(\vec{x})$.

The boundary function between 2 degrees of orange peel, for example strong and intermediate, can be determined by calculating

$$d_{si}(\vec{x}) = d_s(\vec{x}) - d_i(\vec{x}) = 0\,. \tag{3}$$

Therefore, if the vector $\vec{x}$ is composed of either two or three elements, we can represent this boundary as a line and plane respectively, but if it has more than three elements the boundary is given by a hyperplane.

By using this method we classify the samples Ui according to the degree of orange peel estimated from surface parameters (see table 2).

Table 2. Classification of samples.

Degree of orange peel (visual)	Set 1 (prototypes)	Set 2
Weak	W1, W2	U3, U8
Intermediate	I1, I2	U1, U2, U4, U5, U6, U9, U10
Strong	S1, S2	U7

In addition, as a representative way to depict the boundaries between degrees of orange peel, we selected the parameters *S5p* and *Spk* to illustrate it in figure 3.

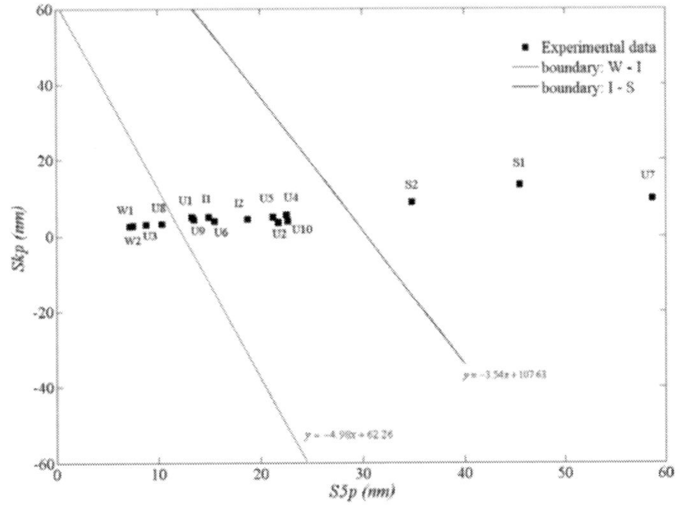

Figure 3. Boundaries of degrees of orange peel for *S5p* and *Spk* parameters.

3. Conclusions

Based on the findings, we can conclude that phase shifting interferometry is an appropriate technique to characterize the orange peel topography of polished metallic surfaces in agreement with the visual classification. In addition, by using a method of recognition of features, as well as the surface parameters calculated from PSI measurements, it is possible to classify samples into the corresponding category of orange peel. Therefore, the implementation of this method in the quality control inspection of polished surfaces can improve and contribute to the standardization of these processes.

4. Acknowledgements

This work was funded by the Austrian COMET Programme (Project K2 XTribology, no. 824187) and carried out at the "Excellence Centre of Tribology". In addition, we thank Berndorf Band GmbH for providing the polished samples.

5. References

[1] Byk additives and instruments, April 30 (2013) www.byk.com/en/instruments
[2] De Groot P 2011 *Optical Measurement of Surface Topography*, ed Leach R (Berlin, Verlag and Heidelberg: Springer) pp 167-185
[3] Leica DCM 3D, "3D Optical Surface Metrology Syistem", January 15 (2011). www.leica-microsystems.com/products
[4] MountainsMap Premium, Digital Surf, France, January 20 http://digitalsurf.com (2011).
[5] ISO 13565-2:1996 Geometrical Product Specification Surface texture Part 2: Height characterization using linear material ration curve.
[6] Leach R 2010 *Fundamental Principles of Engineering Nanometrology*, (UK: Elsevier) 211-258.
[7] Gonzalez R.C 2001 *Digital Imaging Processing*, (New Jersey: Prentice Hall) *693-701*.

Journal of Physics: Conference Series **450** (2013) 012010 doi:10.1088/1742-6596/450/1/012010

Stability comparison between two optical refractometer techniques

S A Chowdhury, E Chehura, R Correia, D Francis, J Hodgkinson and R P Tatam

Dept. of Engineering Photonics, School of Engineering, Cranfield University, Cranfield, Bedfordshire, MK43 0AL, UK

E-mail: j.hodgkinson@cranfield.ac.uk

Abstract. Two optical fibre refractometers are presented; both measure the change in refractive index of an external medium via measurement of the reflectivity of an optical fibre tip. One application of this technology lies in detection of hydrogen using palladium coatings. The first refractometer uses a single source with a probe and reference arm on separate optical fibres, while the second is a dual-wavelength system with a common path for probe and reference beam. We have characterised both systems to compare their sensitivity and long term stability to determine the most suitable configuration in terms of minimising drift and noise.

Journal of Physics: Conference Series **450** (2013) 012011 doi:10.1088/1742-6596/450/1/012011

Development of dithizone based fibre optic evanescent wave sensor for heavy metal ion detection in aqueous environments

K Bhavsar, R Prabhu and P Pollard
IDeaS Research Institute, Robert Gordon University, Aberdeen, UK

E-mail : r.prabhu@rgu.ac.uk

Abstract. Detection of highly toxic heavy metal ions requires rapid, simple, sensitive and selective detection methods in the environment. Optical fibre based sensing facilitates the remote, continuous and in-situ detection approaches in the environment. Herein, we report the development of a dithizone based fibre optic sensor with a simple procedure to detect heavy metal ions in the aqueous environment using an evanescent wave sensing approach. The chromogenic ligand dithizone and its spectral specificity with metal ions has been elaborated in this work.

1. Introduction

Heavy metal ions are highly toxic and are released into the aqueous environment due to the result of industrial activities. These pollutants accumulate in living organisms and create long-term adverse effects on humans, animals and plants [1,2].

Compared to the bulky, expensive, time consuming laboratory based analytical techniques optical fibre based sensing techniques have great potential for remote field sensing applications. Optical fibres are chemically inert, smaller in size, suitable for most of the surface modifications and enable in-situ, real time, distributed, rapid and remote detection with an environment friendly simple setup. One of the optical fibre based sensing approaches is evanescent wave sensing, where light propagating along the fibre generates the evanescent waves at the near surface region of the fibre core. These waves are highly susceptible to changes in the surrounding medium and hence can be used for sensing. This method is sensitive and reliable.

Recent developments on fibre optic sensors to detect metal ions include calixarene molecule based [3], DNA based [4], poly(4-vinylpyridine) based [5] and Al_2O_3 nanoparticles based [6] sensors. Most of these sensors have been reported to detect submicromolar concentrations with fast response times, but lack in procedural simplicity, with complicated design and operation. Among these sensors, use of a single wavelength for detecting multiple metal ions especially based on absorption are known to be highly susceptible to interference and lack selectivity [3].

Dithizone is a well-known ligand used in the spectrophotometric determination of metal ions [7]. Dithizone produces distinct absorption spectra in the presence of metal ions in aqueous solution [8]. Herein, we report a novel dithizone based fibre optic evanescent wave sensor for metal ion detection in aqueous solutions. A simple set up with an evanescent wave spectroscopy based detection approach using dithizone offers increased specificity from other contaminants in the environment compared to the single wavelength based detection approaches reported earlier. In this current research, interaction of chromium, cadmium, copper and mercury ions with dithizone has been elaborated and the detection principle has been demonstrated using mercury ion as model metal ion.

2. Experimental

2.1. Chemicals/Materials

Analytical grade dithizone, 2-propanol and mercury (II) standard solution in nitric acid were purchased from Sigma Aldrich, UK and used without any further purification. All glassware was cleaned in de-ionized water prior to the experiment.

2.2. Principle of metal ions detection and selectivity

Metal ions were adsorbed on the binding site formed by apex S and N atoms, as shown in Figure 1(a), which results in the formation of metal dithizonates by the removal hydrogen atom [7] as shown in Figure 1 (b). Chelation of dithizone with metal ion, results in a change in the absorption

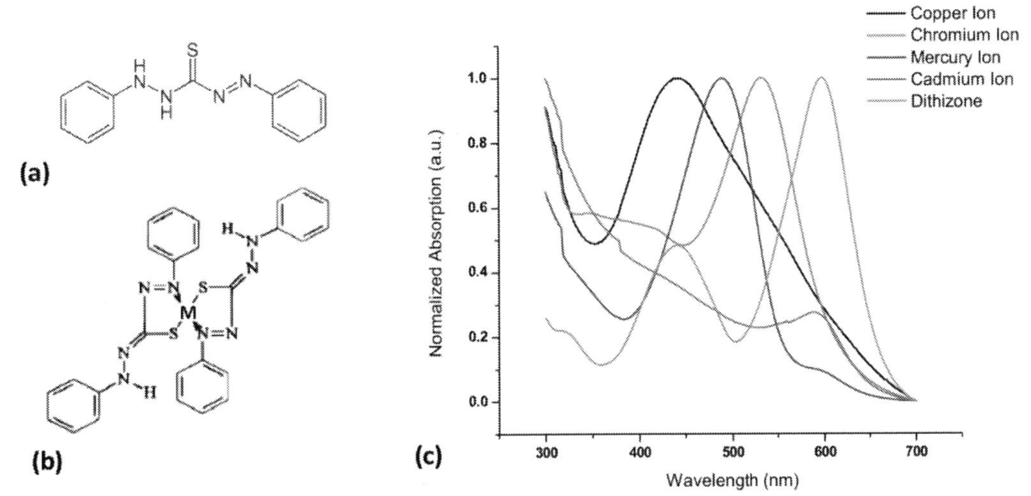

Figure 1(a) Chemical structure of dithizone, (b) metal dithizonate [9] and (c) absorption spectrum of 1ppm Hg, Cu, Cr and Cd ions with dithizone in aqueous solution.

spectra. Spectral response of the metal ions such as copper, chromium, cadmium and mercury with dithizone was evaluated with 1ppm concentration of each with 78μM dithizone and the corresponding absorption spectra are shown in the Figure 1(c). The absorption peak observed for chromium, copper, cadmium and mercury ions were at 532nm, 438nm, 588nm and 490nm respectively. The absorption peak of Hg^{2+} was 43nm, 51nm and 99nm away from the absorption peak of chromium, copper and cadmium respectively. Also, due to the higher toxicity of Hg^{2+} compared to other metal ions, further results are evaluated below using Hg^{2+} as model metal ion.

2.3. Experimental set-up

A simple experimental set up is shown in the

Figure-2. A white light LED (400-700nm) from Electrospell Ltd has been used as light source and the light was coupled to the fibre through an LED-Fibre coupler. For easier light coupling and detection, a multimode polymer fibre of 1mm core diameter has been used. The effect of dithizone to metal ion interaction has been characterised using a Perkin Elmer (Lambda 900) UV/VIS absorption spectrometer. The optical signal has been detected using a Stellarnet (EPP2000C) spectrometer in transmission mode.

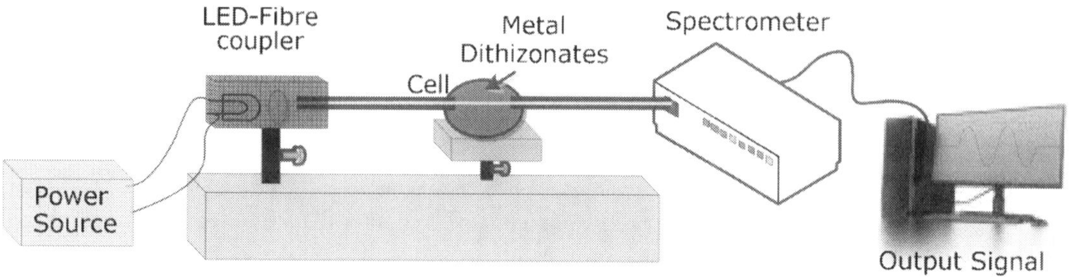

Figure-2 A pictorial view of experimental setup.

2.4. Theory

Light propagating through the fibre generates an evanescent field in the cladding. This exponentially decaying evanescent field in the cladding region will be exploited for the development of the evanescent wave absorption sensor. The phenomenon is achieved by removing the cladding of the fibre. The uncladded fibre acts as an evanescent wave sensor.

The penetration depth (dp) of the evanescent wave is defined as the distance at which the strength of electric field decays to its "1/e" value and is given by [10],

$$dp = \frac{\lambda}{2\pi n_1\sqrt{sin^2\theta - \frac{n_2^2}{n_1^2}}} \quad (1)$$

where, n_1 and n_2 are the refractive indices of the core and cladding, θ is the internal incidence angle normal to the core-cladding interface and λ is the wavelength of light.

Optical power transmitted in the uncladded fibre of length L, is given by [10],

$$P(L) = P(0)e^{-\gamma L} \quad (2)$$

where, P(0) is the initial optical power launched into the fibre, L is the length of uncladded fibre and γ is the evanescent absorption coefficient.

The evanescent absorption coefficient is governed by Beer-Lambert's Law [11] :

$$\gamma = \alpha L \eta \quad (3)$$

where, α is the absorption coefficient, L is the interaction length and η is the fraction of light in the evanescent field.

Therefore, in the case of evanescent wave based sensors, sensitivity depends on the fibre geometry such as diameter, interaction length and other parameters such as incidence angle, critical angle, wavelength of light, refractive index of core and cladding and absorption coefficient of analyte.

3. Results and discussions

Absorption spectra of mercury dithizonates for 78μM dithizone and various Hg^{2+} concentrations were studied using a UV-VIS absorption spectrometer and used as reference data as shown in Figure-3 (a).

A similar experiment was carried out with the fibre optic sensor and the output signal is shown in Figure-3(b). For optimizing the linear working range, the trend of mercury dithizonates absorption peak for 5ppb-20ppm Hg^{2+} concentrations was plotted for both reference data and sensor data as shown Figure-4(a). Results of the sensor follow the same trend as in the case of the reference data. The trend indicates that above 5ppm Hg^{2+} absorption saturates while below 1ppm it drops drastically. It also shows a linear response range within 1-5ppm Hg^{2+} which has been explored further.

To explore the nonlinear trend observed in the mercury dithizonates absorption spectra (>5ppm and <1ppm), the concentration ratio of Hg^{2+}:dithizone was evaluated for a fixed value of Hg^{2+}. 500ppb

Hg^{2+} was used to optimize the Hg^{2+}:dithizone ratio for detection and the corresponding UV-Vis spectra were recorded for 1:1, 1:2, 1:5 and 1:10 Hg^{2+}:dithizone as shown in Figure-4(b). Results

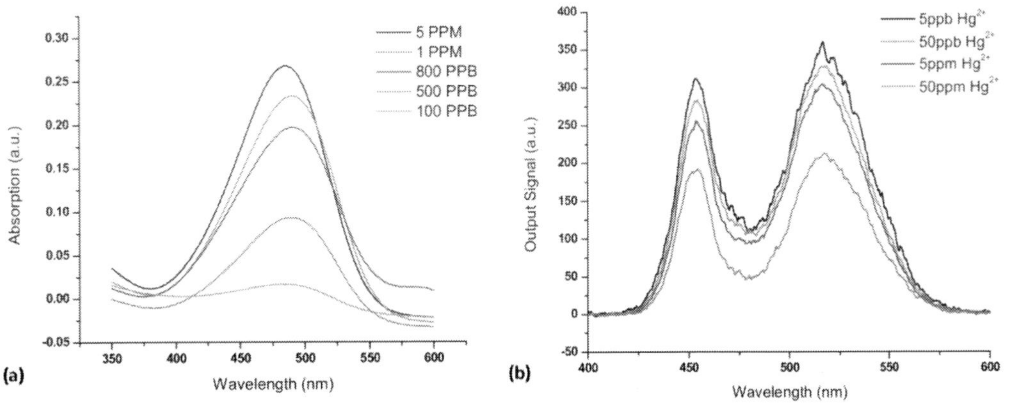

Figure-3 (a) UV-Vis absorption spectra of mercury dithizonates and, (b) output signal from fibre optic sensor at various Hg^{2+} concentrations.

indicate that 2-5 times the dithizone concentration provides optimum value of mercury detection. As discussed, in the Hg^{2+} detection principle with dithizone, each Hg^{2+} requires two dithizone molecules to form the chelate. Therefore, for the concentration ratio <=1 the chelation was poor and lower absorption was observed. The absorption value increases by increasing the Hg^{2+}:dithizone ratio up to a certain extent, thereafter, the dithizone molecules saturates and remains unbound with Hg^{2+}. These unbound dithizone molecules give rise to their absorption peak at 598nm along with the Hg^{2+} peak which can be observed for 1:10 ratio spectrum in the Figure-4(b).

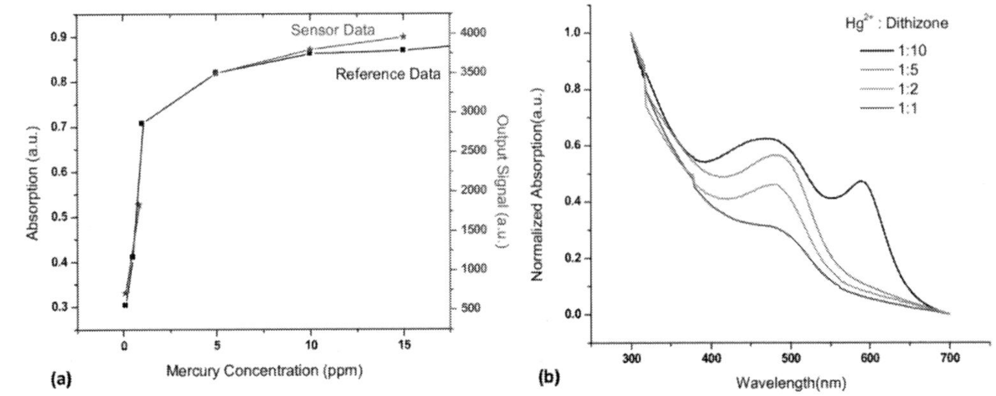

Figure-4 (a) Absorption readings of mercury dithizonates at 490nm with varying concentration of Hg^{2+} and (b) UV-Vis absorption spectra of 500ppb Hg^{2+} with different dithizone concentrations.

As discussed earlier, results in the linear range between 1-5ppm Hg^{2+} obtained from UV/Vis absorption data and the fibre optic sensor are shown in the Figure-5(a) and (b) respectively.

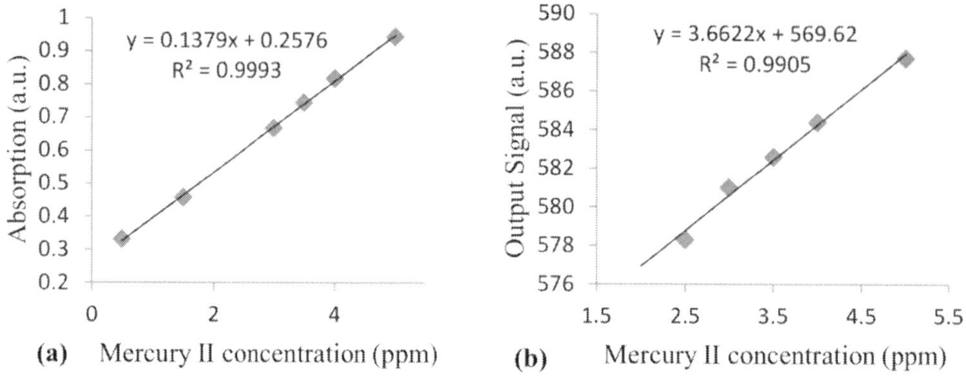

Figure-5 (a) Mercury dithizonate absorption trend measured at 490nm (b) Calibration curve of the fibre optic mercury sensor

4. Summary

Development of a novel dithizone based fibre optic sensor for heavy metal ion detection was demonstrated. Also, the spectral specificity to detect individual ions such as mercury, copper, chromium and cadmium in aqueous solutions using dithizone has been evaluated. The developed sensor demonstrated using Hg^{2+} , features an environment friendly simple procedure and instrumental setup for Hg^{2+} detection with high sensitivity (LOD-5ppb) and adequate specificity. These features demonstrate that the developed sensor can be extended to detect other metal ions from the environment. Further work is being carried out to develop an integrated sensor for real time field applications in onsite monitoring in the agricultural, environmental and marine industries.

5. References

[1] Zahir F, Rizwi SJ, Haq SK and Khan RH. 2005 *Environ.Toxicol.Pharmacol.* **20** 351-60
[2] Tchounwou PB, Ayensu WK, Ninashvili N and Sutton D. 2003 *Environ.Toxicol.* **18** 149-75
[3] Benounis M, Jaffrezic-Renault N, Halouani H, Lamartine R and Dumazet-Bonnamour I. *Mater. Sci. Eng. C*2006 **26**
[4] Long F, Gao C, Shi HC, He M, Zhu AN, Klibanov AM, et al. 2011 *Biosens Bioelectron* **26** 4018-23
[5] Gu B, Yin M, Zhang AP, Qian J and He S. 2011 *Sensor Actuat B. Chem* **160** 1174-9
[6] Perez-Hernandez J, Albero J, Llobet E, Correig X, Matias IR, Arregui FJ, et al. 2009 *Sensor Actuat. B-Chem.* **143** 103-10
[7] White WE. 1936 *J.Chem.Educ.* **13** 369
[8] Litman R, Williams ET and Finston HL. 1977 *Anal.Chem.* **49** 983-7
[9] Yang XQ, Qi SW, Zhang CP, Chen K, Liang X, Yang G, et al. 2005 *Opt.Commun.* **256** 414-21
[10] Sai VVR, Kundu T, Deshmukh C, Titus S, Kumar P and Mukherji S. 2010 *Sensor Actuat B* **143** 724-30
[11] Burke C, StrÃ¡nik O, McEvoy H, MacCraith B. Planar Optical Sensors and Evanescent Wave Effects. In: Baldini F, Chester AN, Homola J, Martellucci S, editors. : Springer Netherlands; 2006 193-215

Investigation of finger reflectance photoplethysmography in volunteers undergoing a local sympathetic stimulation

H. Njoum, P.A. Kyriacou

Biomedical Engineering Research Group, City University of London, UK
haneen.njoum.1@city.ac.uk, p.kyriacou@city.ac.uk

Abstract. Optical sensors used in clinical applications have gained great popularity over the last few decades, especially the photoplethysmographic (PPG) technique used in estimating arterial blood oxygen saturation in the well-known medical devices called pulse oximeters. In this study we investigate the photoplethysmogram further in an effort to understand its origin better, as there is a significant void in the current knowledge on the PPG quantitative measurement. The photoplethysmographic signal provides a heart rhythm pulsating AC component, and a non-pulsating DC component. The signal is commonly believed to originate from tissue volume changes only and hasn't been investigated intensively. This in vivo study examines the source of the PPG signal in relation to pulse amplitude and pulse rhythm while volunteers undergo a right hand ice immersion. It was found that the PPG signal is sensitive in detecting the sympathetic stimulation which corresponds to volumetric and heart rate changes. During the immersion, AC pulse amplitudes (PA) from both hands decreased significantly, while DC levels increased significantly in the right hand and non-significantly in the left hand. Also, a significant decrease in the pulse repetition time (PRT) was observed. Using blood pressure-flow theories, these results suggest that there are possibly other factors in the blood flow regulation that alter the blood optical density which contributes to the detected signal. Further studies need to investigate PPGs in relation to blood optical density and the dynamics of the pulsatile flow effects besides volumetric changes. Such investigations might explore further applications of the PPG in medicine.

1. BACKGROUND

Since the 1930's, the non-invasive optical technique named 'photoplethysmography' has gained broad interest. A photoplethysmographic signal can be obtained at any vascular location on the skin surface using optical probes operating in either reflection or transmission mode [1-2]. During these measurements, light is directed to the tissue, where it travels in complex pathways facing multiple stages of reflection, absorption and scattering. Photoplethysmography is extensively used for the estimation of arterial blood oxygen saturation, a technique known as pulse oximetry. Pulse oximeter sensors use red and infrared optical wavelengths that are sensitive to changes in arterial oxygen saturation because of differences in the light absorption of oxygenated and deoxygenated haemoglobin at these two wavelengths. After the light is transversed or reflected through the vascular tissue, it reaches the photodetector and the current variations are electronically amplified and recorded to provide the photoplethysmographic signal. The detected signal comprises of two PPG components; the pulsatile AC component and the non-pulsating DC component [3-4].

1.1. The origin of the PPG signal

The validity of the PPG quantitative measurement is questioned because most investigators in the field rely on the assumption that the detected light has a linear relationship with the intravascular volume, therefore considering the red cell density to be directly proportional to that space. However, other factors of blood flow behaviour are also believed to contribute to the signal. Findings from some studies on the origin of photoplethysmography have demonstrated that light travelling through blood can diffuse preferentially in the direction of the motion of blood. This pattern of diffusion was suggested to emerge from a combination of the plasma skimming, which occurs at the vessel wall and from the orientation effects of the erythrocytes when in motion [5-7]. In addition, studies have related haematocrit concentration to the PPG pulse amplitude, and attempts to estimate haemoglobin concentration using PPGs have also been initiated [8-9]. The role of such factors and their contribution to the origin of the PPG signals has not been investigated rigorously yet. The dynamics of the pulsatile flow in the blood circulation needs to be considered in relation to the PPG signal and is discussed briefly below.

1.2. The pulsatile flow and the pressure propagation wave

Blood flow distribution controls are initiated by neural, cardiac and respiratory interactions. These maintain a balance between tissue perfusion and heart mechanics in a complex relationship of bioelectrical signals, biochemical reactions and a variety of mechanical forces [10]. The heart impulse is generated at the sinoatrial (SA) and atrioventricular (AV) nodes in the heart, causing a synchronized rhythmic frequency which can be clearly observed in the smaller vascular bed when using the PPG technique. These electrical impulses caused by the SA and AVR nodes result in the contraction of the heart muscle. Therefore, circumferential stresses are generated in the heart, a force normal to the cardiac wall is applied to the fluid, and the intracardiac pressure increases shortly and then falls during the heart relaxation period. The detailed study of biological tissue conductivity in relation to blood flow motion and tissue perfusion is still under investigation [11]. Nevertheless, these forces result in the pressure-flow relationship that has been studied for many years [12-13]. The well-known Hagen-Poiseuille law considers a steady, laminar, viscous and non-compressible fluid in a straight smooth walled tube. The classical approach of haemodynamic points at arterial distensibility and volume change as the primer determinants of pulse pressure. In the presence of the pulsatile blood flow and the complexity of the geometry and elasticity of vessel networks, there are additional forces to be considered. These are inertia and viscosity, which are related to distensibility in a complex manner. These forces are more significant in small capillaries where the blood is thought to obtain shear thinning and viscoelastic properties. This suggestion has been debated, considering the flow to be continuous in a laminar flow. Blood is really a suspension of proteins and cells in an aqueous medium and its dynamics are more complicated than classically assumed. A discussion of non-Newtonian fluid dynamics and its relation to heart forces is beyond the scope of this study [14]. However, the approach adopted considers the complex relationship of pressure and flow dynamics. An attempt to explain PPG signal behaviour in relation to this consideration is discussed later.

Consequently, the volume contained within the arteries is a function of the pressure difference across their wall, called the transmural pressure. Transmural pressure is seen as a haemodynamic signal that exerts a force on the vessel wall. It depends on the intracardial fluid dynamics and the heart morphology. The blood pressure exerted against the walls of an artery and the energy required to move blood at a certain velocity is interrelated by the Bernoulli's principle, where the total energy of blood flow (potential and kinetic energy) at any given point in a system is constant [13]. This approach will be adopted later to explain the change in the pulse repetition time.

1.3. Sympathetic System stimulation

The immersion of one hand in an ice-water bath, the so-called cold pressor test, is a standard autonomic function test to assess the sympathetic activity. Neural and hormonal reactions will allow cutaneous vasoconstriction in the dermis, a rise in both systolic and diastolic blood pressure and therefore, increased heart rates are also associated with the immersion [15-17]. This study was conducted in healthy subjects, who underwent ice immersion of the right hand for 30s, while recording PPG signals and temperature measurements from both hands. Electrocardiogram signals were also recorded.

This *in vivo* investigation was designed to allow local sympathetic nervous stimulation and observe changes in the PPG signal in relation to the following effects; (a) the volumetric changes in the vessels in both hands due to vasoconstriction, followed by vasodilation response, (b) the pulsating wave speed changes due to the heart rate immediate increase during the immersion.

2. MATERIALS AND METHODS

The experimental set-up for this in vivo study is described in Figure 1.

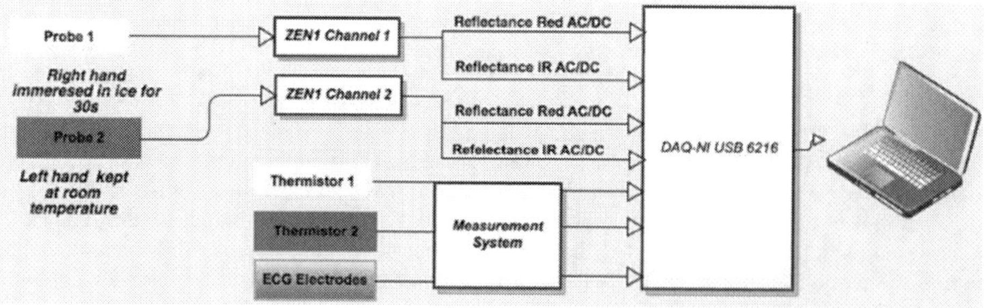

Figure 1: Block diagram of the experimental set up

2.1. Reflectance PPG Probes

A pair of identical custom-built reflectance PPG finger probes (*Probe 1 and Probe 2*) were developed. Each probe comprises of two surface mount ceramic type infrared LEDs with peak emission wavelength of 880 nm, two red LEDs with peak emission wavelength of 660 nm and a photodiode. The LEDs are arranged to maintain a separation distance of 5 mm form the centre of the photodiode. The LED driving current (20 mA) is controlled digitally in the PPG processing system switching the LEDs on and off asynchronously. The duty cycle of the switching pulses is 33.3%.

2.2. Instrumentation

The PPG processing system (*ZEN1*) was developed by the Biomedical Engineering Research Group (BERG) [18].The circuits were constructed on printed circuit boards to pre-process, record and display raw PPG signals from two channels and two wavelengths on a laptop personal computer. The detected currents from the photodiodes are converted into voltage using a differential transimpedance amplifier. The micro-controller, acts as the master clock to generate digital switching signals at a frequency of 500 Hz. These digital signals directly correspond to the LED switching. The mixed PPG signals at the output of the transimpedance amplifier are split into their respective red and infrared PPG signals using a demultiplexer.

The temperature measurement system was developed to contain all necessary temperature linearization circuits. The measurements were obtained utilising two (one for each hand) water resistant rapid response (0.6 s) thermistors (*thermistor 1, thermistor2*).

The Electrocardiogram (ECG) signals were also recorded as a gold standard measurement for heart rate rhythms. The ECG measurement system was developed using an instrumentation amplifier and a sallen-key bandpass filter. Electrodes (Ag/AgCl) were placed using standard limb leads (Lead I).

A 16-bit data acquisition card (6212 DAQ, National Instruments Inc.) was used to acquire the raw analogue signals. All digitized signals were further processed (digital filtering, amplification, etc.) using a virtual instrument implemented in LabView.

2.3. Methods

The experimental protocol for this study was approved by the Senate Research Committee at City University London. Twelve healthy subjects were recruited, 5 females, 7 males, aged 18-45 years (mean age ± SD, 26.9 ± 4.6 years; range, 19-33 years). All subjects gave informed written consents for their participation. The subjects were asked to abstain from smoking and exercising for at least two hours before the experiment. All testing was conducted in a temperature control room with the temperature maintained at 21 ± 2°C. Subjects were required to acclimatize to the testing room temperature upon arrival for a few minutes. The developed reflectance PPG probes were clipped on both index fingers. Skin thermistors were tightly attached on the tip of the middle fingers. The recording started once the volunteer was comfortable and continued for two minutes as baseline monitoring. The volunteer was then asked to slowly immerse their right hand until the wrist was fully in the ice-water bath (maintained at 2 ± 2°C). After 30 s, the volunteer was asked to rest their hand back at room temperature on a dry surface. The recording continued for an average of 9 min as a rewarming period.

3. Results

Good quality PPG signals were obtained from all participants. Changes in infrared PPG components and temperature values for a recording of nine minutes for both index fingers from one of the volunteers is shown in figure 2.

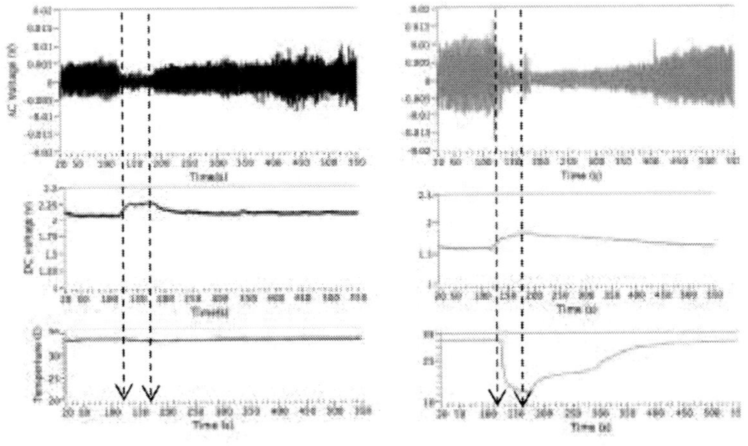

Figure 2: Infrared reflectance AC and DC PPG signals and skin temperatures from a sample volunteer. Left hand signals, resting at room temperature are in black, and right hand signals, undergoing ice-water immersion are in grey. Ice immersion period is marked between arrows.

Journal of Physics: Conference Series **450** (2013) 012012 doi:10.1088/1742-6596/450/1/012012

The increase in the sympathetic tone was apparent once the immersion had started. Temperature values dropped significantly in the right hand and DC PPG levels increased in both hands. The pulse amplitude (AC signal) decreased in both hands. Once the immersion had ended, PPG values gradually returned to baseline in the right hand, while the PPG signals instantly returned to baseline in the left hand (see Fig. 2). Generally, PPG recordings were still useful for cardio assessment in most volunteers during the immersion as can be seen in figure 3.

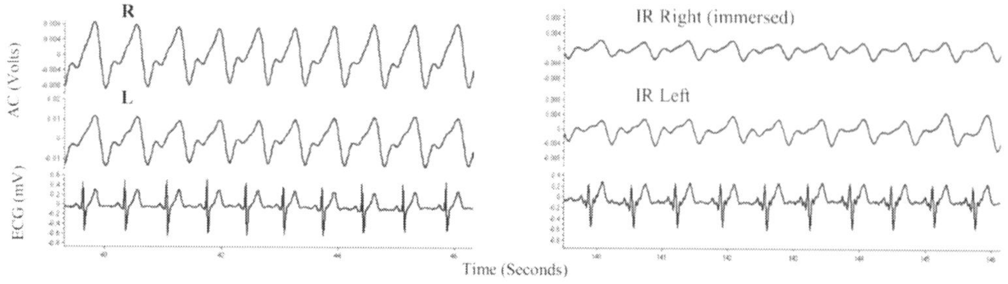

Figure 3A: Baseline ECG and infrared AC PPG recordings from both hands.

Figure 3B: ECG and infrared AC PPG recordings from both hands during the immersion.

Occasionally, volunteers had a drop in PA sufficient to diminish the signal quality and others experienced discomfort at the end of the immersion, which resulted in some movement artefacts.

On average, the temperature values decreased statistically significant in the right hand ($P < 0.005$) for the duration of the immersion. The average decrease in temperature was 20°C (±5°C) at the end of the 30 s of immersion. Thereafter, temperature values increased gradually until the end of the study. The average temperature increase was 12°C (±2°C) after 5 minutes of the immersion. Left hand temperatures had different patterns of response in volunteers, and can be averaged to ±1°C (±0.5°C).

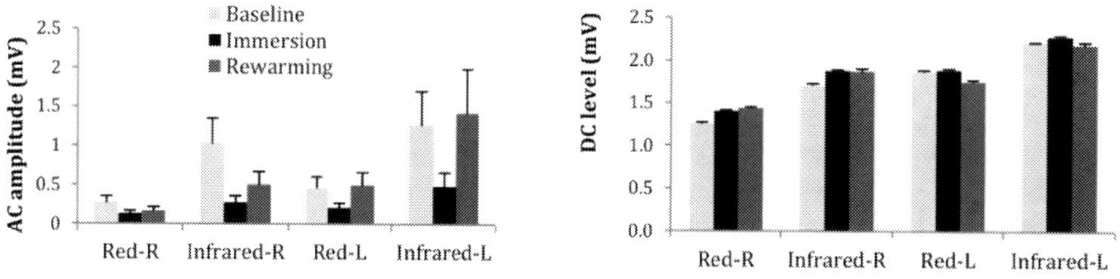

Figure 4: Averaged values for AC and DC amplitudes for the periods; baseline (2 mins), immersion (30 secs) and rewarming (5 mins).

Averaged values for all signals can be seen in figure 4, average right finger pulse amplitude decreased from baseline by 51% (SD ±0.077) for the red wavelength and 73% (SD ± 0.312) for the infrared wavelength. While, the average PA of the left finger decreased from baseline by 55% (SD ±0.138) for the red wavelength and 62% (SD ± 0.427) for the infrared wavelength. All PA values decreased significantly ($P < 0.005$). An immediate increase also occurred in the DC component of the PPG signal in the right hand. The average right DC levels increased significantly ($P < 0.005$) from baseline by 13% (SD ±0.005) and 11% (SD ±0.011) for the red and infrared wavelengths respectively. The average left DC levels increased only by 2% (SD ± 0.018) in the red PPG segment and by 3% (SD ± 0.016) in the infrared PPG segment.

There was an immediate increase in heart rates in all volunteers at the moment of skin-ice bath contact averaged to 10 beats per minute (BPM) as seen in figure 5. Thus, a significant drop ($P<0.005$) in the PRT was observed (see Fig.6). These values returned to baseline once the immersion had ended.

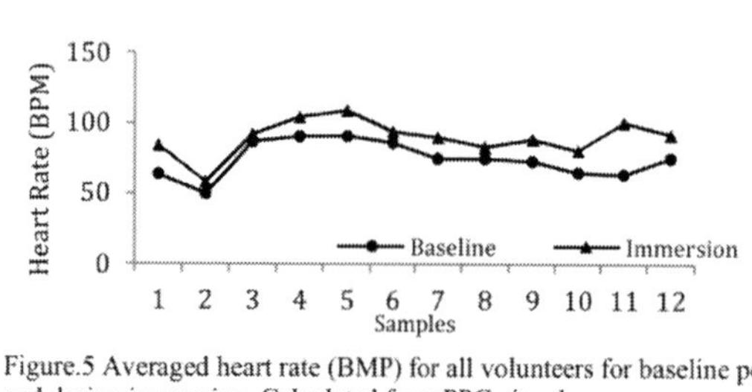

Figure.5 Averaged heart rate (BMP) for all volunteers for baseline period and during immersion. Calculated from PPG signals.

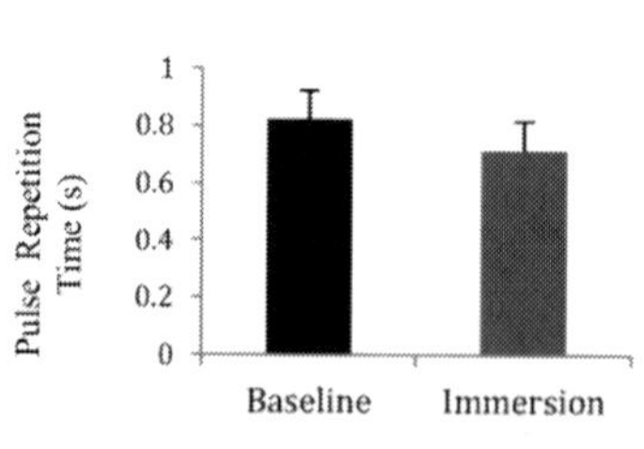

Figure.6 Averaged PPG pulse repetition time for all volunteers.

4. Discussion

This study found that while the right hand temperature values decreased during the immersion in ice water, PPG recordings from both hands have shown that there was an increase in the DC PPG levels, and a decrease in the pulse amplitude of the PPG signal. In addition, there was a decrease in pulse repetition time. The PPG has been used as a tool to measure vasomotor tone under various experimental conditions. The peripheral vasoconstriction and vasodilation affect both the AC and DC components of the PPG signal. Most previous studies have been limited by the inability or lack of interest to access the complete PPG signal. A significant feature of our study is that we recorded the PPG continuously throughout and we analysed both components for different locations under local sympathetic stimulation.

An increase in DC levels was accompanied with a drop in PA in a separate pattern in both hands. This is not completely surprising, because sympathetic stimulation will cause a vasoconstriction in the vascular bed. The vasoconstriction can be seen in the increase of the DC component of the PPG, which perhaps is correspondent to the changes in venous blood fraction. This behaviour is also expected because of the thickening of the dermis layer during the sympathetic response. The dermis is consistent of collagen fibres, which contributes to events of multiple scattering and can be viewed as a turbid media [19]. The different changes in DC levels in both hands confirm that the right hand (undergoing the immersion) is more constricted than that resting at room temperature. This is also expected, due to α-adrenoceptors distribution in the endothelial cells. Endothelial cells release locally acting chemical messengers in response to various changes in their environment, such as changes in oxygen (O_2) or physical stretching of the wall. These local vasoactive mediators act on the underlying smooth muscle to alter its state of contraction. The relative contributions of these various chemical changes in the local metabolic control of blood flow in systemic arterioles are still being investigated [20].

Pulse amplitude decreased significantly in both hands, this can be referred to vasoconstriction effects and the expected drop in the arterial blood volume as described above. Whether PA and DC changes are due to volumetric changes solely is undetermined. The drop in PRT is another observed change to be discussed. The pulsating wave speed is a strong function of pressure as mentioned earlier, due to the elastic properties of the red blood cells and the nonlinear elastic properties of the arterial wall. Previous studies have confirmed that an increase in blood pressure is accompanied with the heart rate increase during the cold pressor test [16]. Others have also observed the gradual drop in PA PPG signals, while volunteers underwent controlled artificially induced hypoperfusion using a blood pressure cuff [21].

Theories admitting the complexity of pressure-flow relationship consider three forces, the friction force, the kinetic force and the force exerted on the wall vessels in relation to its distensibility. These parameters affect each other and the pressure wave is an instantaneous sum of these applied forces. The observed change in the heart rate during the immersion is known to be associated with reduced arterial distensibility [22]. Relying on the Bernoulli's principle, any increase in one form of the energy has to come at the expense of the other; hence the reduced distensibility should be correlated with increased transmural pressure (tension on wall) and therefore, lateral pressure between the red blood cells is expected to decrease. Bernoulli's equation overlooks the viscosity of the fluid so relying on Poiseuille considerations of a viscous fluid, the change in total pressure is expected to result in an increase in their frictional forces. Those forces are affected by the red blood cells count or their phenomenal reversible mechanical properties. These properties are also believed to allow changes in cells' membrane structure, shape and size in order to maintain the required balance. Therefore, an increase in blood viscosity is expected during the increased sympathetic tone and the elevated heart rate [23-24]. These changes are plausibly related to changes in the optical density of blood in terms of concentration, molecules' structure or size [25]. This might explain the similar pulse amplitude behaviour in both hands regardless of the different change in the DC levels. We estimate that blood optical density may vary in relation to pressure and metabolic conditions, which might affect the detected light and will be of direct interest to all investigators in the field of photoplethysmography and its applications. Further *in vivo* and *in vitro* investigations need to explore the pulsatile dynamics in relation to PPG, highlighting blood viscosity and possibly rheology. Potentially, a standardized PPG technique might allow applications related to a variety of blood pathologies (sepsis, sickle cells, anaemia and others) where the blood optical density is altered from the norm.

5. Conclusion

In this experiment, we found that changes in the AC and DC components of the PPG and finger temperature characterize the change in the sympathetic tone during the immersion of the hand in ice (vasoconstriction and increased heart activity) and the parasympathetic tone after the immersion (vasodilation and baseline heart activity). The increased sympathetic tone includes both effects of vasoconstriction and elevated heart rate. Clearly they both contributed to changes in PPG DC levels, pulse amplitude and repetition time. Theories of blood pressure and flow indicate that such effects not only allow changes in the optical path length but might also affect the biological tissues' absorption coefficients during blood circulation. The path to quantify a standardized PPG waveform suggest further exploration of tissue optical properties in relation to the molecular and cellular mechanisms of blood flow control.

References

[1] Hertzman and Spealman, "Observations on the finger volume pulse recorded photoelectrically," 1937. [Online]. Available: http://aqwave.reflexwireless.com/?q=node/15. [Accessed: 10-Dec-2012].
[2] J. A. Nijboer, J. C. Dorlas, and H. F. Mahieu, "Photoelectric plethysmography--some fundamental aspects of the reflection and transmission method," *Clin Phys Physiol Meas*, vol. 2, no. 3, pp. 205–215, Aug. 1981.
[3] J. W. Severinghaus, "Takuo Aoyagi: Discovery of Pulse Oximetry," *Anesth Analg*, vol. 105, no. 6S Suppl, pp. S1–S4, Dec. 2007.
[4] A. V. J. Challoner and C. A. Ramsay, "A photoelectric plethysmograph for the measurement of cutaneous blood flow," *Physics in Medicine and Biology*, vol. 19, no. 3, pp. 317–328, May 1974.
[5] M. Asano, K. Yoshida, and K. Tatai, "Microphotoelectric plethysmography using a rabbit ear chamber," *J Appl Physiol*, vol. 20, no. 5, pp. 1056–1062, Sep. 1965.
[6] L. T. Jespersen and O. L. Pedersen, "The quantitative aspect of photoplethysmography revised," *Heart Vessels*, vol. 2, no. 3, pp. 186–190, 1986.
[7] V. C. Roberts, "Photoplethysmography- fundamental aspects of the optical properties of blood in motion," *Transactions of the Institute of Measurement and Control*, vol. 4, no. 2, pp. 101–106, Apr. 1982.
[8] W. Ochoa and I. Ohara, "The effect of hematocrit on photoelectric plethysmogram," *Tohoku J. Exp. Med.*, vol. 132, no. 4, pp. 413–419, Dec. 1980.
[9] T. K. Aldrich, M. Moosikasuwan, S. D. Shah, and K. S. Deshpande, "Length-Normalized Pulse Photoplethysmography: A Noninvasive Method to Measure Blood Hemoglobin, Bronx, NY," *Annals of Biomedical Engineering*, vol. 30, no. 10, pp. 1291–1298, Nov. 2002.
[10] S. S. Segal, "Regulation of Blood Flow in the Microcirculation," *Microcirculation*, vol. 12, no. 1, pp. 33–45, 2005.
[11] T.-C. Shih, H.-S. Kou, and W.-L. Lin, "Effect of effective tissue conductivity on thermal dose distributions of living tissue with directional blood flow during thermal therapy," *International Communications in Heat and Mass Transfer*, vol. 29, no. 1, pp. 115–126, Jan. 2002.
[12] L. H. Peterson, "The Dynamics of Pulsatile Blood Flow," *Circulation Research*, vol. 2, no. 2, pp. 127–139, Mar. 1954.
[13] H. S. Badeer, "Hemodynamics for Medical Students," *Advan in Physiol Edu*, vol. 25, no. 1, pp. 44–52, Mar. 2001.
[14] H. H. Lipowsky, "Microvascular rheology and hemodynamics," *Microcirculation (New York, N.Y.: 1994)*, vol. 12, no. 1, pp. 5–15, Feb. 2005.
[15] R. G. Victor, W. N. Leimbach, D. R. Seals, B. G. Wallin, and A. L. Mark, "Effects of the cold pressor test on muscle sympathetic nerve activity in humans.," *Hypertension*, vol. 9, no. 5, pp. 429–436, May 1987.
[16] I. Korhonen, "Blood pressure and heart rate responses in men exposed to arm and leg cold pressor tests and whole-body cold exposure," *Int J Circumpolar Health*, vol. 65, no. 2, pp. 178–184, Apr. 2006.
[17] L. Mourot, M. Bouhaddi, and J. Regnard, "Effects of the cold pressor test on cardiac autonomic control in normal subjects," *Physiol Res*, vol. 58, no. 1, pp. 83–91, 2009.
[18] V. Rybynok, J. M. May, K. Budidha, and P. A. Kyriacou, "Design and development of a novel multi-channel photoplethysmographic research system," in *2013 IEEE Point-of-Care Healthcare Technologies (PHT)*, 2013, pp. 267–270.
[19] K. Nielsen, L. Zhao, J. Stamnes, K. Stamnes, and J. Moan, "The Optics of Human Skin: Aspects Important for Human Health," *Solar Radiation and Human Health*, 2008.
[20] D. Ladage, K. Brixius, H. Hoyer, C. Steingen, A. Wesseling, D. Malan, W. Bloch, and R. H. Schwinger, "Mechanisms Underlying Nebivolol-Induced Endothelial Nitric Oxide Synthase Activation in Human Umbilical Vein Endothelial Cells," *Clinical and Experimental Pharmacology and Physiology*, vol. 33, no. 8, pp. 720–724, 2006.
[21] M. Shafique, J. P. Phillips, and P. A. Kyriacou, "Evaluation of a multimode photoplethysmographic sensor during cuff-induced hypoperfusion," in *2010 Annual International Conference of the IEEE Engineering in Medicine and Biology Society (EMBC)*, 2010, pp. 1024–1027.
[22] C. Giannattasio, A. Vincenti, M. Failla, A. Capra, A. Cirò, S. De Ceglia, G. Gentile, R. Brambilla, and G. Mancia, "Effects of heart rate changes on arterial distensibility in humans," *Hypertension*, vol. 42, no. 3, pp. 253–256, Sep. 2003.
[23] H. M. Reims, K. Sevre, A. Høieggen, E. Fossum, I. Eide, and S. E. Kjeldsen, "Blood viscosity: effects of mental stress and relations to autonomic nervous system function and insulin sensitivity," *Blood Press.*, vol. 14, no. 3, pp. 159–169, 2005.
[24] E. Ernst, L. Daburger, and T. Saradeth, "The kinetics of blood rheology during and after prolonged standardized exercise," *Clinical Hemorheology and Microcirculation*, vol. 11, no. 5, pp. 429–439, Jan. 1991.
[25] M. Friebel, A. Roggan, G. Müller, and M. Meinke, "Determination of optical properties of human blood in the spectral range 250 to 1100 nm using Monte Carlo simulations with hematocrit-dependent effective scattering phase functions," *Journal of Biomedical Optics*, vol. 11, no. 3, p. 034021, 2006.

Real-time Microwave Sensor for KCl, $MnCl_2$ and CuCl Solutions Concentration with Ag Patterns Printed on Flexible Substrates

M Ortoneda-Pedrola, O Korostynska[1], A Mason and A I Al-Shamma'a
BEST Research Institute, School of Built Environment, Liverpool John Moores University, Liverpool, UK

E-mail: o.korostynska@ljmu.ac.uk

Abstract. A novel microwave sensor for real-time chloride concentration analysis is reported. The sensor response to deionised water, KCl, $MnCl_2$ and CuCl solutions at 1M and 0.1M concentration was tested. The sensing element, in the form of a silver pattern, was printed on polyimide flexible laminate substrate to suit a broad range of applications. The developed system confirmed the viability of using microwaves for real-time chloride solutions monitoring as the reflected signals represented by S_{11} parameters were unique with clearly observed shifts in the resonant frequencies and amplitude changes when placed in direct contact with 20 µl of each solution.

1. Introduction
The composition of chloride solutions must be determined with high accuracy for many applications, such as food and pharmaceutical industries, chemical processing and environmental monitoring, where the content of impurities in water is an important factor. Salinity is a very important parameter for oceanics, marine environment monitoring, seasonal climate prediction, mariculture, and solar engineering [1]. Notably, the detection of Cu^{2+} may serve as an early warning indicator of the onset of corrosion [2].

Traditional methods of aqueous solutions composition monitoring are lab-based and include standard UV-Vis measurements, flame photometry [3], mass spectrometry, ion-sensitive electrodes and amperometric sensors [4, 5]. On the other hand, a range of modern real-time monitoring approaches exist; for example, fibre-optic sensors, MEMs, lab-on-chip and biosensors. Interestingly, salinity measurement in water environments with a long period grating based interferometer was recently reported [6].

Unfortunately, these techniques are inadequate for real-time continuous monitoring of solution concentrations and a novel approach is required. This paper presents novel methods of water salinity determination based on microwave sensing.

2. Microwave sensing
Microwave sensing is a novel but rapidly developing technology which has been successfully used as a sensing method for various industrial applications including fluid level measurements [7], material

[1] To whom any correspondence should be addressed.

moisture content [8, 9], for continuous process monitoring of biogas plants [10], in the food industry for the verification of the vegetable oil types [11] and in the healthcare industry, for example for real-time monitoring of glucose in diabetic patients [12, 13] and for non-invasive monitoring of bodily fluids [14, 15].

Microwave planar printed patterns for various sensing applications are increasingly used due to their versatility, flat profile and low weight. Their design can be tailored to suit particular applications, coupled with reliability and cost-efficiency. They are easily manufactured using common methods for printed circuit board production, and their impedance can be matched to the input line by altering the micro-strip line feed configuration.

To clarify the principle behind electromagnetic wave sensing suggested in this work for water quality control, it is worth mentioning that microwave sensors in the form of planar printed patterns operate based upon the fact that an object under test, e.g. a water sample, when placed into the vicinity or in direct contact with a microwave sensor, interacts with the electromagnetic waves in a unique manner, which can be specifically correlated with the properties of this material.

By considering how reflected (S_{11}) microwave signals vary at discrete frequency intervals, the change in the signal can be linked to the composition of the object under test.

3. Experimental procedure

3.1. KCl, $MnCl_2$ and CuCl solutions preparation

Potassium chloride KCl (Sigma 60134), manganese chloride $MnCl_2$ (Sigma 2212789) and cuprous chloride CuCl (Sigma C-3294) were dissolved in deionised water at 1M and 0.1M concentrations. All chemical compounds produced clear solutions except CuCl, which is not soluble in water and produced a pale green cloudy suspension.

3.2. Microwave sensor structure

The layout and dimensions of the sensor pattern is shown in Figure 1. Silver was used as the conductive metal material for both the bottom layer, which acted as a ground plane, and the top pattern to maintain chemical neutrality when the device is placed in contact with analyte solution. The thickness of the Ag layers was 35 μm. The microwave sensor was designed on DuPont™ Pyralux® AP Polyimide Flexible Laminate substrate of 2 mil thickness.

A distinct feature of IDE type sensors is their superior sensitivity to change close to the sensor surface, with this sensitivity decaying rapidly with distance away from the surface, as shown in Figure 2. This is advantageous as it reduces significantly the chance of undesirable factors influencing sensor response.

3.3. Measurement setup

The microwave sensor was attached to a Rohde and Schwarz ZVA24 vector network analyser (VNA) via a coaxial cable, as shown in Figure 3. An SMA type Molex connector specified for 50Ω impedance was used. The data (60,000 points for each measurement) was captured in the 0-15 GHz frequency range for the reflected (S_{11}) signals at a constant temperature of 18 °C. The results are repeatable with less than 5% deviation. Notably, average sensor responses are depicted in the graphs shown in following section.

Journal of Physics: Conference Series **450** (2013) 012013 doi:10.1088/1742-6596/450/1/012013

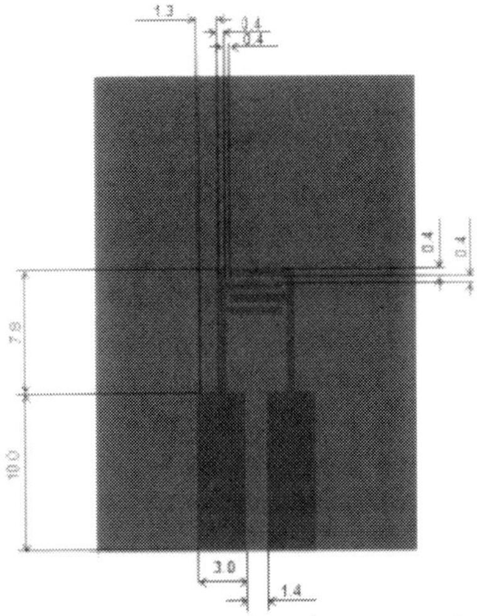

Figure 1. Ag printed pattern layout. All dimensions are in mm.

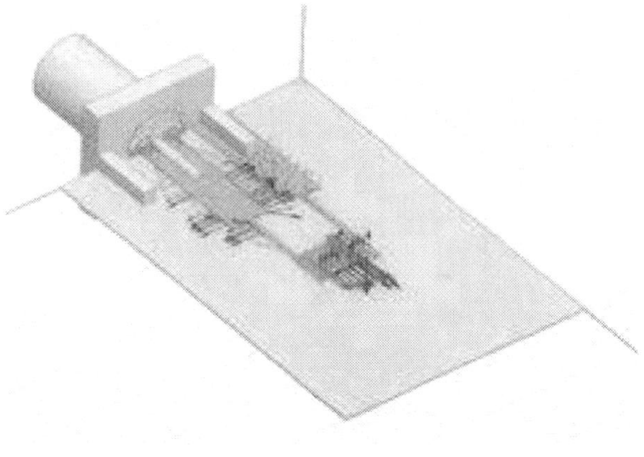

Figure 2. HFSS model of microwave sensor field distribution.

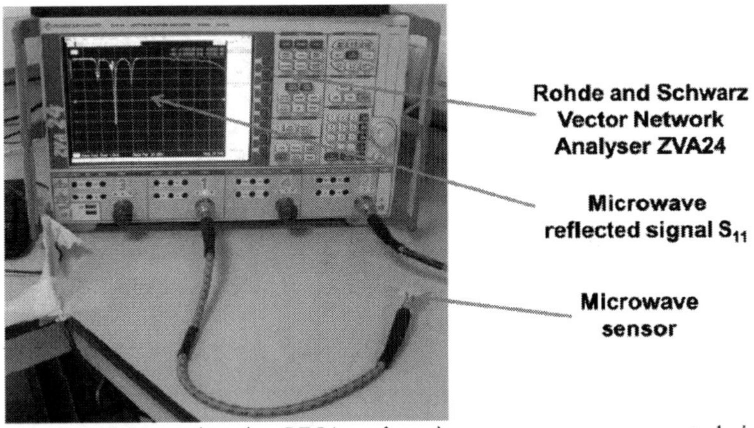

Figure 3. Measurement setup showing VNA and a microwave sensor connected via coaxial cable.

4. Results and discussion

The spectra for the microwave sensor were recorded in the 0-15 GHz frequency range when in contact with 20 µl of deionised water and KCl, $MnCl_2$ and CuCl solutions of 1M and 0.1M concentrations. The ability of this sensor to distinguish the presence and concentration of various chloride solutions was only evident in the change in the amplitude and resonance frequency of the signal occurring at frequencies below 4 GHz.

Figure 4 illustrates the S_{11} signal response of the microwave sensor in the 0-2.5 GHz range when in contact with deionised water and CuCl solutions in 1M and 0.1M concentrations. One may clearly see that the microwave spectrum is different for each solution with both resonant peaks at around 1 and 2 GHz experiencing a frequency shift as well as the change in the signal amplitude.

Similarly, Figures 5 and 6 illustrate the S_{11} signal response of the microwave sensor when in contact with deionised water and KCl and MnCl solutions in 1M and 0.1M concentrations respectively. These graphs also show that the microwave spectra are different for each solution with both resonant peaks experiencing a frequency shift as well as the change in the signal amplitude. Thus, the developed sensor demonstrated a distinct ability not only to detect the presence of various chloride solutions in deionised water (as evident from Figure 7), but to measure their concentrations as well.

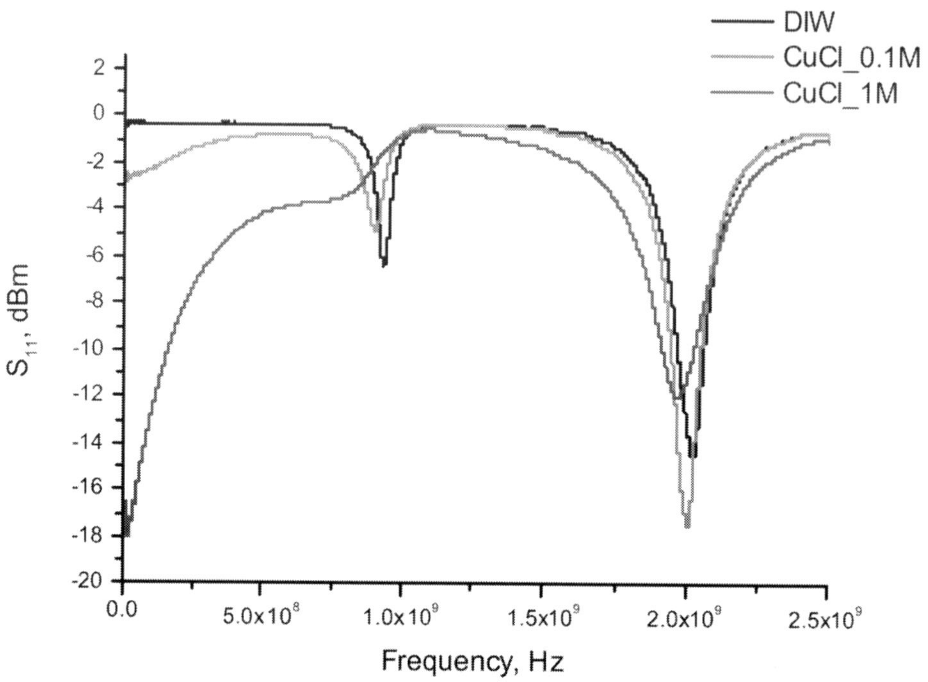

Figure 4. S_{11} signal response of the microwave sensor in the 0-2.5 GHz range when in contact with deionised water and CuCl solutions in 1M and 0.1M concentrations.

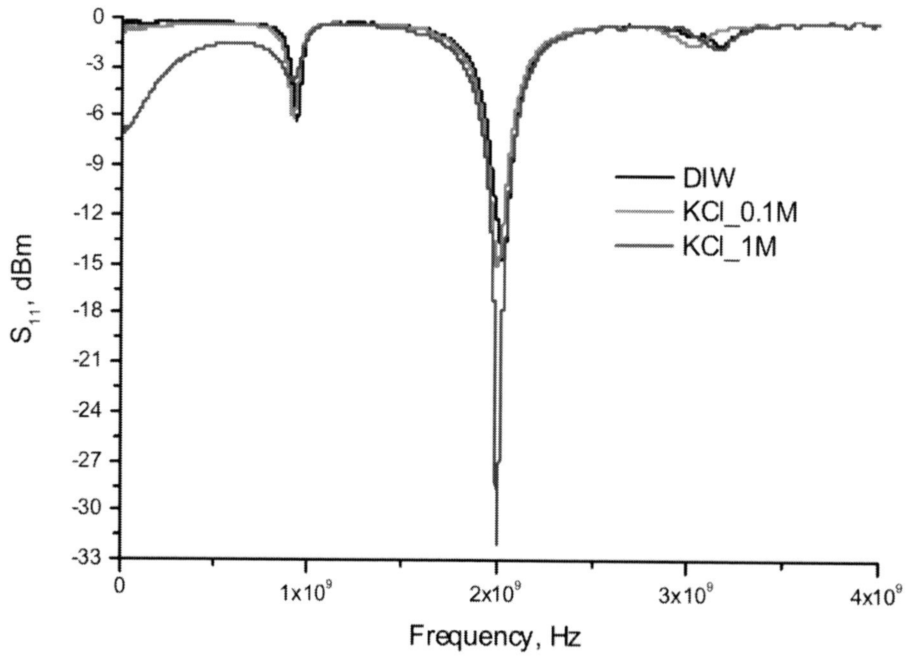

Figure 5. S_{11} signal response of the microwave sensor in the 0-4 GHz range when in contact with deionised water and KCl solutions in 1M and 0.1M concentrations.

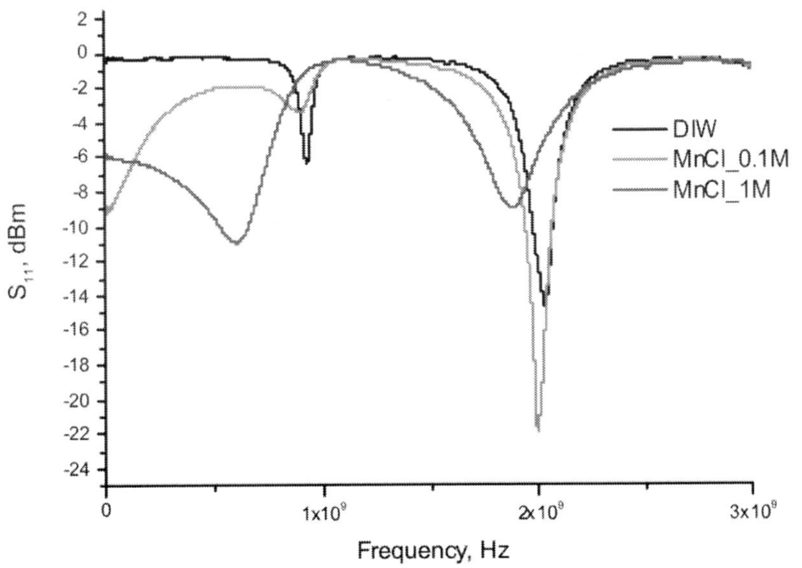

Figure 6. S_{11} signal response of the microwave sensor in the 0-3 GHz range when in contact with deionised water and MnCl solutions in 1M and 0.1M concentrations.

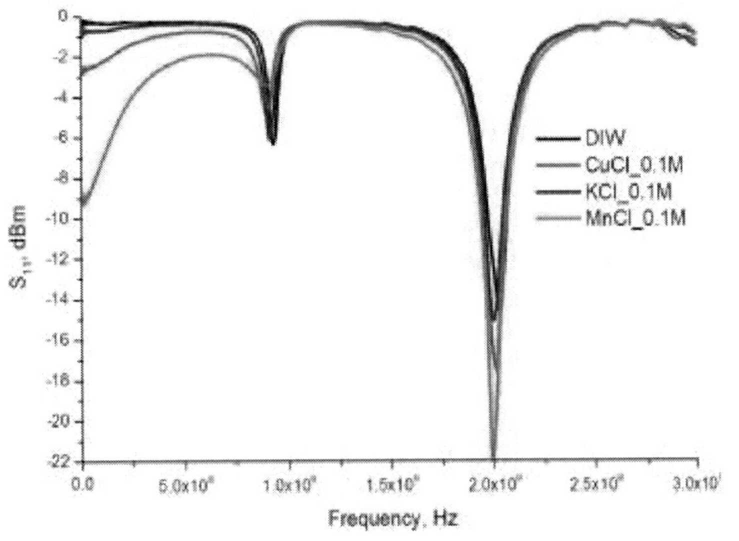

Figure 7. Comparing the microwave sensor responses to different solutions with the same concentration, confirming its ability to distinguish between different types of the solutions as well as their concentrations.

Conclusion

A novel sensor with a silver planar pattern was developed to verify the potential applicability of the microwave sensing approach for real-time monitoring of various chloride solutions presence and concentrations. The sensor responses tested in microwave frequency range revealed that the resonant peaks amplitude and frequency change once the deionised water and KCl, $MnCl_2$ and CuCl solutions of 1M and 0.1M concentrations in 20 μl volumes were placed in contact with the sensor pattern. Each solution produced a characteristic signature spectrum, which would be used for subsequent testing of unknown chloride solutions and concentrations. A distinct advantage of this novel sensor is that minute solution volume is required for the test – a feature not available with counterpart modern methods.

Acknowledgments
This work is financially supported by the European Community's Seventh Framework Programme through the FP7-PEOPLE-2010-IEF Marie-Curie Action project 275201, Water-Spotcheck. The authors would like to express their gratitude to Mike Garner, Garner Osborne Circuits Limited, UK and to Mabruk Farrah, Mechan Controls Plc, for manufacturing sensor prototypes.

References
[1] Z. Yong, L. Yanbiao, Z. Bo *et al.*, "Monitoring technology of salinity in water with optical fiber sensor," *Lightwave Technology, Journal of,* vol. 21, no. 5, pp. 1334-1338, 2003.
[2] A. Cranny, N. R. Harris, N. Mengyan *et al.*, "Sensors for Corrosion Detection: Measurement of Copper Ions in 3.5% Sodium Chloride Using Screen-Printed Platinum Electrodes," *Sensors Journal, IEEE,* vol. 12, no. 6, pp. 2091-2099, 2012.
[3] Z. Daoshan, Z. Zhaohui, and L. Zhijun, "The P, S, Cl Flame Photometric Sensors Fusion and Intelligent Detection." pp. 769-771.
[4] R. Rosen, "Mass spectrometry for monitoring micropollutants in water," *Current Opinion in Biotechnology,* vol. 18, no. 3, pp. 246-251, 2007.
[5] O. Korostynska, A. Mason, and A. I. Al-Shamma'a, "Monitoring of Nitrates and Phosphates in Wastewater: Current Technologies and Further Challenges," *International Journal on Smart Sensing and Intelligent Systems,* vol. 5, no. 1, pp. 149-176, March, 2012.
[6] G. R. C. Possetti, R. C. Kamikawachi, C. L. Prevedello *et al.*, "Salinity measurement in water environment with a long period grating based interferometer," *Measurement Science and Technology,* vol. 20, no. 3, 2009.
[7] J. D. Boon, and J. M. Brubaker, "Acoustic-microwave water level sensor comparisons in an estuarine environment." pp. 1-5.
[8] B. Jackson, and T. Jayanthy, "A novel method for water impurity concentration using microstrip resonator sensor." pp. 376-379.
[9] C. Bernou, D. Rebière, and J. Pistré, "Microwave sensors: a new sensing principle. Application to humidity detection," *Sensors and Actuators B: Chemical,* vol. 68, no. 1–3, pp. 88-93, 2000.
[10] T. Nacke, A. Barthel, C. Pflieger *et al.*, "Continuous process monitoring for biogas plants using microwave sensors." pp. 239-242.
[11] O. Korostynska, R. Blakey, A. Mason *et al.*, "Novel method for vegetable oil type verification based on real-time microwave sensing," *Sensors and Actuators A: Physical*, no. 0, 2013.
[12] O. Korostynska, A. Arshak, P. Creedon *et al.*, "Glucose monitoring using electromagnetic waves and microsensor with interdigitated electrodes." pp. 34-37.
[13] A. Mason, S. Wylie, A. Thomas *et al.*, "HEPA Filter Material Load Detection Using a Microwave Cavity Sensor," *International Journal on Smart Sensing and Intelligent Systems,* vol. 3, no. 3, pp. 16, Sep, 2010.
[14] A. Al-Shamma'a, A. Mason, and A. Shaw, *Patent: Non-Invasive Monitoring Device,* US2012150000 (A1), WO2010131029 (A1), EP2429397 (A1), 2012.
[15] R. T. Blakey, A. Mason, A. Al-Shamma'a *et al.*, "Dielectric Characterisation of Lipid Droplet Suspensions Using the Small Perturbation Technique," *Advancement in Sensing Technology,* Smart Sensors, Measurement and Instrumentation S. C. Mukhopadhyay, K. P. Jayasundera and A. Fuchs, eds., pp. 81-91: Springer Berlin Heidelberg, 2013.

Sensors & their Applications XVII IOP Publishing
Journal of Physics: Conference Series **450** (2013) 012014 doi:10.1088/1742-6596/450/1/012014

Abilities of Raman sensor to probe pollutants in water

K Ben Mabrouk[1,2], TH Kauffmann[1,2] and MD Fontana[1,2]

[1]Université de Lorraine, Laboratoire Matériaux Optiques, Photonique et Systèmes, EA 4423, 2 rue E. Belin, 57070 Metz, FRANCE
[2]Supélec, Laboratoire Matériaux Optiques, Photonique et Systèmes, EA 4423, 2 rue E. Belin, 57070 Metz, FRANCE

E-mail: fontana@supelec.fr

Abstract. The control of water quality requires rapid and efficient monitoring techniques. A Raman sensor is shown to present interest for the analysis of chemical substances dissolved in water. The identification of chemical pollutants and the determination of their content as well are reliably obtained. The main characteristics and abilities of this multi-component Raman sensor are reported.

1. Introduction

Raman spectroscopy (RS) is a widely used technique to know the physico-chemical properties of a solid, liquid or gas, via the study of vibrational characteristics of molecules or ions [1,2,3]. Raman sensors appear only recently [4, 5]. They combine the determination of a physical parameter, as in any sensor, with the physical mechanism associated with this parameter. The position (mode frequency), the width (damping) and the intensity of a Raman line can be exploited in a Raman sensor.

The control of the water quality is a growing challenge in order to insure the access of water in healthy conditions to an always rising number of people through the world. This implies the obligation to detect all pollutants in flow water, rivers, lakes and tanks. Nitrates, sulphates, nitrites, chlorides and phosphates coming either from fertilizers used in agriculture or product manufacturing are specially pursued since they are very water soluble and have a negative incidence in the environment, leading to eutrophication and thus to asphyxia of aqueous ecosystems in rivers and lakes [6,7,8]. Electric conductivity, calorimetry, chromatography and UV spectrophotometry are used [9] to detect different species. Despite some advantages such as their cost and sensitivity, these methods are generally specific to probe one anion only. Moreover they are off-lines and therefore time-consuming techniques. Thus they do not fully respond to the need of continuous monitoring of soil and water. The question is if a Raman sensor could offer an alternative method to detect in situ several contaminants within fairly good accuracy and high speed. As RS is a non-destructive technique analysing only a small volume of a substance without needing any preparation, Raman sensors are in principle suitable for *in situ* and in real-time measurements. Here we demonstrate the abilities of Raman sensor to provide the detection of several chemical substances dissolved in water. In a first stage measurements are performed within laboratory conditions, in order to achieve the calibration and to check the reliability and efficiency of this technique. The characteristics of our sensor are briefly described. The main advantage compared with other techniques is its ability to probe simultaneously or successively several substances. The measurement of nitrates is given as an example, but sulphates, sulphites, phosphates, and nitrites can be detected with the same sensor [10].

2. Identification

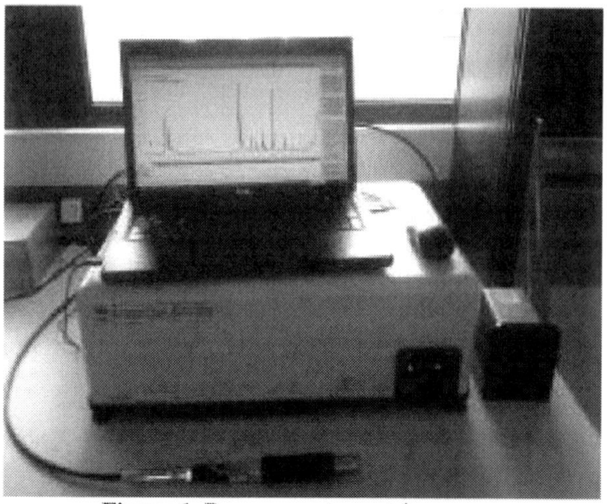

Figure 1. Raman spectrometric sensor.

Measurements were recorded at room temperature using a Raman spectrometer RXN-1 from Kaiser Optical Systems (see Figure 1). This Raman is a portable instrument designed for industrial applications. The spectral resolution obtained with an excitation source at 532nm is about 2cm^{-1}. The Raman detector is a charge coupled device (CCD) camera of 1024 pixels cooled by a Peltier device to -40°C. Measured power at the sample level is 71mW. The Raman scattered light was backscattered collected by the same specific probe, and filtered with a Notch filter before being dispersed by a holographic grating. Spectra were recorded over the 150-4300cm^{-1} frequency range. Figure 2 shows the typical Raman spectrum of an aqueous solution of sodium nitrate compared with the spectrum obtained in pure water. A very intense and relatively sharp peak lying at about 1049cm^{-1} comes out from the nitrate spectrum. This line was shown to be the specific signature of the molecule NO_3^-, since it appears at nearly the same position in several nitrate salts.

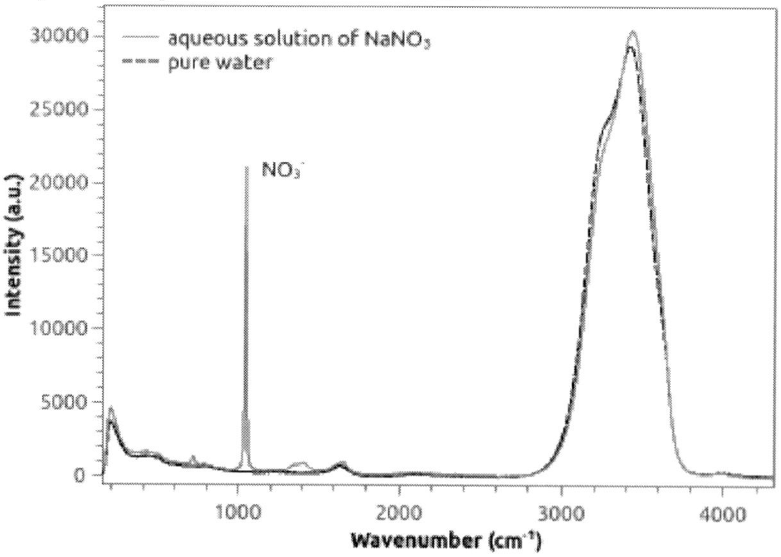

Figure 2. Raman spectra recorded in sodium nitrate aqueous solution, and in pure water.

In addition the line at 1049cm^{-1} can be clearly distinguished from the lines characterizing other salts dissolved in water, as shown in Table 1. The assignment of a Raman peak to a unique anion, consequently provides the simultaneous (or not) identification of several chemical substances in water

[10]. It proves the selectivity of the Raman sensor. Then the Raman peak taken to identify the anion NO_3^- in a mixture is used after an appropriate calibration, to determine its concentration.

Anions	SO_4^{2-}	PO_4^{3-}	HPO_4^{2-}	$H_2PO_4^-$
Main peak position in cm^{-1}	981	936	989	1076

Table 1. Main peak position of several anions in aqueous solution.

3. Calibration and determination of concentration

In Figure 3 is reported the main Raman peak recorded in solutions of sodium nitrates with different concentrations. It is shown that the position of this peak is unchanged whatever the salt content in water is. It is observed that this peak is not fully symmetric since a shoulder appears in the low-frequency side especially for low concentration. Therefore instead of the peak maximum, the intensity integrated over a wide wavenumber range (1020-1070cm^{-1}) was taken for the calibration process of NO_3^-.

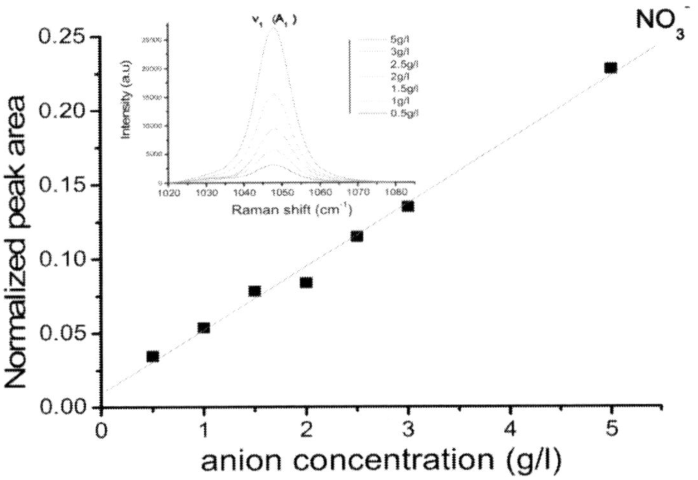

Figure 3. Calibration of the Raman intensity of the peak signature of the nitrate molecule as function of this concentration (expressed in g/l).

As in principle Raman band intensity is proportional to the number of molecules involved in the scattering process, the intensity of the peak signature of the sought substance can provide the substance content. In addition according to the Raman rules the molecule should be non-centro symmetric and highly polarisable, in order to enhance the Raman activity and sensitivity. In fact the absolute intensity is generally difficult to be reached since it needs to account in its calculation the characteristics of the spectrometer, the laser source and detector and all optical components as well. This is impossible in the *in situ* measurement conditions since the intensity depends on the external conditions as the temperature etc. Furthermore even the roughly measured intensity cannot be simply used, since it is affected by any fluctuation of the laser power, or any change in the experimental conditions. As a consequence the relative intensity of the peak normalized to the intensity of an extra band chosen as reference, is utilized in the calibration and determination of a substance concentration. The extra line denotes a peak in the Raman spectrum which is independent of the substance under study, but which is affected by any external perturbation likewise as the signature peak. Extra peak should be relatively intense and well separated from the signature peak. Two kinds of references can be considered. Internal references are provided by the solvent (here the water), or a product added in the mixture, that does not react with initial substances, and possess well known Raman lines. Sometimes external reference is given by the probe itself in the case of contact measurement, since the probe has its own Raman characteristic lines. In our study the OH band located around 1700cm^{-1} is retained, since it holds the above conditions for a suitable reference. It is to mentioned that the upper frequency OH band around 3000cm^{-1}, even it is very intense, cannot be used as reference since affected by of the studied substance (see Figure 2, the intensity of this band changes with the presence of nitrate).

This point is particularly important in the measurement of very small substance quantities. This sometimes as demonstrated below, requires longer recording time. Fluctuations of power intensity or generally experimental perturbations during this time can lead to the same change in Raman intensity, as this induced by a small variation of substance content.

It is remarkable to note that this calibration holds for any nitrate in aqueous solution, regardless the cation. This was checked for dissolved sodium, ammonium and potassium nitrates. Then in order to reinforce the validity of the method, nitrate content in an aqueous mixture containing sulphates and phosphates as well was determined via the above calibration. The results of this blind test are reported in Table 2. They show a fairly good agreement between the expected value and the determined value issued from the calibration. However an error, of about 50mg/l (or 50ppm) arises from the detection of the sought substance (here the nitrate) in a complex mixture. The use of chemometrics techniques [11, 12] in the spectra analysis allows diminishing this error.

[NO_3^-] by weight (mg/l)	200	400	600
[NO_3^-] determined by our sensor (mg/l)	250	397	558

Table 2. Results of blind tests made on several nitrate solutions.

4. Experimental parameters

Here we briefly discuss on experimental conditions. The choice of the exciting wavelength of the laser in the Raman scattering process is of prime importance, since it strongly affects the sensitivity of the sensor. At first the wavelength is taken in the UV, visible (usually 532 and 633nm) or IR (generally 785 and 1064nm) to avoid the large fluorescence of the sample which can hide the Raman spectrum, and to limit the absorption. This is particularly important in the investigation of real samples since waste water which contains substances which obscure Raman signal. This leads to generally perform *in situ* measurements with an IR rather visible excitation. On contrary the use of shorter wavelength allows the enhancement of Raman intensity. In addition the choice of suitable spectrometers and detectors has to be done. A spectrometer with high resolution is necessary for the detection of narrow peaks and/or the separation between different substances in a mixture. Nevertheless, according to the aims required in terms of accuracy and precision, a spectrometer with a poor resolution, (i.e. 5cm^{-1}), and therefore cheaper, can be sufficient. CCD detector also affects the sensor efficiency the equipment. More generally the signal to noise ratio has to be improved. Finally experimental conditions can require the choice of the contact- or contactless-probe, which strongly affects the sensor ability. All these parameters affect the accuracy and precision of the sensor [13].

The main limit in the Raman sensor efficiency in the determination of substance content comes from the fact that the intensity is proportional to the number of molecules scattered by the light, and this number is obviously small if the substance content is weak. Different alternative Raman techniques can partly solve this problem- resonance Raman scattering or surface enhanced Raman scattering [14] – but they are heavier to be implanted and present other drawbacks, such as the nonlinear signal response. However the Raman intensity can be enhanced even within conventional Raman technique. One solution consists into the increase of measurement recording time. Compared with other methods Raman spectrometry has the advantage to provide signal in a few seconds, an increasing measuring time is acceptable. Figure 4 displays the time dependence of the limit of detection LOD, defined by the criterion SNR=2 (Signal to Noise Ratio). LOD can be pushed down by increasing recording time. It is observed that the limit of water drinkability (50ppm) can be reached by a measurement during 2min, which remains reasonable. This result is obtained with a 70mW power source. This time could be reduced if using a larger laser power, but this should induce additional random error related to the intensity instability. Thus compromise between different parameters has to be achieved in order to improve the sensor according to the requirements of the application [10].

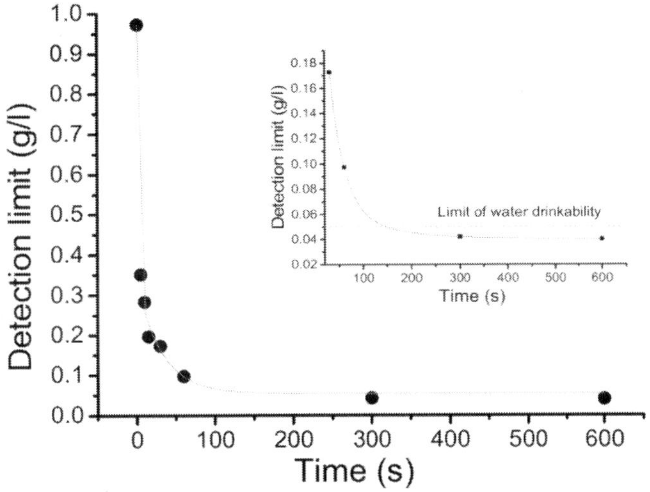

Figure 4. Recording time dependence of the limit of detection. The concentration of nitrate corresponding to the water drinkability is reported for comparison

5. Conclusion

We demonstrated the interest of Raman sensor for the detection of chemical substances dissolved in water. The assignment of well-defined and intense Raman peaks to most frequent anions was achieved, assuring the sensor selectivity. After calibration of various pure substances, the identification of one species in a complex mixture and its concentration determination proves the validity of the technique, firstly within laboratory conditions. Future investigations will concern the measurement of very small quantities by reducing the detection limit, and the application to waste water.

References

[1] Lyon LA, Keating CD, Fox AP, Baker BE, He L, Nicewarner SR, Mulvaney SP and Natan MJ 1998 *Anal. Chem.* **70** 341R.
[2] Colomban Ph and Corset J 1999 *Journal of Raman Spectroscopy* **30** (10)
[3] Lewis R and Edwards HGM 2001 *Handbook of Raman spectroscopy*
[4] White SN, Kirkwood W, Sherman A, Brown M, Henthorn R, Salamy K, Waltz P, Peltzer ET and Brewer PG 2005 *Deep-Sea Research Part I* **52** 2376
[5] Claverie R., Fontana M.D., Durickovic I, Bourson P, Marchetti M and Chassot JM 2010 *Sensors* **10** 3815
[6] Burkholder JM, Tomasko DA, Touchette BW 2007 *Journal of Experimental Marine Biology and Ecology* **350** 46-72
[7] Berenzen N, Hummer S, Liess M and Schulz R 2003 *Bull. Environ. Contam. Toxicol.* **70** 891-897
[8] Jickells T 2005 *Journal of Sea Research* **54** 58-69
[9] Bhadekar R, Pote S, Tale V and Nirichan B 2011 *Am. J. Analytical Chem.* **2**, 1-15.
[10] Fontana M .D., Ben Mabrouk K, Kauffmann T.H. 2013 *Spectroscopic properties of inorganic and organometallic compounds* **44** 40-67
[11] Wold S and Sjöström M 1998 *Chemomectrics and Intelligent Laboratory Systems* **44** 3-14
[12] Roggo Y, Degardin K and Margot P 2010 *Talanta* **81** 988
[13] Kauffmann TH and Fontana MD 2012 *Sensors and Actuators B: Chemical* **161** 21
[14] Faulds K, Hernandez-Santana A and Smith WE 2010 *Spectroscopic properties of inorganic and organometallic compounds: techniques, materials and applications* **41** 1-21

SOLAR POWER WATER DISTILLATION UNIT

Kamran Hameed[1], Muhammad Muzammil.Khan[2], Ijlal Shahrukh.Ateeq[3], Syed Muhammad Omair[4], Muhammad Ahmer[5], Abdul Wajid[6]

[1, 2, 3, 4, 5, 6] Sir Syed University of Engineering and Technology, Department of Biomedical Engineering, Karachi, Pakistan.

kamranhameedpk@hotmail.com

Abstract — **Clean drinking water is the basic necessity for every human being, but about 1.1 billion people in the world lacked proper drinking water. There are many different types of water purification processes such as filtration, reverse osmosis, ultraviolet radiation, carbon absorption, but the most reliable processes are distillation and boiling. Water purification, such as distillation, is especially important in regions where water resources or tap water is not suitable for ingesting without boiling or chemical treatment. In design project It treats the water by combining different methods such as Filtration, Distillation and a technique called concentrated solar power (CSP).Distillation is literally the method seen in nature, whereby: the sun heats the water on the earth's surface, the water is turned into a vapor (evaporation) and rises, leaving contaminants behind, to form clouds. As the upper atmosphere drops in temperature the vapors cool and convert back to water to form water. In this project distillation is achieved by using a parabolic mirror which boils water at high temperature. Filtration is done by sand filter and carbon filter. First sand filter catches the sand particles and the carbon filter which has granules of active carbon is used to remove odor dissolved gases from water. This is the Pre-treatment of water. The filtered water is then collected in a water container at a focus of parabolic mirror where distillation process is done. Another important feature of designed project is the solar tracking of a parabolic mirror which increases the efficiency of a parabolic mirror** [1],[2].

Index Terms— *Water purification processes, Filtration, Reverse osmosis, Ultraviolet radiation, Carbon absorption, Distillation and concentrated solar power (CSP).*

I. Introduction

The atmosphere of the earth absorbs some of the energy and the more air the sun's rays have to travel through to get to the earth's surface, the weaker they become. We can harness the sun's energy to distil water. If there is a full sun, we can safely assume about 100 watts of solar rays per square foot. The key is developing technologies that efficiently convert solar power into usable energy in a cost-effective manner. One way to achieve this is by making a parabolic mirror. The purpose of the parabolic mirror is to collect the sun's energy over as wide an area as possible and focus it onto a smaller area. [2]. Solar powered water purification unit uses the process of distillation to purify the tap as well as contaminated water. Heat by the parabolic mirror at the focus evaporates the water and thus various salts and microorganism will remain left. Then the steam is passes through the condenser which converts the steam into water which is actually called the distilled water. The parabolic mirror is based on a motor which follows the rotation of a sun. The mirror rotation/movement is controlled by a sun tracker circuit. Which follows the sun movement from East to west and slightly north to South because there is a slightly variation in the direction of sun from North to South [4].

A portable distillation unit which can be setup anywhere in remote areas. Solar tracking System's unit which is cost effective. Reduces demand for fuel or firewood. All the material can be obtained from local market. Its main application is in remote areas where pure water is not available for drinking. It is used in various industries and chemical and biological laboratories where highly purified water is essential [1].

II. Design Methodology

A. Purpose of Designing

The water treatment system which is cost effective and consume less energy to purify water, keeping this in mind we decide to utilize solar energy for treatment system. We use solar panel to fulfill the power requirements for the circuits and gear motors. It is a common observation that sun position changes throughout the day for this purpose solar tracker circuit designed which follows the direction of the sun and utilize the solar energy as maximum as possible. Figure 1.1 shown the complete flow of the design project.

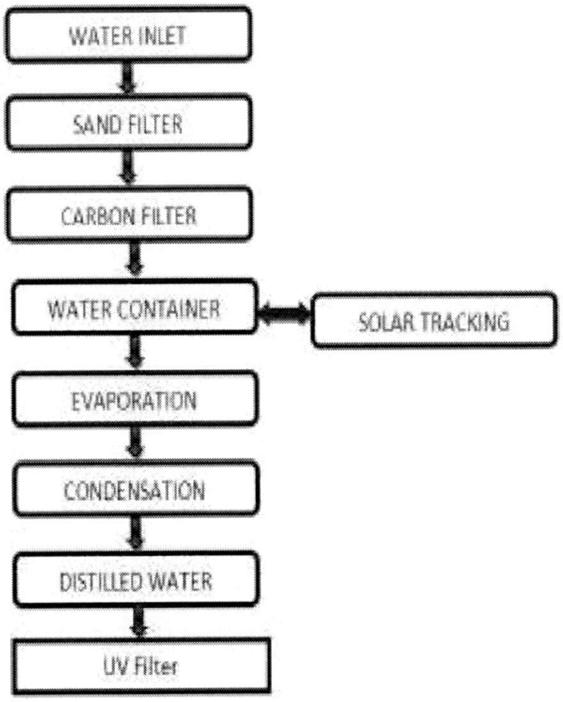

Figure 1.1 System flow chart

B. Material used

In project we used different materials for achieving our desire goals. Which consisted of gear motors which provide torque and low RPM? One gear motor is used for east-west movement and other is use for north-south movement. The project structure made up of iron which provides strength to the structure. Backup Battery is also used which provides power requirement for the solar tracker circuit. Concave mirrors used which reflect inward to one focal point, therefore they are used to concentrate sun rays on a focus point and raise the temperature above 600°C. Condenser is used which allow the steam to convert into purified water.

C. Techniques

In this project we use two major techniques one is concentrated solar power (CSP) and other is solar tracking. Concentrated Solar Power (CSP) is a technology which produces high temperature by concentrating solar energy in a single focal point. This concentrated energy is then used to heat up water or other fluids to produce steam. The focusing of the sun rays can be achieved through different designs such as parabolic trough, parabolic dish. Concave mirror or power tower systems. The steam produces by CSP can be used to drive turbine to produce electricity for desalination plants or it can be directly converted in purified water by passing it through the condenser [5]. Simple and non-polluting CSP technology can be deployed relatively quickly anywhere and can contribute substantially to reducing carbon dioxide emissions. The desert areas with frequent sunshine are ideal places for the deployment of CSP technology. [6].

The basic lay out of solar tracking system is shown in figure 1.2.

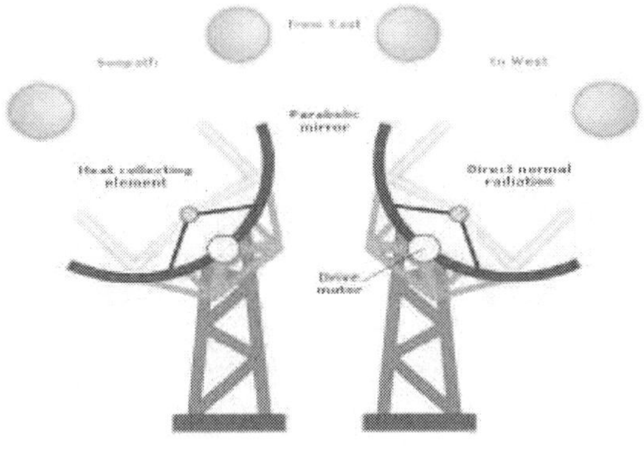

Figure 1.2.Solar Tracking [7]

D. Project phase.

In 1[st]designing phase we made a sand filter by using a PVC pipe of 4 inches diameter and 3 feet in length. In sand filter

In 2nd phase we bought a concave mirror from scrap market which was in two pieces so we join these two pieces to make a single concave mirror of 2 feet's by using fiber cloth and resin.

In last phase we design mechanical structure of a project and modifying the structure from time to time. On the other hand we were also designing the circuit for solar tracking.

III.HARDWARE DESCRIPTION.

A. Charge Controller Circuit.

Charge controller is design to control how much ampere is flowing towards the battery. It is also to prevent overvoltage .Figure 1.3 is shown the complete schematic of charge controller circuit.

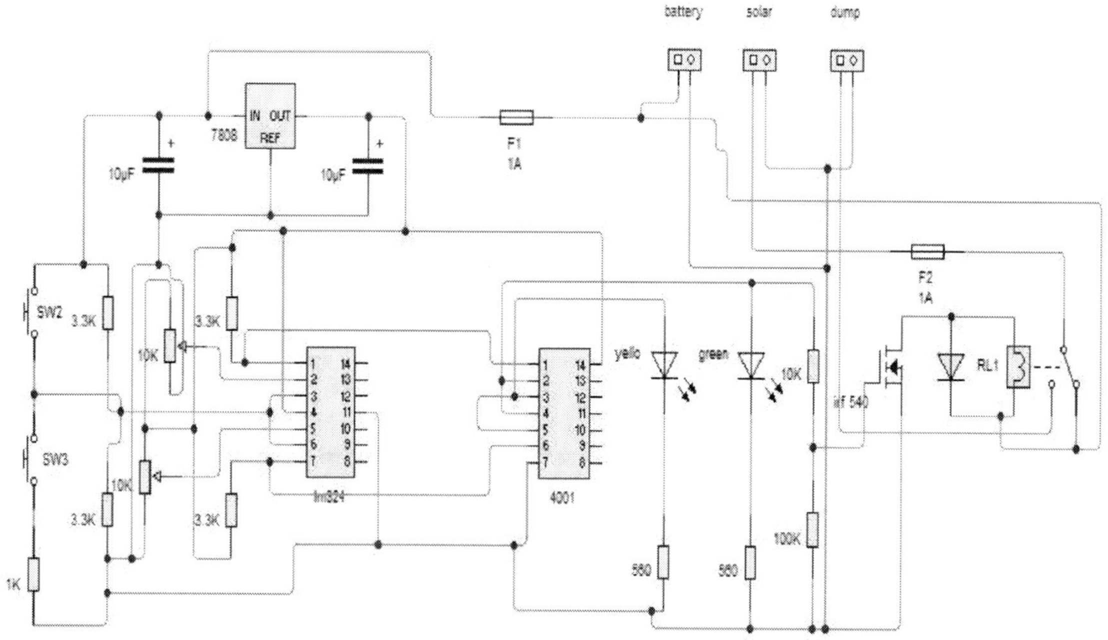

Figure 1.3.Charge controller circuit

A. Solar Tracker Circuit & its description.

It is a circuit which synchronizes the movement of concave mirror and solar panel with the sun. The circuit uses LDR (Light dependent resistance) as a light sensor. One pair of LDR control NORTH-SOUTH movement and other pair control the EAST-WEST movement. Figure 1.4 is shown the complete schematic circuit of solar tracker. With the photocells mounted on the solar collector and in bright sunlight and correctly aimed at the sun each photocell should produce about 3000 ohms of resistance. This value can be obtained by placing different pieces of translucent plastic over the cells; this is not a critical value. The plastic will also protect the cells from the weather. With both photocells seeing the same amount of light trim pot VR1 is adjusted to drive the output of voltage comparator IC1 a low this is indicated by

Journal of Physics: Conference Series **450** (2013) 012015 doi:10.1088/1742-6596/450/1/012015

LED 2 turning off. As long as both cells see the same amount of light the circuit will not respond, even if a cloud has the nerve to pass between your solar collector and the sun.

Now as the sun moves and the vertical barrier causes a shadow to be cast on the west photocell IC1a senses this change and its output will go high, turning on LED 2, and begins the slow charging of C3 through R5. After C3 has charged to the value determined by voltage divider R6 & R7, IC1b's output goes high this turns on transistor Q1 that in turn energizes relay RL1 causing the drive motor to start. The motor then runs turning the collector until the west cell again sees the full light of the sun, this returns the circuit to its initial state and the motor stops. The circuit then waits for the sun to move again and the whole process repeats.

The purpose of the time delay provided by R5, C3 & D1 (D1 is used to discharge C3 at end of cycle).

So now at the end of the day after the sun has been turned off, the west photocell has one more duty to perform. It sends a voltage to IC1c and when compared with the voltage set by trim pot VR2 its output will go high turning on LED 1 and to transistor Q2 that energizes relay RL2 causing the motor to run in reverse, driving the collector back to the east for the start of the next day. Also the output of IC1c goes to IC1d, this turns on the output of IC1d that in turn prevents the possibility of transistor Q1 from being turned on. This then keeps relay RL1 from being energized and prevents lots of smoke and fire that would occur if both relays were allowed to be energized at the same time. Finally there are two switches SW1 & 2. These are limit switches mounted on the collector to control the maximum amount of rotation in the east and west directions.

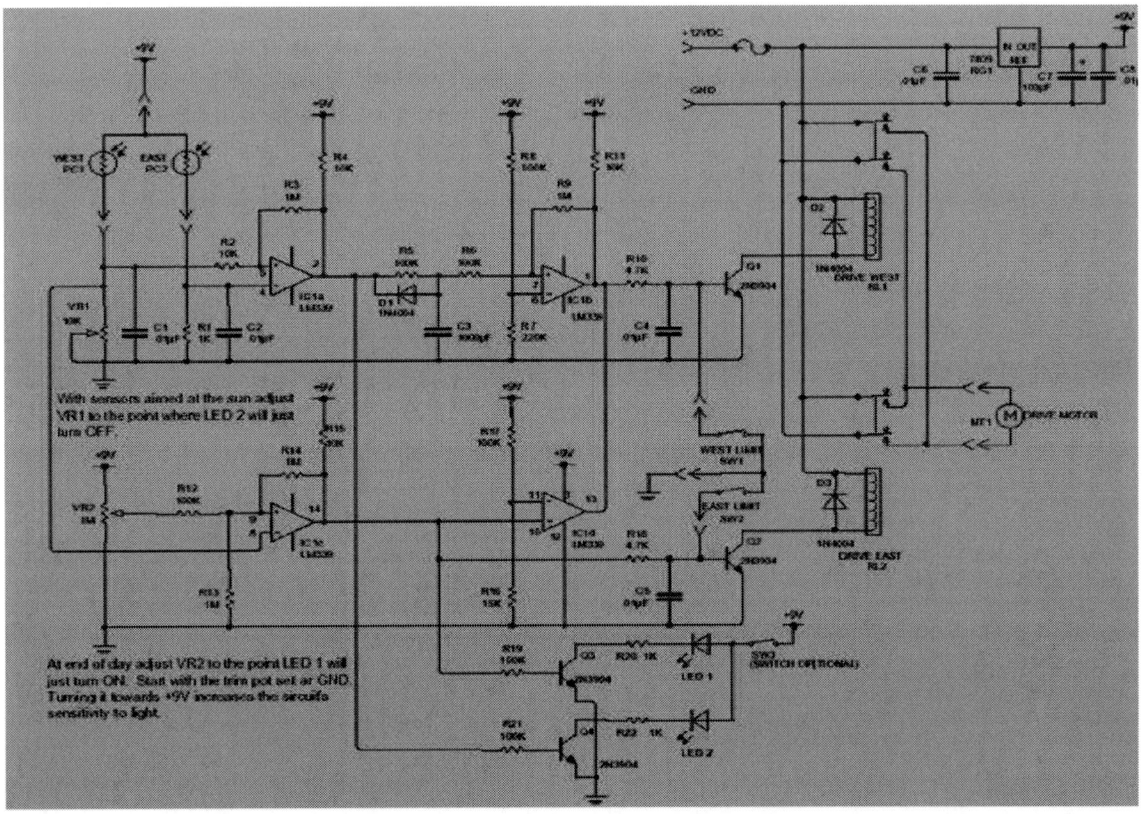

Figure 1.4.Solar tracking circuit

Figure 1.5. Project layout

Conclusion

We observe that concave mirror/reflector use in our project rise the temperature up to 600 – 800 c. which is enough to convert the water into steam and increase the speed of the process. The water obtain from this system is highly purified whose conductivity is in micro Siemens which has many applications in biomedical sciences. The important features are the solar tracking which is a key factor in increasing the efficiency of a system throughout the day. This system is highly cost effective results are obtained at minimum cost, no fuel/line electric supply (220VAC). The only power source is solar. Power requirement is fulfilled by solar panel and battery. Solar panel charging the battery via charge control circuit. This project can be easily installed at any remote area where purified water is not available easily.

References

[1] Frank R Grovesin Journal of Biomedical Engineering Data 1988.

[2] Dr.Lee.t. Rosella and Dr Ronald L.Wathen.MD "Consumption of low TDS water", March 1993.

[3] Gregory, J. "Cryptosporidium in Water: Treatment and Monitoring Methods" Filtration & Separation.

[4] NAS (National Academy of Sciences). National Research Council: drinkingwater and health. Volume 2. National Academy Press, Washington, D.C, 1980.

[5] L. Stoddard, J. Abiecunas, and R. O'Connell Economic, Energy, and Environmental Benefits of Concentrating Solar Power April 2006 David R. Mills and Robert G. Morgan, SOLAR THERMAL ELECTRICITY 2001.

[6] Sargent & Lundy LLC Consulting Group Chicago. Assessment of Parabolic Trough and Power Tower Solar Technology. October, 2003.

[7] P. Schramek, and D.R. Mills. Multi-tower solar array. EuroSun 2000, Copenhagen June 2000.

Journal of Physics: Conference Series **450** (2013) 012016 doi:10.1088/1742-6596/450/1/012016

Real-time Sensing of NaCl Solution Concentration at Microwave Frequencies Using Novel Ag Patterns Printed on Flexible Substrates

M Ortoneda-Pedrola, O Korostynska[1], A Mason and A I Al-Shamma'a
BEST Research Institute, School of Built Environment, Liverpool John Moores University, Liverpool, UK

E-mail: o.korostynska@ljmu.ac.uk

Abstract. A novel microwave sensor for real-time analyte composition analysis is reported. The sensing element in a form of silver pattern was printed on polyimide flexible laminate substrate to suit a broad range of applications where a sensor could be placed in water reservoirs or fluid-carrying pipes for continuous analysis. The developed system confirmed the viability of using microwaves for real-time NaCl monitoring as the reflected signals represented by S_{11} parameters corresponding to deionised water and 0.1M and 0.01M of NaCl were unique with clearly observed shifts in the resonant frequencies when placed in direct contact with 20 μl of each solution.

1. Introduction
The composition of the analyte solutions must be determined with high accuracy for many applications, such as food and pharmaceutical industries, chemical processing and environmental monitoring, where the content of impurities in water is an important factor [1].

Traditional methods of aqueous solutions composition monitoring are lab-based and include standard UV-Vis measurements, mass spectrometry, ion-sensitive electrodes and amperometric sensors [2, 3]. On the other hand, a range of modern real-time monitoring approaches exist; for example, fibre-optic sensors, MEMs, lab-on-chip sensors and biosensors. It was also demonstrated that micro-Raman spectroscopy can allow the determination of the NaCl salt concentration of an aqueous solution with an error close to ±5% [4].

No single system available today can fully address the needs of the customers in its ability to determine, in real time, the composition of water solutions to the desired sensitivity level, bearing in mind the desire for system portability and cost-effectiveness. Therefore, novel real-time monitoring techniques are necessary. One technique potentially capable of meeting this current demand is based on the microwave sensing, as presented in this paper.

2. Microwave sensing
Microwave sensing is a novel but rapidly developing technology which has been successfully used as a sensing method for various industrial applications including fluid level measurements [5], material moisture content [6, 7], for continuous process monitoring of biogas plants [8], for the determination

[1] To whom any correspondence should be addressed.

of moisture content in soil [9] and in the healthcare industry, for example for real-time monitoring of glucose in diabetic patients [10, 11] and for non-invasive monitoring of bodily fluids [12, 13].

Microwave sensors in a form of a cavity resonator for accurate measurements of both organic (sugar, alcohol) and inorganic (NaCl, $KMnO_4$) water solutions concentrations were reported [1]. Notably, the sensitivity of the sensor in determination of NaCl was 0.4 dB/(mg/ml) within 0-1 % concentration range. The sensor was able to detect the concentrations of other water solutions, but its sensitivities are strongly dependent on the type of tested chemical ingredient.

However, the microwave planar printed patterns for various sensing applications are increasingly used due to their versatility, flat profile and low weight. Their design can be tailored to suit particular applications, coupled with reliability and cost-efficiency. They are easily manufactured using common methods for printed circuit board production, and their impedance can be matched to the input line by altering the micro-strip line feed configuration.

The patch antenna represents the frequency-selective element of a phase shift transistor oscillator. The active integrated antenna frequency of operation is determined by both the patch geometry and the electrical loads connected at the two microstrip ends [14]. A convenient termination is normally represented by two open-ended stubs whose electrical length is $\lambda/4$ at the patch resonant frequency, where λ is a complete wavelength. The performance of a microstrip resonator depends on its electromagnetic field distribution, resonant frequency and quality factor (Q).

To clarify the principle behind electromagnetic wave sensing suggested in this work for water quality control, it is worth mentioning that microwave sensors in the form of planar printed patterns operate based upon the fact that an object under test, e.g. a water sample, when placed into the vicinity or in direct contact with a microwave sensor, interacts with the electromagnetic waves in a unique manner, which can be specifically correlated with the properties of this material.

By considering how reflected (S_{11}) microwave signals vary at discrete frequency intervals, the change in the signal can be linked to the composition of the object under test.

3. Experimental procedure

3.1. NaCl solutions preparation

Sodium chloride NaCl (Sigma S6150) was dissolved in deionised water at 0.1M and 0.01M concentrations (equivalent to 5800 ppm and 580 ppm respectively) and produced clear solutions.

3.2. Microwave sensor structure

The layout and dimensions of the sensor pattern is shown in Figure 1. Silver was used as the conductive metal material for both the bottom layer, which acted as a ground plane, and the top pattern to maintain chemical neutrality when the device is placed in contact with analyte solution. The thickness of the Ag layers was 35 μm. Microwave sensor was designed on 2 mil thick DuPont™ Pyralux® AP Polyimide Flexible Laminate substrate.

A distinct feature of IDE type sensors is their superior sensitivity to change close to the sensor surface, with this sensitivity decaying rapidly with distance away from the surface, as shown in Figure 2. This is advantageous as it reduces significantly the chance of undesirable factors influencing sensor response.

3.3. Measurement setup

The microwave sensor was attached to a Rohde and Schwarz ZVA24 vector network analyser (VNA) via a coaxial cable, as shown in Figure 3. An SMA type Molex connector specified for 50Ω impedance was used. The data (60,000 points for each measurement) was captured in the 0-15 GHz frequency range for the reflected (S_{11}) signals at constant temperature of 18 °C. The results are repeatable with less than 5% deviation. Notably, average sensor responses are depicted in the graphs shown in following section.

Journal of Physics: Conference Series **450** (2013) 012016 doi:10.1088/1742-6596/450/1/012016

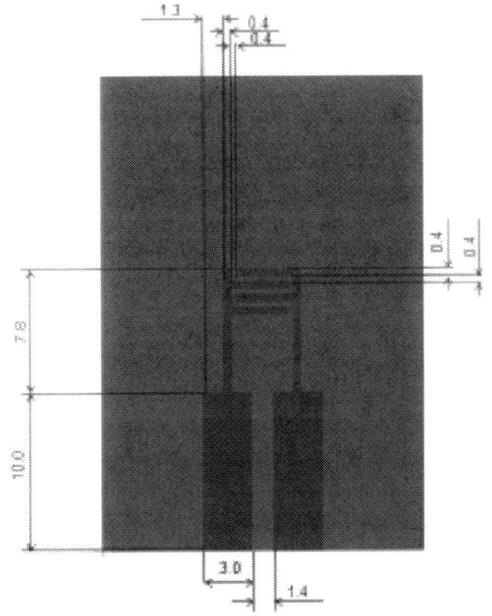

Figure 1. Ag printed pattern. All dimensions are in mm.

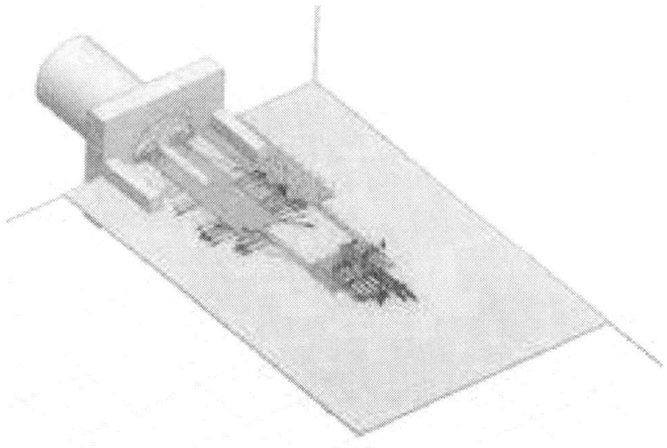

Figure 2. HFSS model of microwave sensor field distribution.

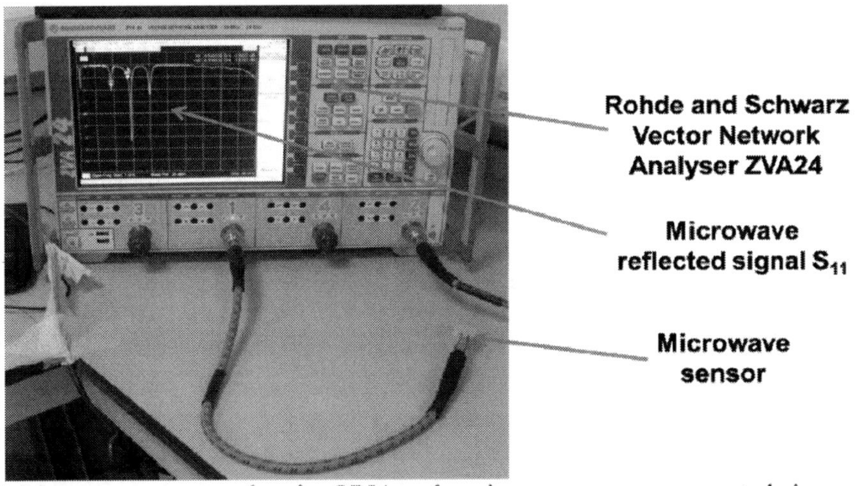

Figure 3. Measurement setup showing VNA and a microwave sensor connected via coaxial cable.

4. Results and discussion

The spectra for the microwave sensor were recorded in the 0-15 GHz frequency range when in contact with 20 µl of deionised water and NaCl analyte solutions in 0.1M and 0.01M concentrations. The ability of this sensor to distinguish the presence and concentration of NaCl is evident due to the change in the amplitude and resonance frequency of the peak occurring at frequencies below 2.5 GHz, as illustrated in Figure 4.

One may choose to monitor the change in the amplitude and frequency shift of a single resonant peak, such as that near 2 GHz, as shown in Figure 5 which depicts the spectra at 1.5-2.5 GHz range.

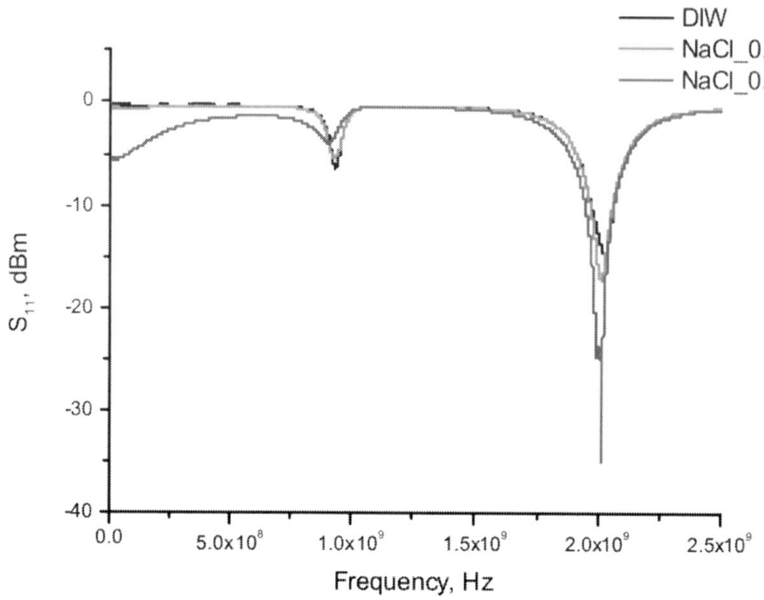

Figure 4. S_{11} signal distribution of the microwave sensor in the 0-2.5 GHz range when in contact with deionised water and NaCl solutions in 0.1M and 0.01M concentrations.

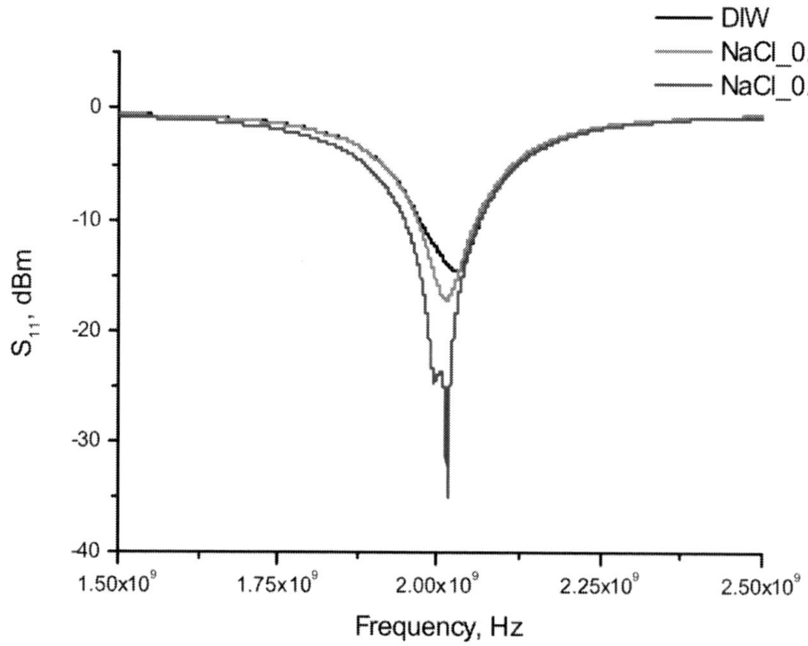

Figure 5. S_{11} signal distribution of the microwave sensor in the 1.5-2.5 GHz range when in contact with deionised water and NaCl solutions in 0.1M and 0.01M concentrations.

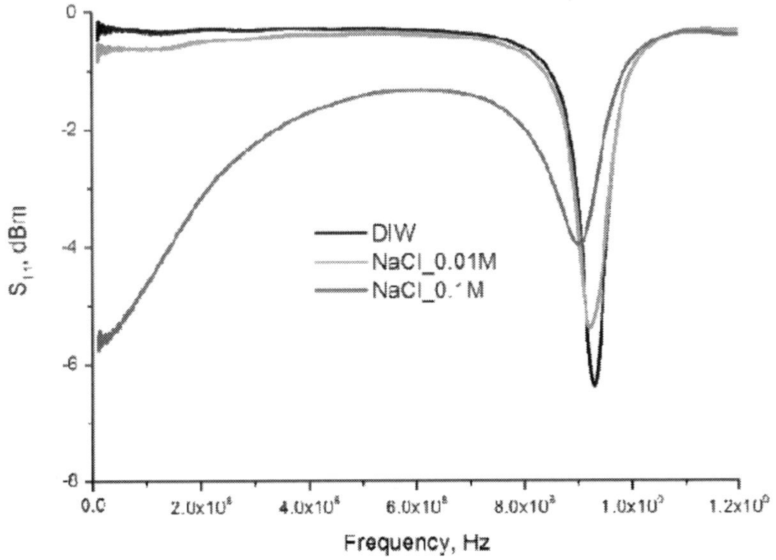

Figure 6. S_{11} signal distribution of the microwave sensor in the 0-1.2 GHz range when in contact with deionised water and NaCl solutions in 0.1M and 0.01M concentrations.

Additionally, lower ranges below 1.2 GHz are characteristic of NaCl presence and concentration, since each spectrum, as shown in Figure 6, is different. It is possible that each spectrum pattern could be used to determine the analyte presented to the sensor. Thus, the developed sensor demonstrates a distinct ability not only to detect the presence of NaCl in aqueous solution, but to measure its concentration as well.

Figure 7. Optical image of a microwave sensor with 20 μl water sample placed over it illustrating hydrophobic nature of the flexible substrate.

Notably, the sensors' responses returned to their original positions, namely air spectra, after each water sample measurement, confirming that the developed microwave sensors are reliable, re-usable and thus a sustainable solution for water quality monitoring. This could be explained by the hydrophobic nature of the DuPont™ Pyralux® AP Polyimide flexible laminate substrate used in the construction of these sensors, as shown in Figure 7.

Conclusion

This research was driven by the need for a novel, versatile, sensitive and cost-effective real-time monitoring method of NaCl presence and concentration in solution. A sensor with a silver planar pattern was developed to verify the potential applicability of the microwave sensing approach for this task. It is believed that the sensor' flexible structure would afford good longevity since this configuration is less prone to failure resulting from mechanical damage than typical alternatives. The sensor responses were tested in the 0-15 GHz frequency range. It was clearly shown that the resonant peaks amplitude and frequency changes once the deionised water and NaCl solutions in 0.1M and 0.01M concentrations (20 μl volumes) were placed in contact with the sensor pattern.

Acknowledgments

This work is financially supported by the European Community's Seventh Framework Programme through the FP7-PEOPLE-2010-IEF Marie-Curie Action project 275201, Water-Spotcheck. The authors would like to express their gratitude to Mike Garner, Garner Osborne Circuits Limited, UK and to Mabruk Farrah, Mechan Controls Plc, for manufacturing sensor prototypes.

References

[1] B. Kapilevich, and B. Litvak, "Microwave sensor for accurate measurements of water solution concentrations." pp. 1-4.
[2] R. Rosen, "Mass spectrometry for monitoring micropollutants in water," *Current Opinion in Biotechnology,* vol. 18, no. 3, pp. 246-251, 2007.
[3] O. Korostynska, A. Mason, and A. I. Al-Shamma'a, "Monitoring of Nitrates and Phosphates in Wastewater: Current Technologies and Further Challenges," *International Journal on Smart Sensing and Intelligent Systems,* vol. 5, no. 1, pp. 149-176, March, 2012.
[4] I. Durickovic, M. Marchetti, R. Claverie *et al.*, "Experimental study of NaCl aqueous solutions by raman spectroscopy: Towards a new optical sensor," *Applied Spectroscopy.* pp. 853-857, 2010.
[5] J. D. Boon, and J. M. Brubaker, "Acoustic-microwave water level sensor comparisons in an estuarine environment." pp. 1-5.
[6] B. Jackson, and T. Jayanthy, "A novel method for water impurity concentration using microstrip resonator sensor." pp. 376-379.
[7] C. Bernou, D. Rebière, and J. Pistré, "Microwave sensors: a new sensing principle. Application to humidity detection," *Sensors and Actuators B: Chemical,* vol. 68, no. 1–3, pp. 88-93, 2000.
[8] T. Nacke, A. Barthel, C. Pflieger *et al.*, "Continuous process monitoring for biogas plants using microwave sensors." pp. 239-242.
[9] K. Y. You, J. Salleh, Z. Abbas *et al.*, "A rectangular patch antenna technique for the determination of moisture content in soil," *PIERS 2010 Cambridge - Progress in Electromagnetics Research Symposium, Proceedings.* pp. 850-854.
[10] O. Korostynska, A. Arshak, P. Creedon *et al.*, "Glucose monitoring using electromagnetic waves and microsensor with interdigitated electrodes." pp. 34-37.
[11] A. Mason, S. Wylie, A. Thomas *et al.*, "HEPA Filter Material Load Detection Using a Microwave Cavity Sensor," *International Journal on Smart Sensing and Intelligent Systems,* vol. 3, no. 3, pp. 16, Sep, 2010.
[12] A. Al-Shamma'a, A. Mason, and A. Shaw, *Patent: Non-Invasive Monitoring Device,* US2012150000 (A1), WO2010131029 (A1), EP2429397 (A1), 2012.
[13] R. T. Blakey, A. Mason, A. Al-Shamma'a *et al.*, "Dielectric Characterisation of Lipid Droplet Suspensions Using the Small Perturbation Technique," *Advancement in Sensing Technology,* Smart Sensors, Measurement and Instrumentation S. C. Mukhopadhyay, K. P. Jayasundera and A. Fuchs, eds., pp. 81-91: Springer Berlin Heidelberg, 2013.
[14] F. Völgyi, "Microstrip Transmission- and Reflection-Type Sensors Used in Microwave Aquametry," *Electromagnetic Aquametry,* K. Kupfer, ed., pp. 243-256: Springer Berlin Heidelberg, 2005.

Journal of Physics: Conference Series **450** (2013) 012017 doi:10.1088/1742-6596/450/1/012017

Interface electronics for an RF resonance-based displacement sensor

E Asua[1], V Etxebarria[1], A García-Arribas[1], J Feutchwanger[1], J Portilla[1], and J Lucas[2]

[1]Departamento de Electricidad y Electrónica. Universidad del País Vasco (UPV/EHU) Apartado 644, 48080 Bilbao, Spain
[2]Elytt Energy. 28046 Madrid, Spain

E-mail: estibaliz.asua@ehu.es

Abstract. We propose, design, and test an electronic interface for a new standalone, affordable and compact displacement transducer based on resonant cavities. The operation of the interface establishes a self-resonance in the cavities and detects the resonance frequency (which is directly related to the position to be measured) by analyzing the attenuation produced by a low pass filter. The results obtained in a first prototype of the interface built with discrete elements show that the obtainable positioning accuracy using this cost-effective solution is about 5 micrometers.

1. Introduction

In many micro- and nano-scale technological applications high sensitivity displacement sensors are needed, especially in ultra precision metrology and manufacturing. There is a great technological interest in methods of measuring displacement at very small scales [1, 2]. Different solutions have been studied: laser interferometry [3, 4], capacitive sensors [5, 6], solutions involving the use of new materials (magnetostrictive TbDyFe [7] or carbon nanotubes [8]), or solutions related to radio frequency (RF) [9] and microwave cavity resonators [10, 11].

In our previous works [13, 14] a displacement transducer based on RF resonant cavities was proposed and tested. It was demonstrated that sub-nanometric displacements can easily be detected and measured using standard but high performance laboratory instrumentation. The principle of operation consists in discrimining resonant frequency variations. The measurements were performed using an RF network analyzer (NA), but this approach limits the use of this kind of sensor in applications where the use of that instrumentation is not feasible, especially for its cost.

We present here a cost-effective prototype of electronic interface to be used together with the resonant cavity transducer in order to replace the NA with minimum loss of resolution and accuracy. The cavity equipped with the proposed electronic interface would result in a standalone displacement sensor, giving a measurable output proportional to displacement or position. It has been experimentally shown that accuracies about 5 micrometers can be achieved.

2. Sensor description

Figure 1 shows a general scheme of the laboratory prototype sensor made of copper, comprising two resonant cavities whose dimensions change in opposite directions as a function of the displacement of a transmission rod. The electromagnetic variations caused by the geometry change of the resonant cavities allow to measure the displacement [13].

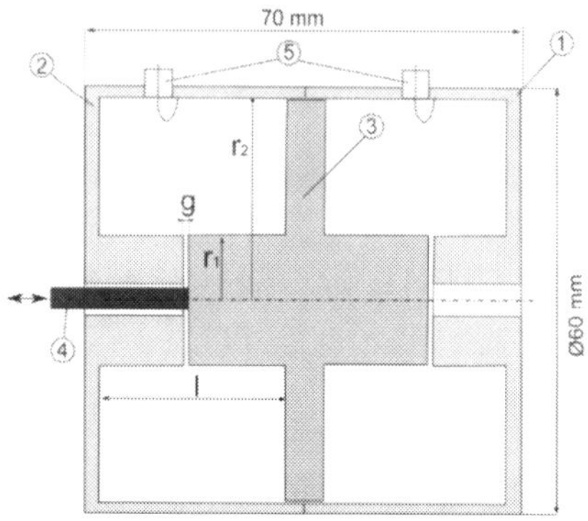

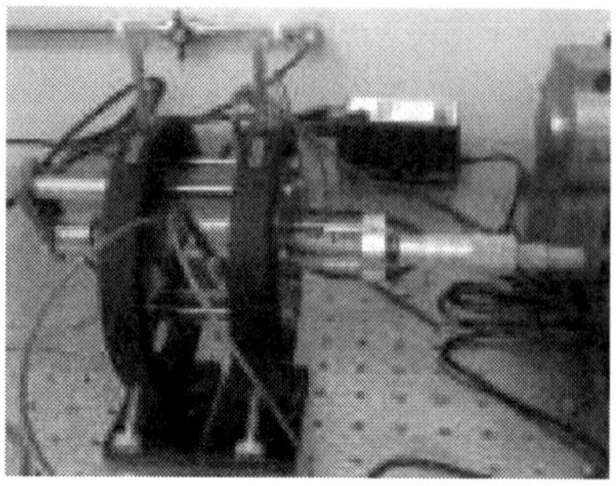

Figure 1. General scheme and photograph of the cylindrical displacement sensor prototype based on resonant cavities. Parts 1 and 2 constitute the body of the sensor. The movement or position to be measured is transmitted to part 3 through the rod 4. Part 5 is an SMA connector for coupling power to the cavity. One resonant cavity is formed on the left side of part 3 and another on the right side. The movement changes the resonant frequency of each cavity in differential form.

A small variation of this gap generates strong variation of the equivalent capacitance, and hence, of the resonant frequency. Let g be the gap, r_1 the radius of the gap region, r_2 the outer radius of the cavity, and l its length. Then, the resonant frequency is:

$$f_0 = \frac{c\sqrt{2g}}{2\pi r_1\sqrt{l\ln(r_2/r_1)}} \tag{1}$$

where c is the speed of light. The variation of frequency with the displacement may be found by taking a logarithmic derivative of the central frequency expression:

$$\frac{\Delta f_0}{f_0} = \frac{1}{2}\left(\frac{\Delta g}{g} - \frac{\Delta l}{l}\right) \tag{2}$$

In order to maximize the sensitivity, the gap dimension (g) must be much smaller than the total cavity length (l), so the first term inside the parentheses dominates.

Each individual cavity changes its frequency from 280 MHz to 730 MHz approximately (one cavity ascending-wise and the other one descending-wise) along the whole displacement range (400 μm) of the transmission rod. It is remarkable the very high sensitivity of the device, whose response changes as much as 2.25 MHz in frequency for each micrometer of displacement (or 1.125 MHz if we use only one cavity). This result confirmed the usefulness of this new way of sensing displacement for sub-nanometric applications.

3. Signal conditioning design and test

In the previous work [13] the frequency response of both cavities was registered using a radio frequency network analyzer (Agilent model E8358A PNA). In this paper an alternative way to measure the resonant frequency is presented. We divide the proposed electronics interface in two stages: the excitation part and the detection one.

3.1. Excitation stage

The purpose of this part is to excite and induce oscillation in the resonant cavity. We have designed a feedback harmonic oscillator using the resonant cavity itself.

IOP Publishing
doi:10.1088/1742-6596/450/1/012017

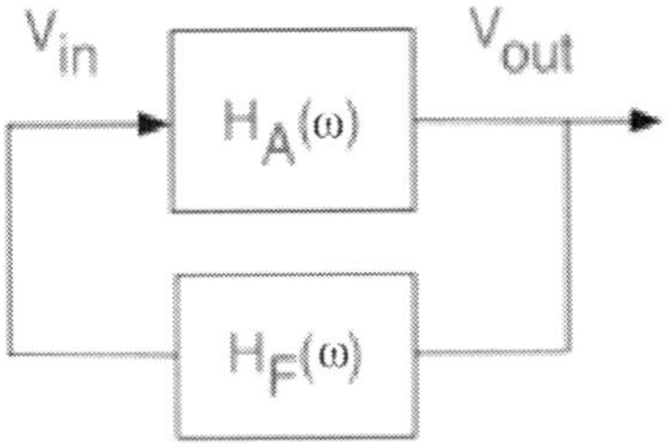

Figure 2. Scheme of the feedback loop to produce the self-oscillation of the resonant cavity. $H_A(\omega)$ is a low noise amplifier. $H_F(\omega)$ comprises the cavity and a phase shifter (see text).

A loop that causes a positive feedback at a selected frequency is the core of any oscillator circuit [15]. The transfer function of the loop shown in figure 2 is described by:

$$\frac{V_{out}}{V_{in}} = \frac{H_A(\omega)}{1 - H_F(\omega)H_A(\omega)} \tag{3}$$

Since there is no input (V_{in}=0), to obtain a nonzero output voltage, a Barkhausen criterion has to be satisfied, that is, the denominator in (3) has to be zero. Thus, these two conditions have to be satisfied:

$$\begin{aligned} &|H_F(\omega)H_A(\omega)| = 1 \\ &\arg[H_F(\omega)H_A(\omega)] = 180° \end{aligned} \tag{4}$$

In the proposed prototype, the $H_A(\omega)$ is a low noise amplifier (ZFL-1000LN, MiniCircuits), whereas the feedback system $H_F(\omega)$ is the cavity itself. The amplifier together with the cavity satisfies the first condition, but to meet the second requirement, it is necessary to introduce a phase shifter (JSPH-661) in the feedback in order to add the necessary phase to the system to produce the self-oscillation of the device. The phase shifter is controlled by a voltage that varies between 0 and 10 V in order to shift the phase between 0 and 270 degrees. A filter (VLF-630) is included after the amplifier to clean its output.

For a given position of the resonant cavities (that is, for a given position to be measured), the cavity oscillates for different values of the phase introduced by the phase shifter. However, the power of the obtained signal and even the resonant frequency depend on the actual value of the phase. We decided to select the phase that delivers the maximum power for each position to calibrate the sensor (see figure 3). There are two positions in which the cavity does not oscillate regardless of the value of the phase. These points (empty squares) are presented in the figure for completeness. From now on we will focus in a 200 μm range, where the response is linear.

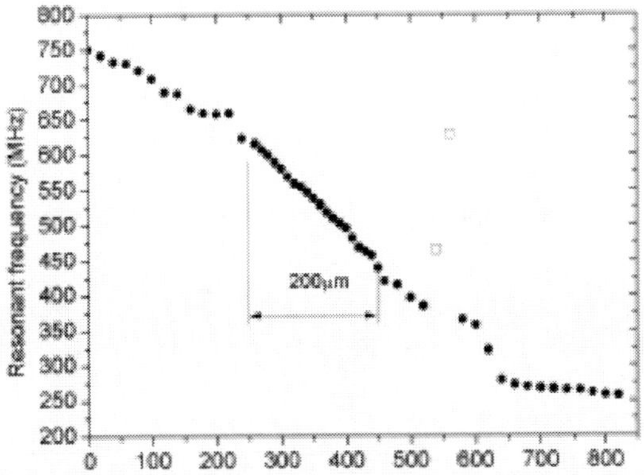

Figure 3. Self-oscillation frequency of the cavity measured using a NA, as a function of position. The phase selected to produce the oscillation is the one that deliver the maximum power.

3.2. Frequency detection

Once the cavity resonates, its output frequency (which is related to the position) has to be measured. To do so, a low pass filter and two RF power detectors are used. A 800 MHz bandwidth low pass filter has been designed to attenuate the signal depending on its frequency. The first RF power (ZX47-50) measures the signal before passing through the low pass filter. The second one (ZX47-50) measures the signal once it has been attenuated. The ratio between those two measurements, that is, the attenuation of the signal, gives us the resonant frequency of the cavity.

Figure 4 shows the elements that constitute the filter, and in figure 5 its performance as measured with a frequency network analyzer. The filter's response is linear in the working frequency of the cavity.

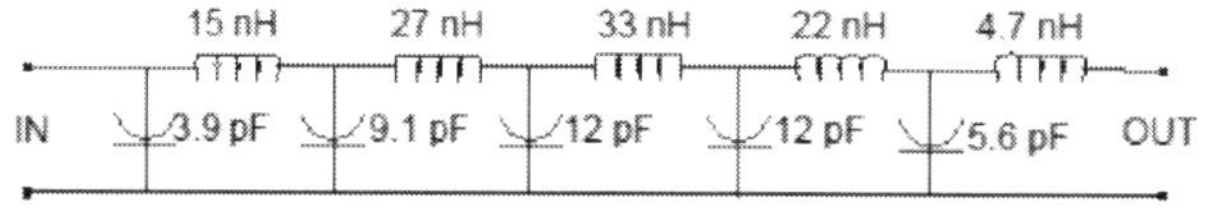

Figure 4. Elements that constitute the low pass filter

The interface used to test the proposed frequency detection procedure have been built using discrete elements. The complete scheme is shown in figure 6. The measurements from power detectors and the voltage to control the phase shifter are managed using a data acquisition board (Model 6011, National Instruments) and LabVIEW software. To determine the accuracy of the electronic interface, the signal produced by the resonant cavity is simultaneously measured by a network analyzer, which is included in the scheme of figure 8. Three ZX30-12-4-S+ couplers have been added when needed to route signals.

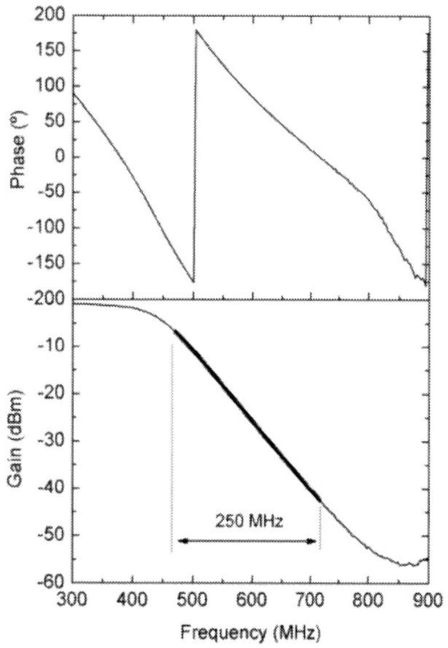

Figure 5. Low pass filter performance. Gain and phase of the transmission parameter S_{21} measured by a network analyzer.

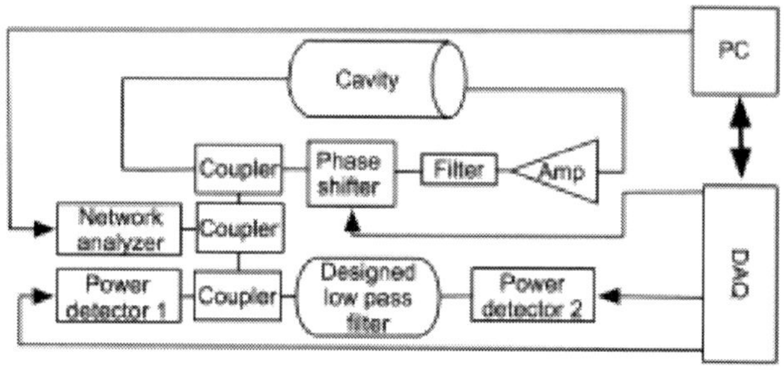

Figure 6. Scheme of the proposed electronic interface. Signals are managed by a data acquisition board connected to a PC. A Network Analyzer is also included to compare the results.

The couplers produce some attenuation, which is independent of the frequency. Thus, the calibration has been done using the ratio between second power detector to the first one when they are connected as the diagram of figure 6 indicates. The results are conveniently fitted to a linear expression:

$$Ratio = 0.00327\,fr(MHz) - 0.0858 \qquad (r=0.996) \qquad (5)$$

It is to be noted again that, in each fixed position, the cavity resonates with different power values depending on the phase shifter. Therefore, the interface is programmed to select the phase shift so the signal power values remain between 0.97 and 0.99 V in the first power detector.

4. Results

Once the cavity oscillates, the resonant frequency has to be measured. As it has been previously mentioned, the method consists of measuring the power of the signal after it passes through the filter. Two nested loops are implemented: the first "coarse" one sweeps the phase shift to achieve oscillation, and the second "fine" one sweeps around this position to reach the correct (0.97-0.99V) power reading. Once this is achieved, the resonant frequency is determined from both power detector readings. In figure 7 the voltages in the first and second power detectors, as well as the calculated frequency based on the ratio between powers and the calibration (equation 5) are shown.

In order to check the performance of the system we compare the frequency as measured with a laboratory calibrated network analyzer and the one given by the proposed electronic interface. As it can be concluded observing figure 8, difference between both readings is about 5 MHz, which corresponds to 5 micrometers according to the transducer intrinsic sensitivity as explained in the introduction section. This error could be improved if we restricted the measurement range of the sensor, or even better if we further constrain the allowable power readings in the first power detector. This latter approach would require a longer time to stabilize the second "fine" loop described above, so there is a clear trade-off between measurement time, precision and range. In any case, from the above experimental results, it can be concluded that the proposed electronics interface is a cost-effective, yet accurate solution for the displacement and position transducer based on resonant cavities.

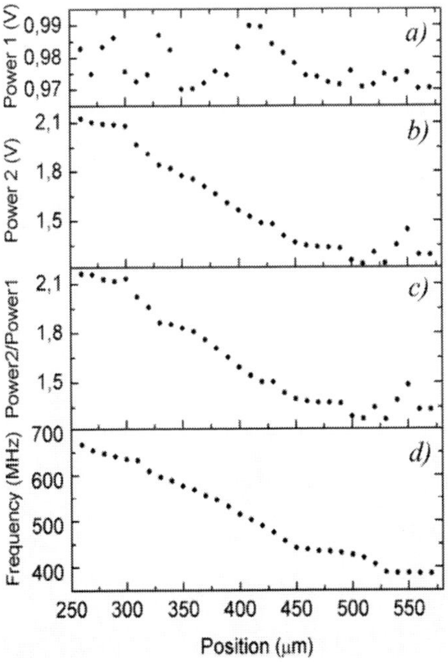

Figure 7. Experimental results. a, b) Voltage measured by power detectors. c) Ratio between both voltages. d) Calculated frequency based on the ratio of powers and calibration (equation 5).

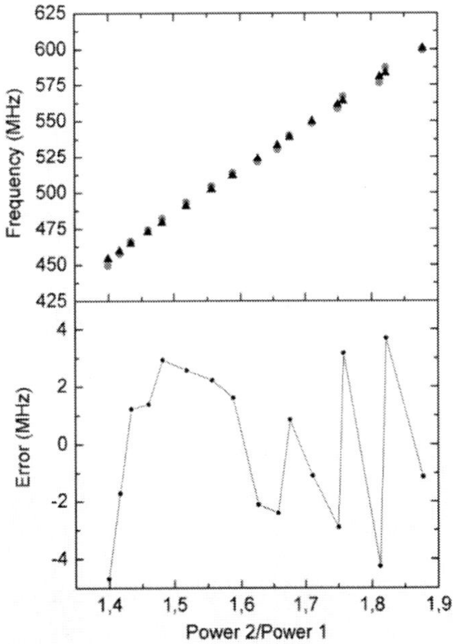

Figure 8. Experimental results. On the top resonant frequency as given by a network analyzer (in black triangles) and as given by the proposed electronic interface (in red points). On the bottom the difference between them.

5. Conclusion and future works

We have designed an electronic interface in order to use it together with a novel position and displacement transducer based on radio frequency resonant cavities, so that it can be used as a standalone positioning sensor. The complete system has been presented and a laboratory prototype has been experimentally tested. It has been shown that the transducer response is linear and exhibits very high sensitivity (1.125 MHz for each micrometer). Besides, it has been experimentally tested that if the proposed electronic interface based on low pass filters, power detectors and low noise amplifier is used, accuracies around 5 MHz in frequency are achieved (corresponding to less than 5 micrometers in displacement). This performance could be further improved restricting the range or increasing the measurement time.

Alternative interface approaches are being considered for testing in order to take better advantage of the sensitivity of the transducer. One of those alternatives implies using the RF cavity to implement a variable low pass filter, so the displacement to be measured produce changes in the attenuation of the filter. Then, a fixed frequency oscillator would be used to supply a fixed amplitude RF signal and the magnitude of the signal after the filter will give the position measurement. This approach has been successfully used in sensors based in self-inductance changes [16].

As a general conclusion, it can be stated that the proposed electronic interface together with the designed displacement transducer based on resonant cavities constitute a cost-effective, yet accurate displacement sensor.

Acknowledgement

The authors are grateful to CICYT and to the Basque Government for partial support of this work though projects DPI2011-24821 and IT-381-10, respectively.

References

[1] Nyce D S 2004 Linear *Position Sensors: Theory and Application* (John Wiley & Sons).
[2] Leach R 2009 *Fundamental Principles of Engineering Nanometrology*. (Norwich, NY: Elsevier-William Andrew Publishing).
[3] Hariharan P 2003 *Optical Interferometry* (2nd Ed. San Diego, CA: Academic Press).
[4] Friedman S J, Barwich L B and Batellan L H 2005 *Review of Scientific Instruments* 76: 123106. DOI:10.1063/1.2130667.
[5] Baxter L K 1997 *Capacitive Sensors: Design and Applications* (John Wiley & Sons).
[6] Kuijpers A A 2004 *Micromachined Capacitive Long-Range Displacement Sensor for Nano-Positioning of Microactuator system*. Thesis.
[7] Yinxiang Li, Jiming An, Zhang, Y and; Qiang Li 2001 *International Conference on Sensor Technology (ISTC 2001)*, Proc. SPIE Vol. 4414, p. 270-272.
[8] Stampfer C , Jungen A, Linderman R, Obergfell D, Roth S and Hierold C 2006 *Nano Lett.*, 6 (7), pp 1449–1453.
[9] Kitagawa A 2011 *Journal of Sensors* ID 360173J.
[10] Tsubono K, Hiramatsu S and Hirikawa H 1977 *J. Appl. Phys*., vol. 16, pp.1641-1645.
[11] Pfizenmaier H and Voigtlaender K 2002 Patent US 6359445.
[12] Song B, Yu Y, Yang W and Ge Y 2008 *IEEE International Conference on Robotics and Biomimetics*, ROBIO 2008, 269 - 275.
[13] Etxebarria V, Lucas J, Feuchtwanger J, Sadeghzadeh A, Hassanzadegan H, Garmendia N and Portilla J 2010 *IEEE Sensors Journal*, vol. 10, p. 1335-1336.
[14] Etxebarria V and Lucas L 2013, ES patent ES 2 377405 B1.
[15] Ludwig R, Bretchko P 2001 *RF Circuit Design: Theory and Applications* (Prentice Hall).
[16] de Cos D, García-Arribas A, and Barandiarán J M 2004 *Sensors and Actuators* A, v. 112, pp. 302–307.

Sensor Interface for Multimodal Evaluation of Capacitive Sensors

Thomas Schlegl and Hubert Zangl

Institute for Electrical Measurement and Measurement Signal Processing, Graz University of Technology, 8010 Graz, Inffeldgasse 23/II, Austria

E-mail: thomas.schlegl@ieee.org, hubert.zangl@tugraz.at

Abstract. Capacitive proximity sensing can be done in different modes. The single ended mode usually offers a higher signal to noise ratio (SNR) and - in conjunction with active guarding - high robustness. However, it can be blind for objects with low permittivity close to the sensor surface. The differential mode usually has a worse SNR but has the capability to detect objects in situations where the single ended mode is blind. Thus, we propose a measurement circuitry to combine both modes. It is compared to state of the art sensors and the benefits of this approach are demonstrated by means of experimental investigations.

1. Introduction - Differential and Single Ended Capacitive Sensing

Capacitive proximity sensing is usually done in one of two measurements modes [1]. They can e.g. be denoted as

- Differential mode (also called mutual capacitance mode) and
- Single ended mode (also called self capacitance mode).

With the differential mode, the capacitance between two conductive areas (called electrodes) is determined. By applying a voltage to a first electrode (called transmitter) and measuring the displacement current induced on the second electrode (called receiver) this capacitance can be measured.

The single ended mode measures the displacement current originating from an electrode (called transmitter). This current corresponds to the capacitance between the transmitter and the distance ground (if measuring to the open environment) as shown in Figure 1(a). A difficulty associated with the single ended mode is the fact that the sensitivity is quite high at the edges of the electrode in particular when conductive objects resides in the vicinity, e.g. as the carrier of the electrode. In this case moisture and contamination may significantly affect the measurement and no reliable proximity determination is possible. A commonly used method to cope with this problem is active guarding where a guard is placed between the actual electrode and a metallic carrier. Thus, the sensitivity moves away from the edges of the electrodes. However, this also leads to a reduced sensitivity with respect to small objects in the vicinity of the electrodes.

Electrical Capacitance Tomography (ECT) represents a technology that is usually based on mutual capacitance. It is a noninvasive imaging system aiming on the determination of the spatial material distribution in some region from capacitive measurements (in differential mode) taken from the outside [2]. This is done by solving a so called inverse problem with heavy signal processing. In process tomography, the region of interest is typically the interior of a pipe. Typical measurement tasks are the determination of the gas content in oil exploration. Recently in [3] we showed that the

ECT approach can be transferred to the open environment. With a differential mode capacitive measurement system we demonstrated the basic functionality, but we also found limitations. For example, it may not be possible to detect small conductive objects. This is due to two competing effects referred to as coupling and shielding effect [4]. Signal cancelation may occur at transitions between these effects. As the shielding effect is the main effect in the single ended mode, this can be used to identify the presence of both effects. Thus, we propose to combine single ended and differential mode to avoid the blind spots.

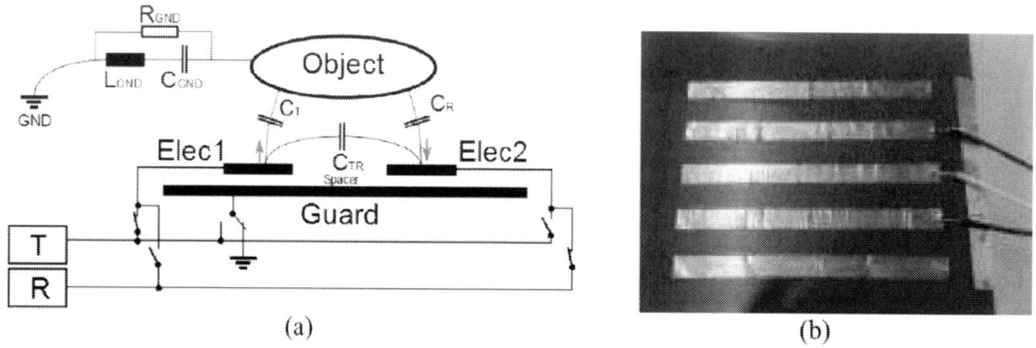

Figure 1 Conductive areas (electrodes) are used as sensing elements. (a) Sketch of the sensor front end with two electrodes. Electrode 1 is used as a transmitter and electrode 2 is used as a receiver. The red arrows indicate the two displacement currents originating from electrode 1 and entering electrode 2, which are measured when using the single ended mode and the differential mode, respectively. The guard electrode can be set to ground or to the excitation signal (i.e., active guarding). An objects parasitic connection to ground is indicated by the equivalent circuit of R_{GND}, C_{GND} and L_{GND}. (b) Picture of the used sensing elements. Thin (approx. 100 mm) and flexible copper strips are used as electrodes. A 2 mm thick PVC material is used as spacer between the electrodes and the guard at the backside.

2. Measurement System

The proposed measurement system (shown in Figure 3(a)) is able to work in the differential mode and the single ended mode. It provides a high measurement rate (> 1 kHz). The measurement frequency can be changed between 10 kHz and 1 MHz to any frequency value. Thus, it is possible to obtain additional information about parasitic effects, due to their frequency dependency as shown in Figure 1(a). Furthermore, a change in the measurement frequency can be used to deal with electromagnetic compatibility (EMC) problems as shown in [5].

Figure 2 gives an overview of the presented measurement system. A sinusoidal signal, generated by a direct digital synthesizer (DDS), is applied to one or more electrodes through a switch circuitry. The displacement current originating the electrodes used as transmitters is measured by a transmitter circuitry (single ended mode). Each electrode is also connected to a receiver circuitry. If an electrode is not used as a transmitter, the receiver circuitry measures the displacement current entering this electrode (differential mode). Since each electrode can be used as a transmitter or a receiver a total of $\frac{N_{elec}(N_{elec}-1)}{2}$ independent measurements, where N_{elec} is the number of electrodes, can be obtained with the differential mode. With the additional single ended mode a total of $\frac{N_{elec}(N_{elec}-1)}{2}+N_{elec}$ independent measurements can be obtained. The backside of the sensor can be connected to ground (differential mode) or to the excitation signal (i.e., active guarding in single ended mode).

A low noise amplifier (LNA) is used to amplify and connect the output of the receiver and transmitter circuits to the I/Q demodulator (where I and Q represent the in-phase and the quadrature component, respectively). With the I/Q demodulator it is possible to get amplitude and phase information of the measured signals with respect to the excitation signal. These DC signals are digitized and stored in a microcontroller. The microcontroller also controls the signal generation, the switch circuitry and the I/Q demodulation (not shown in Figure 2). The processed measurement data can be sent to a host computer to store or post process the data (e.g. signal processing for imaging applications [6]).

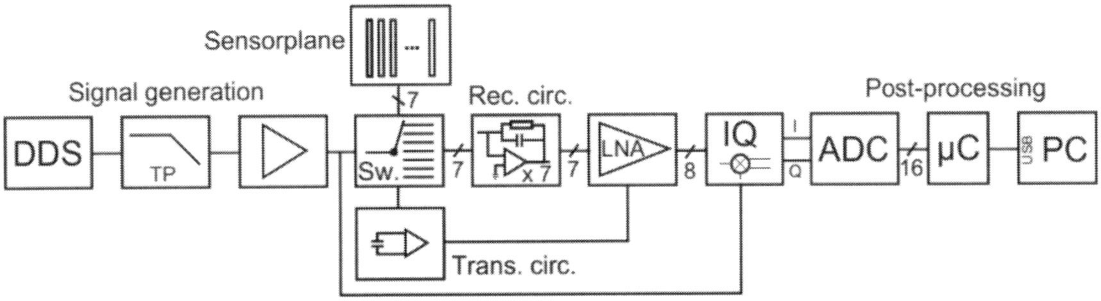

Figure 2 Overview of the presented measurement system. A DDS is used to generate a sinusoidal signal between 10 kHz and 1 MHz. Switches are used to connect the electrodes to the signal generation, receiver and transmitter measurement circuitries. An I/Q demodulator is used to get amplitude and phase information of the measured signal. The post processing consists of an analog digital converter (ADC), a microcontroller (μC) board and a host computer (PC).

Table 1 gives an overview of the proposed measurement system (shown in Figure 3(a)) and the two commercial available systems, which were used in this work.

(a) (b)

Figure 3 Picture of the sensor and the approaching objects. (a) The measurement system consists of three stacked printed circuit boards (PCB): A commercial available μC board at the top, the digital part comprising the DDS, ADC and clock generation positioned in the middle and a PCB at the bottom comprising all amplifier circuits, the switches and the I/Q demodulator. (b) A human hand and two approaching objects were used to evaluate the sensor. The metallic rod has a diameter of 5 mm and the plastic case has a size of 200 x 300 mm.

3. Measurement Results and Discussion

Several experiments (Figures 4 to 6) are carried out with the proposed measurement system and compared to a commercial available singled ended sensor and a differential sensor (Analog Devices AD7148 and AD7746 [7]).

In the first experiments shown in Figure 4 a human hand approaches the sensor surface and leaves again. The human hand can be detected in single ended mode as well as in differential mode with both measurement systems (proposed sensor and commercial available ones). Although, due to shielding and coupling effects (described in Section 1) at a certain distance to the sensor surface the measured capacitance increases (marked with arrows in Figure 4). This effect (decreasing and increasing capacitances) yields to ambiguities in proximity determination.

Table 1 Properties of the Proposed Measurement System compared to state of the art sensors

	Proposed sensor	AD7746	AD7148
Excitation signal	Sinusoidal signal	Square wave	Square wave
Frequency	Tunable from 10 kHz to 1 MHz	32 kHz	250 kHz
Measurement rate	1.25 kHz (max. 6.25 kHz @ 1 MHz)	10 Hz to 90 Hz	40 Hz
Measurement method	Single ended and Differential mode	Differential Mode	Single ended Mode
Shielding	Active guarding and Grounded shielding	Grounded shielding	Active guarding
Number of electrodes	$N_{elec} = 7$	$N_{elec} = 3$	$N_{elec} = 8$
Number of independent measurements	28 (for each frequency)	2	8

As can be seen in Figure 5 a metallic rod (compare Figure 3(b)) can also be detected with a good SNR. Again, the two competing effects shielding and coupling can be observed for the sensors working in the differential mode (marked with arrows).

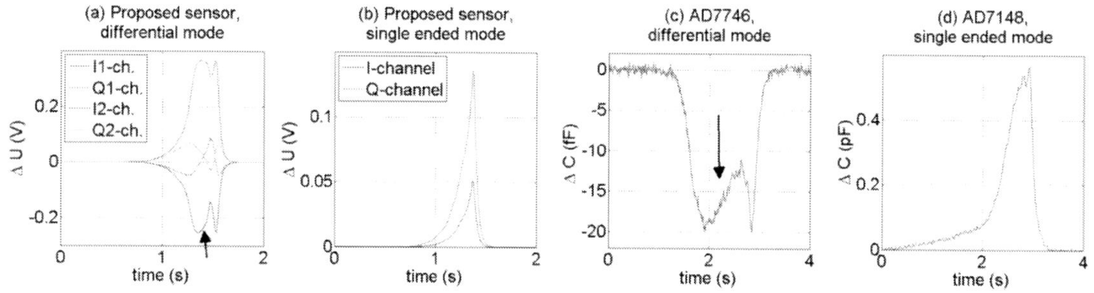

Figure 4 Human hand approaching and leaving the sensor surface. (a), (c) Measurement (I- and Q-channel in the case of the proposed sensor) for the differential mode sensors. The arrows indicate the transitions between the shielding and coupling effects. (b), (d) Measurements with the single ended mode.

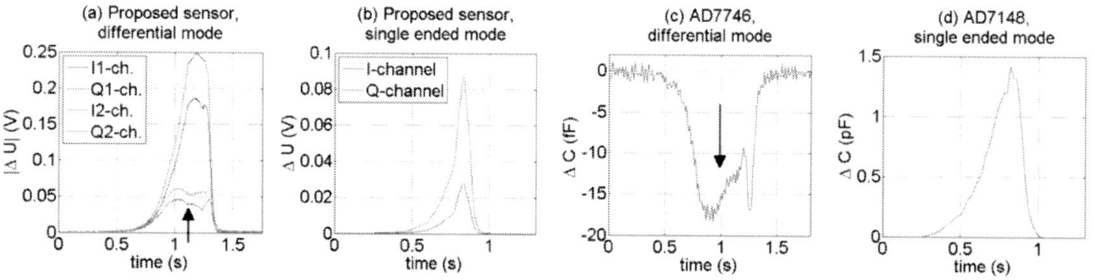

Figure 5 Metallic rod approaching and leaving the sensor surface. (a), (c) Measurement for the differential mode sensors. The arrows indicate the transitions between the shielding and coupling effects. (b), (d) Measurements with the single ended mode.

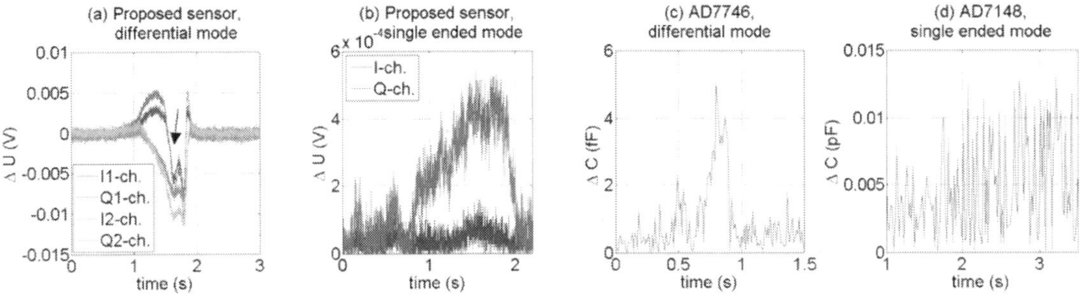

Figure 6 Plastic case approaching and leaving the sensor surface. (a), (c) Differential mode results. Although the approaching object has a small influence on the electric field, it can reliably be detected by the proposed sensor. (b), (d) In the single ended mode it is very difficult to detect this kind of objects.

In Figure 6 an empty plastic case (compare Figure 3(b)) approaches the sensor surface. Due to the small volume and the low permittivity (ε_r close to 1) this object has a small influence on the electric field. Thus, the capacitance change is very small. As can be seen, the plastic case cannot be detected by the commercial sensor working in the single ended mode. In this case the differential mode outperforms the single ended mode. The plastic case can reliably be detected with the proposed sensor working in the differential mode as shown in Figure 6(a).

4. Conclusion

In this paper a versatile and fast multichannel capacitive measurement system has been presented. The measurement system is able to work in the differential as well as the single ended mode. The ability of the measurement system has been demonstrated by means of experimental investigations. The new developed multichannel system features the design and application of new capacitive measurement techniques as it provides a flexible measurement platform for arbitrary sensor arrangements. Thus, the research and development can focus on the frontend, i.e., the electrode topology, and the reconstruction, i.e., imaging methods.

5. Acknowledgement

The research was partially funded by the Austrian Science Fund (FWF): P21855-N22.

References

[1] L. Baxter, *Capacitive Sensors, Design and Applications.* IEEE Press, 1997.

[2] W. Q. Yang and L. Peng, "Image reconstruction algorithms for electrical capacitance tomography," *Measurement Science and Technology*, vol. 14, no. 1, pp. R1 - R13, 2003.

[3] T. Schlegl, M. Neumayer, S. Muhlbacher-Karrer, and H. Zangl, "A pretouch sensing system for a robot grasper using magnetic and capacitive sensors," *Instrumentation and Measurement, IEEE Transactions on*, vol. 62, no. 5, pp. 1299 - 1307, 2013.

[4] T. Schlegl, T. Bretterklieber, M. Neumayer, and H. Zangl, "Combined capacitive and ultrasonic distance measurement for automotive applications," *Sensors Journal, IEEE*, vol. 11, no. 11, pp. 2636 –2642, nov. 2011.

[5] G. Brasseur, "Design rules for robust capacitive sensors," *IEEE Transactions on Instrumentation and Measurement,* vol. 52, pp. 1261–1265, August 2003.

[6] A. M. Olmos, E. Castillo, F. Martinez-Marti, D. P. Morales, A. Garcia, and J. Banqueri, "An imaging method for electrical capacitance tomography based on projections multiplication," *Journal of Physics: Conference Series*, vol. 307, no. 1, p. 012032, 2011. [Online]. Available: http://stacks.iop.org/1742-6596/307/i=1/a=012032

[7] *http://www.analog.com/en/analog-to-digital-converters/capacitance-to-digital-converters/products/index.html*, May 2013

Journal of Physics: Conference Series **450** (2013) 012019 doi:10.1088/1742-6596/450/1/012019

Smart-phone based electrocardiogram wavelet decomposition and neural network classification

N Jannah[1], S. Hadjiloucas[1], F. Hwang[1] and R K H Galvão[2]
[1]School of Systems Engineering the University of Reading
[2]Divisão de Engenharia Eletrônica, Instituto Tecnológico de Aeronáutica, São José dos Campos, SP, 12228-900 Brazil

E-mail: s.hadjiloucas@reading.ac.uk

Abstract. This paper discusses ECG classification after parametrizing the ECG waveforms in the wavelet domain. The aim of the work is to develop an accurate classification algorithm that can be used to diagnose cardiac beat abnormalities detected using a mobile platform such as smart-phones. Continuous time recurrent neural network classifiers are considered for this task. Records from the European ST-T Database are decomposed in the wavelet domain using discrete wavelet transform (DWT) filter banks and the resulting DWT coefficients are filtered and used as inputs for training the neural network classifier. Advantages of the proposed methodology are the reduced memory requirement for the signals which is of relevance to mobile applications as well as an improvement in the ability of the neural network in its generalization ability due to the more parsimonious representation of the signal to its inputs.

Introduction

Early diagnosis of heart diseases enables patients to improve their quality of life through more effective treatments [1]. Analysis and classification of ECG signals can be particularly helpful to identify the initiation of heart conditions such as atrial fibrillation or flutter, multifocal atrial tachycardia, palpitations, paroxysmal supraventricular tachycardia, reasons for frequent fainting, slow heart rate (bradycardia) or ventricular tachycardia. Normally patients will be given a Holter monitor and wear the monitoring electrodes over a period of 24-48 hours. The data-logged signals are post-processed and examined for cardiac beat abnormalities by doctors over the following days. There are restrictions, however, to how often one should perform such measurements. It is not uncommon for patients to complain to their doctors that the arrhythmias they suffered prior to the examination period were not present during the monitoring process, making early diagnosis more difficult. If one is prepared to wear the appropriately placed electrodes more often, it would be possible to use mobile phones as data logging devices directly. Bluetooth emitters such as the RN-42 from Microchip can provide a direct input from a small footprint battery-operated mobile data acquisition card to which the electrodes are interfaced. With the ever-increasing capabilities of smart-phones, portable ECG tele-monitoring is likely to become a common feature for these devices, performing data-logging functionality for the ageing population. Beyond their data-logging functionalities, smart –phones can also use Multimedia Messaging Services (MMS) to enable the recorded signals to be sent directly for diagnosis by experts through current mobile networks or perform directly classification tasks. With the number of patients increasing due to current population sedative lifestyles and unhealthy eating habits, an increased expectation by patients for personalized medical treatment, as well as the envisaged wider proliferation of ECG data-logging devices, it is widely anticipated that there will soon be an overwhelming requirement for expert advice in healthcare service systems worldwide. As a consequence, there are pressures for devising automatic classifiers that could quickly and reliably pre-screen for abnormalities before a referral to experts is made. A further aim of the proposed algorithms is to encourage the data-logging, analysis and classification of heart-beat patterns of subjects at regular intervals throughout their lifetime. Patterns from the same patient when they were healthy should provide a more accurate input to the classifiers for the early detection of abnormalities, direct comparison of benefits from diet and fitness regimes in the cardio-vascular system, or the identification of potentially harmful side-effects to the cardiovascular system from different treatments associated to specific drugs.

The current work discusses efficient decompositions of the ECG signals in the wavelet domain using Discrete Wavelet Transforms (DWTs) and investigates their further processing using neural network based classifiers. There are additional benefits in terms of memory storage requirements as well as a possible reduction in energy and transmission bandwidth requirements which are important from a smart-phone applications perspective that can also be explored through the proposed signal decomposition process. In the next two sections, identification of the QRS complex and ECG signal wavelet decompositions are discussed in more detail whereas in section 4, candidate neural network classifiers are considered.

Identification of the QRS complex in Electrocardiograms (ECG)

ECGs provide a graphic representation of the electrical activity of the heart muscle. It can be seen that the contraction of any heart muscle is associated with electrical changes (depolarization) and these changes can be detected by electrodes attached to the surface of the body and the ECG monitor [2]. The electrocardiogram (ECG) consists of the measurement of electrical activity on the body surface associated with myocardial contraction with respect to time. The standard ECG consists of 12 different leads that record the same electric events but from different viewpoints. Each cardiac cycle in the ECG is normally characterized by a sequence of deflections that make waveforms that are known as the P wave, the QRS complex, and the T wave. Contraction of the atria is associated with the ECG wave called the P wave and that wave represents the sequential activation (depolarization) of the right and left atria. There is a large deflection of the ECG signal when the ventricles are depolarised, that is commonly known as the QRS complex wave. The muscle mass of the atria is small when compared with that of the ventricles due to the P wave being smaller than the QRS complex. The T wave is associated with the return of the ventricular mass to its resting electrical state, which is called repolarisation. Figure 1 shows the basic shape of the normal ECG wave. The ECG can provide evidence and information that can help doctors to identify abnormalities in heart beat rates, arrhythmias, myocardial infarctions, atrial enlargements, ventricular hypertrophies, and bundle branch blocks. ECG records are essential for diagnosing heart diseases such as abnormal cardiac rhythm, abnormal cardiac conduction, ischemia of myocardium, and cardiac hypertrophy [3]. The diagram below shows typical records of a normal ECG signal taken from the European ST-T Database, depicting the presence of the characteristic QRS complex.

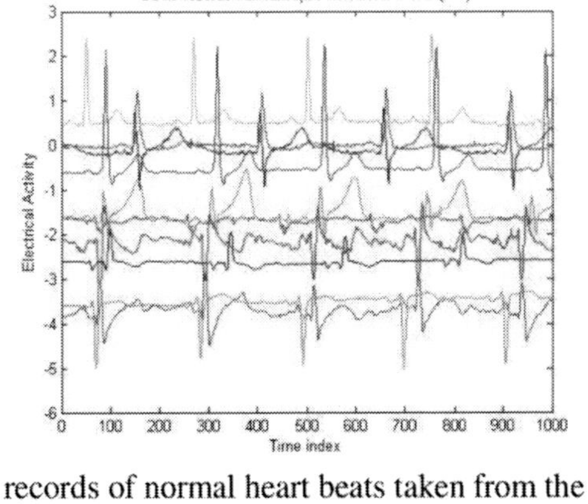

Figure 1. Typical records of normal heart beats taken from the European ST-T Database.

ECG signal decomposition using Discrete Wavelet Transforms (DWTs)

It is convenient to decompose ECG signals into time-frequency representations by using Discrete Wavelet Transforms (DWTs). The main advantage of the DWT is its great time and frequency localization ability, which enables it to reveal the local characteristics of the input signal. Discrete wavelet transforms have already been used by the ECG community because of their multi-resolution capabilities in detecting specific ECG characteristics e.g., the P wave, Q wave, or QRS complex and for cardiac beat analysis [4].

In the signal decomposition using the DWT, both a low pass filter bank (LPF) and a high pass filter bank (HPF) are used to generate time domain responses, these are convolved with the time domain ECG signal. Convolving the response function of the chosen filter (corresponding to a particular mother wavelet) with the signal provides an output which has different energy at different scales. Approximation coefficients relate to the low frequency components of the signal whereas detail coefficients relate to the higher frequency components in the signal. Wavelet decomposition using the DWT provides essentially a multi-resolution representation of the input signal. The user normally retains coefficients up to a particular scale whereas more detailed decompositions become redundant as their incorporation have a negligible effect on the signal. The convolution operation may be conveniently performed in the frequency domain where it is implemented through a simple multiplication process [5].

The ECG signals considered in this study were taken from the European ST-T Database [6]. The features in the ECG signal were extracted using DWTs from the sym3, db4 and db6 wavelet families. Six decomposition levels are more than sufficient to faithfully represent the ECG signals. As shown clearly in figure 2, blocks h[n] and g[n] represent the low-pass and high-pass filter responses (i=1,... 6) respectively, and the ↓2 operator denotes dyadic down-sampling. Approximation A_i and Detail D_i coefficients at each decomposition step are also shown. Only perfect reconstruction quadrature mirror filter banks (orthogonal transforms) are considered in this study because they fully preserve the information content in the signal. This is important from an algorithm certification perspective which is normally associated with the introduction of new algorithms for bio-medical software applications. The discrete wavelet transform can be calculated in a fast manner by using finite-impulse-response (FIR) filter banks.

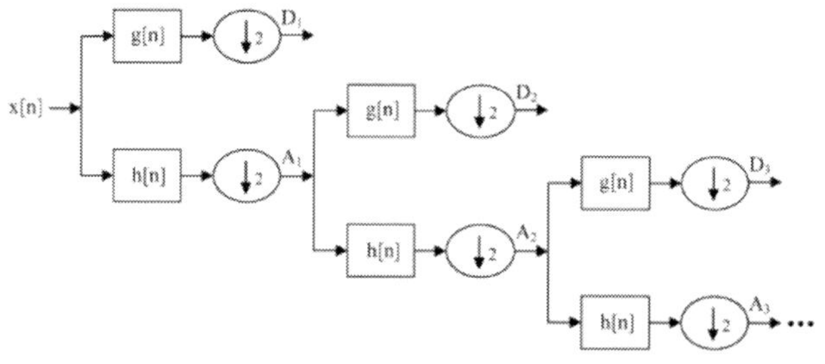

Figure 2. Block diagram of wavelet decomposition.

The filter bank transform can be regarded as a change of variables for the ECG signals x_n:

$$t_j = \sum_{n=0}^{J-1} x_n \, v_j(n), \quad j = 0, 1, \ldots\ldots, J-1$$

where t_j is a transformed variable and $v_j(n) \in \Re$ is a transform weight. The transfer function of the low-pass filter in the z-domain can be written as:

$$H^{(N)}(z) = \sum_{n=0}^{2N-1} h_n^{(N)} z^{-n} = H_0^{(N)}(z^2) + z^{-1} H_1^{(N)}(z^2)$$

where superscript *(N)* denotes that the filtering sequences have length *2N* and $H_0^{(N)}(z)$ and $H_1^{(N)}(z)$ denote polyphasic components of $H^{(N)}(z)$. Further algorithmic details of the proposed filter banks and a discussion of adaptive filter banks for the purpose are discussed in [7]. Results from multi-level decompositions are shown in Figure 3.

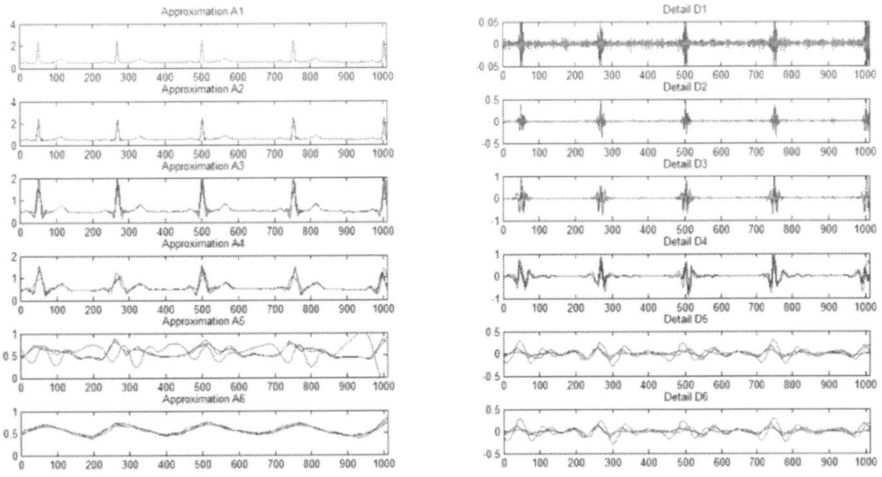

Figure 3. ECG signal multilevel decomposition using sym3, db4 and db6 wavelet families.

Of particular interest to the current study is to identify the most parsimonious representation of the ECG signals in the wavelet domain so that a non-linear neural network classifier can perform the classification task directly in the wavelet domain. Further parameterization of the signals using adaptive wavelets using various adaptive filter banks [7] from the wavelet transform literature are considered in this project.

Discussion of candidate neural network classifiers

Classification of arrhythmias is a complex problem because of a strict requirement for avoiding false-positive or false-negative results. There are many different approaches that can be used to analyze and classify ECG signals. There have been suggestions to use linear discriminant analysis [9], back propagation neural networks [10], self-organizing maps (SOM), learning vector quantization (LVQ) schemes [11], support vector machines (SVM) [12] and fuzzy or neuro-fuzzy algorithms [13].

A Recurrent Neural Network (RNN) was implemented and used as a basis for detection of the variability of ECG signals. RNN can be used to classify different types of ECG beats, such as normal beat, congestive heart failure beat, ventricular tachyarrhythmia beat and atrial fibrillation beat that are obtained from different ECG databases. A particular feature of all neural networks is their multi-layered architecture. Multi-layered networks can be classified as feed-forward or feedback networks, according to their connectivity and the direction of information flow. The recurrence allows the network to remember cues from the past without complicating the learning excessively.

An Elman RNN is a network which in principle is set up as a regular feed-forward network. This means that in this type of network, all neurons in one layer are connected with all neurons in the next layer. Figure 4 depicts the architecture of this type of network, it can be seen that the neurons in the context layer (context neurons) hold a copy of the output of the hidden neurons. Moreover, the output of each hidden neuron is copied into a specific neuron in the context layer.

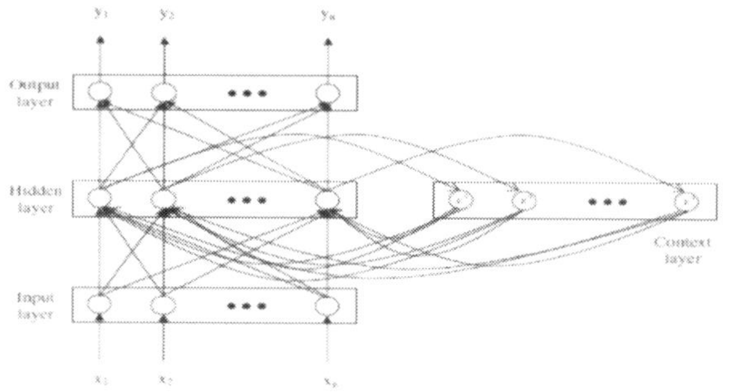

Figure 4. Elman recurrent neural network structure.

RNNs can perform highly nonlinear dynamic mappings and thus, have temporally extended applications; whereas multi-layer feed-forward networks are confined to performing static mappings. This is an important advantage over other topologies for the current application. The strength of all connections between neurons are indicated with a weight, which is similar to a regular feed-forward neural network. At the start point, all weight values are chosen randomly and are then optimized during the training stage. The weights from the hidden layer to the context layer are set to one and are fixed because the values of the context neurons have to be copied exactly in an Elman network. Furthermore, the first output weights of the context neurons are set to be equal to half the output range of the other neurons in the network. Similar to regular feed-forward neural networks, the Elman network can be trained with gradient descent back propagation and optimization methods [14-16]. Elman networks are, therefore, good candidates for ECG signal classification.

In recent years, there has been a growing interest in Continuous Time Recurrent Neural Networks (CTRNN). The popularity of these networks has been increasing because of their simplicity in simulating non-linear dynamical processes. Gallagher *et al.*, (2005) suggested that CTRNNs should be seen as Hopfield type networks with unconstrained connection weight matrices [17]. The unconstrained connectivity provides an improvement in the generalization ability of the networks in the learning process, thus improving its classification ability. Furthermore, CTRNNs are capable of faithfully emulating neuronal activity, and as such they are a natural platform upon which a classifier can be built for ECG diagnostics. CTRNNs are made up of neurons and each neuron's activity can be described by the following expression:

$$\tau_i \frac{dy_i}{dt_i} = -\, y_i + \sum_{j=1}^{N} w_{ji}\,\sigma\big(y_j + \theta_j\big) + s_i I_i(t), \;\; i = 1 \,.....\, N$$

In the above expression, y_i is the internal state of neuron i , τ_i is a time constant of neuron i, N is the total number of neurons, w_{ji} is the strength of the connection from neuron j^{th} to neuron i^{th} , θ_j is a threshold/bias term, $\sigma(x) = 1 / (1+e^{-x})$ is the standard non-linear (sigmoid) logistic activation function and $I_i(t)$ represents a weighted sensory input with strength s_i [17].

The main difference of a CTRNN over other kinds of neural networks is that the neurons could propagate a signal back through the network. Other neural networks considered in ECG are feed forward, which means that neuron signals could only be unidirectional. In addition, CTRNNs are more dynamic in terms of mimicking biological neuronal signal discharge processes. Moreover, CTRNNs are also deemed to be more efficient than other neural networks in terms of computations since they could be used to directly simulate each spike [18]. The CTRNN has additional advantages and computational efficiency over other discrete formulations. For example, using a discrete-time RNNs there is a considerable dependence of the resulting models on the sampling period used in the process, whereas for CTRNNs this can be varied without the need for re-training. Even in the presence of measurements noise the RNNs are capable of providing long-range predictions. Another advantage of CTRNNs is that if they are compared with other NN types such as feed forward neural networks (FFNNs), they have been shown to be more efficient in terms of the number of neurons required to model a dynamic system [19].

Conclusion

The current study proposes the use of smart-phones and a wireless communication network for monitoring patients at home or on the go over a prolonged period of time for the purpose of providing early diagnosis of heart conditions. Different wavelet decomposition schemes of ECG signals assuming different levels of approximation and detail for each wavelet family are considered and their suitability as inputs to neural network classifiers is investigated. The efficiency in parametrizing the wavelet coefficients using adaptive structures from the perspective of improving parsimony in each decomposition step and improving the reliability of the classification task is currently investigated and will be discussed at the conference. The results from different classifiers where the classification process is performed in the wavelet domain are also being considered.

References

[1] Emanet, N 2009 ECG beat classification by using discrete wavelet transform and Random Forest algorithm *Soft Computing, Computing with Words and Perceptions in System Analysis, Decision and Control(ICSCCW 2009),Fifth International Conference (Famagusta, 2-4 September 2009)* pp. 1–4

[2] Vijaya, V, Rao, K K and Rama,V 2011 Arrhythmia Detection through ECG Feature Extraction using Wavelet Analysis *European Journal of Scientific Research* **66** (3) 441–48

[3] Hampoton, J R 2003*The ECG Made Easy*. 6th GH: Churchill Livingstone pp1–10

[4] Adib, A and Haque, MA 2010 ECG beat classification using discrete wavelet coefficients *Health Informatics and Bioinformatics (HIBIT), 5th International Symposium* 1–6

[5] Sadaphule, M M, Mule, S B and Rajankar, S O 2012 ECG Analysis Using Wavelet Transform and Neural Network *International Journal of Engineering Inventions* **1**(12)1–7

[6] Physiobank Archieve Index 2013 *The European ST-T Database*, http://www.physionet.org/physiobank/database/edb.

[7] Froese, T, Hadjiloucas, S, Galvao, R K H, Becerra V M and Coelho, C J 2006 Comparison of extrasystolic ECG signal classifiers using Discrete Wavelet Transforms, *Pattern Recognition Letters* **27**(5) 393-407.

[8] Ubeyli, E D 2009 Combined neural network model employing wavelet coefficients for EEG signals classification *Digital Signal Processing*. **19** (2) 297-308.

[9] Yeh Y C, Wang W J and Chiou C W 2009 Cardiac Arrhythmia Diagnosis Method Using Linear Discriminant Analysis on ECG Signals *Measurement* **42** 778–89.

[10] Yeap T H, Johnson F, and Rachniowski M, 'ECG Beat Classification by a Neural Network *Proceedings Annual International Conference of the IEEE EMBS Society*, pp.167–173, 1990.

[11] Palreddy S, Tompkins W J, and Hu Y H 1995 Customization of ECG Beat Classifiers Developed using SOM and LVQ *Engineering in Med And Biol Soc.*

[12] Ulbeyli E D 2007 ECG Beats Classification Using Multiclass Support Vector Machines with Error Correcting Output Codes *Digital Signal Processing* **17** 675–684.

[13] Özbay Y Ceylan R and Karlik B A 2006 Fuzzy Clustering Neural Network Architecture for Classification of ECG Arrhythmias *Comput Biol Med.* **36** 376–88.

[14] Ulbeyli E D 2010 Recurrent neural networks employing Lyapunov exponents for analysis of ECG signals *Expert Systems with Applications* **37** 1192–99.

[15] Ulbeyli E D 2009 Combining recurrent neural networks with eigenvector methods for classification of ECG beats *Digital Signal Processing* **19** 320–29.

[16] Ulbeyli E D 2008 Recurrent neural networks with composite features for detection of electrocardiographic changes in partial epileptic patients *Computers in Biology and Medicine* **38** 401–10.

[17] Gallagher J C, Boddhu, S K and Vigraham, S 2005 A Reconfigurable Continuous Time Recurrent Neural Network for Evolvable Hardware Applications. *IEEE Computer Society NASA/DoD Conference of Evolution Hardware (EH'05)* 247–20.

[18] Fiore, J M and Gallagher, J C 2004 Continuous time recurrent neural networks: a paradigm for evolvable analog controller circuits *IEEE National Aerospace and Electronics Conference (NAECON) 2000,* 299–304.

[19] Al Seyab, R K and Cao, Y 2008 Nonlinear system identification for predictive control using continuous time recurrent neural networks and automatic differentiation, *Journal of Process Control,* **18** 568–81.

Hydrogen sensing using reduced graphene oxide sheets supported by Pd nanoparticles

Roman Yatskiv and Jan Grym
Institute of Photonics and Electronics, Academy of Sciences of the Czech Republic, Prague, Chaberska 57, Czech Republic

E-mail: yatskiv@ufe.cz

Abstract. We investigated Schottky diode hydrogen sensors prepared by the deposition of reduced graphene oxide functionalized by nanocrystals of Pd on InP substrate. Schottky diodes were investigated by the measurement of current voltage characteristics and further tested for their sensitivity to hydrogen in a cell with a through-flow gas system. Pd nanocrystals which are in direct contact with the semiconductor substrate serve to dissociate hydrogen molecules into atomic hydrogen, lowering the work function of Pd, and resulting in the decreased Schottky barrier height.

1. Introduction

Hydrogen gas has been widely used in chemical industry, medicine, or hydrogen-fuelled vehicles. However, hydrogen is very volatile and extremely flammable. A small leakage of high concentration of hydrogen-containing gases can cause explosion. Thus, development of hydrogen sensors with high sensitivity, short response time, small size and low cost is in great demand. One of the promising sensing materials is graphene because of its unique and outstanding mechanical, thermal and electrical properties [1]. Graphene can be extracted by various chemical and physical methods. Mechanical cleavage of graphite was the initial method to produce single-layer grapheme sheets [2]. Unfortunately it is underproductive process that is not suitable for large scale production. An alternative method is growth of graphene on silicon carbide [3], which allows to prepare high quality mono and multilayers of graphene, however, the ultrahigh vacuum requirement limits its technological applicability. Another suitable methods for production of large area graphene films is chemical vapor deposition [4], but the sheets obtained in this manner suffer from bad structural homogeneity over larger areas. A potentional low-cost method is to first produce graphene oxide and then reduce it to obtain graphene for device applications [5].

Recently, mechanically exfoliated graphene sheet and reduced graphene oxide have been shown to exhibit sensitivity for different sensing gas molecules [6,7]. The gas sensing mechanism is based on the change in the electrical conductivity of graphene due the charge transfer from molecules absorbed on its surface. Unfortunately pristine graphene layers do not show much sensitivity towards hydrogen. Therefore, functionalization of graphene with catalytically active metals such as Pd or Pt can be used to enhance its hydrogen sensitivity.

Lately, it has been demonstrated that highly oriented graphite or graphene form high-quality Schottky diodes (SD) on various n-doped semiconductor such as Si, Ge, InP, SiC, GaN and ZnO [8-10]. A high quality SD formed by colloidal graphite is a good starting point for the fabrication of high sensitivity hydrogen sensor. In our previous works [11-13] we presented high sensitivity hydrogen sensor, which was created by inserting Pd or Pt nanoparticles between graphite Schottky contact and the semiconductor substrate. The sensor elements prepared in this way are capable of detection of low concentration (about 1 ppm) of hydrogen at room temperature.

In this work we propose using the reduced graphene oxide supported by Pd nanocrystals for the fabrication of SD hydrogen sensors.

2. Experiment and results

For our experiment we used commercially available reduced graphene oxide functionalized by nano Palladium (rGO@Pd was supplied by NanoInnova Technologies [14]). The rGO@Pd powder was sonicated in ethanol to produce a uniform suspension. This suspension was deposited on n-type InP substrate by microcapillary tube and let dry. The contact area was checked by optical microscope to maintain the diameter close to 1mm. Prior to the to the rGO@Pd deposition, the InP surface was cleaned in boiled methanol for 5 min. An ohmic contact on the back side was formed by rubbing liquid gallium with a tin rod. The diodes were characterized by current-voltage (I-V) characteristic and further tested for their sensitivity to hydrogen in a cell with through-flow gas system. The measuring system consisted of a test chamber, a sensor holder, a Keithley current/voltage source measure unit, mass flow controllers (Bronkhorst High-Tech B.V.) and of a data acquisition system (LabView, National Instruments).

SEM investigation of the rGO@Pd layer deposited onto the InP substrate showed that it does not form a solid layer, but it consists of irregular sheets of sizes in the range of few μm (figure 1). Graphite oxide sheets were covered by Pd nanocrystals with size from 10 to 30 nm (figure 1a). The rGO@Pd was characterized by energy dispersive x-ray analysis (EDX), which proved the presence of C and Pd.

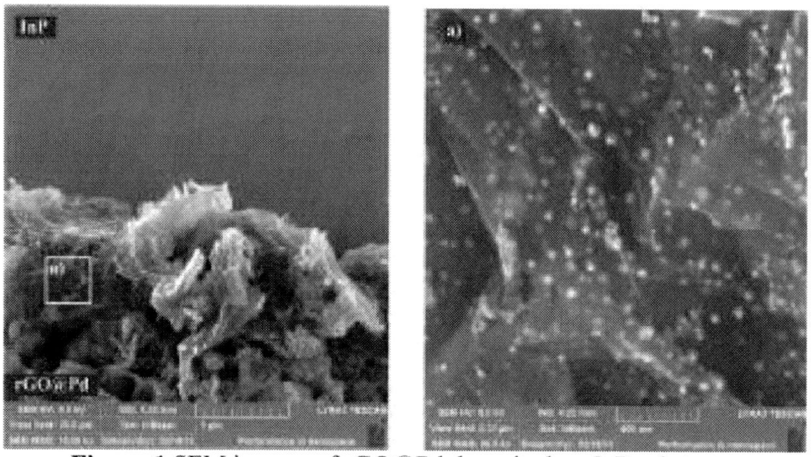

Figure 1 SEM images of rGO@Pd deposited on InP substrate.

The I-V characteristic of the Schottky diodes in a dark chamber at room temperature is presented in figure 2. For metal-semiconductor diodes, according to the thermionic emission (TE) theory, the forward I-V relationship of a Schottky diode at V>3kT/q can be expressed as:

$$I_s = I_0 \exp(\mathrm{q}V/\eta kT)\,, \tag{1}$$

where

$$I_0 = AA^{**}T^2 exp(q\phi_B/kT). \tag{2}$$

where A^{**} is the Richardson constant, which has theoretical value of 9.6 for InP [15], A is the contact area, T is the absolute temperature, k is the Boltzmann constant, ϕ_B is the barrier height and η is the ideality factor. By fitting the forward I-V curves, the barrier height and ideality factor were determined. The ideality factor was found to be higher than one for rGO@Pd/InP Schottky diodes. Nonideal behaviour can be attributed to either (i) defect states in the band gap of the semiconductor providing other current transport mechanism such as barrier tunnelling or generation recombination in the space charge region [16], or (ii) laterally inhomogeneous contacts [17]. A summary of the basic electrical parameter calculated from I-V characteristics is given in table 1.

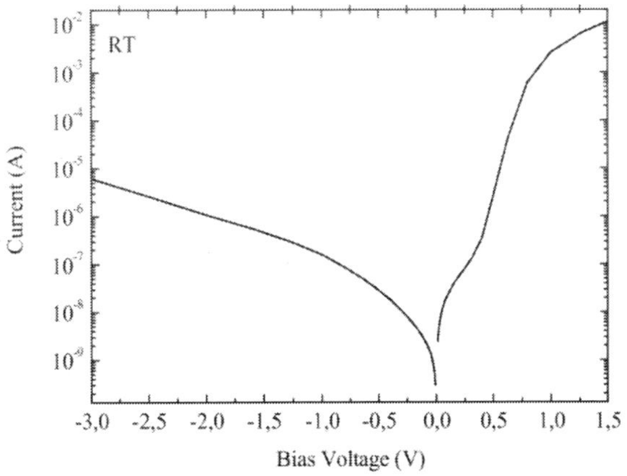

Figure 2 Current-voltage characteristics of the rGO@Pd/InP Schottky diodes.

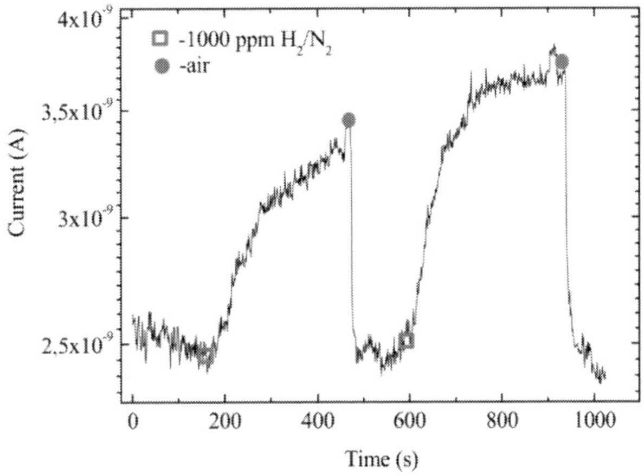

Figure 3 Current-transient characteristics of the rGO@Pd/InP Schottky diodes hydrogen sensor measured at -0.1 V at room temperature.

The rGO@Pd/InP Schottky diodes were tested for their sensitivity to hydrogen in a cell with a through-flow gas system. Our experiments were performed with a mixture of H_2/N_2 containing 1000 ppm of hydrogen in nitrogen (figure 3). The current increase characterized by the sensing response S = 1.6 was observed.

$$S = \{J_H - J_{air}\}/J_{air}, \qquad (3)$$

where J_H is the saturation current density under exposure to hydrogen and J_{air} is the same for air. The mechanism of sensing can be explained by the dissolution and dissociation of hydrogen molecules into atomic hydrogen at the Pd nanocrystals, lowering the work function of Pd (palladium hydride [PdH_x] is formed), and resulting in the decreased Schottky barrier height of the rGO@Pd/InP Schottky diodes. The rGO@Pd/InP Schottky diodes sensor has lower sensitivity in comparison with the result obtained in our previous work, where Pd or Pt nanoparticles were inserted by electrophoretic techniques between the graphite contact and the semiconductor substrate [11]. This result indicates that a key role in the dissociation of hydrogen and consequently in the change of the Schottky barrier height is played by Pd nanocrystals which are in close contact with a semiconductor substrate. The rGO@Pd had layered structure and only small amount of Pd nanocrystals are in direct contact with the semiconductor substrate resulting in relatively low sensitivity response.

Table 1 Electrical parameters calculated from I-V and I-t characteristics for the rGO@Pd/InP Schottky diodes. R-rectification ratio at 1.5V, η-Ideality factor, ϕ_β-Schottky barrier height, S-sensing response at -0.1V (in 0.1% H2/air).

	R	η	ϕ_b, (eV)	S
rGO@Pd/InP	2.1×10^4	2.07	0.87	1.6

3. Conclusion

We have demonstrated the fabrication of Schottky diode hydrogen sensors by the deposition of reduced graphene oxide decorated with Pd nanocrystals on n-type InP substrates. These Schottky diodes are capable of the detection of 1000 ppm hydrogen at room temperature. The sensing mechanism can be explained by the dissolution and dissociation of hydrogen molecules into atomic hydrogen at the Pd nanocrystals which are in direct contact with semiconductor substrate, lowering the work function of Pd, and resulting in the decreased Schottky barrier height of the rGO@Pd/InP Schottky diodes.

Acknowledgments

This work has been supported by the project COST LD12014 of the Ministry of Education CR and by the international collaboration project M100671201 of the ASCR.

Reference

[1] Novoselov KS and Geim AK 2007 *Nature Materials* **6** 183 – 191
[2] Novoselov KS, Geim AK, Morozov SV, Jiang D, Zhang Y, Dubonos SV, Grigorieva IV and Firsov AA 2004 *Science* **306** 666-669
[3] Berger C, Song ZM, Li XB, Wu XS, Brown N, Naud C, Mayou D, Li TB, Hass J, Marchenkov AN, Conrad EH, First PN and Heer WA 2006 *Science* **312** 1191-1196
[4] Reina A, Jia XT, Ho J, Nezich D, Son HB, Bulovic V, Dresselhaus MS and Kong J 2009 *Nano Lett.* **9** 30-35
[5] Stankovich S, Dikin DA, Dommett GHB, Kohlhaas KM, Zimney EJ, Stach EA, Piner RD, Nguyen ST and Ruoff RS 2006 *Nature* **442** 282-286
[6] Schedin F, Geim AK, Morozov SV, Hill EW, Blake P, Katsnelson MI and Novoselov KS 2007 *Nature Materials* **6** 652 - 655
[7] Johnson JL, Behnam A, Pearton SJ and Urall A 2010 *Advanced Material* **22** 4877 – 4880
[8] Tongay S, Schumann T, Hebard A F 2009 *Applied Physics Letters* **95** 222103
[9] Yatskiv R and Grym J 2012 *Appl. Phys. Lett.* **101** 162106
[10] Yatskiv R and Grym J 2013 *Semicond. Sci. Technol.* **28** 055009
[11] Grym J, Prochazkova O, Yatskiv R and Piksova K 2011 *Nanoscale Research Letters* **6** 392.
[12] Yatskiv R, Grym J, Zdansky K and Piksova K 2012 *Carbon* **50** 3928-3933
[13] Zdansky K and Yatskiv R 2012 *Sens. Actuator B-Chem.* **165** 104-109
[14] http://www.nanoinnova.com/Uploads/Features/2385654.pdf
[15] Brillson L J, Brucker C F, Katnani A D, Stoffel N G, Daniels R, and Margaritondo G 1982 *J. Vac. Sci. Technol.* **21** 564-569
[16] Klein A, Sauberlich F, Spath B, Schulmeyer T and Kraft D 2007 *J. Mater. Sci.* **42** 1890-1900
[17] Werner JH and Guttler HH 1991 *J. Appl. Phys.* **69** 1522-1533

Journal of Physics: Conference Series **450** (2013) 012021 doi:10.1088/1742-6596/450/1/012021

Detection of the Short-Circuit Faults in the Stator Winding of Induction Motors based on Harmonics of the Neighboring Magnetic Field

V. Fireteanu[1]
[1]EPM_NM Laboratory, POLITEHNICA University of Bucharest,
313 Splaiul Independentei, 060042, Bucharest, Romania

E-mail: virgiliu.fireteanu@upb.ro

Abstract. Based on the time domain finite element analysis of the electromagnetic field, this paper studies the signature of the short-circuit faults inside the stator winding in the magnetic field outside induction motors. The detection of the such a fault is based on the evaluation of the output voltage of coil sensors placed in the motors neighbouring and the comparison of amplitudes of harmonics of this voltage for the healthy and faulty operation states.

1. Introduction

The study based on finite element models offers much more information on the phenomena characterizing the operation of electrical machines than the classical analytical models. This explains the increase of the interest for the finite element investigations in electrical machines [1-10].

Based on finite element models [5], this paper studies the influence on the magnetic field outside induction motors of a short-circuit fault in the stator winding. The neighbouring magnetic field is investigated through the harmonics in the time variation of the output voltage of coil sensors in case of healthy and faulty motor states, for no-load and for loaded motor operation.

2. Dedicated time domain finite element field-circuit model of induction motors

The geometry and mesh of the electromagnetic field 2D computation domain, Fig. 1(a), (b) and the circuit model, Fig. 1 (c), belong to a four poles squirrel cage induction motor of 11 kW, 3 x 380 V, $f_n = 50$ Hz supplied. This domain contains the stator and the rotor cores, which are magnetic and nonconductive regions, the 48 stator slots - nonconductive, nonmagnetic and source regions, the 32 bars of the rotor squirrel cage aluminum made, the nonmagnetic motor frame of 8 mm thickness and the motor shaft, which are regions of solid conductor type, the motor airgap and an infinitely extended air region outside the motor. In order to investigate the neighboring magnetic field, the computation domain includes two coil sensors with a common magnetic core, the radial sensor *SensorOx*, blue colored, and the azimuth sensor *SensorOy*, in red, Fig. 1 (b).

The Phase B and the Phase C of the stator winding are represented in the circuit model, Fig. 1 (c), by four coil components, which reflects the four groups representing each the four go- and the four return

sides of the elementary coils, Fig. 1 (a). Each elementary coil occupies two stator slots and four elementary coils series connected represent one pole of the stator winding.
It is the Phase A, Fig. 1 (c), of the stator winding where the short-circuit fault is considered. The section of this winding representing one stator pole, which regroups four elementary coils in red in Fig. 1 (a), is represented in the circuit model, Fig. 1 (c), through eight coil components. These are the go-sides and the return sides of four elementary coils. The resistor R in the circuit model, Fig. 1 (c), is used to simulate the short-circuit of the one of the elementary coils of the Phase A. The go-side of the elementary coil in short-circuit is placed around the Ox axis, Fig. 1 (b), (a).

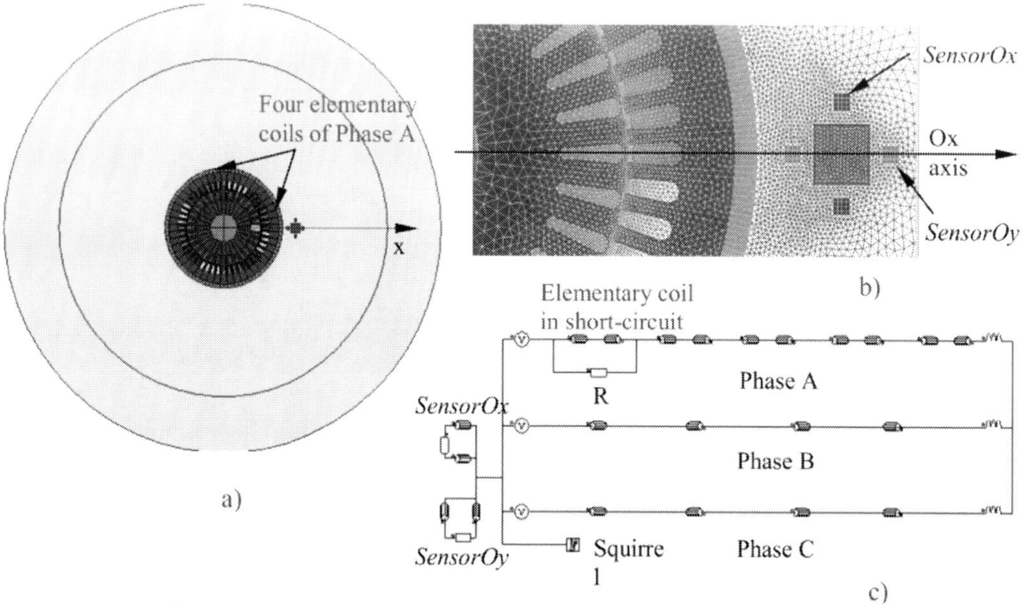

Figure 1. Geometry (a) and mesh (b) of the computation domain and the circuit model (c)

The circuit model includes for each of the two coil sensors, *SensorOx* and *SensorOy*, Fig. 1 (c), two coil components and a resistor of high resistance for the evaluation of sensors output voltage.
The magnetic vector potential $A(x,y,z,t)$, which is the state variable of the electromagnetic field, satisfies the differential equations:

$$\operatorname{curl}\left[1/\mu\cdot\operatorname{curl}\boldsymbol{A}\right]+\left(\partial\boldsymbol{A}/\partial t\right)/\rho=\boldsymbol{J}_s\left(x,y,z,t\right);\ \operatorname{div}\boldsymbol{A}=0 \qquad (1)$$

where μ is the magnetic permeability, ρ is the resistivity and $\boldsymbol{J}_s$ is the current density in the stator slots. The term $(\partial\boldsymbol{A}/\partial t)/\rho$ reflects the density of the induced current in the regions of solid conductor type.
In the 2D field model considered in this paper the source current density has the structure $\boldsymbol{J}_s[0,\ 0,\ J_s(x,y,t)]$. As consequence, the vector potential $\boldsymbol{A}[0,\ 0,\ A_s(x,y,t)]$ is oriented normal on the xOy plane and does not depends on the third coordinate of the Cartesian coordinate system (x, y, z).
The investigation of the electromagnetic field inside and outside the motor uses a transient magnetic model of induction motors [10, 11]. This time domain analysis of the electromagnetic field considers the imposed constant rotor speed, 1450 rpm for the rated load motor operation, and the synchronous speed 1500 rpm for the ideal no-load motor operation.

3. Signature of the short-circuit fault in the magnetic field outside the motors

This section analyses results related the time variation of the output voltage of *SensorOx* and *SensorOy* coil sensors when pass from the healthy state (HE) to the short-circuit faulty state (FA) of motor operation. The changes in amplitude of harmonics of this voltage in the range [500 … 1000] Hz are evaluated.

<u>*No-load motor operation*</u>. The amplitude of the main harmonics of the *SensorOx* voltage in the range [500 ... 1000] Hz, Fig. 2, have an important increase when pass from the healthy state (HE0) to the faulty state (FA0). The ratio FA0/HE0 of the amplitudes, Table 1, shows that the most important increase is reflected by the 650 Hz harmonic, followed by the harmonic with the frequency 850 Hz.

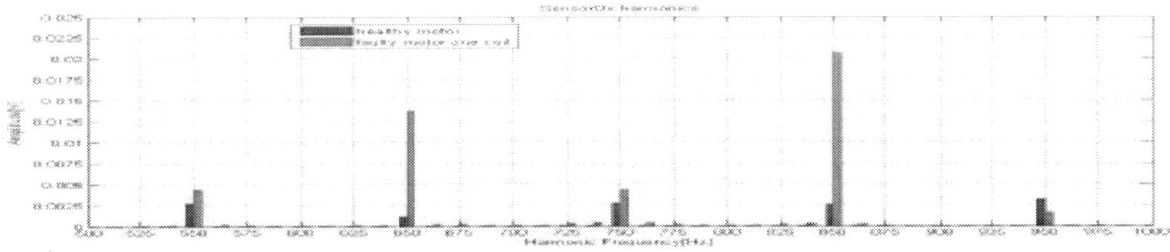

Figure 2. Amplitude of harmonics of *SensorOx* voltage.

Table 1. Amplitude of harmonics of *SensorOx* voltage, no-load operation (1500 rpm)

f [Hz]	550	650	750	850	950
HE0 [mV]	2.729	1.163	2.726	2.52	3.146
FA0 [mV]	4.382	13.72	4.355	20.75	1.625
FA0/HE0	1.61	11.8	1.60	8.23	0.516

In case of the *SensorOy* voltage, Fig. 3, the most important increase, 9.37 times, characterizes the amplitude of the same 650 Hz harmonic. Consequently, *SensorOx* and *SensorOy* offer similar efficiency of fault detection based on the criterion FA0/HE0. For the most important harmonics this ratio increases around 10 times.

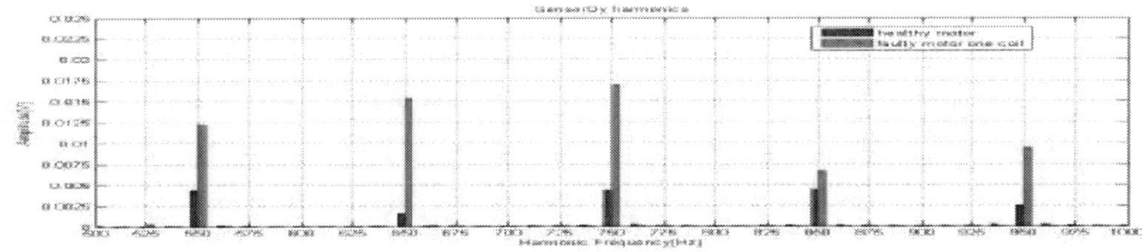

Figure 3. Amplitude of harmonics of *SensorOy* voltage.

<u>*Loaded motor operation*</u>. The amplitude of the main harmonics of the *SensorOx* voltage generally decreases. It is the case of the 625 Hz harmonic, with the value 0.79 of the ratio FA1/HE1, Table 2. In case of the *SensorOy* voltage the decrease of the amplitude of the 925 Hz harmonic is much more important, around 8 times from the healthy state to the faulty state of the motor, Table 2.

Table 2. Ratio FA1/HE1 of amplitude decrease, loaded motor operation (1450 rpm)

f [Hz]	525	625	725	825	925
SensorOx	0.84	0.79	0.99	1.05	0.92
SensorOy	0.47	0.734	0.36	0.134	0.130

4. Detection of the short-circuit fault with increased efficiency

In accordance with the results in the section 3, the on-line detection of the short-circuit fault in the stator winding consists in the survey of the amplitude of a particular harmonic of the output voltage of *SensorOx* or of *SensorOy* coil sensors. For the no-load motor operation the appropriate harmonic has the frequency 650 Hz. In case of loaded motor operation, the harmonic of 925 Hz of the *SensorOy* voltage can be used.

A much more efficient solution for short-circuit fault detection uses two groups of two coils placed at the same distance from the motor axis. The first group consists in the coils *SensorOx*, *SensorOy*, and the second group in the coils *Sensor2Ox*, *Sensor2Oy*, Fig. 4 (a). The second group of two coils is shifted 90 degrees around the motor axis with respect the first group. The two coils *SensorOx*, *Sensor2Ox* and the two coils *SensorOy*, *Sensor2Oy* are series connected. The new output voltages, denominated *SensorsOx+2Ox* and *SensorsOy+2Oy*, Fig. 4 (b), will be further evaluated.
The azymuth position of the four coils system with respect the motor is defined by the angle Alpha, Fig. 4 (a), between the horizonthal Ox axys and the radial line that connect the center of the *SensorOx*, *SensorOy* assembly and the center O of the motor.

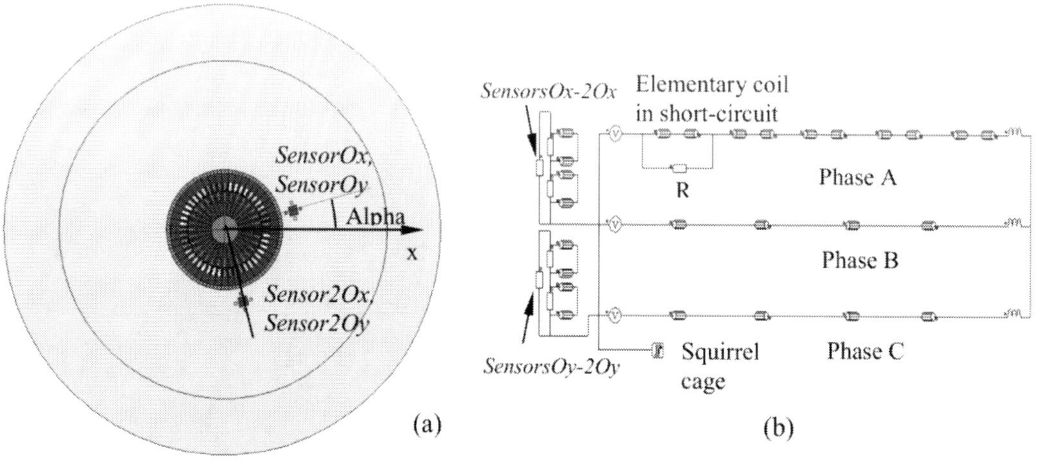

Figure 4. Computation domain geometry (a) for Alpha = 15 degrees and the new circuit model (b)

The numerical results in Tables 3, 4 for Alpha = 15 degrees related the ratio between the amplitude of the main harmonics of *SensorsOx+2Ox* and *SensorsOy+2Oy* voltages, FE0/HA0 for no-load motor operation and FE1/HA1 for loaded motor operation, offers the following findings:

- the efficiency of fault detection through the *SensorsOx+2Ox* and *SensorsOy+2Oy* voltages is much more higher than through the *SensorOx* and *SensorOy* voltages;
- for the no-load motor operation, the efficiency of fault detection through the *SensorsOy+2Oy* voltage is higher than through the *SensorsOx+2Ox* voltage. The maximum of this efficiency, FE0/HA0 = 5115 corresponds to the harmonics of 650 Hz;
- for loaded motor operation, the efficiency of fault detection is lower. The *SensorsOx+2Ox* is the best solution and the maximum efficiency, FE1/HA1 = 730.7 it is obtained with the 625 Hz harmonic.

Table 3. Ratio FA0/HE0 of amplitudes, no-load motor operation (1500 rpm)

f [Hz]	550	650	750	850	950
SensorsOx+2Ox	310.9	522.8	757.0	1155.2	818.5
SensorsOy+2Oy	2935	5115	2668	3249	2043

Table 4. Ratio FA1/HE1 of amplitudes, loaded motor operation (1450 rpm)

f [Hz]	525	625	725	825	925
SensorsOx+2Ox	574.4	730.7	396.4	476.6	367.4
SensorsOy+2Oy	45.31	187.0	110.1	123.2	91.99

5. Identification of the short-circuit position and maximum of the fault detection efficiency

The next tables contain results related the value of the amplitude of the 650 Hz, 750 Hz and 850 Hz harmonics of *SensorsOx+2Ox* and *SensorsOy+2Oy* voltages, which are the main harmonics in case of

no-load motor operation and of the 625 Hz, 725 Hz and 825 Hz harmonics, which are the main harmonics in case of loaded motor operation. The maximum in the dependence of the harmonics amplitude on the azimuth position of coil sensors corresponds to Alpha = 0 degrees, if the *SensorOx+2Ox* voltage is used for the no-load motor operation, Table 5, and if the *SensorOy+2Oy* voltage is used for the loaded motor operation, Table 8.
The minimum in the dependence of the amplitude of the main harmonics with respect the azimuth position of the coil sensors corresponds to the value Alpha = 90 degrees, Tables 7, 8, for the loaded motor operation in both cases, *SensorOx+2Ox* and *SensorOy+2Oy*.
For the detection of the position of the elementary coil in short-circuit remember that the go-side position of this coil corresponds to Alpha = 0, Fig. 4 (a). In this context, the maximum of the amplitude of the 750 Hz harmonic of *SensorOx+2Ox* for no-load motor operation, Table 5, and of the 725 Hz harmonic of *SensorOy+2Oy* for loaded motor operation, Table 8, corresponds also to Alpha = 0 degrees. The minimum of the amplitude of these harmonics corresponds to Alpha = 90 degrees.

Table 5. Amplitude of *SensorOx+2Ox* harmonics, no-load motor operation (1500 rpm)

Alpha[degrees]	-15	0	15	30	45	60	75	90	105
Harm_650 Hz ampli mV]	16.5	19.9	8.78	4.73	5.03	2.73	1.07	0.59	0.56
Harm_750 Hz ampli [mV]	7.85	23.7	16.2	6.23	3.69	2.71	1.14	0.54	2.62
Harm_850 Hz ampli [mV]	6.51	29.9	14.4	4.05	4.25	3.51	1.74	1.99	3.68

Table 6. Amplitude of *SensorOy+2Oy* harmonics, no-load motor operation (1500 rpm)

Alpha[degrees]	-15	0	15	30	45	60	75	90	105
Harm_650 Hz ampli [mV]	33.7	18.7	11.9	2.66	3.54	2.45	0.78	0.89	1.12
Harm_750 Hz ampli [mV]	40.0	12.8	9.56	11.7	6.50	2.19	1.48	2.47	3.62
Harm_850 Hz ampli [mV]	40.5	6.78	12.0	10.0	5.03	1.36	1.48	2.10	3.65

Table 7. Amplitude of *SensorOx+2Ox* harmonics, loaded motor operation (1450 rpm)

Alpha[degrees]	-15	0	15	30	45	60	75	90	105
Harm_625 Hz ampli [mV]	27.5	5.97	25.9	7.49	3.30	1.89	1.08	0.61	1.81
Harm_725 Hz ampli [mV]	33.8	3.48	35.4	8.25	2.32	1.69	1.14	0.30	1.05
Harm_825 Hz ampli [mV]	41.3	1.79	41.2	13.8	4.97	2.38	0.84	0.72	2.01

Table 8. Amplitude of *SensorOy+2Oy* harmonics, loaded motor operation (1450 rpm)

Alpha[degrees]	-15	0	15	30	45	60	75	90	105
Harm_625 Hz ampli [mV]	2.81	64.3	7.46	6.27	3.77	1.86	1.52	1.04	1.04
Harm_725 Hz ampli [mV]	0.53	81.0	8.39	10.0	4.10	1.20	1.64	0.82	1.55
Harm_825 Hz ampli [mV]	2.29	95.9	5.38	9.04	5.88	4.18	0.97	0.42	2.49

In the context of on-line survey of the healthy state of the elementary coil considered for the short-circuit fault investigation, the results in Tables 9 and 10 show that the maximum efficiency of the faulty state detection based on the amplitude of the 750 Hz harmonic of the *SensorOx+2Ox* voltage for no-load motor operation, Table 9, and of the 725 Hz harmonic of the *SensorOy+2Oy* voltage for loaded motor operation, Table 10, corresponds to Alpha = 0 degrees.

Table 9. Efficiency of fault detection through the 750 Hz harmonic of *SensorOx+2Ox* voltage, no-load motor operation (1500 rpm)

Alpha[degrees]	-15	0	15	30	45	60	75	90	105
FA0/HE0	351.8	1296.2	757.0	311.6	182.5	165.3	51.3	30.0	117.1

Table 10. Efficiency of fault detection through the 725 Hz harmonic of *SensorOy+2Oy* voltage, loaded motor operation (1450 rpm)

Alpha[degrees]	-15	0	15	30	45	60	75	90	105
FA0/HE0	6.94	1169.0	110.1	137.3	55.5	16.6	21.5	11.7	20.6

6. Conclusions

The detection of the short-circuit in the stator winding of induction motors can be the result of the comparison of the amplitudes for healthy and faulty motor operation states of harmonics of the output voltage of a neighbouring coil sensor, radial or azimuthally orientated with respect the motor.

An increased efficiency of the fault detection and the possibility to detect the azimuth position of the short-circuit fault inside the stator are offered by a system of two coil sensors series connected, one pole shifted around the motor.

7. References

[1] J. Penman, H.G. Sedding and W.T. Fink, "Detection and location of inter-turn short circuits in the stator windings of operating motors," *IEEE Trans. Energy Conversion.*, vol. 9, No. 4, pp. 652-658, Dec. 1994.

[2] H. Henao, C. Demian, and G.A. Capolino, "A frequency-domain detection of stator winding faults in induction machines using an external flux sensor", *IEEE Trans. Ind. Appl.*, vol. 39, pp. 1272–1279, Sept/Oct. 2003.

[3] T. Assaf, H. Henao, G.A. Capolino, "Simplified axial flux spectrum method to detect incipient stator inter-turn short-circuits in induction machine", in *Proc. IEEE ISIE. 2004*, Ajaccio, France, pp. 815–819, vol. 2, 2004.

[4] M. D. Negrea, "Electromagnetic Flux Monitoring for Detecting Faults in Electrical Machines", doctoral disertation at Helsinki University of Technology, Laboratory of Electromechanics, November, 2006.

[5] V. Fireteanu, P. Taras, "Teaching induction machine through finite element models", *Proc. of XVIII ICEM Conf.*, Sept. 6-9, 2008, Vilamoura, Portugual.

[6] B. Vaseghi, N. Takorabet, and F. Meibody-Tabar, "Transient finite element analysis of induction machines with stator winding turn fault," *Progress In Electromagnetics Research (PIER) Journals,* vol. 95, pp. 1-18, 2009.

[7] A. Ceban, "Methode globale de diagnostique des machines electriques" , PhD Thesis, LSEE Lab., Université d'Artois, France 2012.

[8] L. Frosini, A. Borin, L. Girometta, G. Venchi, "A novel approach to detect short circuits in low voltage induction motor by stray flux measurement", *Proc. of ICEM Conf.,* September 2-5, 2012, Marseille, France.

[9] R. Pusca, R. Romary, A. Ceban, "Detection of inter-turn short circuits in induction machines without knowledge of the healthy state", *Proc. of ICEM Conf.,* September 2-5, 2012, Marseille, France.

[10] A-I. Constantin, V. Fireteanu, V. Leconte, "Effects of the short-circuit faults in the stator winding of induction motors and fault detection through the magnetic field harmonics, *Proc. of ATEE Conf.,* May 23-25, 2013, Bucharest, Romania.

[11] V. Fireteanu, A-I. Constantin, R. Romary, R. Pusca, S. Ait-Amar, "Finite element investigation of the short-circuit fault in the stator winding of induction motors and harmonics of the neighboring magnetic field", *Proc. of SDEMPED Symp.,* August 27-30, 2013, Valencia, Spain.

Journal of Physics: Conference Series **450** (2013) 012022 doi:10.1088/1742-6596/450/1/012022

Rigorous modelling of light's intensity angular-profile in Abbe refractometers with absorbing homogeneous fluids

A. García-Valenzuela*, **H. Contreras-Tello, R. Márquez-Islas, C. Sánchez-Pérez**

Centro de Ciencias y Desarrollo Tecnológico, Universidad Nacional Autónoma de México, Apartado Postal 70-186, Distrito Federal, 04510, México.

*E-mail: augusto.garcia@ccadet.unam.mx

Abstract. We derive an optical model for the light intensity distribution around the critical angle in a standard Abbe refractometer when used on absorbing homogenous fluids. The model is developed using rigorous electromagnetic optics. The obtained formula is very simple and can be used suitably in the analysis and design of optical sensors relying on Abbe type refractometry.

1. Introduction.

There is an increasing need for reliable small-scale refractometers for chemical and biological microfluidic devices [1-3]. However, all refractometers developed to date are assumed will be used with transparent fluids and their use with absorbing liquids is not always possible. On the other hand, one of the more robust ways to measure a refractive index of liquids is by means of an Abbe refractometer introduced many years ago [4]. Actually, most bench-top refractometers in use today are of the Abbe type. Nevertheless, to our knowledge, the Abbe refractometer has not yet been exploited in microfluidics. We believe that this is in part due to the lack of rigorous optical models for measuring with Abbe type going beyond geometrical ray optics. By Abbe type refractometers we mean refractometers in which diffuse light is refracted from a sample fluid into a prism of higher refractive index, making apparent the critical angle from which the refractive index of the fluid can be obtained. It is not difficult to realize that the basic principle of Abbe refractometers could be developed into compact optical sensors for applications in microfluidics. As most refractometers, the Abbe refractometer was thought for use with transparent materials. Nevertheless, in practice it is often used on substances with some degree of opacity without precautions. Very few works have been devoted to understand the use of an Abbe refractometer with opaque media. The use of standard Abbe refractometers with turbid colloidal fluids was investigated in Ref. [5] many years ago. But many questions regarding the correct determination of a critical angle were left open at that time and no optical model was derived then. Only recently we initiated new research on the measurement of the effective refractive index of highly turbid fluids with Abbe type refractometers [6, 7]. In the latter references we did not consider the case of samples of complex refractive index without optical scattering. Thus, the model derived in Ref. [7] is not applicable to homogeneous absorbing fluids, which are actually involved in many chemical and biological sensing applications. In standard Abbe refractometers, the sample is in the form of a fluid film at the base of an optical prism of higher refractive index. Diffuse light is transmitted through the sample to the prism. Light transmitted into the prism is confined to angles of refraction smaller than the critical angle. Thus locating de edge of the light's cone yields the sample's refractive index. However when the sample is not optically transparent, the edge becomes fuzzy and the determination of the refractive index may not be straightforward. In this work we develop a model for the light's intensity angular-profile around the critical angle in Abbe refractometers with homogeneous absorbing fluids. We suppose that light scattering within the sample fluid is negligible. Our aim is to provide a simple formula for the angular intensity profile around the critical angle in a standard Abbe refractometer that could be related to the complex refractive index of the sample.

2. Theoretical model.

In standard Abbe refractometers the sample fluid is placed between the base of a glass prism and a diffusing translucent surface. Light is incident to the diffusing surface from the side opposite to the prism and scattered within the sample fluid film as shown in Fig. 1. The light transmitted into the prism is refracted within an angular cone limited by the critical angle. In the standard methodology, the refractive index of the sample fluid is obtained from locating the critical angle from the edge of the light cone. When the sample has a complex refractive index, the angular profile around the edge of the light cone is modified with respect to that when the sample has a real refractive index. Nevertheless, from the angular intensity profile it can be possible to retrieve the real and imaginary parts of the refractive index of the sample fluid or simply sense variations of their value.

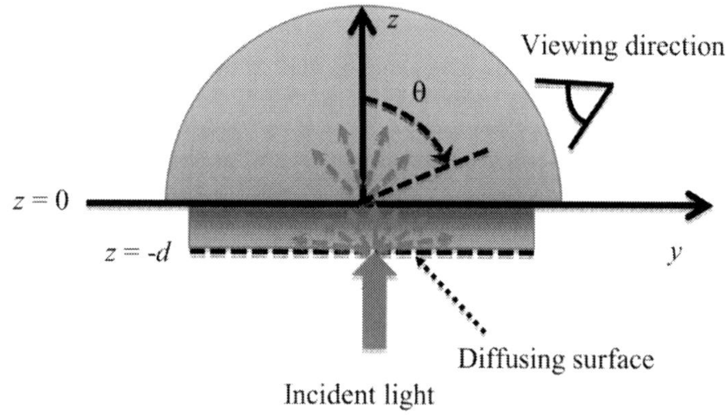

Figure 1: Schematic illustration of the basic operation of a standard Abbe refractometer.

We can follow closely the model developed in [7] for the intensity angular-profiles of light scattered within a turbid colloidal fluid and transmitted into a transparent medium of higher refractive index. The difference here is that the effective radiation sources due scattering are not distributed through the sample but all are located on a plane parallel to the interface between the absorbing liquid sample and the transparent glass of higher refractive index as illustrated in Fig. 1.

Let us start by considering an opaque rough surface, flat on the average, placed at a distance d from the bottom surface (flat) of a transparent glass prism. Let us assume that the space in-between the irregular surface and the bottom of the prism is filled with an absorbing, homogeneous liquid of complex refractive index n_m. Let us place the origin of coordinates on the interface between the prism and the liquid sample with the z-axis pointing into the prism as shown in Fig. 1. Let us assume that monochromatic light of frequency ω is incident to the rough surface from $z < -d$ and scattered in all directions. We may regard the illuminated rough surface as a random array of sources emitting incoherently in all directions. The sources can be modelled as localized electrical currents embedded in the sample liquid of refractive index n_m and distributed within a very thin film just below the liquid sample. Let us divide the radiating current into a large number of independent localized currents. The electric field radiated by the *nth* localized current can be expressed as,

$$\mathbf{E}_n^s(\mathbf{r}) = i\omega\mu_0 \int_{V_n} d^3r' \ddot{\mathbf{G}}(\mathbf{r},\mathbf{r}') \cdot \mathbf{J}_n(\mathbf{r}'), \quad (1)$$

where V_n is the volume occupied by the *nth* current and $\ddot{\mathbf{G}}(\mathbf{r},\mathbf{r}')$ is the Green's function dyadic of the vector Helmholtz equation in the sample medium at frequency ω. The time dependence is implicit in all equations. For observation points above all the radiating currents, we can use the following plane wave expansion of the Green function dyadic [8],

$$\ddot{\mathbf{G}}(\mathbf{r},\mathbf{r}') = \frac{i}{8\pi^2} \int dk_x\, dk_y \frac{(\ddot{\mathbf{I}} - \hat{\mathbf{k}}_+\hat{\mathbf{k}}_+)}{k_z} \exp[i\mathbf{k}_+ \cdot (\mathbf{r} - \mathbf{r}')], \quad (2)$$

where $\mathbf{k}_\pm = k_x\hat{\mathbf{a}}_x + k_y\hat{\mathbf{a}}_y + k_z\hat{\mathbf{a}}_z$ and $k_z = (k_0^2 n_m^2 - k_x^2 - k_y^2)^{1/2}$. Making the change of variable $\mathbf{r}' = \mathbf{r}_s + \mathbf{r}_n$, where $\mathbf{r}_s$ is the position vector relative to the centre of the volume V_n containing the *nth* localized current, we get for points above the rough surface,

$$\mathbf{E}_n^s(\mathbf{r}) = \frac{-\omega\mu_0}{8\pi^2} \int dk_x\, dk_y \exp[i\mathbf{k}_+ \cdot (\mathbf{r} - \mathbf{r}_n)]\, \mathbf{s}_n(k_x,k_y), \quad (3)$$

where

$$\mathbf{s}_n(k_x,k_y) = \frac{(\ddot{\mathbf{I}} - \hat{\mathbf{k}}_+\hat{\mathbf{k}}_+)}{k_z} \cdot \int_{V_n} d^3r_s \exp(-i\mathbf{k}_+ \cdot \mathbf{r}_s) \mathbf{J}_n(\mathbf{r}_s). \quad (4)$$

Since we are assuming n_m is a complex quantity, Eq. (3) is a linear superposition of inhomogeneous plane waves in the sample medium. Note that the kernel is divided by k_z which is zero when $k_x^2 + k_y^2 = k_0^2 n_m^2$. This corresponds to plane waves traveling parallel to the interface between the sample liquid and the prism (the plane $z = 0$).

Now, let us propagate these plane waves through the sample up the prism's bottom surface. For the sake of simplicity we will assume the detection plane is within the prism in the far zone. In general it is relatively simple to consider the actual case in which the detection plane is outside the prism. Let us assume that the detection plane is at a large distance from the interface and viewing towards the interface with the liquid sample. First we propagate the fields to just before the interface by evaluating Eq. (3) at $z = 0^-$. We get,

$$\mathbf{E}_n^s(\mathbf{r})\Big|_{z=0^+} = \frac{-\omega\mu_0}{8\pi^2} \int dk_x\, dk_y \exp[ik_x(x - x_n) + ik_y(y - y_n) - ik_z z_n] \mathbf{s}_n(k_x,k_y). \quad (5)$$

Then, the field just on the other side of the interface at $z = 0^+$ is obtained by splitting the fields into its transverse electric, TE, and transverse magnetic, TM, components and multiplying by the corresponding transmission coefficient. Then we change $\mathbf{s}_n(k_x,k_y)$ by,

$$\ddot{\mathbf{t}}(k_x,k_y) \cdot \mathbf{s}_n(k_x,k_y) = t^{TE} s_n^{TE}(k_x,k_y)\hat{\mathbf{s}}_{TE} + t^{TM} s_n^{TM}(k_x,k_y)\hat{\mathbf{s}}_{TM},$$

where t^{TE} and t^{TM} are the transmission coefficients for TE and TM polarized waves respectively, $s_n^{TE}(k_x,k_y)$ and $s_n^{TM}(k_x,k_y)$ are the projections of $\mathbf{s}_n(k_1,k_2)$ on the unit vectors $\hat{\mathbf{s}}_{TE}=(\mathbf{k}_+\times\hat{\mathbf{a}}_z)/|\mathbf{k}_+\times\hat{\mathbf{a}}_z|$ and $\hat{\mathbf{s}}_{TM}=(\mathbf{k}_+\times\hat{\mathbf{s}}_{TE})/|\mathbf{k}_+\times\hat{\mathbf{s}}_{TE}|$. Since only the z-component of $\mathbf{k}_+$ can be complex, we can see that $\hat{\mathbf{s}}_{TE}$ is a real unit vector. However, $\hat{\mathbf{s}}_{TM}$ is in general a complex unit vector.

Now we can introduce the factor $\exp(ik_z^p z)$ with $k_z^p=(k_0^2n_p^2-k_x^2-k_y^2)^{1/2}$ in the kernel of Eq. (5) to propagate the field into the incidence medium of refractive index n_p. Then, we can propagate the transmitted field to the far zone and use the method of stationary phase to evaluate the integrals [9]. To this end, we assume that the interface transmits light through an aperture with a large but finite area A on the plane $z=0$, and bring the observation point far away compared to the lateral dimensions of the aperture. We get,

$$E_n^s(\mathbf{r})=i\omega\mu_0 k_p\cos\theta\frac{\exp(ik_p r)}{4\pi r}\exp(-i\mathbf{k}_+\cdot\mathbf{r}_n)\vec{\mathbf{t}}(k_1,k_2)\cdot\mathbf{s}_n(k_1,k_2) \tag{6}$$

where $k_1=k_p\sin\theta\cos\phi$ and $k_2=k_p\sin\theta\sin\phi$ are the coordinates of the stationary phase point in the $(k_x,\ k_y)$ space and $(r,\ \theta,\ \phi)$ are spherical coordinates of the observation point $\mathbf{r}$. Also, now $\mathbf{k}_+=k_1\hat{\mathbf{a}}_x+k_2\hat{\mathbf{a}}_y+k_z\hat{\mathbf{a}}_z$ with $k_z=\sqrt{k_0^2n_m^2-k_1^2-k_2^2}$, which is in general complex. That is, now we have $k_z=\mathrm{Re}(k_z)+i\,\mathrm{Im}(k_z)$. Light intensity is given by the magnitude of the time average of Poynting's vector, which in the case of self-propagating plane waves is given by $I_n=\frac{1}{2}(k_p/\omega\mu_0)|\mathbf{E}_n^s|^2$. To obtain the intensity profile in the detection plane we must add the contribution of all localized currents emitting incoherently light towards the detection plane. So we add the intensity at the detection plane due to all localized currents (we do not add the fields for incoherent sources).

For simplicity we assume that all currents are located within a very thin film of thickness h and lateral dimensions of area A. Adding up the intensity due to all localized currents and taking the limit that the volume occupied by any of the localized currents tends to zero, yields,

$$I(\mathbf{r})=\frac{1}{2}\beta D\left(\frac{\cos\theta}{r}\right)^2\iint_A dx_n\,dy_n\int_{-d-h}^{-d}dz_n\exp[2\,\mathrm{Im}(k_z^{eff})z_n]\left|\vec{\mathbf{t}}(k_1,k_2)\cdot\mathbf{s}_n(k_1,k_2)\right|^2, \tag{7}$$

where $D=\omega\mu_0k_p^3/16\pi^2$, β is the density of incoherently-emitting localized currents, and we assumed the area A is large compared to the sample's depth d. Since the unit vectors $\hat{\mathbf{s}}_{TE}$ and $\hat{\mathbf{s}}_{TM}$ are orthogonal to each other and we assume that light is randomly polarized we have,

$$\left|\vec{\mathbf{t}}(k_1,k_2)\cdot\mathbf{s}_n(k_1,k_2)\right|^2=\left|t^{TE}\right|^2\left|s_n^{TE}(k_1,k_2)\right|^2+\left|t^{TM}\right|^2\left|s_n^{TM}(k_1,k_2)\right|^2. \tag{8}$$

Recall that k_1 and k_2 are functions of the viewing angles θ and ϕ. The intensity profile is measured over a very small range (compared to 2π) along the polar angle θ around the direction of the edge of the light cone and about the plane of incidence, that is around $\phi=0$. Note that $\mathbf{s}_n(k_1,k_2)$ is inversely proportional to k_z, and the magnitude of k_z is minimum when $\sqrt{k_0^2n_m^2-k_1^2-k_2^2}=\sqrt{k_0^2n_m^2-k_0^2n_p^2\sin^2\theta}$ is minimum. We may refer to the viewing angle θ that minimizes the latter expression as the critical angle. If the effective refractive index approaches a real number, the minimum of the magnitude of k_z

approaches zero and $\mathbf{s}_n(k_1,k_2)$ becomes singular. However, the transmission coefficients t^{TE} and t^{TM} are in the form of Fresnel transmission coefficients. Then, both t^{TE} and t^{TM} should be proportional to k_z. Thus, Eq. (8) does not have a singularity when $k_z = 0$.

Thus, in general $\left|\vec{\mathbf{t}}(k_1,k_2)\cdot\mathbf{s}_n(k_1,k_2)\right|^2$ should be a smooth function of k_1 and k_2, and therefore, for a fixed $\mathbf{r}_n$ we can approximate it as a constant within the small range of interest of viewing directions. Note that the magnitude of $\mathbf{s}_n$ must be proportional on the average to the intensity of the incident light I_0. The integral in Eq. (7) simplifies to,

$$I(\theta) = I_0 C\cos^2\theta\exp[-2\,\mathrm{Im}(k_z^{eff})d], \qquad (9)$$

where $k_z^{eff} = k_0(n_{eff}^2 - n_p^2\sin^2\theta)^{1/2}$ and C is an instrumental constant determined experimentally. Note that we assumed h very small. This is the main result of this paper. We must draw attention to the fact that θ in Eq. (9) is the angle of travel of light inside the prism. In a practical device the detector will be placed outside the prism and one must take into account refraction of light when exiting the prism.

3. Numerical examples

In Fig. 2a we plot the angular intensity profile of light refracted near the critical angle assuming the sample thickness is d = 100 μm and the fluid has a refractive index n_m = 1.33 + ix. We show plots for x = 10^{-6}, 10^{-5}, 10^{-4} and 10^{-3}. Note that a substance with x = 10^{-3} would be completely opaque in a 1 mm thick cuvette. For the plots in Fig. 3 we assumed the refractive index of the prism is 1.515 and the wavelength in vacuum is 635 nm.

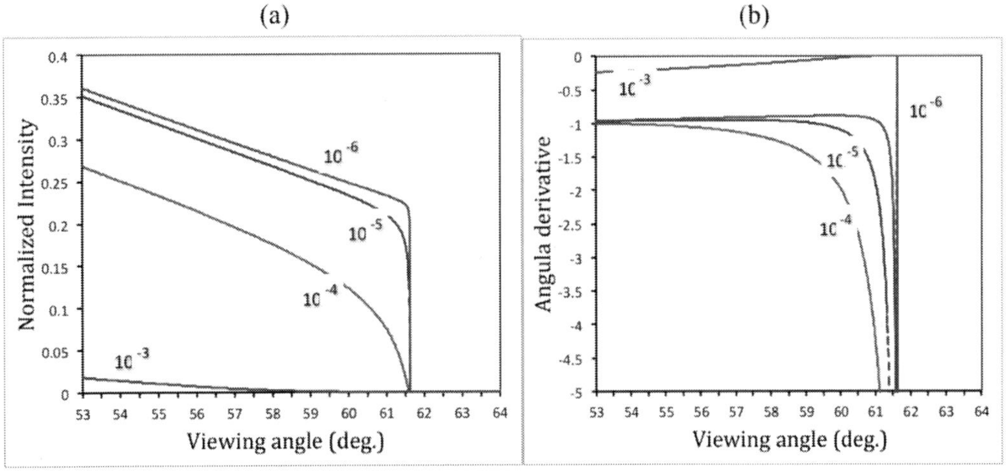

Figure 2: (a) Plots of the intensity versus the angle of refraction (inside the prism) near the critical angle for a sample fluid of refractive index n_m = 1.33 + ix with x = 10^{-6}, 10^{-5}, 10^{-4} and 10^{-3}. The refractive index of the prism was assumed to be 1.515, the wavelength 635nm and the sample thickness = 100 μm. (b) plots of the angular derivative of the curves in (a).

We can see that for a small imaginary part of the refractive index, the intensity profile is "step-like" at the critical-angle determined by the real part of the refractive index and the refractive index of the prism. As the imaginary part of the refractive index of the sample fluid increases, the angular profile becomes smoother, but it sinks to zero at the same angle. In the plots for x = 10^{-6}, 10^{-5} and 10^{-4} one can se there is an inflexion point. One should take the inflexion point as the critical angle, θ_c. From it one may obtain the real part of the refractive index of the sample using the usual formula $n_s = n_p \sin^{-1}(\theta_c)$ where n_p is the refractive index of the prism and n_s would be in this case the real part of the refractive index of the sample.

In Fig. 2b we plot the angular derivative of the curves shown in Fig. 2a. Clearly the first three plots for $x = 10^{-6}$, 10^{-5} and 10^{-4} show a sharp maximum at basically the same angle. However, the curve for $x = 10^{-3}$ does not show a maximum and therefore the corresponding plot in Fig. 2a has no inflexion point. We may then conclude that standard Abbe refractometry can no longer be performed. Nevertheless, by adjusting Eq. (9) to the angular intensity profile one could retrieve the real and imaginary parts of the refractive index of the sample. We may point out here that if d is reduced the angular intensity profile for $x = 10^{-3}$ changes and an inflexion point appears. However if d is reduced too much, then the intensity profile starts to extend noticeably beyond the inflexion point. Anyhow, depending on the particular design of a refractometric device based on refraction of diffuse light near the critical angle one can use Eq. (9) as long as scattering within the sample is negligible.

4. Conclusions

We derived a simple formula for the angular intensity profile of light refracted from a homogeneous absorbing fluid into a transparent prism of higher refractive index. The formula could be used to model refractometric devices based on refraction of diffuse light from an absorbing fluid sample near the critical angle. It could also be advantageously used to retrieve the real and imaginary parts of the complex refractive index of a sample fluid from angular intensity profiles around the critical angle in Abbe type refractometers.

Acknowledgments
We acknowledge financial support from Dirección General de Asuntos del Personal Académico from Universidad Nacional Autónoma de México through grant PAPIIT IN-106712.

References

[1] Lei L, Li H, Shi J, Chen Y., "Microfluidic refractometer with integrated optical fibers and end-facet transmission gratings", Rev Sci Instrum. 2010 Feb;81(2):023103. doi: 10.1063/1.3280226.
[2] Sang Youl Yoon[a] and Sung Yang, "Microfluidic refractometer with micro-image defocusing", Lab on a Chip 11, pp. 851-855, 2011.
[3] Sergio Calixto, Martha Rosete-Aguilar, David Monzon-Hernandez, and Vladimir P. Minkovich, "Capillary refractometer integrated in a microfluidic configuration", Applied Optics, Vol. 47, Issue 6, pp. 843-848 (2008).
[4] Malacara, Daniel, Geometrical and Instrumental Optics (Boston: Academic Press, 1988).
[5] G H Meeten and A N North. "Refractive index measurement of turbid colloidal fluids by transmission near the critical angle". Meas. Sci. Technol 2 441, 1991.
[6] H Contreras-Tello and A García-Valenzuela, "Experimental evidence of an effective medium seen by diffuse light in turbid colloids", Journal of Physics: Conference Series 274, 012129 (2011).
[7] A. García-Valenzuela and H Contreras-Tello, "Optical model enabling the use of Abbe-type refractometers on turbid suspensions", Opt. Lett. 38 (5), pp. 775-777 (2013).
[8] W. C. Chew, *Waves and Fields in Inhomogeneous Media* (Piscataway NJ, 1995), Chapter 7.
[9] Robert E. Collin, Antennas and Radiowave Propagation, (McGraw Hill NY, 1985), Chap. 4.

Journal of Physics: Conference Series **450** (2013) 012023 doi:10.1088/1742-6596/450/1/012023

Long range magnetic localization- accuracy and range study

J Vcelak[1],A Zikmund, J Kral[2]
[1]CTU-University Center for Energy Efficient Buildings, Nam. Sitná 3105, Kladno, CZ
[2]CTU-Faculty of electrical engineering, Technická 2, Prague 6, CZ

E-mail: vcelakj1@fel.cvut.cz

Abstract. Undersurface localization systems are accuracy demanding application. Localization methods used for common position estimation in open space or in building cannot be used since they are usually based on RF signal transmission or satellite navigation. Magnetic localization methods are usually the only usable solution. Horizontal directional drilling is one of accuracy demanding application where the operator needs to know exact position of the underground unit with respect to a given point on the surface. Long range surface magnetic localization system will be presented in this paper. This paper will summarize achievable accuracy of magnetic localization and maximal range with respect to given localization error. First results measured with presented system will be presented as well as results from FEM modeling. The influence of target distance, magnetic sensors noise, orientation sensor accuracy and surrounding material will be evaluated and considered in this work.

1. Introduction

Increasing number of horizontal directional drilling jobs caused demand for high accuracy systems that are able to follow exactly planned track of the drill also in industrial or urban agglomerations. There are several systems for precise drill-head underground guidance. In principal we can differ following navigation and guidance approaches:

1.1. Electronic compass + inclinometers with dead reckoning algorithm

This method is not very precise since the accuracy of magnetometers is strongly limited[1][2]. Such system usually measures geomagnetic field of the Earth and estimates actual azimuth, pitch and roll. The other disadvantage is that vicinity of any soft or hard iron near the magnetic sensor will cause unpredictable error which is dependent on mutual alignment of Earth's magnetic field vector and the drill head. Advantage is that these systems can be designed for very small bore diameters down to 2 inches and they are quite cheap in comparison to others. Such systems are used for short drills with low demand on accuracy of the tract and exit point position.

1.2. Optical gyro + inclinometers and dead reckoning algorithm

Because of the high accuracy of optical gyroscopes they can be easily used for high accuracy orientation measurement (azimuth 0.02°)[3]. Optical gyros are used for orientation estimation and can be recalibrated (northseeking calibration) during the job several times so long time drift will not cause a problem. The accuracy of such system is very high and on the drill with total length of 2.5km the

[1] To whom any correspondence should be addressed.
[2] To whom any correspondence should be addressed.

accuracy of exit point might be better then 2.5m. That disadvantage of these systems is price of optical gyroscope and dimensions. These systems are used for longer drills or drills in industrial or urban agglomerations where high accuracy is necessity.

1.3. Surface unit localization
The unit is localized with respect to given point on the surface. The working principle is very similar to localization of two underground systems[5]. The accuracy of unit itself doesn't need to be that high since the precise position is estimated during the drill job and drill track is corrected in order to match the planned route. Localization methods are mainly based on magnetic methods with surface coil or low frequency radio beacons in downhole unit and radio receiver in walkover surface unit. The beacons have limited accuracy and range and in higher depths the localization is very inaccurate. In this contribution we will focus on the localization using a surface coil for localization of downhole unit with magnetometer sensors.

This paper describes third configuration with surface excitation coil and downhole magnetometer system. System localizes downhole unit with respect to surface coil.

2. Method and unit description
The localization system consists of a surface transmitting coil and downhole unit with two magnetometers. The excitation surface coil generates square wave magnetic field with high amplitude (N=1500, d=20cm, I=20Ap-p, 1Hz) and given time pattern. This signal is measured by both magnetometers in downhole unit. According the known model of surface excitation coil processing algorithm estimates the position of magnetometer with respect to the surface coil (X,Y,Z). Model geometry is shown in figure 1.

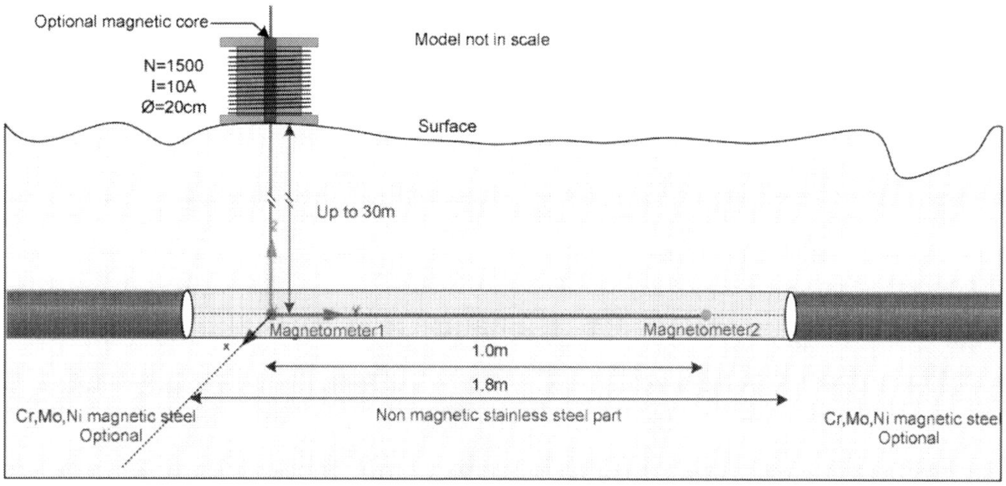

Figure 1. Measurement geometry.

The model of the coil is in far distance described using classical equation of magnetic dipole[4](1).

$$\boldsymbol{B}(\boldsymbol{r}) = \frac{\mu_0}{4\pi}\left(\frac{3\boldsymbol{r}\cdot(\boldsymbol{M}\cdot\boldsymbol{r})}{r^5} - \frac{\boldsymbol{M}}{r^3}\right) \quad (1)$$

Where ***B(r)*** is magnetic field vector, ***r*** is vector representing position in the space with respect to the magntic the magnetic field source with magnetic moment ***M***.

Magnetic field drops with the cube of the distance to magnetic dipole source (surface coil). We can identify following sources of potential errors in magnetic localization according described method and perform sensitivity analysis of each source

Input parameters to the data processing algorithm are following:

Magnetometer data – accuracy and noise of the magnetic sensor

Orientation sensors (roll, pitch & azimuth) – accuracy and noise of the orientation sensor

The influence of these parameters to final position estimation algorithm were investigated.

3. Results

The accuracy of magnetometer orientation is important parameter which can lead to unpredictable errors that are related to mutual orientation of the surface coil and magnetometers. Figure 2 shows position error dependency with respect to overall distance to magnetic source and orientation sensor precision. Three different error values of orientation sensor accuracies are considered 0.01 deg in green, 0.1 deg in blue and 0.5 deg in red color line. Accelerometers used to estimate pitch and roll are quite precise today and can achieve accuracy of 0.5°with appropriate system calibration [1],[2]. The situation with azimuth is much worse. Azimuth accuracy bellow 0.5 deg is not easily achievable with magnetic sensors [1] and therefore optical gyroscopes are much better choice as azimuth orientation sensor.

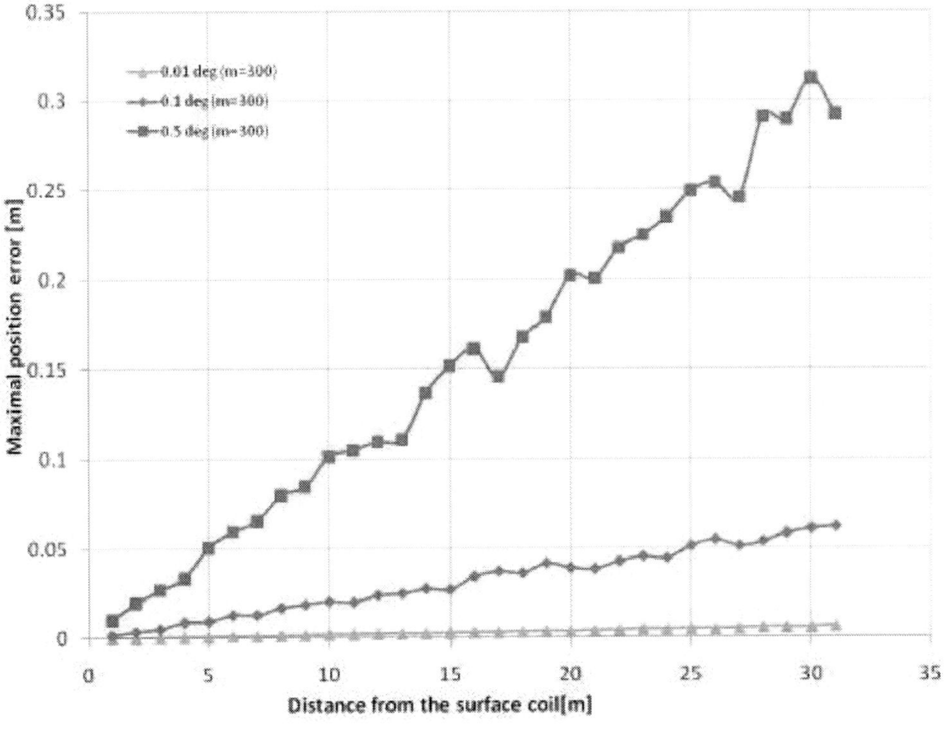

Figure 2. Localization accuracy with respect to distance from target and accuracy of orientation sensor.

The influence of the magnetic sensor noise is shown in figure 3. Also three different sensor noise levels were considered $0.01nT_{p-p}$ in green, $0.1nT_{p-p}$ in blue, $1nT_{p-p}$ in red line. It is clear that noise level of magnetic sensor has crucial influence on accuracy of the whole system. High accurate sensors (fluxgates) with low noise has to be used in order to minimize the contribution to final position error. Presented system uses three axis fluxgate magnetometer with noise level of 12 pT RMS/Hz. Nevertheless if we consider the whole 24-bit ADC magnetic data processing chain we will end-up with noise value closer to 30pT. Obviously the influence of the sensor noise increasing with distance to surface coil (decreasing of magnetic signal to noise ratio).

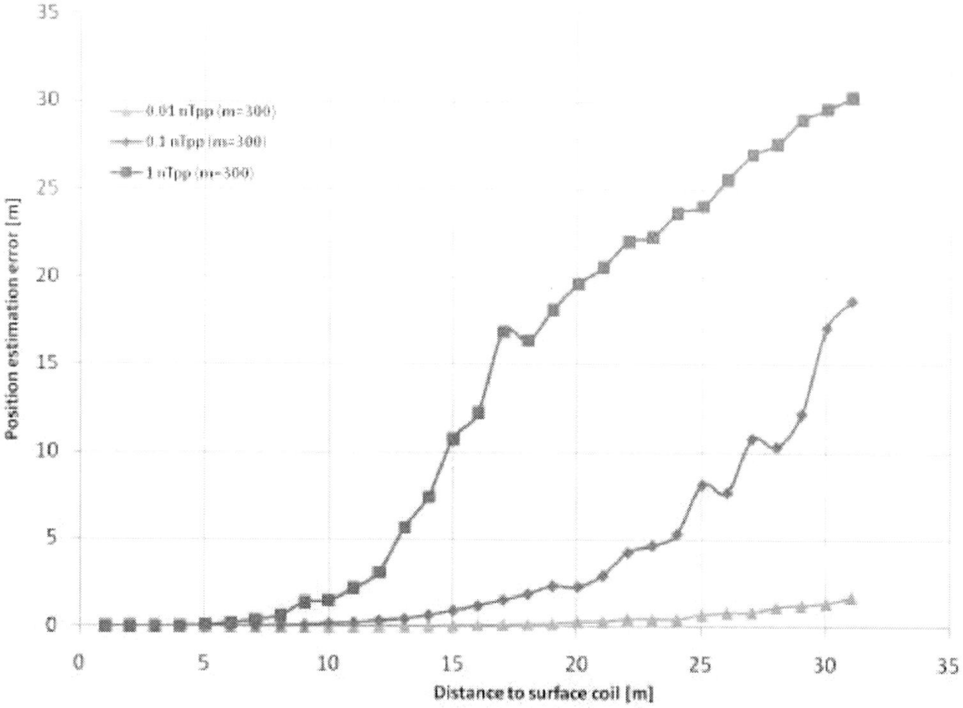

Figure 3. Infuence of magnetic sensor noise to final positioning accuracy.

It has to be pointed out that presence of any magnetic material in close vicinity of the magnetic sensors can ruin the position accuracy. Threrefore it is necessary to keep magnetic sensors in non-magnetic environment also in case of horizontal underground drilling even if all other components of the drill chain are magnetic. The model situation is shown in figure 4 where presense of magnetic material (long steel magnetic pipes) 40 cm far from magnetic sensors cause deviation of magnetic field vector in order of degrees. In such situation the estimation of position using general dipole model accordign equation (1) fails or the accuracy of position drops dramatically. Detailed characterization of this influece and possible compensation of the magnetic material presence is above the scope of this paper and will be evaluated separately.

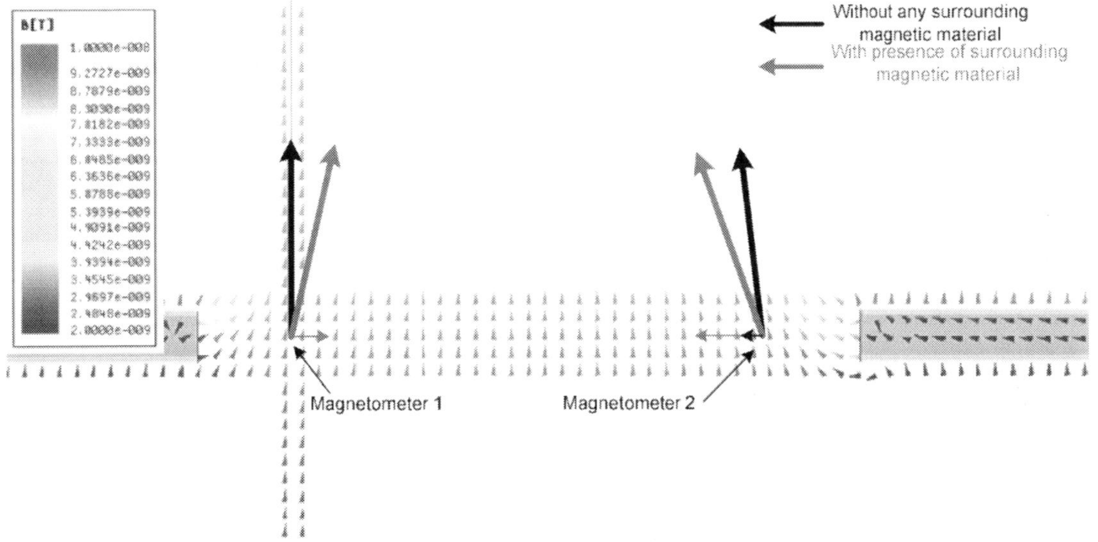

Figure 4. Model situation - Influence of magnetic material in close vicinity of the magnetometer.Same geometry as in figure 1

The accuracy and overall performance of the nonmagnetic system without presence of disturbing magnetic material was verified during real measurement. The mesurement was done in area with low magentic disturbances. The inspected depth Z range was up to 28 m, Y range was up to 8m and X range was 2m. Different mutual orientation between magnetometers and surface coil were tested also. We compare the position calculated from magentic localization with the real measured position. Total error vector ***E*** is calculated (2) and its magnitude (3) presented in figure.

$$\overline{E} = \overline{R_{real}} - \overline{R_{calc}} = [X_{err}, Y_{err}, Z_{err}] \quad (2)$$

$$|\vec{E}| = \sqrt{X_{err}^2 + Y_{err}^2 + Z_{err}^2} \quad (3)$$

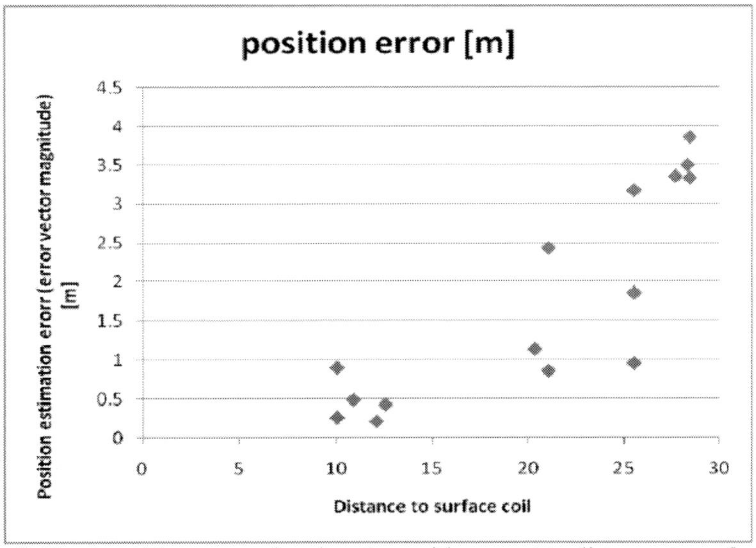

Figure 5. Total position error of real system with respect to distance to surface coil.

4. Conclusion

We use magnetic localization system simulation model to estimate the influence of sensor noise as well as orientation sensor accuracy to final accuracy of position estimation. It was shown that orientation sensor accuracy has much lower influence compared to sensor noise influence which is a range limiting factor. The influence of the magnetic material was shown in one model situation and was pointed out that it can ruin the algorithm precision if not compensated. It has to be noted that orientation sensor influence as well as magnetic material influence strongly depends on mutual position of the excitation coil and magnetometer. Real system was used to verify precision also in long distances (28m range). Surprisingly also in 28m range the precision of estimated position is still bellow 4m.

References

[1] Včelák J 2009, "Application of Magnetic Sensors for Navigation Systems", Shaker Verlag GmbH Germany, ISBN 978-3832277383

[2] Včelák J, Ripka P, Platil A, Kubík J, Kašpar P 2006, Sensors & Actuators A, Errors of AMR Compass and Methods of Their Compensation, Vol. 129, pages 53-57

[3] Ledroz AG., Pecht E., Cramer D., Mintchev MP 2005, IEEE TRANSACTIONS ON INSTRUMENTATION AND MEASUREMENT, FOG-based navigation in downhole environment during horizontal drilling utilizing a complete inertial measurement unit: Directional measurement-while-drilling surveying, , VOL. 54, NO. 5

[4] Tomek J, Mlejnek P, Janasek V, et al., 2007, Sensor Letters: Gastric motility and volume sensing by implanted magnetic sensors, 276-278

[5] Vcelak J, Zikmund A, Ripka P, 2012, Journal of Electrical Engineering, Long Range Magnetic Localization system, , Vol 63, NO 7, p19-22

Acknowledgments

This work has been supported by the European Union, OP RDI project No. CZ.1.05/2.1.00/03.0091 – University Centre for Energy Efficient Buildings.

A Fluidic Cell Embedded Electromagnetic Wave Sensor for Online Indication of Neurological Impairment during Surgical Procedures

R.T. Blakey[1], **A. Mason and A.I. Al-Shamma'a**

BEST Research Institute, School of Built Environment, Liverpool John Moores University, Liverpool, UK.

Email: r.t.blakey@2010.ljmu.ac.uk

Abstract. Lactate is known to be an indicator of neurological impairment during aortic aneurysm surgery. It is suggested that cerebrospinal fluid removed during such surgery could provide useful information in this regard. Medical professionals find the prospect of online detection of such analytes exciting, as current practice is time consuming and leads to multiple invasive procedures. Advancing from the current laboratory based analysis techniques to online methods could provide the basis for improved treatment regimes, better quality of care, and enhanced resource efficiency within hospitals. Accordingly, this article considers the use of a low power fluidic system with embedded electromagnetic wave sensor to detect varying lactate concentrations. Results are promising over the physiological range of 0 – 20 mmol/L with a calibration curve demonstrating an R^2 value > 0.98.

1. Introduction

It is well known that patients undergoing surgical or endovascular aneurysm repair (EVAR) of acute and chronic thoraco-abdominal aortic disease [1, 2] have an inherent risk of paraplegia. This is caused by restriction of the spinal cord blood flow and lack of oxygen during the procedures, typically referred to as spinal cord ischemia [3]. The current preventative measure practiced by surgeons is to collect cerebral spinal fluid (CSF) by inserting a spinal drain. In doing so, pressure upon the spinal cord is reduced, and thus, so too is the inherent risk of paraplegia.

The authors have been working for some time toward developing a method which could make use of the collected CSF as a marker of some underlying patient condition. At the moment any CSF collected would normally be discarded as biological waste. However, researchers at Liverpool Heart and Chest Hospital (LHCH) are particularly active in looking for ways to identify the onset of paraplegia in patients. Part of this work involves CSF analysis in the search for indicators of sub-clinical cord ischemia and compromise. This notion at that CSF might harbour such important information is supported separate research [4] which observed that lactate concentration in CSF increased with the onset of patient paraplegia. One would normally expect to see physiological levels of lactate in blood around 1 mmol/L; in pathology this may rise to 10-15 mmol/L, or higher in the case of exercise [5].

[1] r.t.blakey@2010.ljmu.ac.uk

Analysis of CSF would be most practical immediately after it has been drained from a patient, since in many cases this would allow the surgeon to take immediate corrective action in addition to improved diagnostic accuracy. Other authors have considered optical [6, 7] and magnetic resonance [8] techniques for the quantification of analytes in CSF. In many cases, the *in-situ* analysis of CSF could remove the need for subsequent surgical procedures, thus improving the efficiency of hospital resources and arguably improving patient care. Previous work by the authors [9-12] has considered *microwave spectroscopy* using a resonant microwave cavity as a potential method to analyse CSF in real-time as it is extracted from a patient during surgery.

Microwave spectroscopy has seen use for a variety of purposes, including monitoring of water quality [13, 14], filter monitoring [15] and also for use monitoring oil, gas and water concentrations [16]. It is thought that the use of microwave spectroscopy, as in the previously mentioned applications, will provide affordable and rapid diagnostics for the healthcare scenario described in this work.

This work goes a step further through the application of a novel sensor system, patented by the authors [17-19], which makes use of a planar electromagnetic structure rather than a cavity. This structure could conceivably be placed in-line with a spinal tap and, more importantly, is much cheaper to manufacture than a cavity system. Thus, it serves as a more practical option than the cavities used in previous "proof-of-concept" work.

The microwave sensing technique is robust, requiring low power and having good depth of penetration in respect of analyte materials. Also, of particular interest in medical applications is the non-ionising nature of microwave radiation; the technique used in this paper has low power output at around 1 mW (0 dBm) but has good penetration depth and equipment could be made portable for use at the bedside. These are features which practitioners see as a significant benefit over existing technologies, particularly X-ray imaging [20].

The multi-parameter nature of broadband microwave analysis can provide unique signal spectrum signatures. In this work, captured microwave signals are presented in the form of scattering parameters (commonly referred to as S-parameters), with measurement of the reflected (S_{11}) and transmitted (S_{21}) microwave signal being possible. These signals vary depending upon properties of the analyte presented to the sensing structure, such as conductivity and permittivity [21]. Conductivity is a measure of a material's ability to conduct an electric current. Permittivity is a measure of how an electric field is affected by a dielectric medium, which is determined by the ability of a material to polarise in response to the field, and reduce the total electric field inside the material. Therefore, permittivity (ε_r) as defined in (1) relates to a material's ability to transmit an electric field and is a complex value which varies with frequency, and accounts for both the energy stored by a material (ε') as well as any losses of energy (ε'') which might occur.

$$\varepsilon_r = \varepsilon' + j\varepsilon'' \qquad (1)$$

As a material alters in concentration or type, it is likely that its permittivity will change leading to a change in response if the material is the target of microwave radiation. By measuring this response over a range of frequencies, one can characterise materials in order to infer their properties.

2. Materials and Methods

The novel structure used in this work is based upon a co-planar design. This design was chosen in order to minimise losses and interferences which might occur from external sources. Co-planar devices are well known for exhibiting this vital property, and therefore in this application offers a great deal of control in relation to how the sensor responds to analyte materials [22]. In particular, the sensor is constructed to ensure that only a small area is sensitive to dielectric change, which enhances significantly its robustness for the aforementioned use (i.e. in-line with a spinal tap). In addition, and as demonstrated in Figure 1, the electric field created by the sensor is dissipated significantly within 2 mm of its surface.

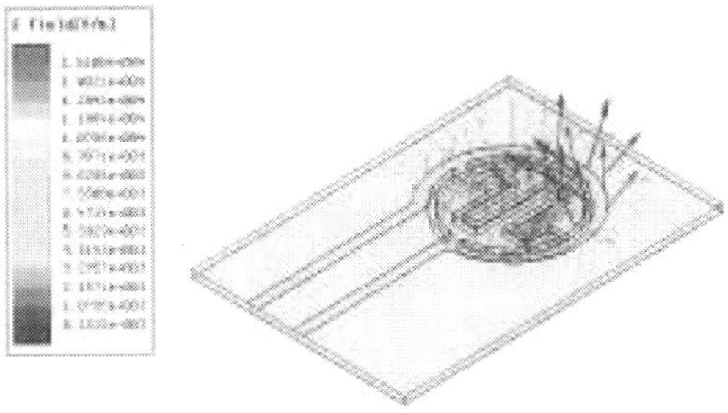

Figure 1. Illustrating the dissipation of the sensor electric field only a short distance from the sensor surface, which is particularly beneficial in removing environmental effects and interference.

The sensor is constructed in-house using an end mill process. An FR4 epoxy substrate is used which is coated on both sides with a 35 μm copper layer. After the milling process, tin plated finish is applied to the copper, and a side mount SMA connector is attached to provide connectivity to a Vector Network Analyser (VNA), discussed later, or other appropriate hardware.

The sensor is embedded in a bespoke fluidic cell constructed of polymethyl-methacrylate (PMMA), which has a high performance liquid chromatography (HPLC) compatible inlet and outlet ports in order to allow fluid to pass through the cell and come into contact with the sensor device itself. Although this system represents a laboratory environment, it is feasible that a fluidic cell such as this could be placed in-line with a spinal tap, or similar depending on the application, and thus the CSF analyte could easily be introduced to the sensor during a surgical procedure. Testing of this principle will be the aim of future work. The sensor and fluidic cell combination is shown in Figure 2.

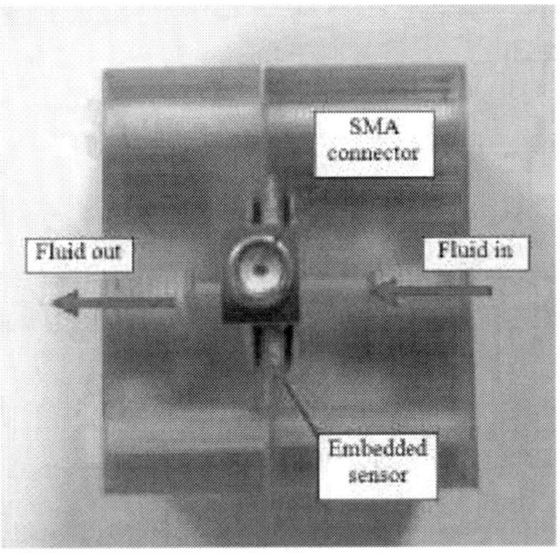

Figure 2. The sensor and fluidic cell combination used for the experimental work.

In order to test the capability of this sensor system, the work involves the use of a synthetic CSF, comprising deionised water combined, in varying quantities, with L-(+)-Lactic acid solution (Sigma-Aldrich, 27714). In particular, the following concentrations (mmol/L) were measured: 0, 0.2, 0.4, 1, 1.6, 2, 4, 10 and 20. The experimental work was repeated 6 times and an average of the results taken. Between each repetition the fluidic cell was cleaned with deionised water in order to eliminate possible contamination. A quaternary pump (Perkin Elmer) was used to introduce samples to the fluidic cell, and this was used to mix the lactic acid and deionised water to the concentrations previously detailed. The deionised water and lactate solutions were stored in an incubator at 30 °C, and feed tubes between the solutions and fluidic cell were kept short to minimise temperature loss. Each measurement took 3 minutes; 1 minute was allowed for the system to be flooded with the analyte and a further 2 minutes settling time was allowed. A low (2 ml/min) pumping speed was selected to minimise the build-up of air in the fluidic cell resulting from turbulence, particularly in the area contacted by the sensor. In the real world, this speed could easily be emulated *in-situ* although the speed of pumping itself has been found to have little direct effect on the response of the sensor beyond issues with air build-up.

Dielectric measurements were performed using an Agilent Technologies (Hewlett Packard) 8720 ET Vector Network Analyser (VNA). The instrument was set to generate a signal between 100 MHz and 1 GHz over 1601 data points for 10 linear frequency sweeps and calculate the S_{11} (reflected signal) parameters. Data was collected using bespoke LabView software and later analysed, as shown in Section 3 of this paper. The complete experimental setup is shown in Figure 3.

Figure 3. Overview of the experimental setup, including quaternary pump, vector network analyser and sensor system.

3. Results

The S_{11} measurements taken during the experimental work were used in order to identify change which occurred due to the varying concentration of lactic acid. Two resonant peaks where noted as responding to the change in lactate concentration. When water was present in the cell, these peaks were at approximately 394 MHz and 573 MHz; the latter however was found to give more pronounced response to lactate and so results for the fundamental resonance only are presented in this paper.

In Figure 4, the response of the sensor is shown as captured from the VNA, with the shift in electromagnetic spectrum clearly demonstrated. In Figure 5, the sensor response is shown in the form of a calibration curve over the full range of concentrations tested. Deionised water is used as a background reading against which all other readings are normalised; thus the results emphasise the change in response of the sensor at the indicated frequency. The sensor shows a logarithmic response, with the greatest sensitivity shown up to concentration levels of 2 mmol/L. From Figure 5 it is clear that there is a good correlation ($R^2 > 0.98$) between the sensor response and the lactate concentration.

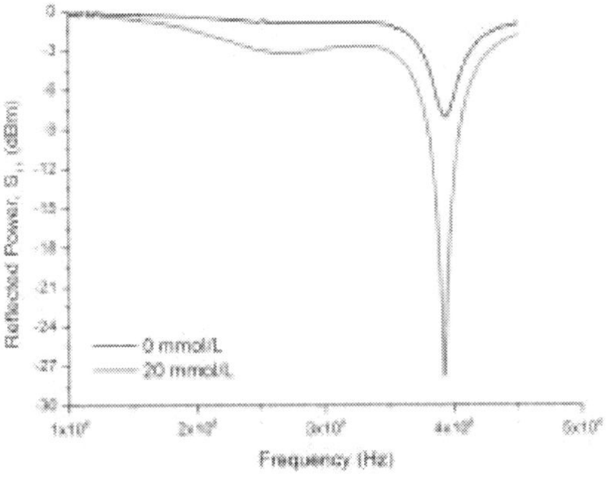

Figure 4. Illustrating the shift in peak frequency at approx. 394 MHz for 0 and 20 mmol/L concentrations of lactate.

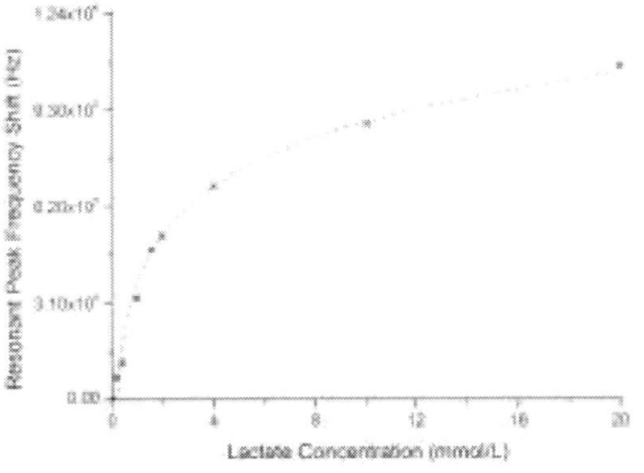

Figure 5. Illustrating the shift in peak frequency at approx. 394 MHz with different concentrations of lactic acid, with good correlation between results and the logarithmic best fit curve ($R^2 > 0.98$).

These results are an important step forward for this work, which has for a number of years strived to develop a method for real-time CSF analysis which would be simple to implement and not require any significant change to current surgical procedure. While previous work has shown the potential for electromagnetic wave techniques (e.g. cavities and planar structures), this shows a technique which could be applied in-line to current spinal tap procedures.

The technique uses only a small amount of power in the microwave region (approx. 1 mW). It is therefore non-destructive and, more importantly in a medical setting, non-ionising. It is important also to mention that the sensor response, after cleaning, returned to its original baseline output when flooded with deionised water which is promising in terms reliability and robustness for long term use. Longevity may not be a pressing concern in the healthcare field, as it is likely the sensor system would be single use due to potential biohazard risks. However, in other applications long term use of the sensor would certainly be preferable, and the authors are looking to apply this technique in other areas including water quality monitoring, industrial process monitoring and for biomedical screening.

One feature of note with the sensor, namely with the substrate material chosen, was the effect of water absorption. Over a period of time, the sensor substrate could be observed to absorb water, which was evident as drift in the sensor output signal. This was easily compensated however by allowing a significant pre-experimental period where the sensing system was exposed continuously to deionised water for a period > 40 hours. After this time, the change in sensor output as a result of drift during a single experiment (i.e. maximum change is 6.17×10^4 Hz) is considerably smaller than the change resulting from change of lactate concentration (i.e. 1.07×10^6 Hz). This drift could be compensated during manufacture, and would not be a feature of any sensor used for *in-situ* measurements. Figure 6 shows the variation of sensor output over time, with the largest change in resonant frequency apparent within the first 20 hours of operation. After this time the sensor output stabilises significantly, with small variations being accounted for by small changes (± 2 °C) in the laboratory temperature as a result of the thermostat controlled heating and ventilation system.

The variation of sensor output due to the choice of substrate is currently under investigation, and it is proposed that using alternative substrates (e.g. ceramic) will resolve current issues with sensor drift. In addition, temperature variation is likely to be resolved by using a differential measurement method. Once implemented, these remedial methods are expected to greatly improve the sensor usability.

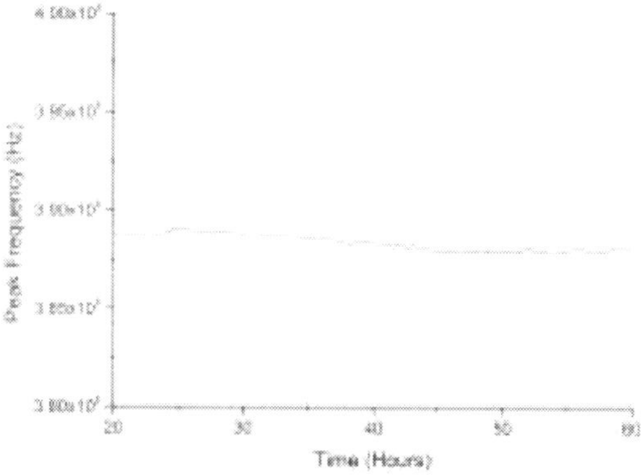

Figure 6. Sensor drift with time 20-60 hours after the sensor is first exposed to deionised water. Measurements are taken at approx. 40 hours, when sensor drift is minimal.

This work has successfully proven the concept that novel electromagnetic sensor embedded into a fluidic cell is an attractive technique capable of providing real-time information on the composition of the CSF. In previous work a microwave cavity technique was used to perform offline measurements of lactate suspended in a phosphate buffered saline (PBS) solution, since this is considered more representative of blood than deionised water. Due to the added conductivity of the media, it is proposed that this is a far more challenging environment for the sensor to operate in. Therefore further work with this sensor will consider the use of both PBS and blood background media spiked with similar concentrations of lactate as demonstrated in this paper in order to determine a more realistic idea on the sensitivity of this sensor for the intended application. An important aspect however is that the purpose of this sensor is not for precise quantification of lactate in blood, but rather to determine if the level is too high. This alone would indicate if there is some underlying issue with the patient; a precise value would be of little use to a surgeon in theatre. If the sensor indicates the elevated lactate level, its exact value could later be determined using standard laboratory procedures.

Alongside this work, the response of the sensor to other analyte solution types and concentrations is being explored and a database of these microwave signature spectra is being compiled, which can later be used for online process control in a broad range of industrial applications in the wastewater industry, chemical and pharmaceutical production lines.

4. Conclusion

An electromagnetic wave sensor embedded within a fluidic cell has been reported. The system demonstrates an improvement over previous work by the authors, and is a step toward achieving the goal of a truly online system for the indication of lactate level (i.e. *is it abnormally high?*) in a patient under-going a procedure such as surgical or endovascular aneurysm repair. The paper demonstrated sensitivity of the sensor system to varying levels of lactic acid, in the range 0 to 20 mmol/L in a deionised water solution. The measured results (i.e. resonant peak shift) at 394 MHz from the electromagnetic wave sensor show an excellent correlation with the varying concentration. Further work will consider other background media (e.g. blood), in addition to other potential applications in the biomedical, process and water industries where real-time monitoring of various fluidic analyte materials is becoming ever more necessary.

References

[1] M. Sellman, T. Ivert, G. Ronquist, K. Caesarini, L. Persson, and B. K. Semb, "Central nervous system damage during cardiac surgery assessed by 3 different biochemical markers in cerebrospinal fluid," *Scand J Thorac Cardiovasc Surg*, vol. 26, pp. 39-45, 1992.

[2] R. E. Anderson, A. Winnerkvist, L. O. Hansson, O. Nilsson, L. Rosengren, G. Settergren, and J. Vaage, "Biochemical markers of cerebrospinal ischemia after repair of aneurysms of the descending and thoracoabdominal aorta," *J Cardiothorac Vasc Anesth*, vol. 17, pp. 598-603, Oct 2003.

[3] F. Wadouh, C. F. Arndt, E. Oppermann, H. G. Borst, and R. Wadouh, "The mechanism of spinal cord injury after simple and double aortic cross-clamping," *J Thorac Cardiovasc Surg*, vol. 92, pp. 121-7, Jul 1986.

[4] B. Drenger, S. D. Parker, S. M. Frank, and C. Beattie, "Changes in Cerebrospinal Fluid Pressure and Lactate Concentrations during Thoracoabdominal Aortic Aneurysm Surgery," *Anesthesiology*, vol. 86, pp. 41-47, 1997.

[5] R. Dawson, D. Elliott, W. Elliott, and K. Jones, *Data for Biochemical Research*, 3rd Edition ed. USA: Oxford University Press, 1986.

[6] A. Curulli, "Electrochemical Synthesis of Poly(1,2-diaminobenzene) Microstructures: A New Material for Biosensing Applications," *Sensor Letters*, vol. 6, pp. 682-689, 2008.

[7] F. Beyette, B. Booher, J. Drennan, L. Carraher, J. Butler, P. Bowman, J. Clark, and P. Wilsey, "Device for Quantification of Bilirubin in Cerebral Spinal Fluid," *IEEE Transactions on Biomedical Engineering*, vol. 58, pp. 773-776, 2011.

[8] A. Subramanian, A. Gupta, S. Saxena, R. Kumar, A. Nigam, S. K. Mandal, and R. Roy, "Proton MR CSF analysis and a new software as predictors for the differentiation of meningitis in children," *NMR Biomed*, vol. 18, pp. 213-25, Jun 2005.

[9] J. H. Goh, A. Mason, A. I. Al-Shamma'a, M. Field, and P. Browning, "Lactate Detection Using Microwave Spectroscopy for In-Situ Medical Applications," *International Journal on Smart Sensing and Intelligent Systems*, vol. 4, pp. 338-352, September 2011.

[10] J. H. Goh, A. Mason, A. I. Al-Shamma'a, S. Wylie, M. Field, and P. Brown, "Lactate Detection Using a Microwave Cavity Sensor for Biomedical Applications," presented at the Fifth International Conference on Sensing Technology, Palmerston North, New Zealand, 2011.

[11] J. H. Goh, A. Mason, A. I. Al-Shamma'a, M. Field, M. Shackloth, and P. Browning, "Non Invasive Microwave Sensor for the Detection of Lactic Acid in Cerebrospinal Fluid (CSF)," *Journal of Physics: Conference Series*, vol. 307, 2011.

[12] J. H. Goh, A. Mason, M. Field, P. Browning, and A. I. Al-Shamma'a, "Using a Microwave Sensor as an Online Indicator of Neurological Impairment during Surgical Procedures," *Key Engineering Materials*, vol. 543, pp. 368-371, 2013.

[13] N. Al-Dasoqi, A. Mason, R. Alkhaddar, and A. I. Al-Shamma'a, "Use of Sensors in Wastewater Quality Monitoring - A Review of Available Technologies.," presented at the 2011 World Environmental & Water Resources Congress, Palm Springs, California, 2011.

[14] O. Korostynska, A. Mason, and A. I. Al-Shamma'a, "Monitoring of Nitrates and Phosphates in Wastewater: Current Technologies and Further Challenges," *International Journal on Smart Sensing and Intelligent Systems*, vol. 5, pp. 149-176, March 2012.

[15] A. Mason, S. Wylie, A. Thomas, H. Keele, A. Shaw, and A. I. Al-Shamma'a, "HEPA Filter Material Load Detection Using a Microwave Cavity Sensor," *International Journal on Smart Sensing and Intelligent Systems*, vol. 3, pp. 322-337, 2010.

[16] S. R. Wylie, A. I. Al-Shamma'a, and A. Shaw, "RF sensor for multiphase flow measurement through an oil pipeline," *Measurement Science and Technology*, vol. 17, p. 2141, 2006.

[17] A. I. Al-Shamma'a, A. Mason, and A. Shaw, "Non-Invasive Monitoring Device," 2010.

[18] D. P. Evenson, J. E. Parks, M. T. Kaproth, and L. K. Jost, "Rapid Determination on Sperm Cell Concentration in Bovine Semen by Flow Cytometry," *Journal of Dairy Science*, vol. 76, pp. 86-94, 1993.

[19] J. Gabrielson, M. Hart, A. Jarelöv, I. Kühn, D. McKenzie, and R. Möllby, "Evaluation of redox indicators and the use of digital scanners and spectrophotometer for quantification of microbial growth in microplates," *Journal of Microbiological Methods*, vol. 50, pp. 63-73, 2002.

[20] A. Sultana, N. Safavian, M. H. Izadi, K. S. Karim, and J. A. Rowlands, "Characterization of bias induced metastability of amorphous silicon thin film transistor based passive pixel sensor switch and its impact on biomedical x-ray imaging application," in *Medical Imaging 2009: Physics of Medical Imaging, February 9, 2009 - February 12, 2009*, Lake Buena Vista, FL, United states, 2009.

[21] D. Kajfez, "Temperature characterization of dielectric-resonator materials," *Journal of the European Ceramic Society*, vol. 21, pp. 2663-2667, 2001.

[22] C. A. Balanis, *Antenna Theory: Analysis and Design*, 3rd ed.: John Wiley & Sons, 2005.

Journal of Physics: Conference Series **450** (2013) 012025
doi:10.1088/1742-6596/450/1/012025

Detection of γ-radiation and heavy metals using electrochemical bacterial-based sensor

M. Al-Shanawa[1,3], A. Nabok[1], A. Hashim[1], T. Smith[2], S. Forder[1]
[1] Sheffield Hallam University, Materials & Engineering Research Institute
[2] Sheffield Hallam University, Biomedical Research Centre
City Campus, Howard Street, Sheffield, UK
[3]Faculty of Science, University of Basra, Basra, Iraq
E-mail: a.hashim@shu.ac.uk

Abstract. The main aim of this work is to develop a simple electrochemical sensor for detection of γ-radiation and heavy metals using bacteria. A series of DC and AC electrical measurements were carried out on samples of two types of bacteria, namely *Escherichia coli* and *Deinococcus radiodurans*. As a first step, a correlation between DC and AC electrical conductivity and bacteria concentration in solution was established. The study of the effect of γ-radiation and heavy metal ions (Cd^{2+}) on DC and AC electrical characteristics of bacteria revealed a possibility of pattern recognition of the above inhibition factors.

1. Introduction

This work is a part of ongoing research targeting the development of novel, simple, and cost effective methods for monitoring environmental pollutants, particularly radionuclides and heavy metals being common contaminants of water resources [1, 2]. It is known that micro-organisms such as bacteria are very sensitive to γ-radiation produced by radionuclides as well as to heavy metals [3, 4]. The use of micro-organisms for assessment of general toxicity of aqueous environment was reported previously [5]. Identification of the types of pollutants in the environment and the evaluation of their concentration is much more difficult task which is impossible to solve using a single sensor. However, the sensor array approaches [6] utilising several types of bacteria being inhibited differently by different types of pollutants could solve the above problem.

Our early experiments with *Escherichia coli* (*E. coli*) established a correlation between optical properties of liquid bacteria samples and bacteria concentration [7]. The effect of γ-radiation on concentration of live E coli bacteria was studied using three different optical techniques: (i) fluorescent microscopy with which yields directly the ratio of live/dead bacteria, stained, respectively, with "green" and "red" fluorescent dyes, (ii) optical density measurements at 600 nm, and (iii) fluorescent spectroscopy. The results obtained were encouraging however, the use of expensive and bulky optical instrumentation limits the sensor development.

In this work, we used simple electrical (electrochemical) measurements for the same purpose, i.e. first, establishing the correlation between electrical properties (conductivity) of liquid bacteria samples and live bacteria counts, after that studying the effect of γ-radiation and heavy metal ions (Cd^{2+}) on bacteria. Electrochemical measurements were successfully used for studying electrical properties of cells deposited on metal electrodes and showed great prospects of using such cell-based sensors for detection of various analytic [8, 9]. In our study, the principles of cell-sensors were extended further to more complex objects such as bacteria. In this work, in addition to *E.coli* bacteria, we used another type of bacteria, *Deinococcus radiodurans* known by its high resistance to γ-radiation [10]. The use of two types of bacteria may lead to pattern recognition of inhibition factors, in our case γ-radiation and Cd^{2+} ions.

2. Experimental methodology

2.1. Bacteria sample preparation

Two types of bacteria were selected for this work: Gram-negative *Escherichia coli* (*E. coli*) and Gram-positive *Deinococcus Radiodurans* (*D. Radiodurans).* Non-pathogenic HD5α strain of *E. coli* was used in LB (Luria-Bertain) broth [11] as a medium for *E. coli* cell culture. Anderson R1 strain of *D. Radiodurans,* which is extremely resistant to ionizing radiation, ultraviolet light, desiccation, oxidizing and electrophilic agents [10] was used in this work with 1-Nutrient agar (oxoid cm3). Both types of bacteria and respective growth media were acquired from SIGMA-ALDRICH CO. and OXOID LTD. Other chemicals, i.e. $CdCl_2$, salt were purchased from SIGMA-ALDRICH CO.

Cultivation of bacteria was performed in several stages. The first step was to cultivate a specific strain of bacteria in Petri dish containing solid broth agar, in order to use it as a bacteria source in future. In the second stage, one colony of bacteria was added into a sterile flask containing 50ml of liquid broth. Finally, the flask containing the bacterial culture was placed inside shaking incubator operating at 150 rpm shaking speed. The incubation temperatures were 30 C° for *D. Radiodurans* and 37 C° for *E. coli.* Bacteria start growing after 16 hours for *E. coli* and 24 hours for D. *Radiodurans.*

2.2. Experimental procedures

The cultivated bacteria density and changes in the live bacteria counts after exposed to pollution were recorded with Optical Density Photometer (6715 UV/Vis Spectrophotometer JENWAY OD600).

Solid radioactive source (Co-57) with the activity of 330 MBq was used for irradiation of bacteria samples. This source emits Gamma radiation (136, 14) keV with the equivalent dose of about 2000 mSv/h and the half-life 271.79 days [12]. The radiation dose (for bacteria samples) was varied using different exposure times from 1h up to 360h. $CdCl_2$ salt (from Sigma Aldrich) was used for testing bacteria samples. Solutions of different concentrations of $CdCl_2$ (down 0.1mM) were prepared by multiple dilution of 1M stock solution of $CdCl_2$ in de-ionised water (MQ Elga). Bacteria samples were mixed with $CdCl_2$ solutions in 1:1 ratio and kept incubated from 1 hour till 384 hours.

All electrical measurements were performed in a simple, sandwich-type, two-electrode cell. The cell schematically shown as inset in Figure 1 has PTFE body (1) and contains platinum counter electrode (2); gold-coated glass slide acting as working electrode (3) was sealed against the cell via rubber O-ring (4), the cell has inlet and outlet tubes (5, 6) allowing the injection of liquid medium containing bacteria. DC electrical test were performed using 6517A Keithley electrometer in voltage range ±0.5V. It has to be noted that preliminary electrochemical measurements were made in a standard three-electrode cell using DropSens microSTAT200 equipment (results are not presented here) and showed no difference to the measurements done in two-electrode cell in the ±0.5V voltage range which was selected in order to avoid oxidation and reduction electrochemical reactions on metal electrodes. AC electrical measurements were performed using (hp-4284A PRECISION LCR METER) in the frequency range from 20 Hz to 1 MHz with the amplitude of AC voltage of 100mV and no DC bias applied. The spectra of two parameters Gp and Cp corresponding to a parallel connection of conductance and capacitance were recorded.

3. Experimental results and discussion

3.1. DC electrical measurements

Typical I-V characteristics of E-coli samples having different concentrations of bacteria are shown in Figure1. Cathode current appeared to be much higher than anode current. The increase in bacteria concentration in solution leads to decrease of cathode current. This could be explained by adsorption of bacteria on the working electrode surface and thus the reduction in DC conductivity. Samples of *D. radiodurans* bacteria showed very similar DC characteristics. For detection purposes, the values of cathode current at –0.5V will be discussed throughout this paper.

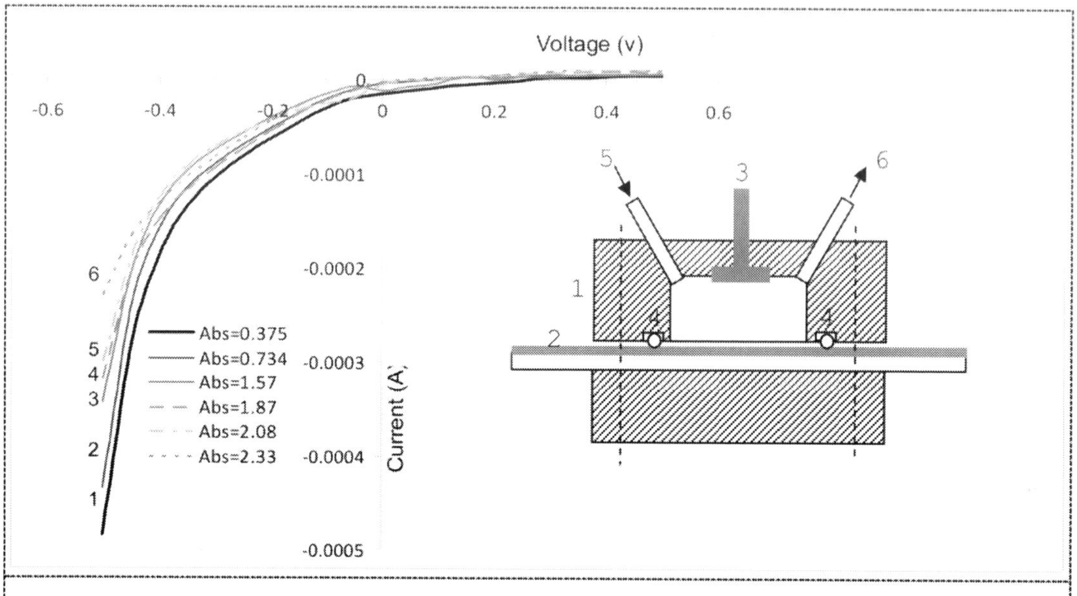

Figure 1. Typical DC I-V characteristics of E-coli samples of different concentrations measured in Abs units of optical density (OD_{600}): (1) 0.375, (2) 0.734, (3) 1.57, (4) 1.87, (5) 2.08, (6) 2.33 Inset shows the design of two-electrode cell.

The effect of γ-radiation and heavy metal ions (Cd^{2+}) on cathode DC current is illustrated in Figures 2 and 3. As one can see in Fig.2, DC conductivity of *E. coli* samples goes down exponentially with the radiation dose increase, while for *D. radiodurans* DC conductivity increases a little at low doses then decreases linearly at high doses. The effect of Cd^{2+} ions is similar on both bacteria (see Fig.3).

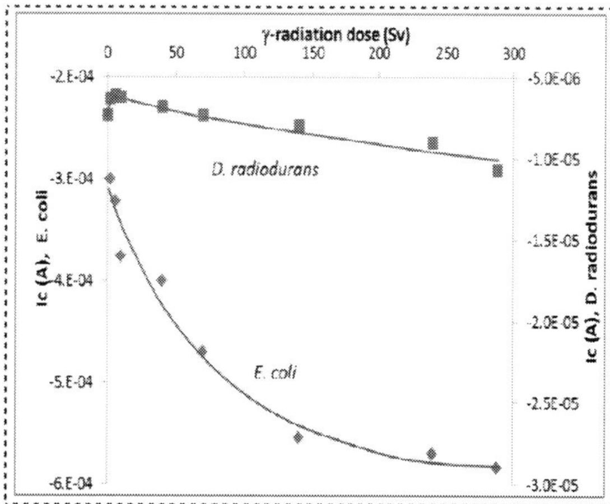

Figure 2. Dependence of cathode current of *E. coli* and *D. radiodurans* bacteria samples on γ-radiation dose.

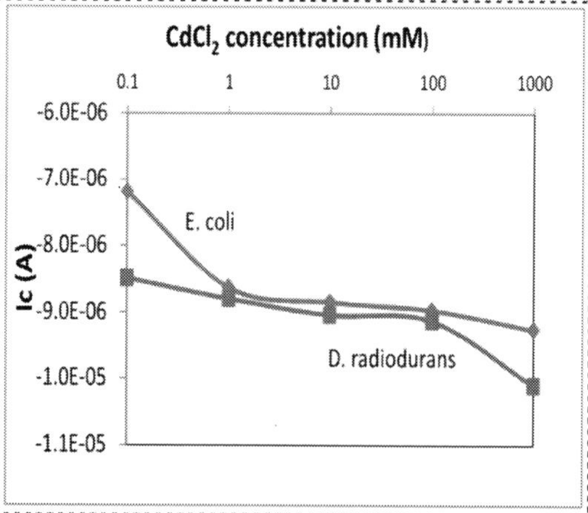

Figure 3. Dependence of cathode current of *E. coli* and *D. radiodurans* bacteria samples on $CdCl_2$ concentration

3.2. AC electrical measurements

Typical spectra of AC conductance (Gp) and capacitance (Cp) for bacteria samples are shown in Figure 4. The conduction values increase with the frequency and reach saturation at about 900kHz; Such behaviour is similar to that observed for cells [1] and can be explained by the dominant contribution from the material (bacteria in our case) absorbed on the surface of metal electrode. For sensing purposed, it would be useful to analyse the values of Gp at the saturation, i.e. at 900 kHz.

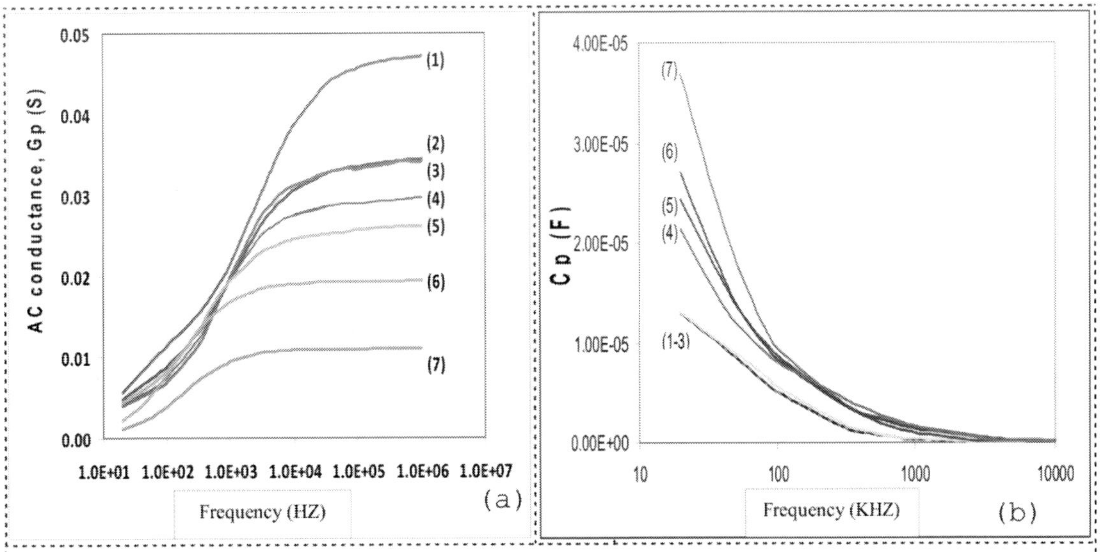

Figure 4. Spectra of Gp (a) and Cp (b) for E. coli bacteria samples of different concentrations presented in optical density Abs units: (1) clear broth, Abs =0.0, (2) Abs =0.237, (3) Abs =0.532, (4) Abs =0.801, (5) Abs =1.39, (6) Abs =1.57, (7) Abs =2.08

The capacitance, in contrary, maximal at low frequency and decreases in 4-5 orders of magnitude at 1MHz. The increase in bacteria concentration leads to larger capacitance values. These facts are a complimentary proof of the key role of bacteria adsorbed on the surface of metal.

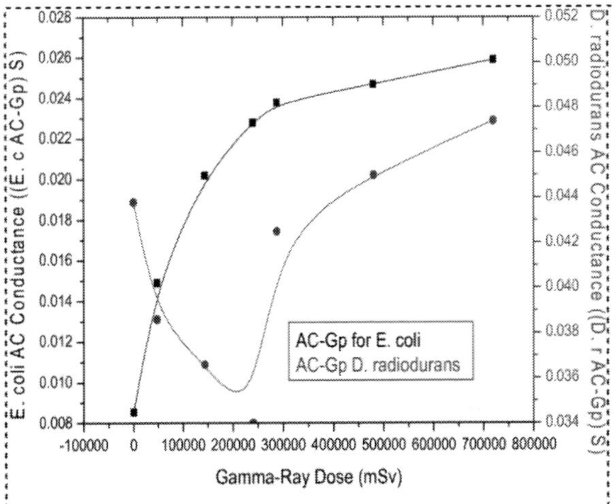

Figure 5. The effect of γ-radiation on Gp values at 900 kHz for *E. coli* and *D. radiodurans* bacteria samples.

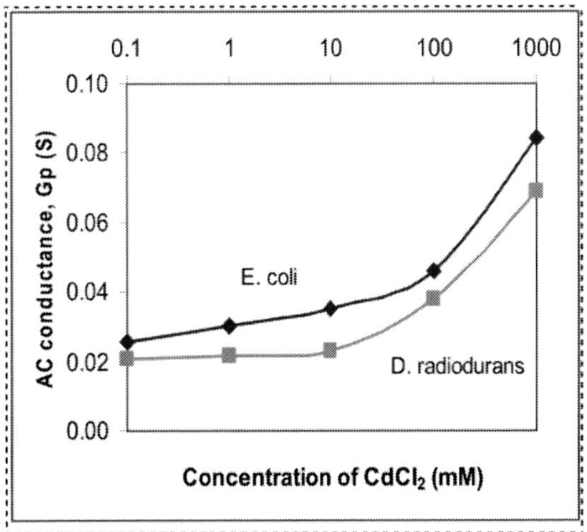

Figure 6. The effect of $CdCl_2$ on Gp values at 900 kHz for *E. coli* and *D. radiodurans* bacteria samples.

The effect of γ-radiation and $CdCl_2$ salt on Gp values at 900 kHz is demonstrated in Figures 5 and 6. AC conduction of *E. coli* increases with the increase in radiation dose which correlate well with the decrease in bacteria concentration. In case of *D. radiodurans,* Gp decreases at low radiation doses and increases at high doses. Cd^{2+} ions cause similar effect on both *E. coli* and *D. radiodurans* bacteria with Gp rising on the increase of $CdCl_2$ concentration. Detailed analysis of AC data (which is currently underway) should be based on a simple equivalent circuit model consisting of parallel conductance and capacitance most-likely associated with adsorbed layer of bacteria on the metal surface.

3.3. Data analysis: Pattern recognition of inhibiting factors

The data of DC and AC electrical measurement correlate and complement each other; they are also in a good agreement with the results of optical study [7]. The difference in responses of *E. coli* and *D. radiodurans* bacteria to γ-radiation and heavy metal ions (Cd^{2+} in our case) allows the application of principle of pattern recognition solving the "reverse problem", i.e. identification and quantification of pollutants in the unknown environment. Figure 7 shows the plot of relative DC responses (Ic current values were normalised by Ic_0 corresponding to untreated bacteria samples) of *E. coli* and *D. radiodurans* bacteria as X- and Y- axes respectively. The "open square" data points correspond to

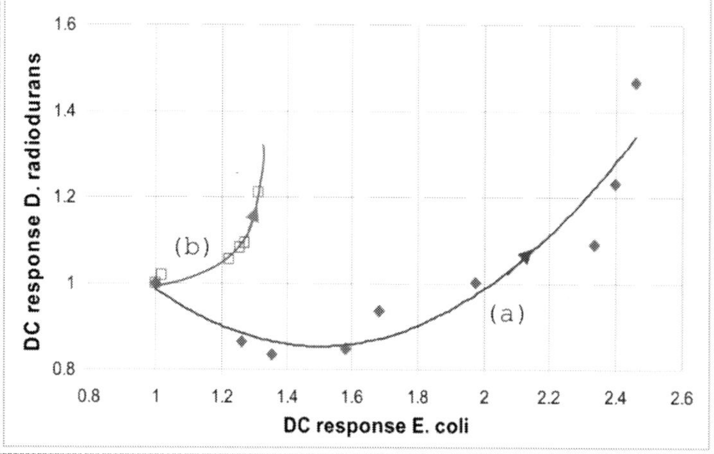

Figure 7. Pattern recognition of the effects of γ-radiation (a) and Cd^{2+} ions (b) using relative response of DC current measurements of *E. coli* and *D. radiodurans* bacteria. The arrows show, respectively, the directions of increase in γ-radiation dose from 0 to 288 (Sv) and $CdCl_2$ concentrations from 0 to 1M.

$CdCl_2$ exposure, while "filled diamond" points correspond to γ-radiation exposure. The separation of the two curves is clear which offers great possibility of pattern recognition of the above inhibiting factors.

4. Conclusions

DC and AC electrical study of two types of bacteria *E. coli* and *D. radiodurans* was a success. The obtained data correlated between themselves as well as with the data of optical study carried out previously. The measurements of DC current and Gp and Cp spectra can be used for quantification of live bacteria concentrations and thus the effect of γ-radiation and heavy metal ions on bacteria. Moreover, the difference in responses of *E. coli* and *D. radiodurans* bacteria to γ-radiation and heavy metal ions allows the application of principle of pattern recognition for identification and quantification of pollutants. This work proved the concept of a simple and cost effective electrical bacteria-based sensor and sensor array for preliminary assessment of the presents of toxins in water.

Acknowledgements

The authors would like to thank the Iraqi government for sponsoring PhD project.

References

[1] D. G Jones 2001, Journal of Environmental Radioactivity, Volume 53, Issue 3, Pages 313-333
[2] Adam K. Wanekava, Wilfred Chen and Ashok Mulchandain 2008, Journal of Environmental Monitoring, Issue 6, 703-712
[3] Kira S. Makarova, L. Aravind, Yuril. Wolf, Roman L. Tatusoy, Kenneth W. Minton, Eugene V. Koonin and Michael J. Daly 2001, Microbiol Mol Biol Rev; 65(1): 44–79.
[4] H. Chua, P. H. F. Yu, S. N. Sin, M. W. L Cheung 1999, Elsevier (Chemosphere). Vol 39, Issue 15, 2681-2692.
[5] Starodub N.F., Katzev A.M., Levkovetz I.A., 2003, Boston Transducers: Digest of Technical Papers 1- 2 1197-1200
[6] Sensor array concept, any text book on sensors
[7] Al-Shanawa M, Nabok A, Hashim A, Smith T 2013 Sensors & Transducers 149 179-186
[8] Md. Abdul Kafi, Tae-Hyung Kim and Jeong-Woo Choi 2001, SENSORDEVICES 2011: The International Conference on Sensor Device Technologies and Application, 147-150
[9] Yúfera A., Cañete D., Daza P., 2001 Technologies and Application, 143-146.
[10] Daly M J, Gaidamakova E SENSORDEVICES 2011: The International Conference on Sensor Device
K, Matrosova V Y, Vasilenko A, 2004, Science (AAAS) 306. 5698: 1025-8.
[11] BJ Rembacken, AM Snelling, DM Chalmers, ATR Axon, 1999 THE LANCET, Vol 354, Issue 9179, 635-639.
[12] G. Audi, O. Bersillom, J. Blachot, A. H. Wapstra, 2003, Nuclear Physics A, Vol. 729, Issue 1, 1-28.

Journal of Physics: Conference Series **450** (2013) 012026 doi:10.1088/1742-6596/450/1/012026

Arm and wrist surface potential mapping for wearable ECG rhythm recording devices: a pilot clinical study

W.D. Lynn[1], O.J. Escalona[2], D.J. McEneaney[3]

[1] Faculty of Computing and Engineering, University of Ulster, UK
[2] School of Engineering, University of Ulster, UK
[3] Cardiovascular Research Unit, Craigavon Area Hospital, UK

E-mail: lynn-wd@email.ulster.ac.uk; oj.escalona@ulster.ac.uk; david.mceneaney@southerntrust.hscni.net

Abstract. This study addresses an important question in the development of a ECG device that enables long term monitoring of cardiac rhythm. This device would utilise edge sensor technologies for dry, non-irritant skin contact suitable for distal limb application and would be supported by embedded ECG denoising processes. Contemporary ECG databases including those provided by MIT-BIH and Physionet are focused on interpretation of cardiac disease and rhythm tracking. The data is recorded using chest leads as in standard clinical practise. For the development of a peripherally located heart rhythm monitor, such data would be of limited use. To provide a useful database adequate for the development of the above mentioned cardiac monitoring device a unipolar body surface potential map from the left arm and wrist was gathered in 37 volunteer patients and characterized in this study. For this, the reference electrode was placed at the wrist. Bipolar far-field electrogram leads were derived and analysed. Factors such as skin variability, 50Hz noise interference, electrode contact noise, motion artifacts and electromyographic noise, presented a challenge. The objective was quantify the signal-to-noise ratio (SNR) at the far-field locations. Preliminary results reveal that an electrogram indicative of the QRS complex can be recorded on the distal portion of the left arm when denoised using signal averaging techniques.

1. Introduction

Cardiovascular disease remains globally the single most common cause of death in developed and developing nations. Sudden death, usually caused by lethal arrhythmias, accounts for 50% of these deaths [1]. Furthermore, patients with palpitations or loss of consciousness (syncope) account for a large proportion of attendances at hospital outpatients and emergency rooms [2]. Many of these patients have transient abnormal heart rhythm (arrhythmias) which are important in the diagnosis of heart disease and have prognostic significance. As there are many different cardiac arrhythmias accurate detection and recording is essential and thus continuous monitoring of the patients heart rhythm is required for periods lasting from several days up to a number of weeks or more. A portable heart rhythm monitor (Holter monitor, Fig.1) detects and records electrical signals generated by the heart from the body surface. Two or more self-adhesive electrodes are applied to the body surface (usually on the chest wall or arms) and after signal amplification and filtering an electrical signal from the heart is reconstructed (electrocardiogram, ECG). However, interference from other bioelectrical activity in the body, particularly artefact from muscle activity may render recording of cardiac electrical activity difficult. In general, the strength of the cardiac electrical signal detectable at the body surface will depend on the distance between the electrodes and on their positions relative to the heart. Electrodes are thus usually placed on several standard locations on the the chest wall or on both arms to maximise the signal. This technique is unsuitable for recording for more than a few days due to unreliability of electrode adhesion and interference with the patient's activities. Positioning of electrodes in positions more convenient for the patient such as at the wrist would be desirable but the ECG signal at this location is undetectable due to distance from the heart and interference (artefact) from arm muscle activity which obscures the signal. The current alternative approach to long term ECG monitoring is the implantable loop recorder (Fig.2). This is a solid state device similar in size to a memory stick containing sensors and monitoring/recording apparatus. The implantable loop recorder is placed under the skin of the chest wall near the heart during a surgical procedure. The loop recorder can monitor the heart rhythm for up to two years or more. However the technology is expensive (€2000) and requires a surgical procedure with risks including infection and scar formation. The combined cost of device, hospitalisation, surgery and follow up are thus considerable [5].

The diagnostic benefits of the clinical electrocardiogram (ECG) have been exploited by clinicians for over a century. The reproduction of the heart's electrical activity in the dv/dt domain is used to define abnormal heart conditions such as myocardial damage and arrhythmias on millions of cardiac patients globally [3]. The ECG uses 10 or more non-invasive skin mounted voltage probes to detect cardiac electrical activity. Positioning of the probes is used to build up a series of two dimensional pictures of the electrical activity in the myocardium, allowing the clinician to isolate damage and expedite treatment. A typical 12-lead ECG is recorded over a time period of less than 1 minute [3]. This time span would be sufficient for detecting new or established myocardial ischaemia and persistent arrhythmias. However, it is highly unlikely that infrequently occurring intermittent arrhythmias will be detected.

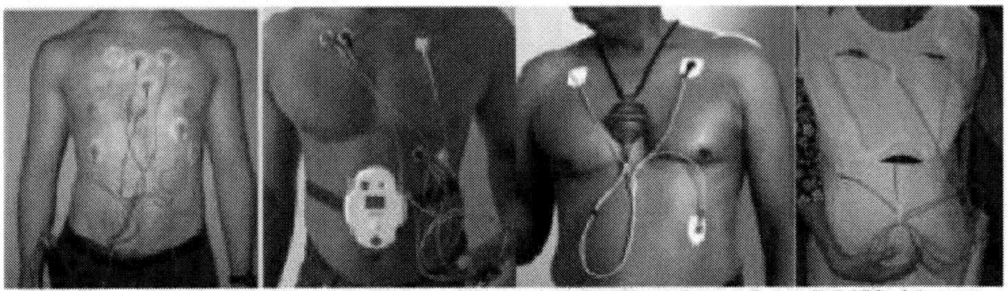

a) Traditional Holter Monitor b) NASA's Lifeguard c) Cardionet's monitor d) Lifebelt's system

Figure 1. Holter monitor type systems

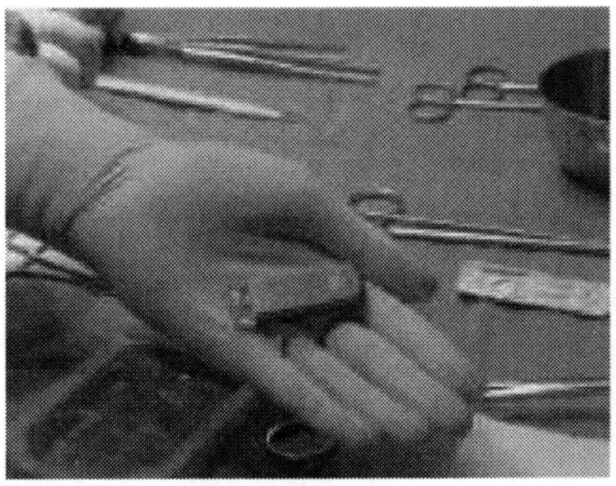

Figure 2. Implantable loop recorder.

The Holter system (Fig. 1) [4] uses a less aggressive, slow reacting, skin worn wet electrode, physically similar to that used during the diagnostic 12-lead ECG, but chemically less aggressive. Electrode choice is important as the more aggressive the reaction the less likely the patient is to tolerate long term exposure. The Holter recording typically utilises 2 or more electrodes across the chest area. The electrode array has a good signal to noise ratio due to the ECG electrodes physical distance across the cardiac signal source, providing a good quality ECG signal from the surface of the chest. Although the Holter has a high capability to provide quality signals, it is often confounded by patient interaction, for example, tampering with electrodes, percussion noise due to the piezo electric response of the wiring harness and electrode contact degradation. Skin adhesive systems are prone to degradation at differing rates, dependent on the patent skin type. In addition, the patients often suffer from localised irritation if the electrode is worn for an extended period.

Another, non surgical, alternative would be thus desirable and hence, the drive to develop a recording technique for a body location remote from the heart. Gemperle et al.[6] have suggested several spaces on the human body, Figure 3, where both solid and flexible forms could be comfortably and reasonably unobtrusively be located without interfering significantly with the subjects range of motion. Although they were interested in the application of wearable computers, their guidelines are also relevant to the positioning of bioelectric sensing devices in the form of patches and bands.

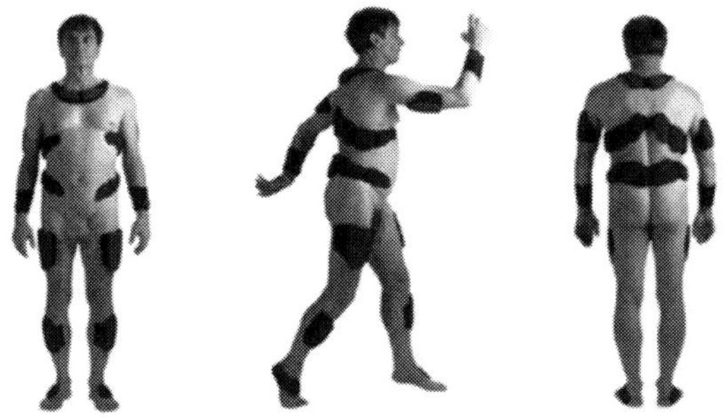

Figure 3. Body sites for comfortable "wearability" (Gemperle et al. 1998)[6].

2. Methods

In order to establish a method for the recovery of ECG rhythm information from far field sites, such as the wrist, a study of the ECG signal levels and their prevalence over the background noise level was characterised after the application of conventional ECG denoising techniques. The study maps the gradient of signal degradation as distance from the heart increases along the extremity of the subject's left limb. A conventional high-resolution ECG signal averaging technique as described in [7] was used for the ECG denoising process; as it is a widely accepted method for small cardiac QRS synchronised potential recovery.

2.1. Clinical ECG recordings

Ethical approval for the investigation was obtained from the Office of Research Ethical Committees Northern Ireland (ORECNI). An information sheet was provided and informed signed consent obtained. Each subject was connected to the Biosemi data recorder using the electrode position protocol described in Figure 5a and 5b. The electrode positions correspond to the comfort zones established by Gemperle et al.[6].

The subjects were selected from the general intake of the cardiology department at Craigavon Area Hospital. The subjects (age range 42-78 years) Body mass index was recorded. Recordings were carried out on different days using the same room environment and recording equipment. A database of 38 recordings from the left arm was created and processed using Mathworks Matlab programming environment.

Recording duration was 500 seconds at a sample rate of 2048Hz and 24 bit resolution. This gives a measurement resolution of 31.25e-9V per bit.

2.2. Arm and wrist unipolar mapping leads

The unipolar mapping leads/electrodes of importance are CH1, a reference high level ECG channel used for temporal alignment in the signal averaging process. Electrodes for the bipolar lead (CH11- CH12) are place across the bicep on the upper arm. CH10 is placed on the upper forearm and electrodes for the bipolar lead (CH6 - CH2) are located on the wrist. The reference electrode, REF, electrode as depicted in Figure 5(a) and 5(b), is also located on the subject's wrist interproximal to CH2 and CH6. The electrodes used were the Ambu Blue Sensor N type, designed for recovery of neonatal ECGs. This electrode type was chosen due to its physical size as a typical adult electrode made adhesion to the wrist area cluttered.

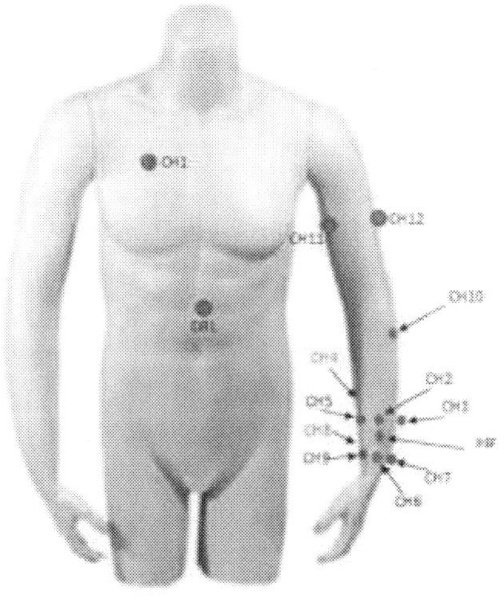

Figure 5(a). Electrodes positions.

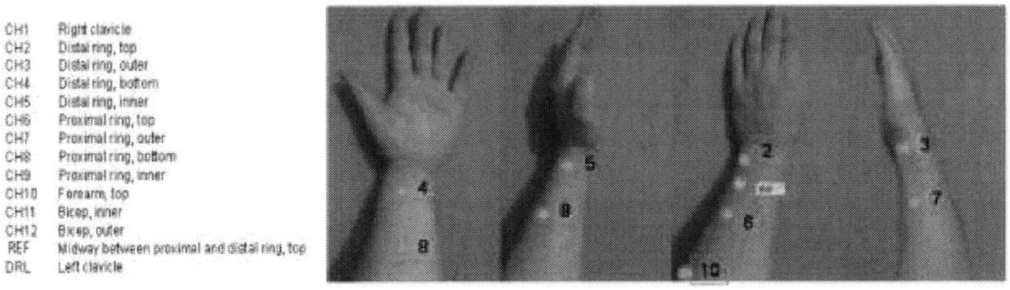

Figure 5(b). Electrodes positions on left arm & wrist.

2.3. ECG denoising process

Each recording was denoised using an industry accepted technique of signal averaging. The recovery of a strong ECG from Channel 1 allowed the remote channels to be aligned and averaged consistently. The beat recovery was consistently greater than 98% with *n* ranging from 348 to 512, dependant on the subjects heart rate. Noise figures for each subject were measured using a 40mS window located on the S-T segment. This positioning was chosen for its regional stability. Signal figures were measured in a 120mS window centred on the QRS complex – Figure 6.

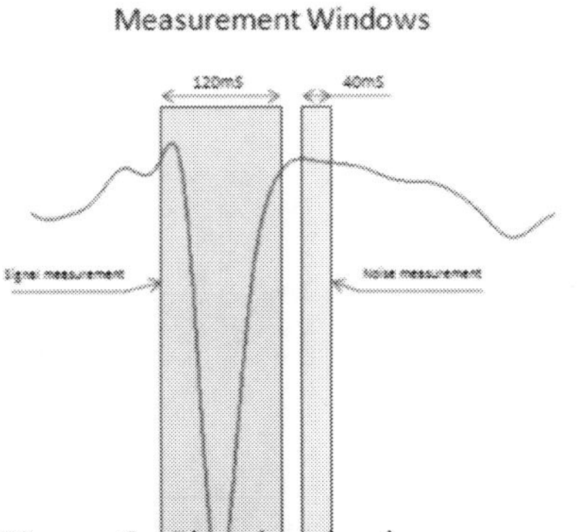

Figure 6. Signal and noise measurement windows locations

The standard deviation of the data points was calculated and subsequently the signal-to-noise ratio (SNR). Table 1a shows the SNR across 5 channel positions on 6 subjects. Table 1b-e shows the individual subject data that was used to construct the mean averages in table 1 a.

Channel 1 is the reference electrode, channel 11-12 is a mathematically produced bipolar channel across the bicep, channel 10 is on the forearm, Channel 4 is on the wrist and channel 6-2 is a mathematically produced bipolar channel located on the wrist. These channels have been chosen due to their progressive displacement from the signal origin. They best map the ECG signal degradation along the arm.

3. Results

Five recordings were rejected (Subjects 12, 20, 27, 31 and 34) due to elevated levels of coherent noise from power line interference. Six recordings were classified as having ideal recording quality. Table 1: (a) to (f) present the signal and noise levels characterization results, namely: original QRS noise level before ECG denoising, post ECG denoising signal level in rms (μV), post denoising background noise level and the SNR (S/N) after the denoising process. Table 1(g) presents the SNR *vs* selected distant far-field channels from the heart (Ch11-12, Ch10, Ch4 and Ch6-2). The best six channels signal and noise characterization analysis. Figure 7 plots the mean SNR of the best six channels along the arm, in distancing order from the hearty. The graph shows a progressive reduction in signal strength (SNR) as distance from the heart increases.

Figures 8 and 9 show the breakdown of the data. Figure 8 shows the inclusion of the remaining 25 acceptable channels. Figure 9 isolates the 5 rejected channels and shows how the mean SNR is affected by the inclusion of coherent power line noise. The mean SNR is reduced by almost 3dB due to the poor condition of the signals recorded in these instances.

Table 1(a). Best recordings: **Patient 1**.
Denoising: averaged beats, N = **434**;
Percentage of rejected beats = **1.2%**

Patient 1	Original QRS Noise Level (μV)	Post Denoising Signal RMS Level (μV)	Post Averaging Noise Level (μV)	Post Averaging S/N Ratio
CH1	73.10	67.18	3.20	20.97
CH2	0.61	0.10	0.00	25.85
CH3	7.19	0.19	0.06	2.96
CH4	11.88	0.26	0.12	2.23
CH5	3.44	0.15	0.22	0.68
CH6	12.50	0.14	0.11	1.38
CH7	12.50	0.23	0.19	1.17
CH8	9.38	0.17	0.04	4.51
CH9	24.34	0.65	0.34	1.88
CH10	13.13	0.11	0.10	1.08
CH11	28.13	1.26	1.80	0.70
CH12	7.81	1.44	0.32	4.49
CH11-12	29.12	2.65	2.09	1.27
CH10-2	13.14	0.09	0.10	0.92
CH6-2	12.51	0.05	0.11	0.48

Table 1(b). Best recordings: **Patient 2**.
Denoising: averaged beats, N = **559**;
Percentage of rejected beats = **1.3%**

Patient 2	Original QRS Noise Level (μV)	Post Denoising Signal RMS Level (μV)	Post Averaging Noise Level (μV)	Post Averaging S/N Ratio
CH1	150	105.11	7.65	13.75
CH2	2.8	0.15	0.03	4.72
CH3	8.44	0.18	0.01	14.39
CH4	8.13	0.16	0.02	10.18
CH5	8.13	0.16	0.04	4.10
CH6	12.8	0.14	0.06	2.52
CH7	12.03	0.17	0.01	15.68
CH8	13.4	0.17	0.05	3.49
CH9	5.9	0.15	0.04	3.44
CH10	3.91	0.17	0.04	4.67
CH11	3.44	1.26	0.36	3.54
CH12	12.19	3.29	0.56	5.83
CH11-12	12.67	4.54	0.92	4.93
CH10-2	4.81	0.09	0.03	2.81
CH6-2	13.1	0.05	0.03	1.81

Table 1(c). Best recordings: **Patient 3**.
Denoising: averaged beats, N = **511**;
Percentage of rejected beats = **5.94%.**

Patient 3	Original QRS Noise Level (μV)	Post Denoising Signal RMS Level (μV)	Post Averaging Noise Level (μV)	Post Averaging S/N Ratio
CH1	40.10	108.99	1.97	55.29
CH2	3.75	0.10	0.06	1.66
CH3	12.50	0.23	0.17	1.38
CH4	13.75	0.10	0.05	2.00
CH5	6.25	0.19	0.07	2.65
CH6	18.44	0.13	0.02	6.16
CH7	7.50	0.11	0.02	6.11
CH8	8.44	0.08	0.01	5.23
CH9	7.75	0.08	0.12	0.63
CH10	8.13	0.13	0.01	15.67
CH11	24.12	4.65	0.06	72.40
CH12	9.24	7.89	0.16	48.95
CH11-12	25.83	7.93	0.22	36.86
CH10-2	8.95	0.09	0.06	1.43
CH6-2	18.82	0.15	0.08	1.85

Table 1(d). Best recordings: **Patient 10**.
Denoising: averaged beats, N = **506**;
Percentage of rejected beats = **0%.**

Patient 10	Original QRS Noise Level (μV)	Post Denoising Signal RMS Level (μV)	Post Averaging Noise Level (μV)	Post Averaging S/N Ratio
CH1	80.00	67.18	1.18	20.97
CH2	267.30	0.10	0.04	25.85
CH3	43.00	0.19	0.02	2.96
CH4	76.00	0.26	0.01	2.23
CH5	225.70	0.15	0.05	0.68
CH6	68.00	0.14	0.05	1.38
CH7	267.40	0.23	0.02	1.17
CH8	73.00	0.17	0.01	4.51
CH9	61.00	0.65	0.02	1.88
CH10	193.40	0.11	0.01	1.08
CH11	142.60	1.26	0.00	0.70
CH12	118.00	1.44	0.02	4.49
CH11-12	185.1	2.65	0.02	1.27
CH10-2	330	0.09	0.04	0.92
CH6-2	275.8	0.05	0.02	0.48

Table 1(e). Best recordings: **Patient 17**. Denoising: averaged beats, N = **461**; Percentage of rejected beats = **0%**

Patient 17	Original QRS Noise Level (μV)	Post Denoising Signal RMS Level (μV)	Post Averaging Noise Level (μV)	Post Averaging S/N Ratio
CH1	73.20	93.46	2.48	37.68
CH2	176.30	0.13	0.37	0.34
CH3	53.00	0.14	0.57	0.25
CH4	77.00	0.70	1.05	0.67
CH5	121.70	0.14	0.38	0.36
CH6	78.00	0.16	0.58	0.27
CH7	102.40	0.17	0.55	0.30
CH8	198.00	0.30	0.64	0.47
CH9	75.10	0.13	0.40	0.33
CH10	109.40	0.12	0.49	0.25
CH11	188.60	4.09	2.14	1.91
CH12	118.00	2.98	6.04	0.49
CH11-12	222.5	6.44	4.31	1.49
CH10-2	207.5	0.09	0.86	0.11
CH6-2	192.8	0.15	0.95	0.16

Table 1(f). Best recordings: **Patient 36**. Denoising: averaged beats, N = **434**; Percentage of rejected beats = **1.22%**

Patient 36	Original QRS Noise Level (μV)	Post Denoising Signal RMS Level (μV)	Post Averaging Noise Level (μV)	Post Averaging S/N Ratio
CH1	96	91.59	0.93	98.85
CH2	164	0.25	0.24	1.06
CH3	146	0.22	0.28	0.79
CH4	93	0.17	0.39	0.44
CH5	152	0.27	0.33	0.83
CH6	125	0.30	0.25	1.17
CH7	132	0.21	0.45	0.48
CH8	120	0.31	0.38	0.82
CH9	88	0.25	0.29	0.86
CH10	157	0.28	0.27	1.01
CH11	97	2.85	1.03	2.78
CH12	143	2.52	0.23	10.78
CH11-12	172.8	4.57	1.25	3.67
CH10-2	227	0.16	0.17	0.91
CH6-2	206.2	0.09	0.02	4.99

Table 1(g). Six best case recordings: SNR mean values along four distancing bipolar leads along the left arm, to the wrist.

Subject	ch1	ch12-11	ch10	ch4	ch6-2
1	20.97	1.27	1.08	2.23	0.48
2	13.75	4.93	4.67	10.18	1.81
3	55.29	36.86	15.67	2.00	1.85
10	60.59	197.05	22.22	21.23	2.15
17	37.68	1.49	0.25	0.67	0.16
36	98.85	3.67	1.01	0.44	4.99
Mean	**47.85**	**40.88**	**7.48**	**6.12**	**1.91**

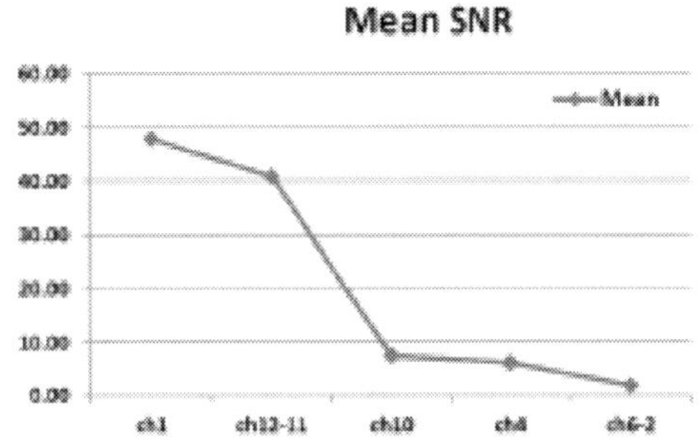

Figure 7. Six best case recordings: plot of mean (N=6) SNR values along the left arm, as calculated in Table 1 (g).

Table 2. All acceptable case recordings SNR mean values on channels along left arm.

Subject	Ch1	Ch12-11	Ch10	Ch4	Ch6-2
1	20.97	1.27	1.08	2.23	0.48
2	13.75	4.93	4.67	10.18	1.81
3	55.29	36.86	15.67	2.00	1.85
5	13.23	7.24	4.87	7.35	1.71
6	11.36	13.48	2.28	0.41	0.24
7	48.35	0.81	1.76	3.00	0.60
8	48.78	33.45	0.84	2.83	4.70
9	6.80	11.17	10.50	12.39	0.86
10	60.59	197.05	22.22	21.23	2.15
11	16.37	20.25	3.63	2.58	4.55
13	1087.9	2.83	1.68	0.98	2.25
14	16.56	8.73	2.09	3.35	0.37
15	42.00	4.20	1.08	0.17	0.30
16	15.68	6.74	15.73	4.27	0.69
17	37.68	1.49	0.25	0.67	0.16
18	78.67	15.16	3.18	2.69	1.72
19	146.33	0.78	0.67	0.45	0.32
21	206.33	106.25	1.78	5.80	2.22
22	20.43	13.62	0.24	0.23	0.61
23	17.64	3.40	0.90	0.96	0.34
24	18.78	17.63	15.93	2.09	2.53
25	121.56	20.63	8.92	7.68	5.15
26	16.17	1.04	0.74	3.75	2.64
28	20.86	8.85	5.99	3.73	0.64
29	38.03	1.00	0.77	2.12	0.99
30	341.61	15.88	3.91	9.29	1.29
32	30.64	24.51	0.74	0.31	0.25
33	29.99	1.24	6.13	2.05	0.57
35	22.98	3.21	2.99	1.74	1.66
36	98.85	3.67	1.01	0.44	4.99
37	69.00	0.96	0.83	0.72	2.39
38	51.14	45.50	8.49	6.22	2.35
Mean	**88.26**	**19.81**	**4.74**	**3.87**	**1.67**

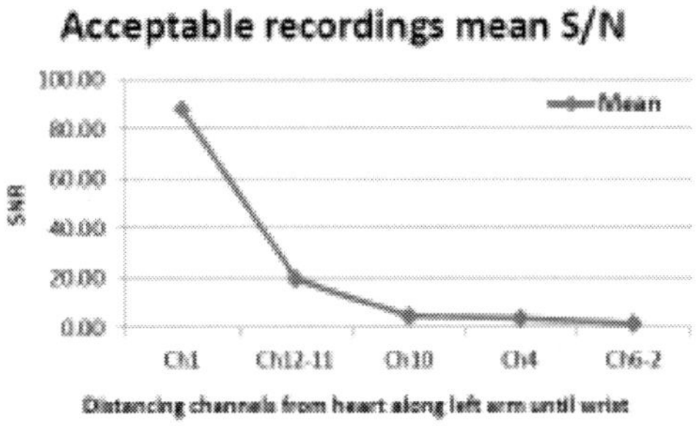

Figure 8. All accepted recordings: plot of mean (N=37) SNR values along the left arm, as calculated in Table 2.

Table 3. Five rejected recordings: SNR mean values along four distancing bipolar leads along the left arm, to the wrist.

Subject	Ch1	Ch12-11	Ch10	Ch4	Ch6-2
12	11.08	1.00	3.35	1.19	2.11
20	14.92	2.00	0.20	1.31	0.32
27	8.32	2.18	2.76	1.87	1.32
31	3.73	2.01	1.23	1.69	2.49
34	7.20	3.08	0.76	1.04	1.62
Mean	9.05	2.05	1.66	1.42	1.57

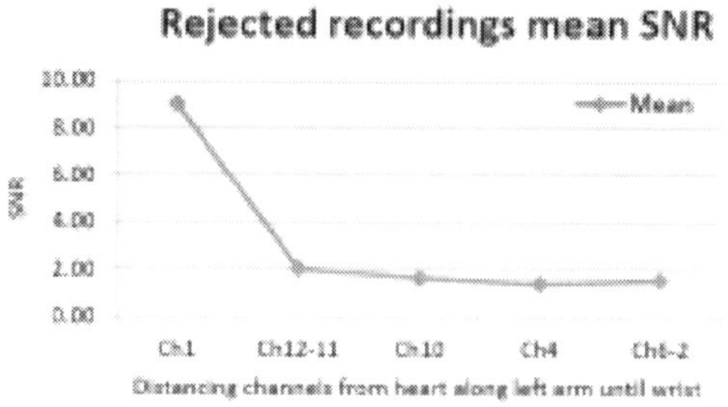

Figure 9. Rejected 5 recordings: plot of mean (N=5) SNR values along the left arm, as calculated in Table 3.

4. Discussion

Much work has been carried out on mapping the ECG across the torso. New sensors in wearable ECG monitoring system technologies [8] and available smart ECG denoising techniques, have renewed interest on the feasibility of acquiring the electrical activity of the heart from stand alone wearable devices placed along the arm or wrist of the patients requiring continuous long term monitoring of their rhythm. Currently, there is limited characterisation data about how the electrogram diminishes as it propagates along the arms.

The basic principle of the work carried out was to determine if signal levels of sufficient magnitude for real-time detection are present at the wrist and other 'comfort zones' along the left arm. Secondary to this, a map of signal level along the arm was created using subjects selected from a clinical environment. The information presented in this paper shows a variance in the ability to measure detectable levels of signal at the extremities of the arm. This in itself is promising.

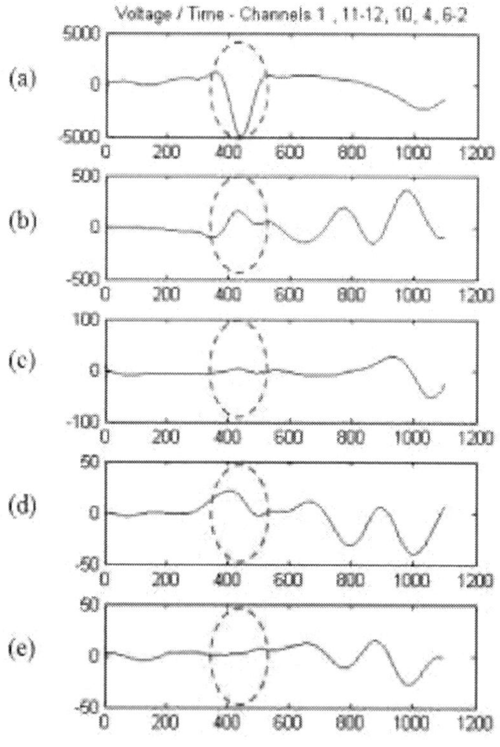

Figure 10. Subject 1 voltage-time frames after ECG denoising process (434 averaged beats).

Figure 10 shows Subject 1 voltage time waveforms from each of the 5 channels of interest. The waveforms are 537mS wide and centred on the QRS fiducial marker. This scale of window allows for the entire PQRSTU waveform to be included in the signal averaging process. The QRS complex has the highest magnitude and greatest rate of change and is most likely to be detected across the body model. It is also easier discriminated against a background of low frequency noise; therefore, it is the event of interest in this case. Using subject 1 as an example - channel 6-2 has the worst SNR of 0.47. It can be seen in Fig. 10 (e) that once the SNR drops below 1 it is not possible to visually resolve a ventricular event. Once the SNR increases above 1 the QRS inflection is clearly visible.

5. Conclusions

This pilot study has set out to investigate the signal strengths of the ECG as it propagates along the arm. The mapped left arm surface potential measurements of ECG synchronised signal levels obtained, after a conventional signal averaging denoising process, and their comparison with the remaining background noise level, provide objective and quantitative evidence of expected electrogram activity signal strength distribution along the arm to the wrist. The results presented in the various tables and graphs, were reasonably in line with the theoretical expectations of bipolar ECG signals rapid attenuation as the bipolar electrode segment distance from the heart increases. A simple dipole model of electric field propagation would suggest that field strength decrease squarely with distance. The exponential decay curve shape of the average (N=32) SNR data of bipolar channels along the arm in clinical cases approximates this model. Nevertheless, the reality is largely more complex as each person's unique human body impedance characteristics and other model parameter, affect the path of the electromagnetic field propagation.

For any bipolar ECG channel data to be of use for cardiac rhythm monitoringthe SNR must be at least >1, after an adequate ECG denoising process. The data provided in this pilot study in a cardiac clinical group of 37 cases, reveals the broad conclusion that 88% of the 32 recordings that were deemed acceptable, had a SNR greater than 1 in the mathematically derived bipolar channel Ch12-11 (bicep). The distal bipolar channel Ch6-2, had about 48% of recordings with a SNR of greater than 1. Other distant bipolar channels near the wrist can be expected to present SNR figures less than 1.

6. References

[1] Heikki V, et al., 2001 Sudden death due to cardiac arrhythmias. *New England Journal of Medicine* **345** 1473-1482.

[2] Zimetbaum P, et al. 1998 Evaluation of patients with palpitations, *New England Journal of Medicine* **338** 1369.

[3] Mark, R. G. 1990 Biological Measurement: Electrical Characteristics of the heart, in Systems and Control Encyclopedia, Singh, M. G., (ed), Oxford, U.K.. Permagon Press, P400 -456

[4] McAdams, E, et al. 2010 *Wearable electronic systems: Applications to Medical Diagnostics/Monitoring* 14.

[5] Krahn AD, et al. 2004 Insertable loop recorder use for detection of intermittent arrhythmias. *Pacing and Clinical Electrophysiology* **27** 657 – 64.

[6] F. Gemperle, C. Kasabach, J. Stivoric, M. Bauer, R. Martin 1998 Design for Wearability, *ISWC Second International Symposium on Wearable Computers (ISWC'98)*, 116.

[7] Escalona OJ, Mendoza M, Villegas G, Navarro C 2011 Real-time system for high-resolution ECG diagnosis based on 3D late potential fractal dimension estimation. *Computers in Cardiology* **38** 789-792.

[8] McAdams ET, et al. 2011 Wearable Electronic systems, *In Wearable Monitoring Systems, Eds. Bonfiglio A. and DeRossi D. Springer Science and Business Media*, 179-204.

Optical sensor for heat conduction measurement in biological tissue

A Gutierrez-Arroyo, C Sanchez-Perez*, N Aleman-Garcia

Centro de Ciencias Aplicadas y Desarrollo Tecnológico, Universidad Nacional Autónoma de México, AP 70-186, 04510 México DF, México.

*E-mail: celia.sanchez@ccadet.unam.mx

Abstract. This paper presents the design of a heat flux sensor using an optical fiber system to measure heat conduction in biological tissues. This optoelectronic device is based on the photothermal beam deflection of a laser beam travelling in an acrylic slab this deflection is measured with a fiber optic angle sensor. We measure heat conduction in biological samples with high repeatability and sensitivity enough to detect differences in tissues from three chicken organs. This technique could provide important information of vital organ function as well as the detect modifications due to degenerative diseases or physical damage caused by medications or therapies.

1. Introduction

Biomedical science and practical medicine demand innovative and cutting edge techniques for early and accurate diagnosis of diseases. Assessing the state of a tissue may be carried out to characterize the heat conduction therein[1,2]. The thermal properties can determine the physical characteristics of biological tissues[3] and thus may be associated with morphological and compositional changes caused by various degenerative processes. This work is aimed to demonstrate the feasibility of this technique to be applied in the area of biophotonics for characterizing heat conduction in biological tissue using a device based on the photothermal deflection phenomenon in an optical integrated device offering an alternative technique to characterize tissue properties with respect to actual qualitative microscopic tests performed in clinical ambiances for identifying alterations in tissues.

2. Design and characterization of optical sensor

The proposed device is based on the phothermal deflection of a laser beam propagating in a slab of thermo-optical material, acrylic for instance, in which is applied a temperature gradient that causes the beam to be deflected to the coldest section of the slab. The deflection is a function of the temperature gradient so it is also a function of the heat applied to one of the block sides perpendicular to the axe of propagation. The deflecting angle is measured with an intensity modulated fiber optic angle sensor. This optoelectronic device quantifies the power change of a light beam caused by a decoupling of the modal profile in optical fibers, due to the deflection that arises when it travels through the thermo-optic material exposed to a temperature gradient. In this case the lateral deflection angle θ may be approximated as,

$$\theta = \frac{1}{n}\frac{dn}{dT}\int \nabla T(x,t)\cdot dz\,, \tag{1}$$

where n is the refractive index and dn/dT is the thermo-optical coefficient of the acrylic block respectively. ∇T is the temperature gradient given by:

$$\nabla T(x,t) = \frac{\partial Q}{\partial x} - \rho c\frac{\partial T}{\partial t}\,, \tag{2}$$

In this expression ρ is the density and c is the specific heat capacity of the block material.

The experimental setup of the heat flux sensor is shown in Figure 1. It is composed of a laser source at the wavelength of 1550 nm, an optical fiber for signal transmission at the input and output with a collimator at its butt (two identical collimator lens) to ensure negligible divergence angle along the optical path. Located between the collimating lens pair is a thermo-optic material, specifically a clear acrylic block, selected because of its high thermo-optic coefficient (*dn/dT*), easy manufacturing and low cost. The device uses a pair of plates to align the input and output sides of the block to be perpendicular to the beam path in order to achieve normal incidence and avoid light reflection. Light at the output is then collected by the output fiber and measured with a Si photodetector interfaced with a computer.

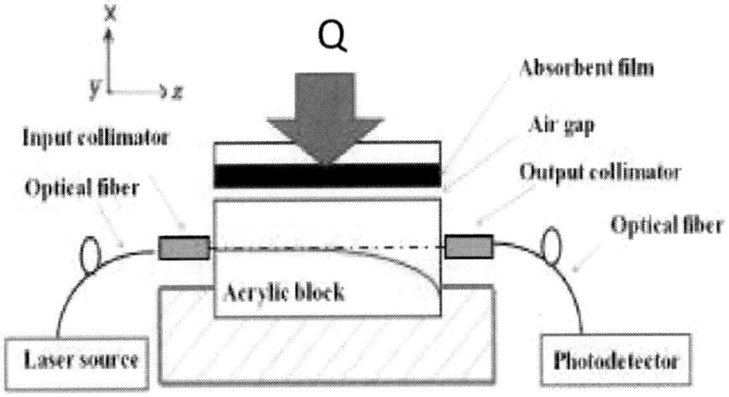

Figure 1. Heat flux optical sensor experimental setup.

Initially we characterize the optical sensor response applying a heat pulse of Q = 5.03 [mW/cm^2] during 410 [s]; for this, the heat source is turned on for 80 [s] and turned off at 490 [s]. Each measurement was registered until 1050 [s] elapsed to register the recovery of the signal during cooling. We consider a time before t = 20 [s] in order to monitor the stability of the system. Then we make a repeatability test with three monitoring of the sensor response under the same conditions, as we show in the Figure 2.

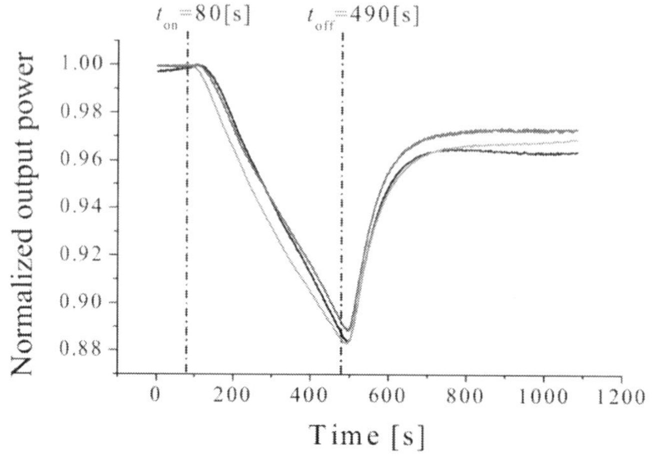

Figure 2. Repeatability test in sensor's calibration.

We made a calibration test monitoring the output power response for three different heat values applied as source Q_s,. In Figure 3, we show the experimental calibration curves of the sensing system, when the medium at the boundary acrylic block is an air gap in between. As in the previous test the heat source is turned on at time 10 [s] and turned off at time 420 [s] and registered until 1050 [s] elapsed.

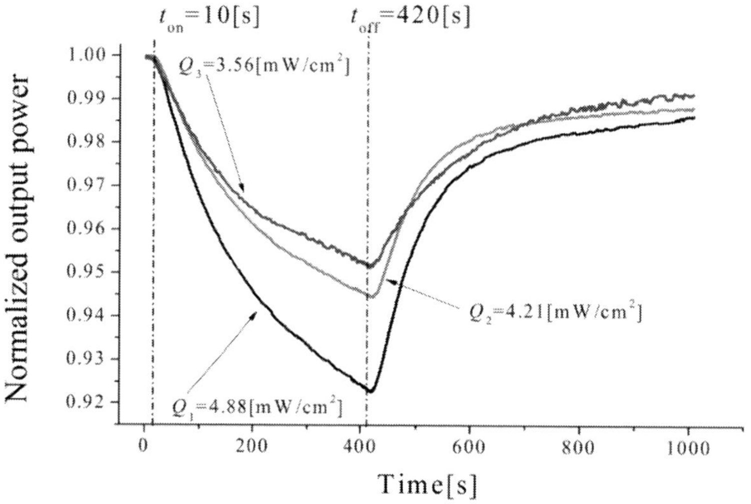

Figure 3. Optical sensor's response for three heat flow values.

In results shown in Figure 3, at the moment when the white light lamp is turned on, it causes the deflection of the beam as heat flows through the acrylic slab. The optical output power undergoes a progressive decrease even after the heat is turned off at time t_{off} = 420 [s], as a result of thermal inertia presented in acrylic material. Then, the beam begins to correct the skew angle and causes the gradual increase of the output power. Using these results to characterize the thermal source, we calculated the power decrease ΔP in the output signal by the heat flow Q_s applied by the thermal source, measured in the range of 10 [s] to 250[s] of curves in Figure 3. We plotted ΔP values versus Q_s shown in Figure 4:

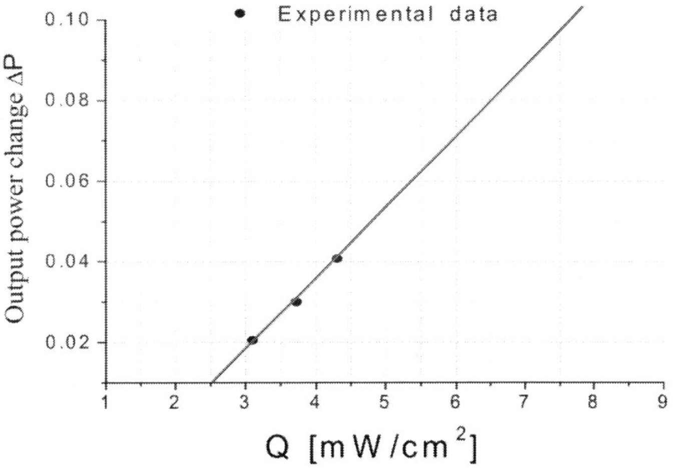

Figure 4. Output power difference as a function of heat flow applied.

3. Measurement of heat conduction in biological tissues

To demonstrate the sensitivity of our device to characterize thermal properties of biological tissues, we measure heat conduction in three different commercial chicken tissue (heart, liver, gizzard), applying a heat pulse of Q=4.221 [mW/cm^2] during 420 [s]. A thin tissue layer was placed between the heat source and the acrylic slab. The thickness of the tissue layer was about 1 mm. Results are shown in Figure 5.

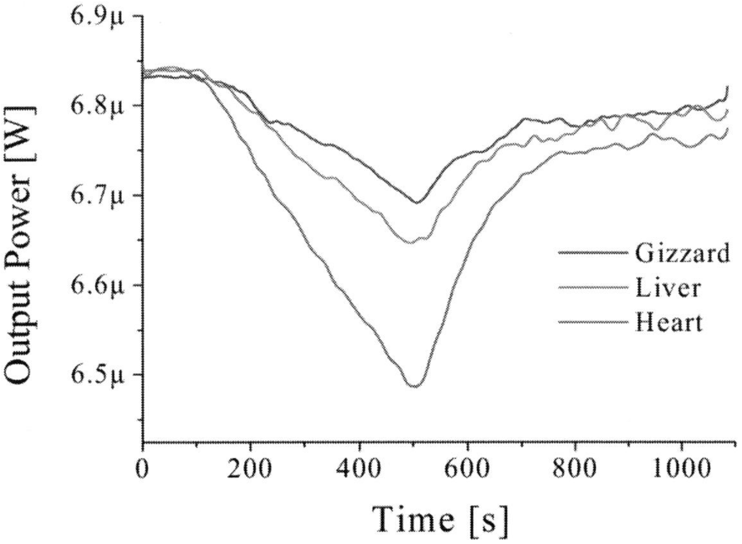

Figure 5. Heat conduction test in chicken tissues for a heat flux of Q=4.221 [mW/cm^2].

One can see in Figure 5 that the three tissues have different thermal conduction behaviour that is expressed as a different slope and power decrease. The chicken heart had the highest heat conduction rate while the liver had a medium rate and the gizzard had the lowest rate. We consider these variations could be caused by their composition and structure, it was also been reported the strong relation of water content with thermal conductivity[4]. The heat flux sensor´s has a good sensitivity to detect differences in thermal behaviour that characterizes each type of tissue once it has been exposed to a certain heat flux.

4. Discusion

Heat flow measurement technique proposed in this work operates with high sensitivity and resolution to detect differences in thermal properties in biological tissues through the measurement of heat conduction using an integrated optical beam deflection device. This device could identify structural or compositional changes in tissues dues to a degenerative disease or certain medications. The potential of this optical integrated device is to quantify from the heat conduction curves, with the aid of a heat diffusion numerical method thermal properties, as effusivity, diffusivity and thermal conductivity of biological tissues that could be used in clinical studies.

References

[1] Bhaavaraju N C, Cao H and Yuan D Y 2001 Measurement of directional thermal properties of biomaterials 2001 IEEE Trans. on Biomed. Engin. **48** 261
[2] A Yoshida, K Kagata and T Yamada Measurement of thermal effusivity of human skin using the photoacoustic method 2010 Int. Journal of Thermophys. **31** 2019.
[3] C Sanchez-Perez, A Gutierrez-Arroyo and N Aleman-Garcia Heat flux measurement in biological tissues by photothermal beam deflection 2012 AIP Conf. Proc. **1494** 62.
[4] R L Micintosh, V Anderson A comprehensive tissue properties database provided for the thermal assessment of a human at rest 2010 Biophys. Rev. and Lett. **5** 129.

Acknowledgments

Authors would like to thank Dr. Garcia-Valenzuela for valuable discussions. We acknowledge finantial support from DGAPA UNAM througth grant PAPIIT IT100612.

Journal of Physics: Conference Series **450** (2013) 012028 doi:10.1088/1742-6596/450/1/012028

Empowering smartphone users with sensor node for air quality measurement

Dinko Oletic and Vedran Bilas

University of Zagreb, Faculty of Electrical Engineering and Computing, Unska 3, Zagreb, HR-10000

E-mail: dinko.oletic@fer.hr

Abstract.
We present an architecture of a sensor node developed for use with smartphones for participatory sensing of air quality in urban environments. Our solution features inexpensive metal-oxide semiconductor gas sensors (MOX) for measurement of CO, O_3, NO_2 and VOC, along with sensors for ambient temperature and humidity. We focus on our design of sensor interface consisting of power-regulated heater temperature control, and the design of resistance sensing circuit. Accuracy of the sensor interface is characterized. Power consumption of the sensor node is analysed. Preliminary data obtained from the CO gas sensors in laboratory conditions and during the outdoor field-test is shown.

1. Introduction

Information on air quality in urban environments is typically generated by a limited number of stationary, accurate, but expensive, monitoring stations. In order to increase quality of information, higher spatial resolution of data can be obtained by extending the traditional air quality measurement infrastructure using a larger number of cheap wireless sensor nodes for environmental monitoring. Early designs featuring stationary sensor nodes [1, 2], were later extended into heterogeneous networks including mobile nodes mounted to various vehicles [3, 4]. Particularly interesting are concepts of *participatory sensing*, featuring citizens participating in context-aware collection of information. In such designs, sensors typically use smartphones for geo-tagging and relaying information into the web-service accessed by public [5–8].

Typically, concentrations of atmospheric gasses and pollutants, such as CO, CO_2, NO_2, O_3, SO_2, and volatile-organic compounds (VOC), are measured, along with temperature, humidity, barometric pressure and particulate matter. Concentrations of most of the mentioned gasses are measured either using passive electrochemical sensors, featuring lower power consumption, but higher prices and shorter life-time [7], or using longer-lasting and lower priced semiconductor metal-oxide (MOX) sensors [8, 9]. Requiring heating, minimization of their high power consumption remains an open research topic [10, 11].

In this article, we present an architecture of an air quality measurement sensor node featuring MOX gas concentration sensors and Bluetooth communication, designed for integration into smartphone-based participatory sensing system. We focus on sensor interface circuit design featuring a novel heater implementation, enabling further optimization of power consumption spent for heating.

2. Methods and materials

2.1. Sensor node architecture

We constructed a sensor node for ambient gas sensing, consisting of sensor interface circuitry, signal processing and communication board, and power-supply, as shown in Fig. 1.

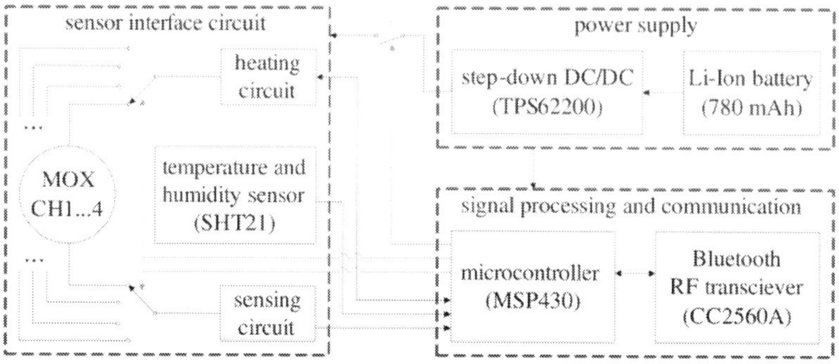

Figure 1. Sensor node architecture.

Sensor interface circuitry features SHT21 ambient temperature and humidity sensor, and a gas sensor interface containing heating and sensing circuit multiplexed by 4 independent MOX sensors. O_3 (MICS-2610), NO_2 (MICS-2710), CO/VOC (MICS-5121), and CO (MICS-5525) sensors are currently used [12].

Signal processing and communication board is adopted from Texas Instruments EZ430-256x development kit [13]. The board contains 16 bit MSP430 microcontroller running FreeRTOS. It controls the following tasks: (1) gas measurement sequence described in detail in Section 2.3, (2) communication, and (3) power management.

Communication is implemented using Ethermind Bluetooth 2.1 stack featuring Serial Port Profile (SPP). A simple request-response application-layer protocol has been developed on top of SPP to enable integration of the sensor node with the *AirSense* smartphone application for participatory sensing [14]. Physical layer is implemented by CC2560 Bluetooth radio transceiver.

Sensor node is powered from a single-cell Li-Ion battery through a 3.3 V step-down DC/DC switching-mode voltage regulator. Power supply of the sensor interface is controlled by the microcontroller. Radio wake-up functionality allows the microcontroller to spend its idle time in LPM3 low power mode, and to be turned on by radio transceiver only upon receipt of an incoming radio message.

2.2. Gas sensor interface

Repeatable temperature regulation of the MOX heater is required, as the sensitivity of the sensing layer depends on its temperature [10, 12]. Heater control can be achieved by regulating the heater's electrical power [11]. Provided that the relationship between temperature and electrical power is known [12], we designed the power-regulated heating circuit shown in Fig. 2.

Heating circuit's main component is a programmable DC current source. It is implemented using a TPS61165 step-up DC/DC switching-mode LED driver, sourcing I_h through heater R_h. Magnitude of I_h is adjustable by duty-cycle D of a PWM signal. Current range can be adjusted by the resistor R_{FB}, as defined by equation (1).

$$U_{FB} = 200mV \times D, I_h = U_{FB}/R_{FB} \tag{1}$$

$$R_s = R_{ref} \times (U_s/U_{ref} - 1) \tag{2}$$

Sensing circuit is implemented using non-inverting operational amplifier. U_s and U_{ref} are digitized using MSP430's on-board 12-bit A/D converter. From there, sensing-layer resistance R_s is calculated according to (2). Expected R_s are in the range of 1 to several 100-s of kΩ [12]. Measurement range is adjusted by programming R_{ref} and U_{ref}.

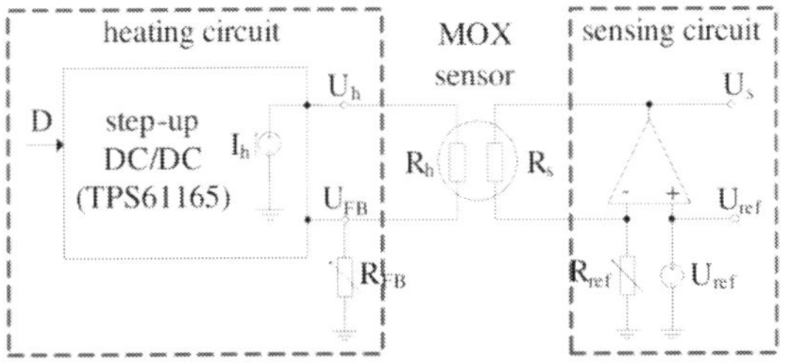

Figure 2. Heating and sensing circuitry.

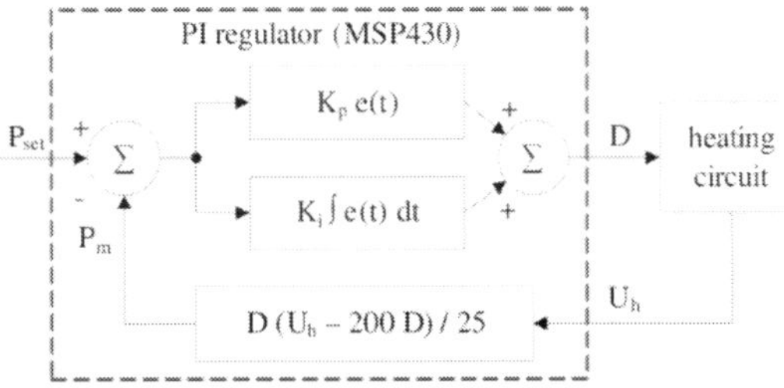

Figure 3. Heating power PI regulator.

2.3. Signal acquisition and processing

Gas concentration measurement sequence consists of heating by the arbitrary heating power function $P_{set}(t)$ and measurement of the sensing layer resistance R_s, which is then related to the gas concentration.

Regulation of heating power to a set-point $P_{set}(t)$ is implemented with PI regulator shown in Fig. 3. Required heating power range is 40 to 100 mW. Actuated variable is duty-cycle (D) of a PWM signal, used for programming the current I_h, as defined by (1). In the feedback loop, high-side voltage of the heater U_h is measured, and the low-side voltage U_{FB} is approximated from (1). From there an error signal is calculated, as defined by (3). PI regulator is implemented in the software. Frequency of the regulation loop is set to $f_{reg} = 50$ Hz. Empirically chosen coefficients of $K_p = 5 * 10^{-4}$ and $K_i = 1 * 10^{-3}$ have proved suitable to ensure stability of the regulator in the required heating power range.

$$P_{set} - P_m = D \times (U_h - 200 \times D)/25 \tag{3}$$

Repeatability of the temperature settle-point during heating is ensured by correcting the heating power by its ambient temperature equivalent. Temperature is measured using SHT21S temperature and humidity sensor. Humidity compensation can also be performed, but is claimed to have less effect on sensor sensitivity [15]. Drift of the sensing layer resistance after prolonged period of sensor shut-down is minimized by keeping the sensor layer free of contaminants which may have deposited [15]. We address the issue by including an interval of overheating prior to the heating. Hysteresis of the heating characteristic exhibited during measurement repetition is addressed by introducing a cool-down interval after every measurement sequence.

3. Testing

3.1. Sensor interface accuracy

Verification of the heater power regulator design was performed by varying heater resistance in the range of $R_h = 40...100\ \Omega$, and by varying heating power set-point in the range of $P_h = 40...100$ mW. Mean relative error as a measure of accuracy, and standard deviation as a measure of repeatability of power regulation, were evaluated for every (P_h, R_h) pair, each iterated 10 times. Also, mean relative error and standard deviation of resistance readout R_s were calculated from time-series of samples of U_s and U_{ref}. Samples were obtained at sampling frequency of f_{reg}, for $R_s = 1...150$ kΩ, at two settings of $R_{ref} = \{43, 100\}$ kΩ, and with fixed $U_{ref} = 0.55$ V.

IOP Publishing
doi:10.1088/1742-6596/450/1/012028

3.2. Power consumption
Durations of stationary operating states and their associated mean current consumption at $V_{bat} = 4.0$ V were measured for MSP430 microcontroller, CC2560 radio chip, and sensor interface board. This enabled construction of power state-machines used for further power consumption modelling.

Autonomy was verified by a battery discharge test. Sensor was powered from a single cell Li-Ion 3.6 V, 780 mAh, charged to 4.15 V. Discharge included alternating periods of sleep and periods of polling the sensor from the Bluetooth peer-device with average frequency of 1 / min. Each polling message caused radio wake-up of microcontroller from LPM3 to the active state, and single measurement sequence after which sensor node returned to sleep.

Measurement sequence consisted of 5 s of overheating by maximum power P_{max}, followed by 10 s of heating by nominal power P_{nom} and 1 s of cool-down. Sensing layer resistance was obtained by averaging the R_s sampled by f_{reg} during last 2 s to minimize deviations of R_s. Such measurement sequence was used in all subsequent tests.

3.3. Validation using gas sensors
Preliminary tests with controlled gas concentrations were performed for CO sensors MICS-5121 and MICS-5525, in order to assess relationship of sensing layers' resistance versus: (1) atmospheric concentration of CO, and (2) 100 ppm of CO. Procedure was conducted using Dräger calibration gasses, at gas volume flow of 0.5 L/s, 23°C and 45 % relative humidity.

The same CO sensors were used in an outdoor test-run performed in Zagreb, Croatia, on 22nd of March, 2013. Sensor node was fastened to a sleeve of a jacket with a velcro strap, and carried along a 5 km route around the town centre. Concentration of CO, along with ambient temperatures and humidities were measured with temporal frequency of 1 minute and sent to a smartphone *AirSense* application. GPS tags and timestamps were added by the application, and information was used by the publish/subscribe-based participatory sensing web-service.

4. Results
Lower power regulation error is exhibited at higher resistor load as lower current needs to be sourced by TPS61165 DC/DC, as verified by Fig. 4. Mean relative error of power regulation of < 0.5 % can be achieved at $R_h > 60\ \Omega$. Standard deviation is below 0.5 mW for the whole testing range of R_h and P_h. Mean value of relative error of resistance measurement of $< 5\%$ can be expected for $R_s > 20$ kΩ with $R_{ref} < 43$ kΩ. Upper boundary of measured R_s is defined by A/D converter's reference voltage. Standard deviation of resistance measurement error increases roughly linearly with measured R_s, being 1.2 kΩ at 20 kΩ, and 2 kΩ at 140 kΩ.

Simplified power state machine showing states contributing the most to sensor node power consumption is depicted in Fig. 6. Mean power consumption in active state is 141-192 mW, depending on the MOX heating power. Consumption in sleep, with microcontroller in LPM3 and radio in the mode enabling wake-up on incoming message, is 9 mW. Results of battery discharge test are depicted in Fig. 7, showing the autonomy of more than 72 hours before battery voltage dropped below 3.3 V. 28 hours of discharge included sensor polling, and 44 were spent in sleep.

Examples of measurements of R_s logged throughout the whole heating interval during the test of CO sensors are shown in Fig. 5. Resistances at atmospheric concentration are $R_0 =$ 279 kΩ for MICS-5121 and 129 kΩ for MICS-5525. Sensitivities to 100 ppm of CO, expressed analogously to data in [12], are $S_{100} = R_{100ppm}/R_0 = 5.94$ and 1.96 for MICS-5121 and MICS-5525, respectively. Drift of atmospheric R_0 after longer period of non-usage is observed.

Assembled sensor node, with battery pack and Samsung Galaxy Mini 2 running AirSense application is shown in Fig 8. Results of outdoor test as seen through web-interface are shown in Fig. 9. Dots indicate GPS locations where sensor data was acquired. Details can be visualized by clicking the dots [14].

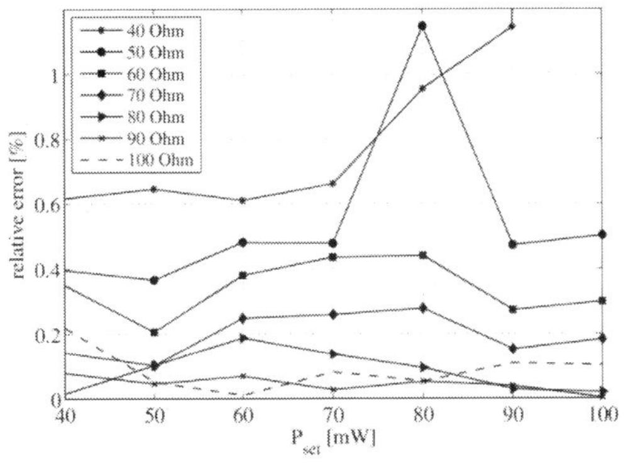

Figure 4. Power regulation error in dependency of heating power and heater's resistance.

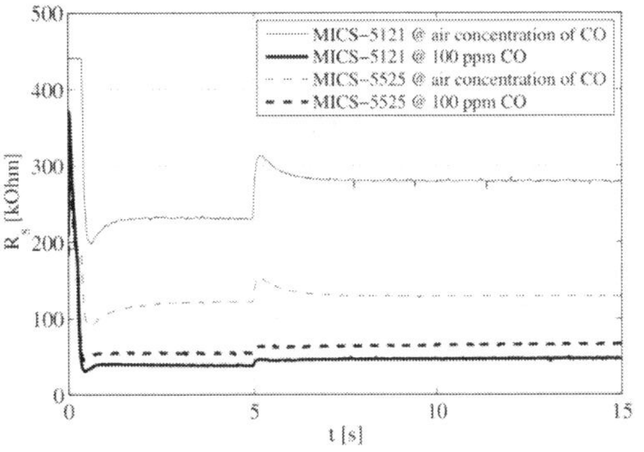

Figure 5. Examples of sensor resistance responses obtained from CO sensors.

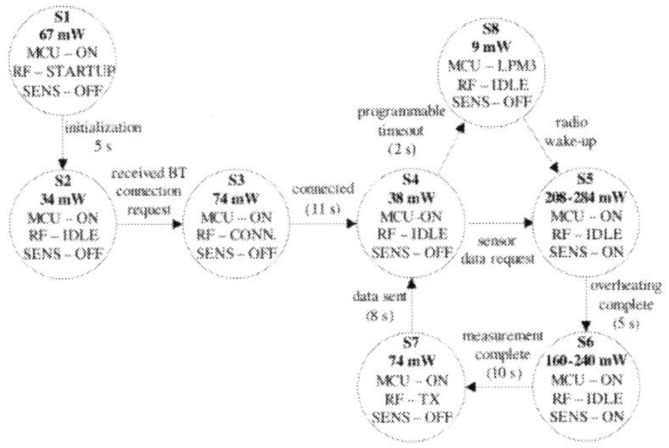

Figure 6. Operating states of a sensor node with associated power consumptions.

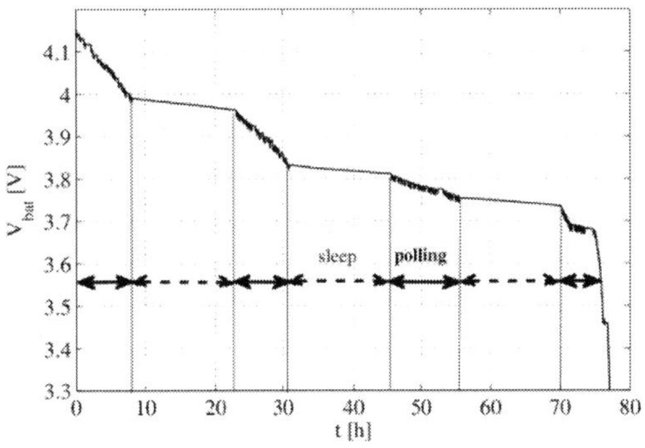

Figure 7. Battery discharge test.

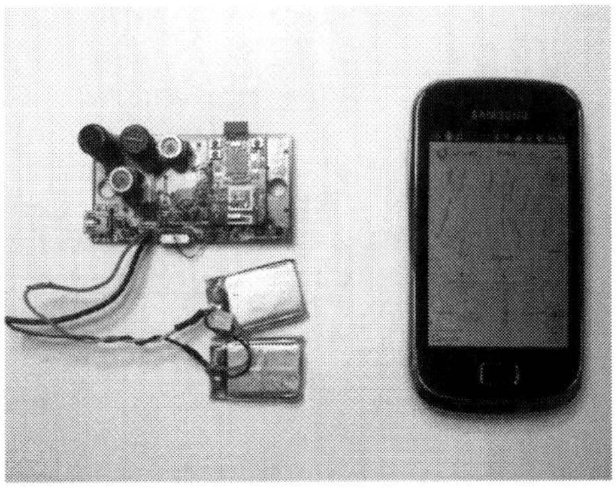

Figure 8. Sensor node with battery. Dimensions of PCB are 35 x 68 mm.

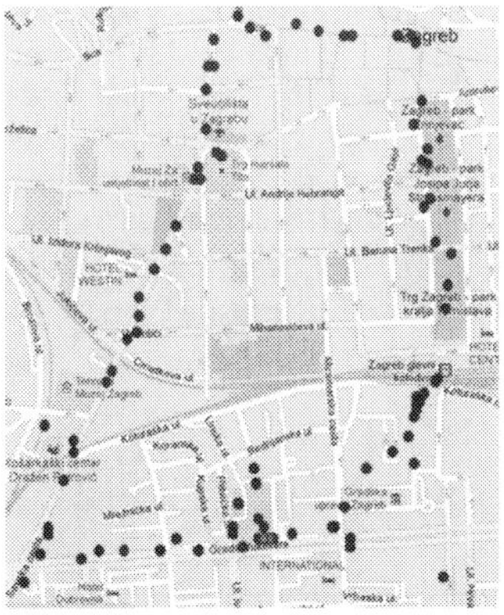

Figure 9. Field-test results, shown in AirSense web-interface [14].

5. Conclusion

In this article we have shown our work-in-progress on the sensor node for gas concentration measurement, suitable for integration with smartphone-based participatory sensing of air quality. We have proposed a heating circuitry design featuring power regulator consisting of programmable current source and software-implemented PI regulator. Results have shown that design enables accurate and repeatable setting of heater power (e.g. temperature).

This allows us to further explore advanced measurement techniques featuring arbitrary heating power functions $P_{set}(t)$ to optimize total heating energy in order to prolong the sensor autonomy. Also, in future work we plan additional interventions into the design of the resistance sensing circuitry to maximize accuracy throughout the entire range of sensor response R_s, exhibiting high-dynamics and non-linearity. Also, drift of R_s when sensor is exposed to prolonged periods of non-activity, temperature compensation, and calibration are to be addressed in detail.

Acknowledgments

Authors would like to thank Josip Biscan for participation in development of the sensor interface, Dario Antolovic for help with testing, Marijan Gorse from Dräger for sensor calibration equipment, and students of professor Ivana Podnar-Zarko at Department of Telecommunications who developed the *Airsense* smartphone application and associated web-services used in the field test.

References

[1] Tsujita W, Yoshino A, Ishida H and Moriizumi T 2005 *Sensors and Actuators B: Chemical* **110** 304–311
[2] Bers J, Gosain A, Rose I and Welsh M 2008 *JJ Proceedings of the 2008 IEEE International Conference on Technologies for Homeland Security, Waltham, MA* (Citeseer)
[3] Dutta P, Aoki P M, Kumar N, Mainwaring A, Myers C, Willett W and Woodruff A 2009 *Proceedings of the 7th ACM conference on embedded networked sensor systems* (ACM) pp 349–350
[4] Elen B, Peters J, Poppel M V, Bleux N, Theunis J, Reggente M and Standaert A 2012 *Sensors* **13** 221–240
[5] Honicky R, Brewer E A, Paulos E and White R 2008 *Proceedings of the second ACM SIGCOMM workshop on Networked systems for developing regions* (ACM) pp 25–30
[6] Miyaki T and Rekimoto J 2008 *Ubicomp 2008 Workshop Programs* pp 187–190
[7] Zappi P, Bales E, Park J H, Griswold W and Rosing T Š 2012
[8] Hasenfratz D, Saukh O, Sturzenegger S and Thiele L 2012 *Proc. 1st Intl Workshop on Mobile Sensing: From Smartphones and Wearables to Big Data*
[9] Choi S, Kim N, Cha H and Ha R 2009 *Sensors* **9** 7970–7987
[10] Lee A P and Reedy B J 1999 *Sensors and Actuators B: Chemical* **60** 35–42
[11] Depari A, Flammini A, Marioli D, Sisinni E, Comini E and Ponzoni A 2012 *Instrumentation and Measurement, IEEE Transactions on* **61** 2374–2383
[12] 2013 SGX Sensortech Sensors Datasheets URL `http://www.sgxsensortech.com/site/sgx-sensortech-products-services/gas-sensors-environmental-monitoring/sensor-selector/`
[13] 2013 EZ430-RF2560 Wiki URL `http://processors.wiki.ti.com/index.php/EZ430-RF256x`
[14] 2013 Osjeti ZG zrak, project website URL `http://croco4.tel.fer.hr/osjetizg/`
[15] 2009 MICS Application Note 3: Using MICS-5525 for CO Detection URL `http://www.sgxsensortech.com/site/wp-content/uploads/2012/10/AN3-Using-MiCS-5525-for-CO-Detection.pdf`

Sensors & their Applications XVII IOP Publishing
Journal of Physics: Conference Series **450** (2013) 012029 doi:10.1088/1742-6596/450/1/012029

Pilot study: Assessing repeatability of the EcoWalk platform resistive pressure sensors to measure plantar pressure during barefoot standing

Martha Zequera, Oscar Perdomo, Carlos Wilches and Pedro Vizcaya
Pontificia Universidad Javeriana, Departament of Electronics, BASPI Research Group - Diabetic Foot, Carrera 7 No. 40 - 62, Bogotá, Colombia

E-mail: mzequera@javeriana.edu.co

Abstract.
Plantar pressure provides useful information to assess the feet's condition. These systems have emerged as popular tools in clinical environment. These systems present errors and no compensation information is presented by the manufacturer, leading to uncertainty in the measurements. Ten healthy subjects, 5 females and 5 males, were recruited. Lateral load distribution, antero-posterior load distribution, average pressure, contact area, and force were recorded. The aims of this study were to assess repeatability of the EcoWalk system and identify the range of pressure values observed in the normal foot. The coefficient of repeatability was less than 4% for all parameters considered.

1. Introduction

The foot is one of the most unique and complex biomechanical structures in the nature [1]. Plantar pressure measurements provide useful information to know the normal condition or disorders of the feet [2]. Several studies were performed on healthy subjects, looking for repeatability, parameters and mathematical relationship between subjects depending age, size, weight, height and BMI [3, 4, 5, 6]. The plantar pressure measurement systems have emerged as popular tools in clinical, ergonomics, and academic research environments.

Unfortunately, the commercially available measurement systems have objectionably sensitive to several disturbances, but this aspect is not reported factually in the literature. Some metrological performances of plantar pressure measurement systems showed the presence of bias and random errors without calibration information or schemes for error compensation by the manufacturer, which leads to uncertainties in the measurements, carried with these plantar pressure measurement systems [7].

EcoWalk system is a relatively new product among the platform pressure measuring devices. Repeatability is one of the elements which may define such dependability [8]. Clinical personal need good repeatability to ensure the best diagnose based their judgments on consistent measures [9]. Therefore, the difference ranges between of normal in-shoe pressure values in a healthy and unhealthy people are needed to identify abnormalities.

The aims of this study are to assess the repeatability of the EcoWalk system and identify the range of pressure values observed in the normal foot.

2. Materials and methods

2.1. Subjects

Ten healthy volunteers were chosen for this study, 5 women and 5 men), aged 18 - 27 years and body mass index below 30. Subjects were excluded if they presented musculoskeletal injuries or biomechanical alterations, which may affect measurements. Ethical approval was granted by the Ethics Committee for Medical Research of the San Ignacio Hospital. All subjects participating in the study signed a written informed consent.

2.2. Equipment

The measurement system consisted on 3 components: the pressure platform EcoWalk (Ecosanit, Anghiari, Arezzo, Italy), the software EcoFoot 4.0, a support, a digital weight scale, and a reference frame.

The pressure platform EcoWalk is composed by 2304 resistive sensors, arranged in an array of 48×48 sensors (1 sensor/cm^2), and a sample rate of 30 Hz. EcoFoot 4.0 was used to measure: lateral load distribution (LLD), antero-posterior load distribution (APLD), average pressure (AP), contact area (PA), and force (F).

The support was given to help the subjects to keep their balance during the measurement, but without affecting seriously the pressure measured. The purpose of the reference frame is to maintain the feet in an anatomical position during measurements. Previous studies revealed that a reference frame allows greater repeatability of pressure measurements [10].

2.3. Protocol

For recruitment, a doctor performed a comprehensive physical exam of the participants, assessing the state of their feet, and selecting those who didn't present musculoskeletal injuries or biomechanical alterations. The subjects approved were asked to be barefoot, with their pockets empty, jewelry and watches removed, and wearing comfortable clothes, in order to avoid changes in the measurement.

Each subject were measured 30 times with a technique for controlling the position of their feet, using the reference frame. The technique consisted on position the feet using the frame, and removing it before measuring the plantar pressure, to avoid noise caused by the presence of the frame on the platform.

The measurement process was supervised by the research group staff, who guided the accommodation of the feet of the subject and body position.

2.4. Data measured

The pressure platform EcoWalk measures plantar pressure distribution, capturing the pressure at each sensor and calculating the different variables. LLD indicates the percentage of load on each foot; APLD is percentage of load in rearfoot and forefoot; AP is the average pressure on each foot; CA are the count of all sensors activated on each foot, since the sensor area is 1cm^2; and FI is the force calculated from the sum of pressures on each sensor on each foot (1):

$$F = \sum P_{\text{sensor}} \cdot CA \tag{1}$$

2.5. Data processing

After measuring the 10 subjects, each measurement (300 measurements in total) were processed to calculate CA and F, while LLD, APLD, AP were given by the software. On each variable, an statistical analysis was carried out in order to assess its performance and the least variability within-subjects, was selected.

The variable selected was analyzed in order to select the most significant measurements from each subject.

2.6. Statistical analysis
An ANOVA statistical analysis was performed using the software SPSS, mean, median, mode, variance and standard deviation, with independent variables being gender, foot, and position (anterior or posterior).

3. Results
The set of signals obtained during the study in ten healthy subjects were analyzed using parametric statistical and nonparametric methods with MATLAB software. A significant difference was found within-subject comparison between different parameters under study.

The statistical analysis showed a significant response between subjects with different gender and type of foot analyzed, which is an expected response according to the literature.

In Figure 1 is shown the force variable in the ten subjects studied, the values presented an abnormal distribution in within-subject and between-subject. A larger variance and a tendency to a non-normal data, it can be modeling and represented such as a skewed distribution.

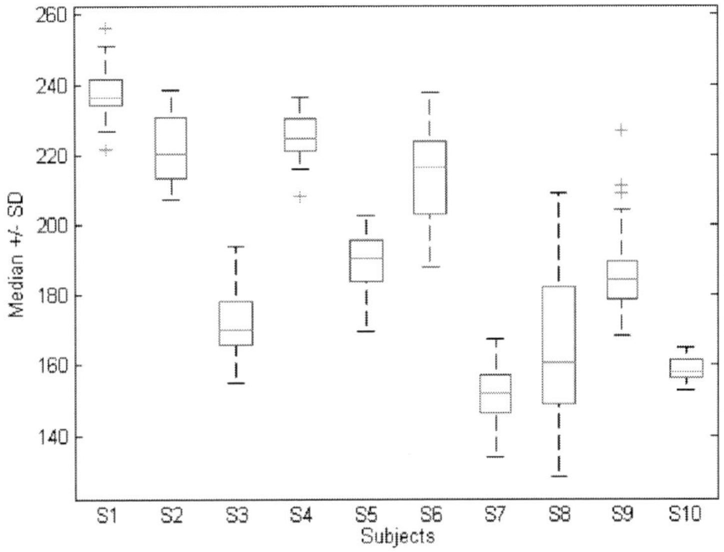

Figure 1. Subjects vs Force

In Figure 2 is shown the contact area variable in the ten subjects. It was observed that there is less variance, which leads to a more consistent data within a smaller range. Statistical tests showed that the distribution was normal because the mean, median and mode were approximately equal.

4. Discussion
The repeatability of the EcoWalk system was assessed in ten healthy subjects. It was identified the range of pressure values and contact area for different age, size, weight, height and BMI. The coefficient of repeatability (CR) for the contact area parameter was less than 4% for all the ten subjects.

According to [3, 4, 5], the repeatability can be estimated through mathematical and statistical analysis of the parameters involved in the measurements of the pressure plantar system. The contact area can be an excellent parameter for helping objectively to choose the most representative sample in a set of measurements.

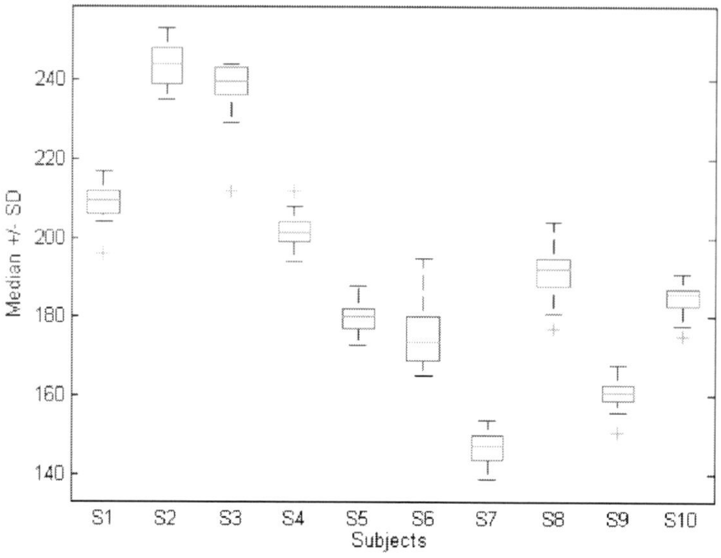

Figure 2. Subjects vs Contact Area

The needs of a metrological performance of the available measurement systems by manufactures have to be reported. The presence of bias and random errors with an appropriated calibration information or schemes for error compensation lead to a totally knowledge of the uncertainty allowing accuracy in the measurements.

Acknowledgments

We sincerely thank Natalia Orjuela and Nicole Ramírez, students from BASPI Research Group - Diabetic Foot Project, for their efforts during the development of this study, and the Department of Orthopedics at Hospital San Ignacio (Bogotá, Colombia), for their help examining and selecting suitable subjects for this study. We are grateful to the Pontificia Universidad Javeriana for having funded the study.

References

[1] Ramanathan A, Kiran P, Arnold G, Wang W and Abboud R 2010 *Foot and Ankle Surgery* **16** 70–73
[2] Saggin B, Scaccabarozzi D and Tarabini M 2013 *IEEE Transactions in Instrumentation and Measurement* **62** 766–776
[3] Zhu H, Wertsch J, Harris G F, Price M B, and Alba H M 1989 *21st IEEE Annu. Int. Conf. Images Eng. Med. Biol. Soc.* **3** 822–823
[4] Bennett P J and Duplock L R 1993 *J. Amer. Podiatr. Med. Assoc.* **83** 674–678
[5] Che H, Nigg B M and de Koning J 1994 *Clin. Biomech* **9** 335–341
[6] Jordan C and Bartlett R 1995 *Gait and Posture* **3** 215–220
[7] Hsiao H, Guan J and Weatherly M 2002 *Ergonomics* **45** 537–555
[8] Putti A, Arnold G, Cochrane L and Abboud R 2008 *Gait and Posture* **27** 501–505
[9] Putti A, Arnold G, Cochrane L and Abboud R 2007 *Gait and Posture* **25** 401–405
[10] Zequera M, Garavito L, Sandham W, Alvarado J A, Rodríguez A, Wilches C A, Villa A C, Bernal J and Quintero S 2011 *5th European IFMBE Conference, IFMBE Proceedings* **37** 830–833

Improvements to Zirconia Thick-Film Oxygen Sensors♣

William C. Maskell[1, 2, ♦], Daniel J. L. Brett[2] and Nigel P. Brandon[1]

[1] *Energy Futures Laboratory, Department of Earth Science and Engineering, Imperial College London, London SW7 2AZ.*
[2] *Electrochemical Innovation Laboratory, Department of Chemical Engineering, University College London, London WC1E 6BT.*
♣ *The work was carried out at Sensox Ltd, a company which subsequently went into liquidation.*
♦ *Corresponding author, e-mail: w.maskell@imperial.ac.uk*

Abstract

Thick-film zirconia gas sensors are normally screen-printed onto a planar substrate. A sandwich of electrode-electrolyte-electrode is fired at a temperature sufficient to instigate sintering of the zirconia electrolyte. The resulting porous zirconia film acts as both the electrolyte and as the diffusion barrier through which oxygen diffuses. The high sintering temperature results in de-activation of the electrodes so that sensors must be operated at around 800 °C for measurements in the percentage range of oxygen concentration. This work shows that the use of cobalt oxide as a sintering aid allows reduction of the sensor operating temperature by 100-200 °C with clear benefits. Furthermore, an interesting and new technique is presented for the investigation of the influence of dopants and of the through-porosity of ionically-conducting materials.

1. Introduction

Oxygen gas sensors have many applications including, and in particular, the monitoring and control of combustion processes. The incorporation of oxygen sensors in the latter systems can enable the air-to-fuel ratio to be maintained at an optimum value with respect to both efficiency and emissions. Domestic systems are cost-sensitive and require low-cost sensors. Screen-printed thick-film sensors [1] do offer the prospect of low cost as they can be produced in substantial numbers per substrate [2]. The high temperature necessary for sintering the zirconia results in deactivation of the electrodes due to grain growth of the platinum metal particles in the electrodes. Deactivated electrodes, combined with insufficiently low through-porosity, necessitate high sensor operating temperatures in order to support sufficient currents to drive the sensor into the diffusion-limited region. Power input rises with increasing operating temperature but also operating life is expected to be reduced. It is shown that the use of cobalt oxide as a sintering promoter reduces the through-porosity of the zirconia for given firing conditions; this allows substantial decrease of the sensor operating temperature without adversely affecting electrochemical performance.

Various studies have shown the importance of matching of thermal expansion coefficient (CTE) of thick-film and substrate. Alumina substrates, which are widely used in the electronics industry, have a CTE of ~7×10^{-6} °C^{-1} while zirconia displays a value of 10-11×10^{-6} °C^{-1}. Hence, upon cooling from the sintering temperature of 1300-1400 °C the zirconia film is prone to crack [1-3]. In this work a substrate was used consisting of a mix of magnesia (MgO) and the spinel, $MgAl_2O_4$. By choosing an appropriate mix proportion the CTE can be selected between 9-14×10^{-6} °C^{-1}.

2. Experimental

2.1 General

All preparative work was done under clean conditions in a Class 100 enclosure. Substrates of MgO-$MgAl_2O_4$ with a CTE to match zirconia, manufactured and supplied by Advanced Ceramics of Stafford, UK, were used throughout.

2.2 Printing, Doping and Firing of Films

A DEK 1202 screen printer was used for all printing operations.
Three types of samples were prepared:

1. A platinum electrode was printed and dried then over-printed with a zirconia electrolyte film and dried.
2. As above then overprinted with a second electrode and dried.
3. A zirconia electrolyte film only was printed and dried with no platinum electrodes.

These samples were then either

A. fired without doping or
B. doped with cobalt (nitrate) and then fired (the cobalt nitrate decomposed to the oxide during the firing process).

In order to dope the films a solution of cobalt nitrate hexahydrate in ethanol (50:50 by weight) was prepared (reagents from Sigma- Aldrich, Analytical Reagent grade) which was then applied to the printed layer with a brush and dried prior to firing to provide a level of cobalt doping in the zirconia of approximately 5 cation %. Films were fired in air using a muffle furnace with a maximum operating temperature of 1600 °C.

2.3 Characterisation of Films

Electrodes of the Type 2 films were electrically connected to a voltage supply and the current flowing in the circuit measured. The sensor was operated in a furnace in a mixture of air and nitrogen, using the experimental arrangement described previously [4], gas flow rates being controlled by mass-flow valves. Substrates were scribed on the reverse side to the film, snapped and the fracture edges of Types 1 and 3 films examined using a scanning electron microscope (Cambridge Instruments 240 Stereoscan operated at 30kV).

3. Results and Discussion

3.1 Micrographs

Types 3A, 3B 1A and 1B zirconia films fired at 1000°C showed no significant differences as this was below the temperature at which significant sintering occurred. The zirconia thick-film (Fig. 1) shows individual particles which remain essentially separate (unsintered) after firing at 1000°C.

Fig. 1 Cross-section of thick-film (◇1) on a substrate (◇2). The thick-film is zirconia doped with cobalt and the sample was fired at 1000 °C for 10 minutes. A similar micrograph (not shown) without cobalt-doping, fired at the same temperature, indicated no significant difference to this.

Figures 2 and 3 show the structure of the zirconia thick films fired at 1300°C without and with cobalt doping respectively. It can be seen that sintering (joining together at the contact points of the particles, shrinking of small particles, grain growth of larger particles and reduction of porosity) had taken place in both cases but that this had occurred to a greater extent where cobalt was present. The zirconia films in Type 2 and 3 samples showed no difference in appearance indicating that the interposition of a platinum thick-film between the substrate and the zirconia did not modify the sintering behaviour of the ceramic film. The micrographs confirm, as has been reported elsewhere [5], though not with respect to thick films, that cobalt acts as a sintering promoter for zirconia.

3.2 Current-voltage characteristics

As the voltage applied to an amperometric oxygen sensor (positive to front electrode, negative to electrode in contact with the substrate) in an oxygen-containing gas is increased from zero, the current rises and then reaches a plateau when the current becomes controlled and limited by the rate of diffusion of oxygen through the diffusion barrier. In the case of the sensors under discussion here, the diffusion barrier is the porous electrolyte. Where diffusion in the pores is normal bulk diffusion, the limiting current in an amperometric sensor is essentially proportional to the oxygen concentration up to about 10%, showing progressive

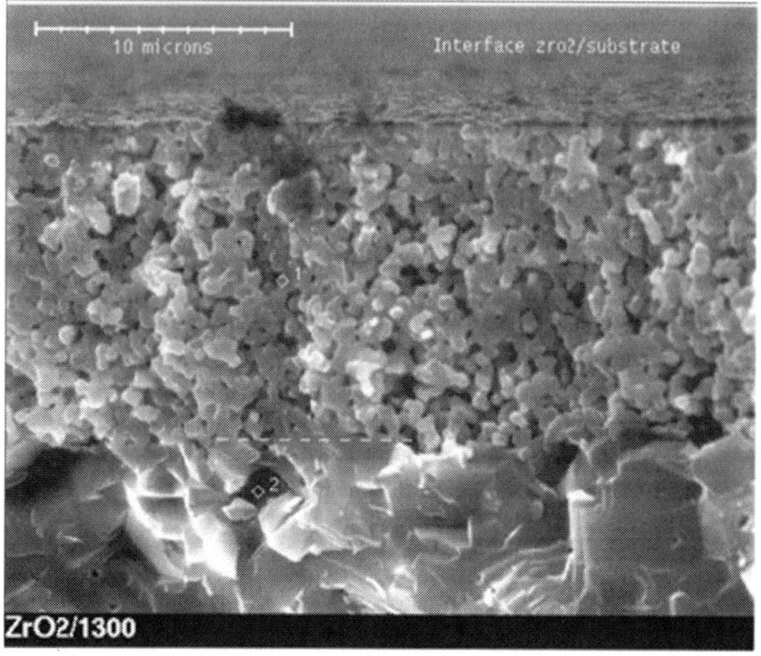

Fig. 2 Cross-section of thick-film (◇1) on a substrate (◇2) with no intervening layer of platinum. The interface between the thick-film and the substrate is shown by the broken line. The thick-film is zirconia, there was no doping with cobalt and the sample had been fired at 1300 °C for 1 hour.

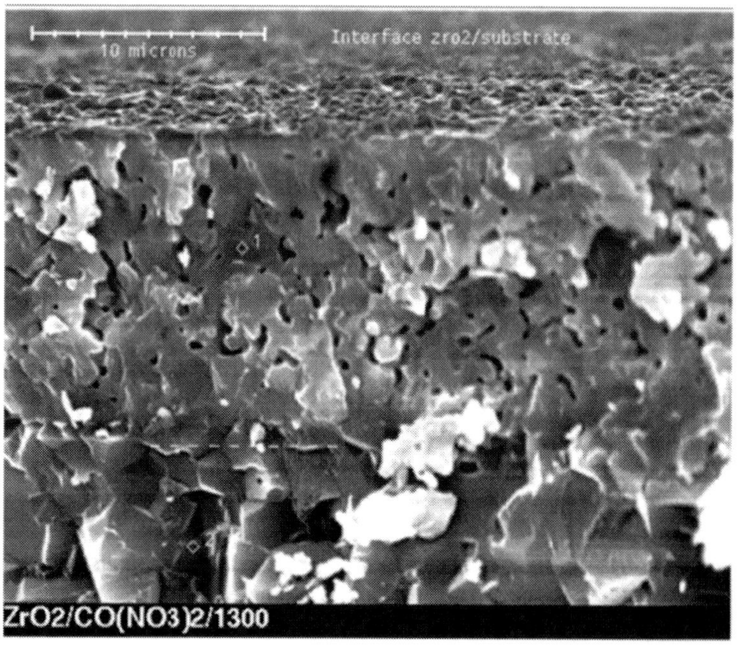

Fig. 3 Cross-section of thick-film (◇1) on a substrate (◇2) with no intervening layer of platinum. The interface between the thick-film and the substrate is shown by the broken line. The thick-film is zirconia, doped with cobalt and fired at 1300 °C for 1 hour.

deviation from linearity towards 21% (the concentration in air) and above. This arises because the differential equation describing transport through the film contains a diffusion term and a flow term. The latter term becomes progressively more important as the oxygen concentration is raised [6]. Where diffusion in the pores is governed by Knudsen diffusion it is less clear whether the flow term should be included; this requires further consideration.

It was found that sensors made without cobalt addition required an operating temperature of 800 °C in order to achieve diffusion limitation in the percentage range of oxygen. Where doping with cobalt was used, this was reduced to 700 °C in air and even to 600 °C in a gas containing 5% oxygen. The marked effect of cobalt addition on the electrochemical (current-voltage) characteristics is shown in Fig. 4 and 5. Substantial reductions in limiting currents are shown to result as a consequence of cobalt addition. So the reduction in operating temperature by use of the dopant was 100-200 °C which is remarkable and provides the important benefits of reduced operating power and expected increase in the operational life of the sensor.

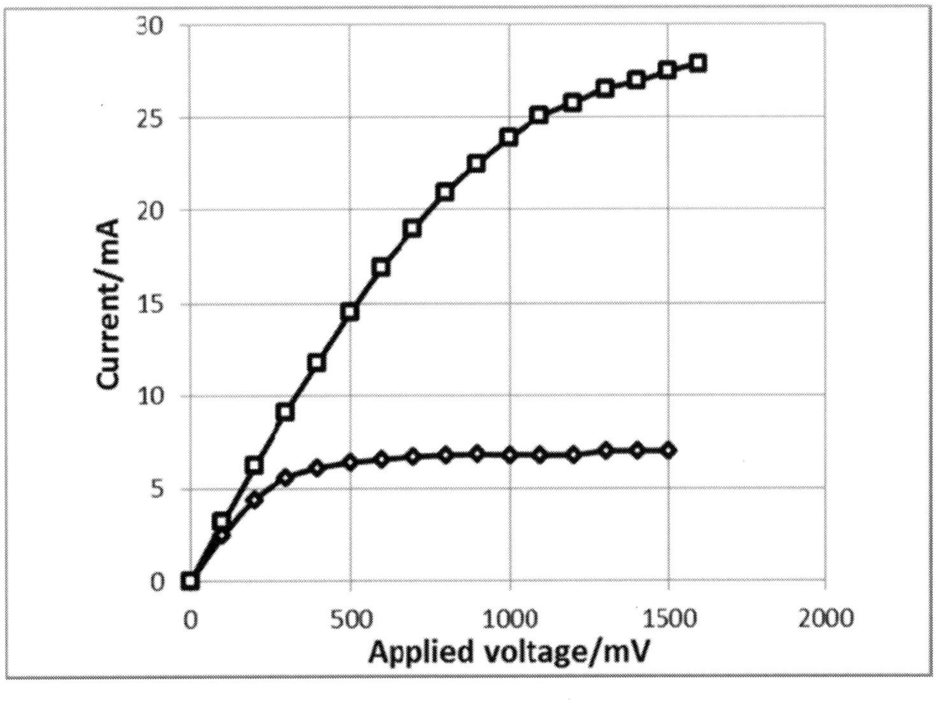

Fig. 4 Current-voltage curves for sensor fired at 1400°C for 2 hours without cobalt addition. Sensor operated at 800°C in (□) 21% oxygen, (◊) 5% oxygen.

3.3 Other aspects

The device (sensor) structure and associated techniques provide interesting and new tools for the preparation and characterisation of porous ionically-conducting materials.

4. Conclusion

The addition of cobalt oxide (formed from the decomposition of the nitrate) to zirconia thick films promoted sintering of the zirconia reducing the through-porosity for given firing conditions or permitted reduced sintering temperature for similar final film structure. The sensors produced using cobalt oxide as dopant operated in the percentage oxygen range at temperatures 100-200 °C lower than those produced in the absence of the cobalt sintering promoter. This has positive benefits in terms of operating power and expected increase in sensor operating life. Furthermore, a new technique for investigating the through-porosity of zirconia and related materials and the influence of sintering aids is proposed and demonstrated.

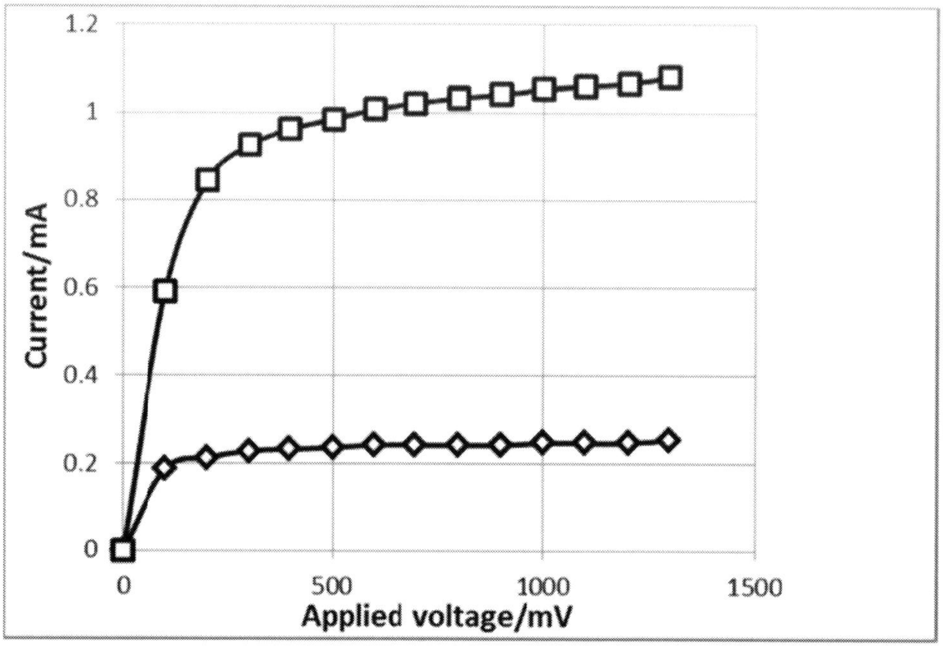

Fig. 5 Current-voltage curves for sensor fired at 1400°C for 2 hours with cobalt addition. Sensor operated at 700°C in (□) 21% oxygen, (◊) 5% oxygen.

5. Acknowledgements

The authors acknowledge a contribution from J Gopaul and assistance from Middlesex University and the Department of Trade & Industry.

6. References

[1] Chu W F, Leonhard V, Erdmann H and Ilgenstein M, *Sensors and Actuators* 1991 **B4** 321
[2] Velasco G and Pribat D, *Proc 2nd Int. Meeting on Chemical Sensors*, Bordeaux, France, 1986 pp 79-94
[3] Gopaul J, Maskell W C and Pitt K E, *J. Appl. Electrochem.* 1999 **29** 93
[4] Kaneko H, Maskell W C and Steele B C H S, *Solid State Ionics* 1987 **22** 161
[5] Lewis G S, Atkinson A and Steele B C H S, *J. Mater. Sci. Lett.* 2001 **20** 1155
[6] Usui T, Asada M, Nakazawa M and Osanai, *J. Electrochem. Soc.* 1989 **136** 534

LOW FREQUENCY NOISE OF ANISOTROPIC MAGNETORESISTORS IN DC AND AC-EXCITED METAL DETECTORS

J Vyhnanek, M Janosek, P Ripka
Czech Technical University in Prague, Technicka 2, 166 27 Prague, Czech Republic
E-mail: vyhnajan@fel.cvut.cz

Abstract. Magnetoresistors can replace induction sensors in applications like non-destructive testing and metal detection, where high spatial resolution or low frequency response is required. Using an AC excitation field the magnetic response of eddy currents is detected. Although giant magnetoresistive (GMR) sensors have higher measuring range and sensitivity compared to anisotropic magnetoresistors (AMR), they show also higher hysteresis and noise especially at low frequencies. Therefore AMR sensors are chosen to be evaluated in low noise measurements with combined processing of DC and AC excitation field with respect to the arrangement of processing electronics. Circuit with a commercial AMR sensor HMC1001 and AD8429 preamplifier using flipping technique exhibited 1-Hz noise as low as 125 pT/√Hz. Without flipping, the 1-Hz noise increased to 246 pT/√Hz.

1. Introduction

Magnetoresistors (MR) fall between Hall sensors and induction sensors in terms of sensitivity and noise. Unlike induction sensors, MRs have the frequency response starting from DC and they are therefore favorite sensors for non-destructive testing devices which detect deeply buried cracks [1]. MRs have small dimensions and high spatial resolution which allows to build array arrangements which can be used for metal detection and object recognition [2]. They are also readily available in commercial packaging as electronic components.

The limiting factors for these applications are the noise of the sensor, gain temperature drift, hysteresis and also offset temperature drift when sensors are used at low frequencies or DC. These parameters limit the detection depth in metal detection and non-destructive testing. Noise can be generally filtered by averaging, however this affects the speed of operation and temperature drifts become more pronounced. There are generally three competing magnetoresistive technologies: giant magnetoresistive (GMR), tunneling magnetoresistive (TMR) and anisotropic magnetoresistance (AMR) sensors. In the case of GMR and TMR, the hysteresis and noise are generally higher [3] than for AMRs, which are subject of this study.

We focus on the sensor noise which disqualifies MRs in favor of induction coils, whereas other parameters speak for MRs – they have small size with high spatial resolution and they are mass produced devices available in packages for assembly in printed circuit boards, so that they can be easily used in arrays [4]. A widely used technique of improving of the AMR sensors parameters is the so called flipping – periodic remagnetization of the sensor by applying large bipolar magnetic field pulses. With the magnetization of opposite polarity the output characteristic is reversed – "flipped". Flipping was shown to improve the offset and gain temperature stability and to reduce crossfield sensitivity of the sensor [5]. Metal detector noise was investigated in three possible circuit arrangements with and without flipping - their effects on AC and also DC detector noise were studied.

2. Measurement setup

For experimental measurements, AMR sensor HMC1001 (Honeywell) was used. This sensor has still the best available noise specifications from the off-the-shelf magnetoresistors. It is a barber-pole sensor with MR elements with 850 ohms resistance arranged in a full bridge, featuring on-chip flipping and compensation coils for feedback operation. The sensitivity is 140 V/T for the selected supply voltage of 5.5 V. As the sensitivity is low, the contribution of the noise of the processing electronics is not negligible. Electronics noise could be removed by the crosscorrelation technique [6], it is however not practical (speed of measurement).
The typical choice for the signal processing of an AMR bridge is a low noise instrumentation amplifier (Figure 1a). We chose AD8429 with a 1 nV/√Hz input voltage noise (gain = 100x) and 1.5 pA/√Hz current noise. Due to the high common mode of the bridge, the instrumentation amplifier cannot be set to the full voltage span therefore another amplifier with the gain 10x was connected as the last stage. A similar arrangement was evaluated in [7] where the high bridge supply of 24 V was applied in order to achieve higher sensitivity and lower noise; however 24V is impractical due to sensor heating.
AMR sensor exhibits two significant types of noise: the 1/f type magnetic noise and the white thermal noise. The white magnetic noise is still some orders of magnitude below the thermal resistive noise of the bridge elements; therefore it is not further taken into account. Whereas the 1/f noise affects low-frequency measurements and depends on the manufacturing process, the white noise influencing AC measurements can be predicted by the bridge resistance and parameters of the instrumentation amplifier.

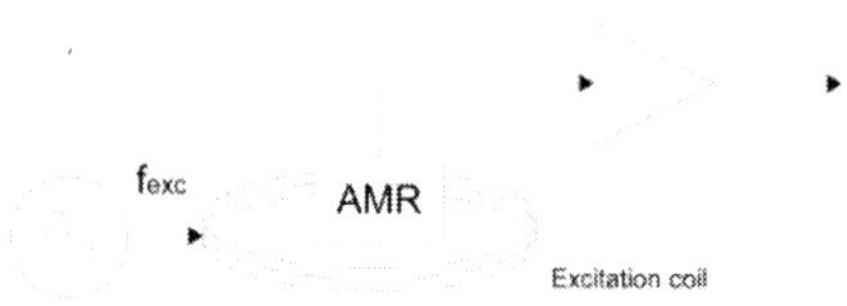

Figure 1a - Direct measurement

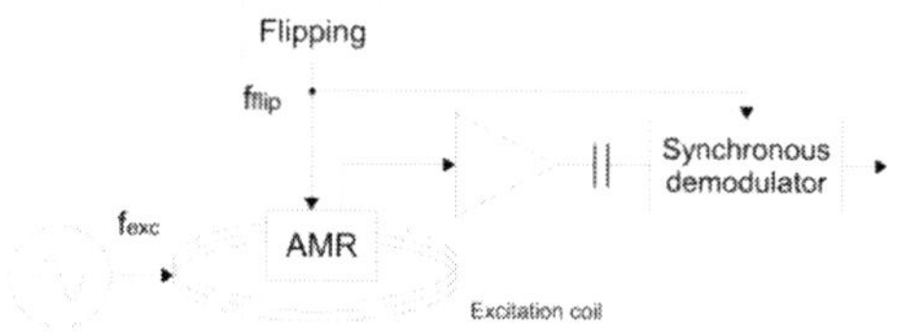

Figure 1b - Measurement setup with flipping

The commonly used method for improving the parameters of an AMR sensor is the so-called "flipping": the sensitive magnetic layer of the AMR is remagnetized in the opposite direction, thus reducing the hysteresis and eliminating the temperature offset drift of the sensor and AC electronics.
For processing the output signal in the flipped mode where the output becomes modulated, a synchronous demodulator is used – Figure 1b. The demodulator in our case includes a switched integrator which eliminates noisy spikes in signal when the sensor is being remagnetized [3] – Figure 2a.
The noise was measured with the Agilent FFT Analyzer 35670A in all cases, without any further amplification, using DC-coupling, 100 averages and a Hanning window. The sensor together with amplifier stage was placed in a 6-layer magnetic shielding can with 100.000x attenuation of the ambient magnetic field noise.

3. Experimental results

3.1. Noise of the electronics

The noise of the electronics was evaluated by connecting a dummy bridge made of resistors of the same value as the MR elements in HMC1001 (850 Ω). Figure 2b shows the noise spectrum obtained at the output of the amplifier (input of the synchronous demodulator in Figure 2a). The 1/f noise with the equivalent of B_E = 95 pT/√Hz at 1 Hz (recalculated using sensitivity S = 140 V/T) is dominantly due to the instrumentation amplifier noise. The white noise with the equivalent of 30 pT/√Hz results both from the bridge thermal noise and the voltage and current noise of the instrumentation amplifier.

The expected white noise of the electronics B_{EW} can be calculated as

$$B_{EW}^2 = (SV_R)^2 + (SV_N)^2 + (SRI_N)^2 \tag{1}$$

$$B_{EW} = S \cdot \sqrt{(V_R)^2 + (V_N)^2 + (RI_N)^2} \tag{2}$$

where V_R is the resistor voltage noise, V_N is the amplifier voltage noise and I_N is the amplifier current noise.

$$B_{EW} = \frac{140V}{T} \cdot \sqrt{\left(\frac{4.12nV}{\sqrt{Hz}}\right)^2 + \left(\frac{1nV}{\sqrt{Hz}}\right)^2 + \left(\frac{1.5pA \cdot 850\Omega}{\sqrt{Hz}}\right)^2} = 31.4\frac{pT}{\sqrt{Hz}}, \tag{3}$$

which matches the measured amplifier noise in Figure 2 (b).

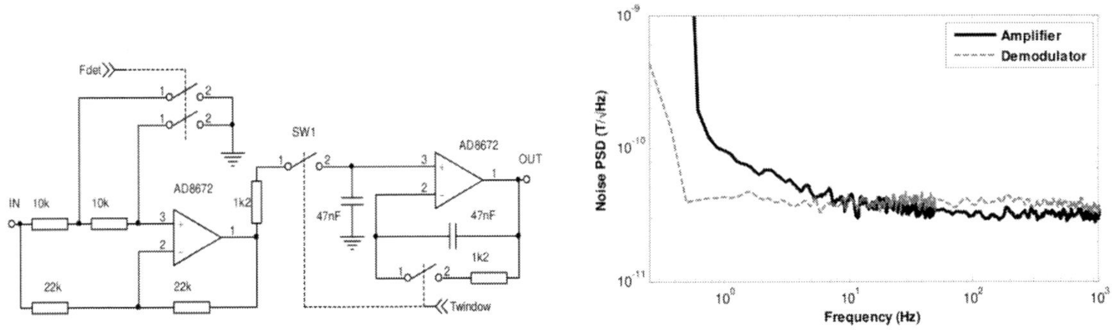

(a) (b)

Figure 2. Synchronous demodulator schematics (a) and comparison of noise of the electronics at amplifier output and demodulator output (b)

The noise of synchronous demodulator was measured at the demodulator output with a 10-kHz reference and the same dummy resistor bridge. The spectrum shows an increased white noise level of 40 pT/√Hz. This was identified as the effect of the switched integrator used in the windowing circuit with the time window set to 70%. With the time window of 100%, the white noise level was 32 pT/√Hz. In the demodulator spectrum there is no 1/f noise of the instrumentation amplifier, due to the fact, that the frequency range was shifted by the 10-kHz demodulation frequency.
Knowing the electronic noise, the noise measurements were done using three different circuit arrangements.

3.2. Direct measurement

This arrangement with simple electronics is depicted in Figure 1a. The output of the AMR bridge is directly amplified. It has the full frequency span limited only by the corner frequency of the amplifier stage. The eventual feedback compensation, which eliminates gain drift and improves linearity, can be realized with a single-opamp PI controller. In this case, the 1/f sensor noise was dominating, the total noise value B_{N1} = 246 pT/√Hz at 1 Hz (Figure 3b).
From the measured values, we can estimate the 1-Hz noise of the sensor itself (B_{S1}) as

$$B_{S1} = \sqrt{B_{N1}{}^2 - B_{E1}{}^2} = \sqrt{246^2 - 95^2}\ \frac{pT}{\sqrt{Hz}} = 226\ \frac{pT}{\sqrt{Hz}} \tag{4}$$

This result is in agreement with values published in [5]. The 1-kHz white noise, which would dominate in the AC application, is approx. the same as the electronic noise - 32 pT/√Hz.
If we use this mode, a single remagnetizing pulse should be performed at least while switching on the device to assure the magnetic state of the sensor.

3.3. Modulation – flipping

The block diagram is shown in Figure 1b, the flipping current and output waveform are shown in Figure 3a. By using flipping at 10 kHz with peak-to-peak amplitude of 3.6 Amps, the output signal was modulated and shifted to the white-noise frequency range of the instrumentation amplifier. The resulting noise of 125 pT/√Hz at 1 Hz (Figure 3b) makes flipping the obvious choice for precise DC field measurements, avoiding the offsets and 1/f noise of the instrumentation amplifier. When compared to 1/f noise of the amplifier and non-flipped sensor (4), the noise clearly further decreased: flipping improved also the low-frequency noise of the magnetoresistor.

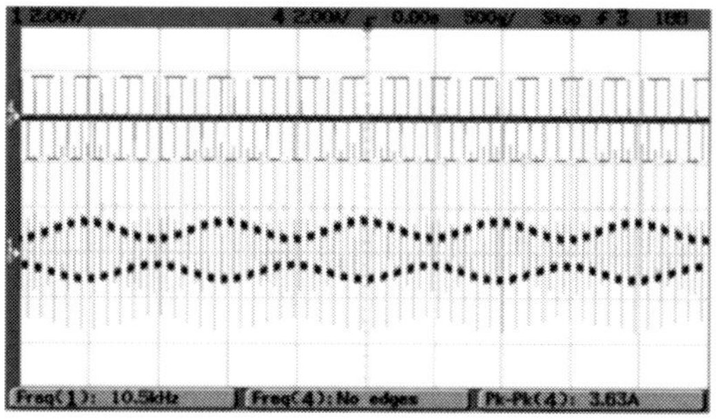

(a)

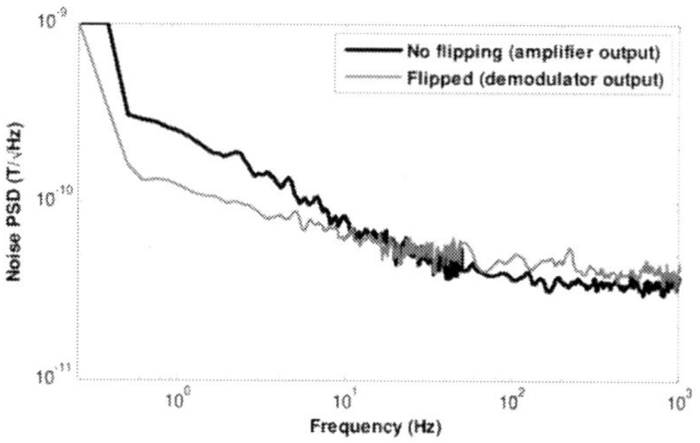

(b)

Figure 3 – Modulated output signal of flipped AMR (a) and comparison of overall noise with direct measurement (b)

However, flipping is power-demanding as the narrow current pulses have to have an amplitude of several amperes in order to properly magnetize the sensitive layer and reduce the 1/f noise [5]. The maximum allowed power dissipation allows maximum flipping frequencies in the order of tens of kHz, limiting the measuring frequency range. It is however possible to use a lower flipping frequency and higher excitation frequency.

3.4. D. Sensor as rectifier

A flipped sensor can be used as a rectifier [8], basic block diagram is presented in Figure 4a. If the flipping frequency is derived from the excitation frequency, the excitation signal is synchronously rectified and the output signal looks like in Figure 4b. The excitation frequency is evaluated with a simple low pass filter connected to the output of the amplifier stage, thus eliminating complex detection circuitry.
A simple feedback compensation is however possible only for a DC range, therefore the sensor should be positioned perpendicularly to the excitation field [1]. However a disadvantage is that the offset drift and the 1/f noise of the amplifier are not eliminated even at the excitation frequency f_{exc}.

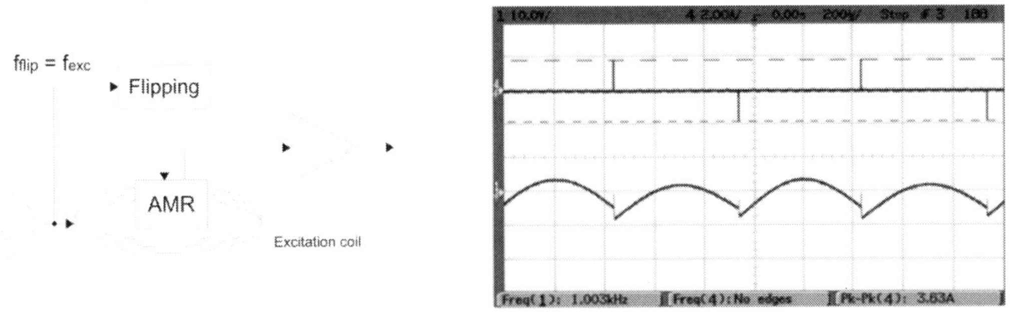

(a) (b)

Figure 4- Special arrangement with f_{flip}=f_{exc} (a), and the flipping current (upper trace) and amplified sensor output (lower trace) (b)

4. Conclusion

When the AMR-based metal detector works with an AC excitation field to sense the eddy current response of metal objects and only the AC frequency response is evaluated, then the noise level is determined by the sensor white noise level and it is approximately the same for either direct measurement or for flipping with demodulator– the 1-kHz noise was about 30 pT/√Hz in both circuit arrangements.

For DC measurements, the 1/f noise of the amplifier and offset drifts of the sensor are best suppressed by flipping modulation technique. For the HMC1001 AMR sensor with 5.5 V_{DC} bridge supply and 10-kHz flipping frequency, we have found an improvement from 246 pT/√Hz 1-Hz noise with direct measurement down to 125 pT/√Hz when using flipping with appropriate demodulation. This improvement was found to be larger than simple effect of shifting the modulated signal out of 1/f amplifier noise: flipping was found to further improve the sensor low-frequency noise.

The commonly used flipping method is however power-demanding with complicated detection electronics; also the measuring frequency range is limited, which is a difficulty in NDE. The possibility of using f_{flip}=f_{exc} was thus investigated: while it allowed to reduce flipping power and to simplify signal processing circuitry, the 1/f noise of the amplifier was however present in this case.

References

[1] Jander A, Smith C, Schneider R 2005 Magnetoresistive sensors for nondestructive testing and evaluation Proc. SPIE. vol. 5770 p. 1-13

[2] Wold RJ, Nordman CA, Lavely EM, Tondra M, Lange E, Prouty M 1999 Development of a handheld mine detection system using a magnetoresistive sensor array, *Proc. SPIE* vol. 3710 pp. 113-123

[3] Popovic RS, Drljaca PM, Schott C 2002 Bridging the gap between AMR, GMR, and Hall magnetic sensors *Proc. 23rd Int. Conf. On Microelectronics* vol. 1 pp. 55-58

[4] Vyhnanek J, Janosek M, Ripka P 2012 AMR Gradiometer for Mine Detection, *Sensors and Actuators,* vol. 186, pp. 100-104

[5] Hauser H, Fulmek PL, Haumer P, Vopálenský M, Ripka P 2003 Flipping field and stability in anisotropic magnetoresistive sensors *Sensors and Actuators*, vol. 106, pp. 1-3

[6] Stutzke NA, Russek SE, Pappas DP 2005, Low-frequency noise measurements on commercial magnetoresistive magnetic field sensors *J. Appl. Phys.*, vol. 97, pp. 10Q107

[7] He DF, Tachiki M, Itozaki H 2009, Highly sensitive anisotropic magnetoresistance magnetometer for Eddy-current nondestructive evaluation, *Review of Scientific Instruments*, vol. 80, pp. 036102-1 - 036102-2

[8] Tumanski S 2001, Thin film magnetoresistive sensors (IOP Publishing), pp. 78-79

Spectral response analysis of PVDF capacitive sensors

B. Reyes-Ramírez[*]**, C. García-Segundo, A. García-Valenzuela**

Centro de Ciencias Aplicadas y Desarrollo Tecnológico, Universidad Nacional Autónoma de México, Ciudad Universitaria, Coyoacán, Apartado Postal70-186, Distrito Federal 04510, México.

Correspondent author: bartolome.reyes@ccadet.unam.mx

Abstract. We investigate the spectral response to ultrasound waves in water of low-noise capacitive sensors based on PVDF polymer piezoelectric films. First, we analyze theoretically the mechanical-to-electrical transduction as a function of the frequency of ultrasonic signals and derive an analytic expression of the sensor's transfer function. Then we present experimental results of the frequency response of a home-made PDVF in water to test signals from 1 to 20 MHz induced by a commercial hydrophone powered by a signal generator and compare with our theoretical model.

1. Introduction.

It is known that the optical absorption of pulsed or modulated light, prompts elastomechanical perturbations via non-radiative processes. This is the so-called photoacoustic effect [1]. For each pulse of light, the energy conversion associated to the non-radiative processes remains confined to the interaction volume, inducing a short-living local increase of pressure (thermo-elastic expansion). Then this propagates through the bulk as an elastomechanical perturbation. I.e. a photoacoustic perturbation propagates through the sample as sound. The frequency dominion of the so-induced perturbations is in the ultrasonic range and can cover from say 100 KHz and up to several hundreds of MHz depending upon the physical properties of the sample. By sample we mean the media where the optical to elastomechanical energy conversion takes place. The registering of this photoacoustic perturbations demands the use of tailored ultrasonic sensors. Due to the high sensitivity and the heavy performance conditions as required, the most frequent sensors used for this purpose are ultrasonic transducers based on PZT-piezoelectric ceramics [2]. These are quite reliable, stable and broadly used in scientific and technical sensing applications [1, 2]. However, they present some limitations in terms of performance and their manufacture process for photoacoustic applications such as: full registering of information at short time constant [3], or for arrays of ultrasonic sensors with specific morphologies and wide bandwidth spectral response as these required in photo-acoustic imagenology [4, 5]. An alternative for manufacturing the required ultrasound sensors and assembling arrays of them is the use of pyroelectric and piezoelectric Polyvinylidene fluoride (PVDF) films in capacitive configuration. The PVDF's piezoelectric component offers a much broader bandwidth window than PZT ceramics [2, 5-7]. With PVDF films at hand, targeted designs of sensing structures are at reach for covering specific sensing needs in pulsed photo-acoustic imaging [4, 5,7]. Indeed one of the first concerns that arise when designing sensors for such a broad bandwidth is their spectral response quality. In general it is very difficult, if not impossible, to have a flat non-negligible response over a spectral range of many MHz. Because as the frequency of sound increases the wavelength decreases, above certain frequency the wavelength inside the PVDF slab is smaller than the thickness and resonances are to be expected. For ultrasound frequencies up to 50 MHz, the wavelength of sound will be smaller than many piezoelectric film sensors. We have not found in the literature explicitly an expression for the frequency dependent transfer function of film piezoelectric capacitive sensors. In this paper we derive an expression for transfer functions assuming ultrasound is incident normally on a piezoelectric film. Then we compare measurements performed on a home-made PVDF capacitive sensor in the frequency range from 1 to 20 MHz.

2. Theoretical analysis

The piezoelectric materials are anisotropic dielectric materials. In these materials an electric field **E** induces a dipolar moment on the constituent molecules, which is proportional to this electric field. The contribution to the total electrical dipolar moment per unit volume (**P**), resulting from the applied electric field and in the absence of mechanical stress, can be written as, $\mathbf{P}_E = \varepsilon_0 \ddot{\chi}_e \cdot \mathbf{E}$. Where $\ddot{\chi}_e$ is the electrical susceptibility tensor and ε_0 is the electrical permittivity of the vacuum. The mechanical stress field, **S**, also contributes to the induced dipolar moment per unit volume in piezoelectric materials. Within the linear response regime of the material and in the absence of an electric field, the contribution to **P** by mechanical stress is written as, $\mathbf{P}_S = \overleftrightarrow{d} \cdot \mathbf{S}$. Here $\overleftrightarrow{d}$ is the piezoelectric tensor. However, in the presence of small electric and stress fields, the electrical polarization per unit volume in the piezoelectric is simply the sum of $\boldsymbol{P_E}$ and $\boldsymbol{P_S}$. As result, in a piezoelectric material we have that,

$$\mathbf{P} = \overleftrightarrow{\mathrm{d}} \cdot \mathrm{S} + \varepsilon_0 \overleftrightarrow{\chi}_e \cdot \mathbf{E}. \tag{1}$$

We assume that the stress field and electric field have amplitudes small enough so that neither field affects the piezoelectric tensor or the susceptibility tensor; see figure 1 for schematic visual guidance. Notice that when the pressure perturbation impinges on the slab on the normal direction respect the piezoelectric surface, the stress field inside the slab has all components equal to zero except the S_{33} component, which turns out to be,

$$S_{33} = \psi_{\mathrm{int}}(z,t). \tag{2}$$

Thus, the stress field acting on the electrical polarization vector of the piezoelectric materials produces component along the z axis; given by $P_{ZS} = d_{33}S_{33}$ while the electrical polarization induces a charge density and a surface charge density according to the following relations, $\nabla \cdot P(r,t) = -\rho_b(r,t)$ and $P(r,t)\cdot\hat{n} = \rho_s(r,t)$. Here $\hat{n}$ is the unit outward normal to the piezoelectric surfaces. Then the stress-induced charge densities are,

$$\rho_b(r,t) = -\nabla \cdot (d_{33}S_{33}\hat{a}_z) = -d_{33}\frac{\partial S_{33}}{\partial z} \quad , \quad \rho_z^{(0)} = -d_{33}\, S_{33}\big|_{z=0} \text{ and } \rho_z^{(d)} = -d_{33}\, S_{33}\big|_{z=d} .$$

As a result one obtains that the field within the slab is,

$$E_z^S(z) = \left| -\frac{d_{33}}{\varepsilon} S_{33}(z) - \frac{d_{33}}{\varepsilon}\left[\frac{S_{33}(h)+S_{33}(0)}{2}\right] \right|. \tag{3}$$

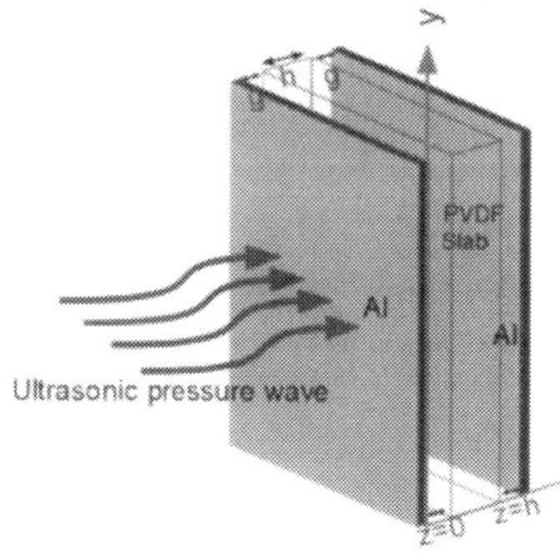

Figure 1. Schematic diagram of transduction from ultrasonic pressure perturbation to electric signal.

For the condition in which the piezoelectric slab is between two flat metallic electrodes, e.g. forming a parallel-plate capacitor (see figure 1), the electrodes are charged with a surface density $\rho_s^{(e)}$ and an electric field is produced between the electrodes. This electric field induces also the polarization of the piezoelectric slab. The electric field and polarization induced by the charges on the electrodes simply adds up with the electric field and with polarization induced by the stress. Within the gaps of width g, shown in the figure above, the electric field due to the electrode charges is normal to the interfaces with amplitude, $E_z^{(e)} = \rho_s^{(e)} / \varepsilon_0$. Within the piezoelectric slab the electric field is also in the z direction and of amplitude $E_z^{(e)} = \rho_s^{(e)} / \varepsilon$. The total electric field inside the piezoelectric slab is,

$$E_z(z) = -\frac{d_{33}}{\varepsilon} S_{33}(z) - \frac{d_{33}}{\varepsilon}\left[\frac{s_{33}(h)+s_{33}(0)}{2}\right] + \rho_s^{(e)} / \varepsilon . \tag{4}$$

By letting the gaps tend to zero, the electric potential difference between the electrodes is then obtained as,

$$E_z(z) = \frac{Q_e h}{A\varepsilon} + \frac{d_{33}h}{\varepsilon}\left[\frac{s_{33}(h)+s_{33}(0)}{2}\right] + \int_0^h \frac{d_{33}}{\varepsilon}\left[S_{33}(z)\right]dz . \tag{5}$$

On the other hand, when a monochromatic plane ultrasonic wave is incident normally to the sensor, the ultrasonic pressure perturbation inside the piezoelectric slab is $\psi_a + \psi_b$, where $\psi_a = A(\omega)\psi_i e^{j(\omega t - k_2 z)}$, $\psi_b = B(\omega)\psi_i e^{j(\omega t + k_2 z)}$, ψ_i the amplitude of the incident pressure $k_2 = 2\pi/\lambda_2 = \omega / c_2$. In the latter expression λ_2 is the wavelength, c_2 is the phase velocity of sound in the piezoelectric and $\omega = 2\pi f$, with f being the frequency of the incident ultrasonic wave. Thus the amplitudes in the perturbation within the slab are given by,

$$A(\omega) = \frac{\left[1 + r_{21}\right]}{1 - r_{21} r_{23} e^{-2jk_2 h}} \quad \text{and} \quad B(\omega) = \frac{-r_{23}\left[1 + r_{21}\right]e^{-2jk_2 h}}{1 - r_{21} r_{23} e^{-2jk_2 h}}, \tag{6}$$

where, $r_{21} = (z_2 - z_1)/(z_2 + z_1)$ and $r_{32} = (z_2 - z_3)/(z_2 + z_3)$, are the reflection coefficients at normal incidence at the interfaces between the slab (medium 2) and the incidence medium (medium 1, for $z < 0$) and the interface between the slab and the transmission medium (medium 3, $z > h$), respectively; where z_1, z_2 and z_3 are the acoustic impedance of medium 1, 2 and 3, respectively. We then have,

$$V_c = \frac{Q_e}{C_0} + \frac{d_{33}h}{\varepsilon}\left[\frac{A(\omega)\left[e^{-jk_2h}+1\right]+B(\omega)\left[e^{jk_2h}+1\right]}{2}\right]\psi_i + \int_0^h \frac{d_{33}}{\varepsilon}\left[\psi_a(z)+\psi_b(z)\right]dz\,, \tag{7}$$

where $C_0 = \varepsilon A/h$ is the electrical capacitance. Performing the integrals yields,

$$V_c = \frac{Q_e}{C_0} + \frac{d_{33}h}{\varepsilon}\left[\frac{A(\omega)\left[e^{-jk_2h}+1\right]+B(\omega)\left[e^{jk_2h}+1\right]}{2}\right]\psi_i + \frac{d_{33}}{\varepsilon}\left[\frac{A(\omega)\left[e^{-jk_2h}-1\right]}{-jk_2}+\frac{B(\omega)\left[e^{jk_2h}-1\right]}{jk_2}\right]\psi_i\,.$$

Notice that the factor $\exp(j\omega t)$ is implicit in all terms. Clearly the equivalent circuit is a capacitor C_0 in series with a voltage source V_s given by,

$$V_s(\omega) = \frac{d_{33}h}{2\varepsilon}\left\{A(\omega)\left[e^{\frac{-j\omega h}{C_2}}+1\right]+B(\omega)\left[e^{\frac{j\omega h}{C_2}}+1\right]\right\}\psi_i + j\frac{d_{33}C_2}{\omega\varepsilon}\left\{A(\omega)\left[e^{\frac{-j\omega h}{C_2}}-1\right]-B(\omega)\left[e^{\frac{j\omega h}{C_2}}-1\right]\right\}\psi_i$$

where we used $k_2 = \omega / c_2$. If we connect the circuit to an oscilloscope we add an impedance in series and close the circuit. The voltage drop across the input impedance of the oscilloscope is, $V_{osc} = IZ_i$, while the voltage drop across the piezoelectric capacitor must be zero. Thus we get,

$$V_{osc} = \frac{-Z_i}{-Z_i + \frac{1}{j\omega C_0}} V_s\,. \tag{8}$$

If we define the transfer function as,

$$H(\omega) = \frac{V_{osc}(\omega)}{\psi_i(\omega)}\,, \tag{9}$$

we get,

$$H(\omega) = \frac{-Z_i(\omega)}{-Z_i(\omega) + \frac{1}{j\omega C_0}}\left\{\frac{d_{33}h}{2\varepsilon}\left[A(\omega)\left(e^{\frac{-j\omega h}{C_2}}+1\right)+B(\omega)\left(e^{\frac{j\omega h}{C_2}}+1\right)\right]+ j\frac{d_{33}C_2}{\omega\varepsilon}\left[A(\omega)\left(e^{\frac{-j\omega h}{C_2}}-1\right)-B(\omega)\left(e^{\frac{j\omega h}{C_2}}-1\right)\right]\right\} \tag{10}$$

This is our result for the transfer function of a film piezoelectric capacitor transducer for normally incident ultrasound. For incidence at angles different from normal, the Eq. (2) would be expressed in terms of the corresponding angular projection and its implications shall be cascaded over the calculation up to reaching an expression equivalent to Eq. (10).

3. Numerical Example

In Fig. 2 we plot the magnitude of the transfer function given in Eq. (10) versus the angular frequency ω form 1 MHz to 20 MHz for a 100 μm thick PVDF film immersed in air, water and in a medium of the same acoustic impedance as that of the PVDF (a perfect impedance match). To generate the curves in Fig. 2 we assumed the speed of sound in air, water and in the PVDF is 343 m/s, 1484 m/s, and 2200 m/s respectively, and we supposed the corresponding densities are 1000 g/cm^3, 1593 g/cm^3 and 1780 g/cm^3, respectively.

We can appreciate that the transfer function predicted by Eq. (10) sinks at about 13 MHz. This is a result of destructive interference of the multiply reflected ultrasound waves within the PVDF film. Note that the difference between the spectral response of the sensor in water and with a perfect impedance match is negligible.

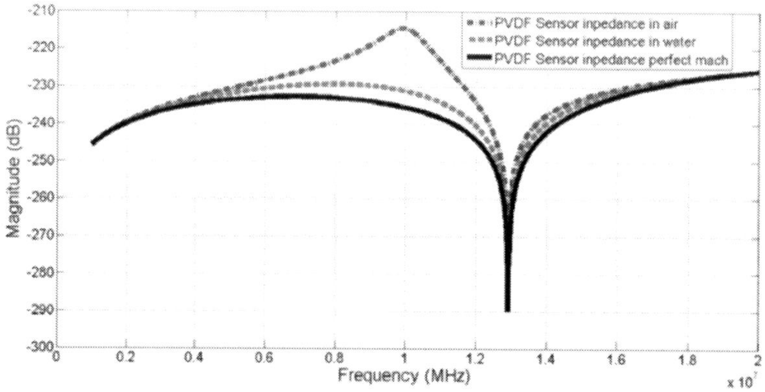

Figure 2. Plot of analytic calculations for $|H(\omega)|$ as a comparison from different media (air, water, PVDF).

4. Comparison with experimental measurements.

We fabricated a PVDF capacitor sensor by depositing very thin layers of aluminum on the surfaces of a 110 μm thick PVDF film and connecting it to a voltage preamplifier (ONDA, model HA-1100). The sensors were sealed with a thick plastic coating for its protection. The sensitive area of the PVDF capacitor was 3 mm^2. The capacitor sensor was immersed in a bucket filled with water and the signal generated by a calibrated commercial hydrophone emitting ultrasound waves towards the sensor was registered with an oscilloscope.

In Fig. 3 we show the sensor's frequency response and the prediction of Eq. (10) above. The experimental response is somewhat different to the theoretical curve, but this is expected since Eq. (10) does not include the effects of the electrodes nor sealant coating that covers on the real sensor. Nevertheless we can see in the experimental curve a drop at 13 MHz as predicted by the model. This is only a preliminary result and of course, a broader analysis and more experiments are needed to improve the sensors response as demanded in photo-acoustic imaging applications. Nevertheless, the model derived here is a necessary tool for developing low-noise ultrasonic sensors with broad frequency response range, as required. We are currently working on a comprehensive experimental characterization of our PVDF sensors of different thicknesses and on the extension of our model.

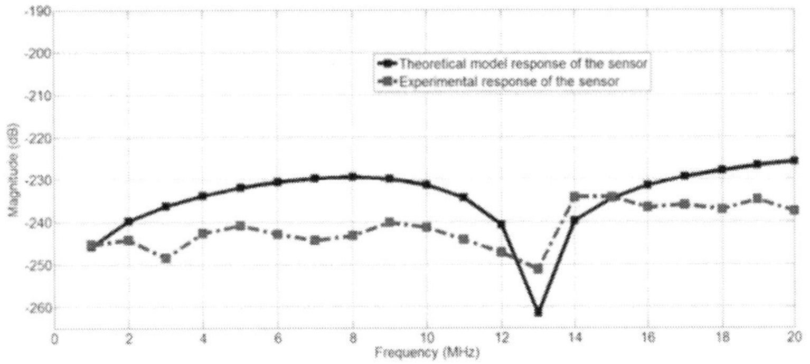

Figure 3. Comparison of the frequency response of the sensor and Eq. (10).

5. Conclusions

We derived an analytic expression for the transfer function of a piezoelectric capacitor sensor taking into account multiple reflections of ultrasonic waves within the sensor. It was found that dips in the transfer function arise due to interference effects. We obtained experimental data on the frequency response of a 110 mm thick PVDF sensor and observed a dip around 13 MHz as predicted by the theoretical model. The possible relevance of these dips on the sensors performance in photo-acoustic imaging is currently under assessment.

Acknowledgments

We acknowledge financial support from Dirección General de Asuntos del Personal Académico of Universidad Nacional Autónoma de México through grant PAPIIT IN-106712 and to Instituto de Ciencia y Tecnología from Ciudad de México for the partial support of this work through a breast cancer research contract.

References

[1]. C. K. N. Patel and A. C. Tam: Pulsed optoacoustic spectroscopy of condensed matter, Rev. Mod. Phys. 53, 517 (1981).

[2]. G. L. Messing, S. Trolier-McKinstry, E. M. Sabolsky, C. Duran, S. Kwon, B. Brahmaroutu, P. Park, H. Yilmaz, P. W. Rehrig, K. B. Eitel, E. Suvaci, M. Seabaugh, and K. S. Oh.: "Templated grain growth of textured piezoelectric ceramics". Critical Reviews in Solid State and Materials Sciences 29, 45–96 (2004).

[3]. "Experimental observations of optically induced non-radiative fast-pulses in metals": C. García-Segundo, A. J. Smith, and J.-P. Connerade. J. Modern Optics. Vol. 51, 233 (2004).

[4]. V. Moock , C. García-Segundo, E. Garduño and F. Arámbula-Cosio: "Signal Processing for Photoacoustic Tomography". Proceedings of the 5th International Conference on Image and Signal Processing and 5th International Conference on BioMedical Engineering and Informatics (5th CISP'12 / IEEE-BMEI'12). Chongqing, China. 16-18 Oct. 2012, pp. 957 – 961. DOI: 10.1109/CISP.2012. 6469836.

[5]. G. Paltauf, R. Nuster, S. Gratt: "Photoacoustic section imaging with integrating detectors". ALT'12 Conference Proceedings. Thun, Switzerland 2-6 Spetember, 2012. DOI: 10.12684/alt.1.93.

[6]. G. Gautschi: Piezoelectric Sensorics: Force, Strain, Pressure, Acceleration and Acoustic Emission Sensors, Materials and Amplifiers. Springer, 5-9 (2012).

[7]. T. DH, Foster FS.: Fabrication and characterization of transducer elements in two-dimensional arrays for medical ultrasound imaging, IEEE Trans Ultrason Ferroelectr Freq Control, 39(4):464-75 (1992).

Journal of Physics: Conference Series **450** (2013) 012033 doi:10.1088/1742-6596/450/1/012033

Hard-field Tomography Imaging in Secondary Contrast

J A Cantoral Ceballos, M P Wood, N Nurgiyatna, M A Banuelos Saucedo, P Wright and K B Ozanyan[1]
School of Electrical and Electronic Engineering and The Photon Science Institute
The University of Manchester, M13 9PL, UK

E-mail: k.ozanyan@manchester.ac.uk

Abstract. Tomography is now a well-established method for indirect imaging, where physical measurements of a set of path integrals of the target function can be inverted to yield its spatial distribution. Traditionally, using low frequency electrical and/or magnetic fields, Tomography has been applied to imaging of cross sections of the real and imaginary permittivity of objects and at high photon energies (x-rays and gamma-rays) – the object density.

The intermediate spectral range, including the visible, infrared and terahertz regions, allows chemically- specific imaging based on resonant absorption, as well as imaging based on the real part of the complex refractive index, e.g. exploiting measurements of the pulse delay and wave-guiding losses introduced by the target function. A common approach based on the simple path-integral description allows to identify the target function in all these cases.

Here we review three cases of hard-field Tomography which allow imaging in secondary contrast, i.e. the recalculation of the target function distribution into another parameter which is not directly measured: a) temperature Tomography, b) refractive index Tomography and c) surface deformation Tomography.

a) Basic thermodynamics does not allow measurements of path integrals of temperature. However, temperature images can be achieved in secondary contrast, whereby reconstructed images of absorption are re-calculated in values of temperature [1]. In the demonstrated case of water vapour secondary temperature contrast is achieved by using spectrally-resolved Landweber reconstruction of absorption, to produce a set of 42 quasi-monochromatic images (at frequencies around two selected transitions in the near infrared, 5263.7 and 6476.1 cm-1, temperature 1600–2000 K, concentration 0.92- 7.86%).) of the local attenuation coefficients, which are then used as data points in a Voigt-shape fitting algorithm (Levenberg-Maquardt-Fletcher) to estimate the most likely values of the linestrengths of the transitions at each pixel. The ratio of these values is then used to calculate the pixel temperature.

b) It is possible to produce a cross-sectional image by "speed-of-light" tomography by measuring the pulse delay across an object [2]. We demonstrate this in the THz region by employing femtosecond THz pulses across objects with refractive index close to near-stoichiometric and sooting flames (n=1.02). We show that measuring the pulse delay is equivalent to taking a line integral of a function containing the refractive index and consequently the set of measurements constitute a well-behaved Radon transform of the spatial distribution of values of the optical density. We use a solid object, with refractive index similar to a flame, to evaluate the achievable spatial resolution. With this approach, images with pixel size as small as 5 wavelengths are reliably reconstructed.

[1] To whom any correspondence should be addressed.

c) The secondary contrast approach can be applied also in the context of Photonic Guided-Path Tomography [3], where path integrals of transmission losses are taken along optical fibres. The variety of optical fibre sensor technologies allows a range of diverse secondary contrasts, if fibre-optic embodiments of path integrals are sensitised to e.g. bending, chemical properties such as humidity, pH, etc. or the concentration of a certain chemical species. We demonstrate deformation tomography realised as a "smart carpet":

http://www.reuters.com/video/2012/10/08/magic-carpet-a-hi-tech-safety-net-for-el?videoId=238243515&videoChannel=5,

where the bending of the optical fibres modulates the transmitted light intensity. In this case, the achieved secondary contrast is based on a relationship between the curvature of the imaged surface and the losses induced in the fibre, which is not straightforward to describe – particularly in one of the possible approaches involving the cutting of transversal grooves reaching the fibre's core. Secondary contrast in deformation also brings out the challenge in the definition of the imaging plane – in this case it is not flat as in conventional Tomography.

Finally, we examine the general approaches to developing other secondary contrasts, by using the formal Radon transform expression satisfying mathematically the description of a measurable physical phenomenon. Alternatively, new target functions are possible to realise by defining the exact analytical relationship with a measurand amenable to Radon transform.

References:

[1]. "Temperature Mapping from Molecular Absorption Tomography". Wood M. P. and Ozanyan K. B. Proceedings of the IEEE SENSORS 2011 Conference, Limerick (Ireland). pp.865-869 doi:10.1109/ICSENS.2011.6127014

[2] "Sensing in Sooting Flames:THz Time-Domain Spectroscopy and Tomography". Darabkhani H.G., Banuelos-Saucedo M.A., Young J.S., Stringer M., Wright P., Wang Q., Zhang Z., Miles B.and Ozanyan K.B. Proc. IEEE SENSORS 2012 Conference, Taipei (Taiwan) doi:10.1109/ICSENS.2012.6411257.

[3] "Smart Carpet for Imaging of Objects' Footprint by Photonic Guided-Path Tomography" Cantoral Ceballos J.A., Nurgiyatna N., Scully P. and Ozanyan K.B. Proc. IEEE AFRICON11, Livingston (Zambia), doi: 10.1109/AFRCON.2011.6071986 .

A Study on Signal Group Processing of AUTOSAR COM Module

Tae Man Han [1], Jeong-Hwan Lee [2], Hyun Yong Hwang [3] and Yong Hak Ahn[4]
[1] Principle Member, *ETRI*, Daejeon, 305-700, South Korea
[2] Senior Member, *ETRI*, Daejeon, 305-700, South Korea
[3] Senior Member, *ETRI*, Daejeon, 305-700, South Korea
[4] Professor, Dept. of C.E., Sejong University, Seoul, 143-747, South Korea

E-mail: {[1] tmhan, [2] jeonghwan.lee, [3] hyhwang}@etri.re.kr, [4]yohans@sejong.ac.kr

Abstract. In vehicle, there are many ECU(Electronic Control Unit)s, and ECUs are connected to networks such as CAN, LIN, FlexRay, and so on. AUTOSAR COM(Communication) which is a software platform of AUTOSAR(AUTomotive Open System ARchitecture) in the international industry standards of automotive electronic software processes signals and signal groups for data communications between ECUs. Real-time and reliability are very important for data communications in the vehicle. Therefore, in this paper, we analyze functions of signals and signal groups used in COM, and represent that functions of signal group are more efficient than signals in real-time data synchronization and network resource usage between the sender and receiver.

1. Introduction

AUTOSAR is an open software platform in order to solve the complexity of software depend on hardware and ECU, improve reusability and exchangeability of automotive electronic software between automobile manufacturers and auto parts makers. It is now the industry standard which more than 170 automobile manufacturers and IT companies participate[1].

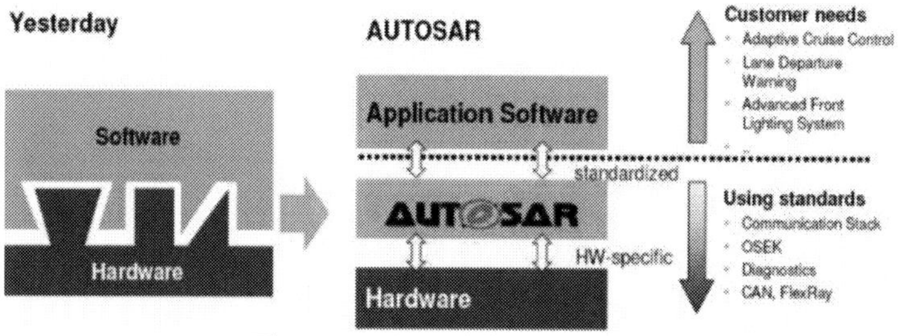

Figure 1. AUTOSAR Vision

AUTOSAR constitute mainly three layers such as AUTOSAR SW-C(Software-Component), RTE(Run-Time Environment), and BSW(Basic Software). BSW composes Service Layer, EAL(ECU

[4] Corresponding author

Abstraction Layer), MCAL(Micro-controller Abstraction Layer), and CDD(Complex Device Driver) as the standard layers[2].

COM belongs to the Service layer in the AUTOSAR software layer, is between PDU Router and RTE, and its main functions are signal processing, signal group processing, I-PDU group communication control, signal gateway, and so on.

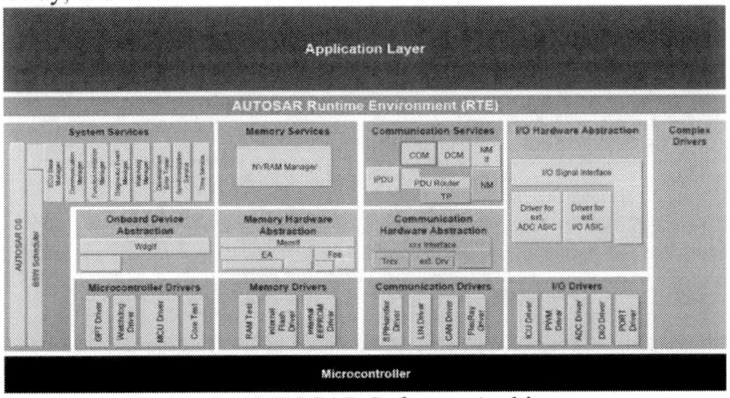

Figure 2. AUTOSAR Software Architecture

In this paper, we analyze signal and signal group functions which are main functions of COM in AUTOSAR and compare the two functions in real-time data synchronization and network resource usage.

2. AUTOSAR COM Functionality

COM makes PDU using signals or signal group received from the higher module RTE and sends it to lower module PDU Router. On the contrary, COM extracts signals and signal group from PDU received from PDU Router and sends them to RTE[3].

2.1. AUTOSAR COM Signal

Because COM provides the higher module RTE with Com_SendSignal function, RTE can send signals to COM. When RTE sends signals calling Com_SendSignal function, COM stores the received signals from RTE in internal I-PDU buffer and sends I-PDU stored in the buffer to the lower module PDU Router by calling PduR_ComTransmit function provided by PDU Router[4]. Also, as COM provides the higher module RTE with Com_ReceiveSignal function, RTE can receive signals from COM. When PDU Router tells to COM that it receives PDU by Com_RxIndication function, COM analyzes signals in PDU, notifies RTE that signal was received by calling callback functions each of which is assigned to each signal.

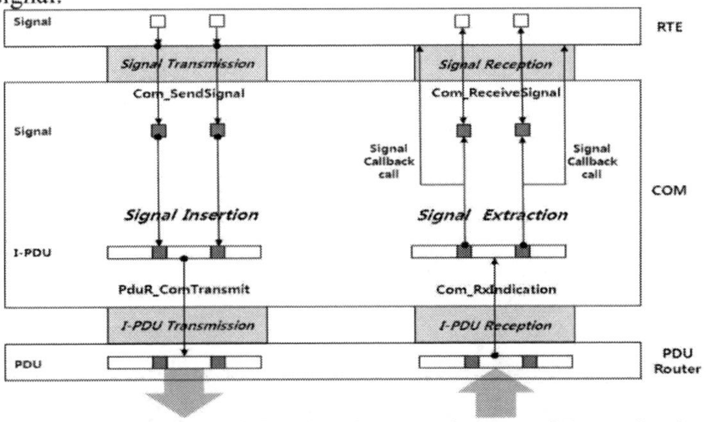

Figure 3. Simplified model for Signal Transmission and Reception in COM

2.2. AUTOSAR COM Signal Group

In AUTOSAR, COM provides the signal group function to support the notion of complex data type. Signals in the signal group are named group signal and they consist of one more group signals. Also, COM synchronizes group signal sent from higher module RTE and PDU transmission by using shadow buffer mechanism. As COM provides higher module RTE with Com_UpdateShadowSignal and Com_SendSignalGroup functions, RTE can send signal group to COM. RTE send group signal in signal group to COM by calling Com_UpdateShadowSignal and COM stores received group signals from RTE in shadow buffer. After RTE sends all group signal to COM, it will call Com_SendSignalGroup function to transmit. At this time, COM reads group signal stored in shadow buffer and writes it in I-PDU buffer. And then COM send I-PDU to lower module PDU Router by calling PduR_ComTransmit function provided by it.

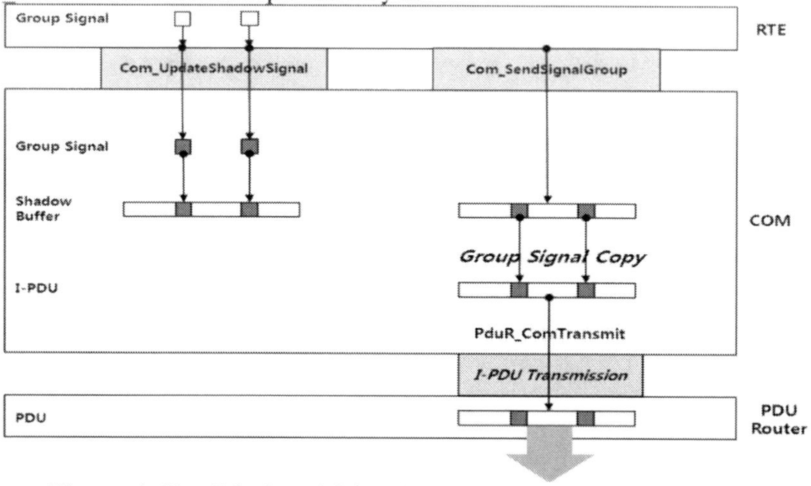

Figure 4. Simplified model for Signal Group Transmission in COM

Also, since COM provides higher module RTE with Com_ReceiveShadowSignal and Com_ReceiveSignalGroup functions, RTE can receive group signals sent from COM. When PDU Router notifies PDU reception by calling Com_RxIndication provided by COM, COM analyzes signal groups in PDU and tells reception of signal groups to RTE by calling callback functions each of which is assigned to each signal group. COM copies group signals stored in I-PDU to shadow buffer by calling Com_ReceiveSignalGroup function, and RTE receives group signals stored in shadow buffer of COM by calling Com_ReceiveShadowSignal function according to group signals.

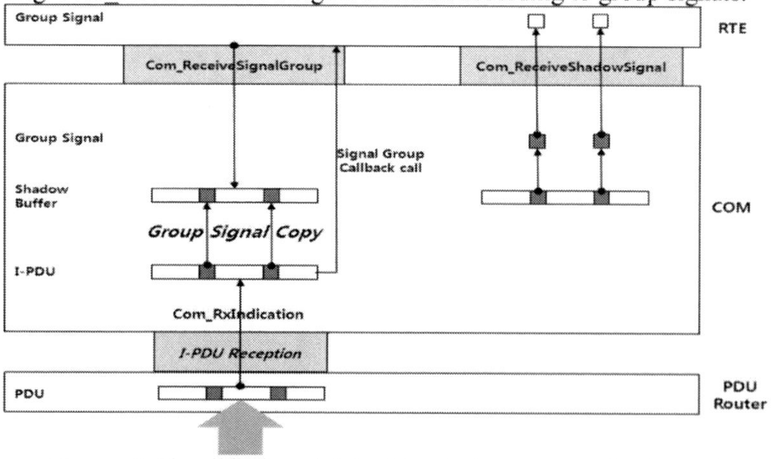

Figure 5. Simplified model for Signal Group Reception in COM

3. Implementation

3.1. Test Design

We designed four signals and one signal group to contain four group signals for comparison and analysis of COM signal and signal group in aspect to real-time data synchronization and network resource, and implemented it. To make the same test environment, we use unsigned char data type for four signals which consist of two 2-bit signals and two 4-bit signals. And group signal also has the same composition as signal and is sent at 10ms periods. In addition, the size of PDU in order to send signals and group signals is designed as 2 bytes.

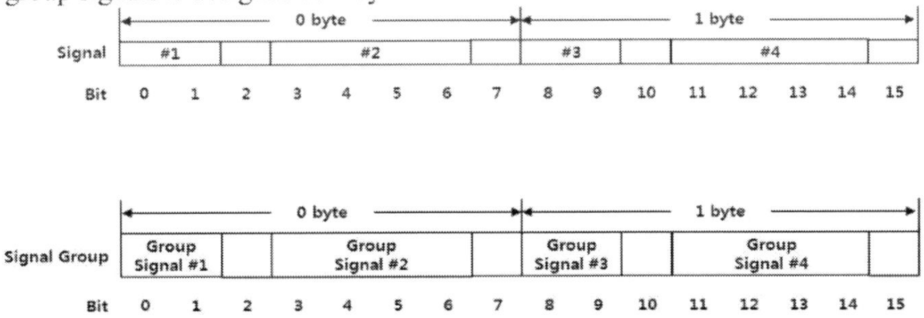

Figure 6. Signal and Signal Group Architecture

3.2. Test Environment

We use C-language to implement, Altium's Tasking VX-toolset for C166 as complier, Vector's CANoe as simulator, and Lauterbach' TRACE-32 as debugging device, and program image results are loaded the target board built in Infineon XC2365A MCU.

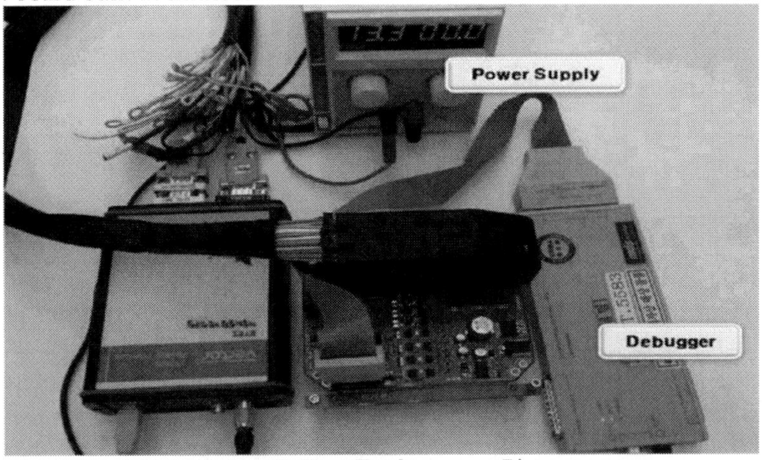

Figure 7. Test Environment Picture

3.3. Test Results

Higher RTE module calls Com_SendSignal function four times to send four signals, and COM also calls PduR_ComTransmit function four times to send them to PduR. Therefore, 4 PDUs are sent via CAN network for four signals and when 4th PDU is sent, for the first time, all up-to-date signals are transmitted via CAN network.

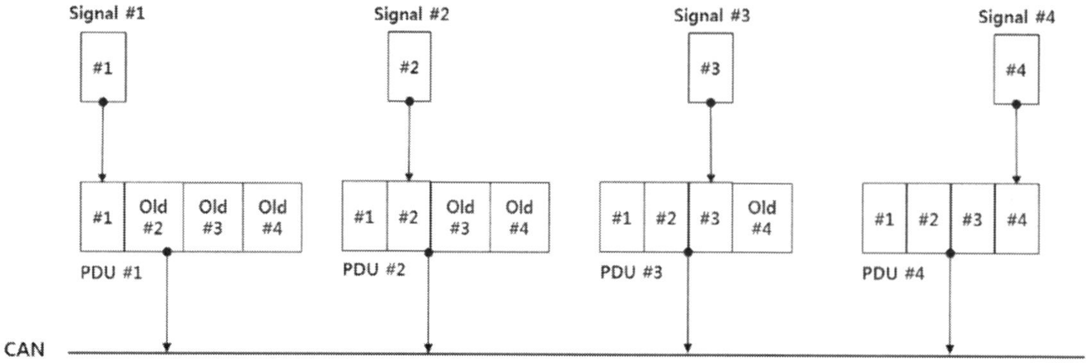

Figure 8. Signal Transmission Procedure

Consequently, as you can see Figure 9, when the last signal is been sending, PDU in which all signals are updated as new signals will be sent via CAN network, and transmission of other PDU except the last PDU results in waste of network resource which is unnecessary. Also, when we transmit 2-byte PDU, CAN bus shows 0.05~0.07% loads and 0.07% peak load.

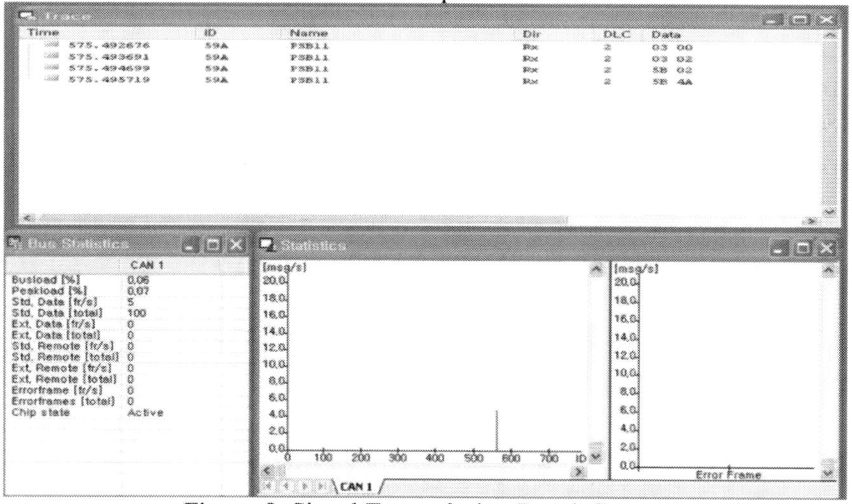

Figure 9. Signal Transmission Result Screen

Unlike transmitting signal, in transmission of signal group, higher RTE module calls Com_UpdateShadowSignal function four times to transmit four group signal in signal group to COM module. But, Com_SendSignalGroup function is called for transmission of signal groups via CAN network, and COM calls PduR_ComTransmit function to send them to PduR.

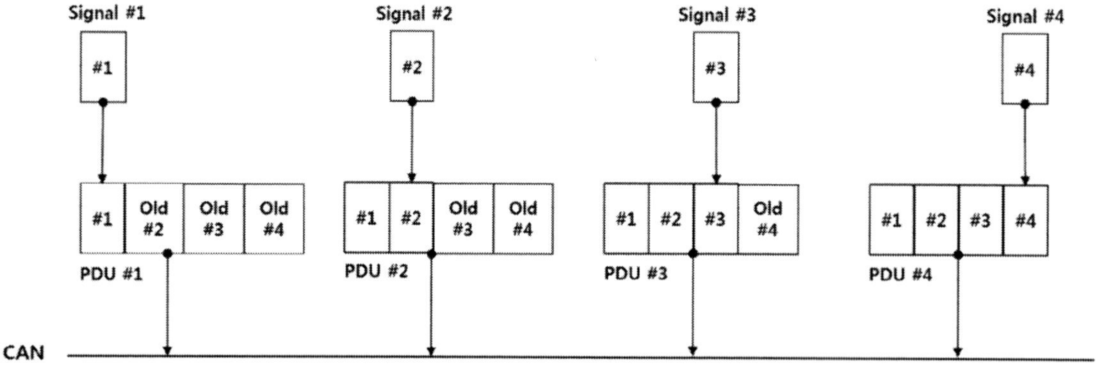

Figure 10. Signal Group Transmission Procedure

Accordingly, As you can see Figure 11, for transmission of four group signals to CAN, all new group signals are sent by transmitting PDU once. The CAN bus shows 0.04~0.06% loads, and 0.07% peak loads. In transmission of 2-byte PDU, it shows difference of at least 0.01% bus loads compared to transmission of signals. As the number and size of PDU increase, it shows remarkable difference of bus loads. Transmission of signal groups doesn't result in waste of network because it doesn't make unnecessary transmission and since transmits new group signals in real-time, it has benefit of synchronization with receiver.

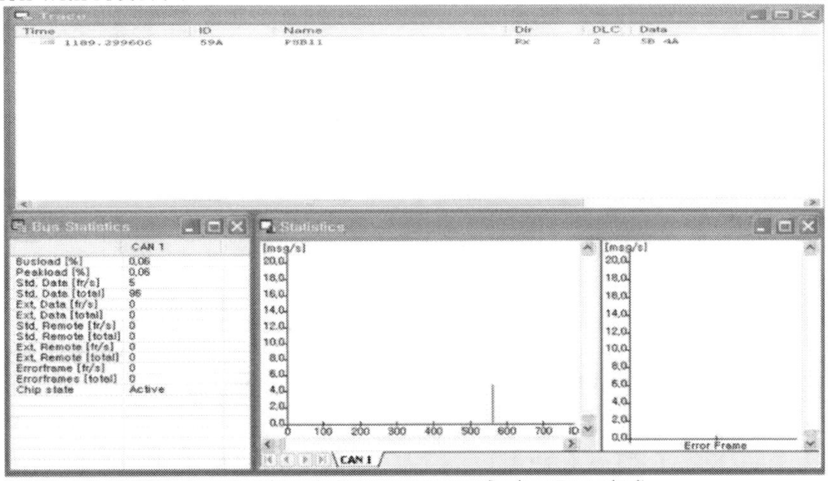

Figure 11. Signal Group Transmission Result Screen

4. Conclusion

In the results of this paper, at transmission of 2-byte PDU, the method of signal groups has at least 0.01% less loads than one of signals. This means that as the number and size of PDU increase, it show remarkable difference of bus loads. Hence, it is turned out that in automobiles for real-time data reliability and synchronization, and efficient usage of network resource, applicant of the method of signal group which is provided by AUTOSAR COM is effective.

5. Acknowledgments

This work was supported by the Industrial Strategic Technology Development Program funded by the Ministry of Knowledge Economy (MKE, Korea) [10041648, Robust Fault-Resilient SW for Vehicle Processors].

6. References

[1] AUTOSAR Home Page, http://www.autosar.org
[2] AUTOSAR , Dec. 2008, AUTOSAR Layered Software Architecture
[3] AUTOSAR, 2010, Specification of Communication V3.2.0 R3.1 Rev 5
[4] AUTOSAR, 2010, Specification of PDU Router V3.2.0 R3.1 Rev 5

Development of device producing electrolyzed water for home care

K Umimoto[1], **S Nagata**[1], **J Yanagida**[2]
[1]Department of Biomedical Engineering, Osaka Electro-Communication University, Osaka, Japan
[2]Department of Medical Technology, Kobe Tokiwa University, Kobe Japan

E-mail: umimoto@isc.osakac.ac.jp

Abstract. When water containing ionic substances is electrolyzed, electrolyzed water with strong bactericidal ability due to the available chlorine(AC) is generated on the anode side. Slightly acidic to neutral electrolyzed water (pH 6.5 to 7.5) is physiological pH and is suitable for biological applications. For producing slightly acidic to neutral electrolyzed water simply, a vertical-type electrolytic tank with an asymmetric structure was made. As a result, a small amount of strongly alkaline water was generated in the upper cathodic small chamber, and a large amount of weakly acidic water generated in the lower anodic large chamber. The pH and AC concentration in solutin mixed with both electrolyzed water were 6.3 and 39.5 ppm, respectively, This solution was slightly acidic to neutral electrolyzed water and had strong bactericidal activity. This device is useful for producing slightly acidic to neutral electrolyzed water as a disinfectant to employ at home care, when considering economic and environmental factors, since it returns to ordinary water after use.

1. Introduction

Electrolysis is well known as being a method of separating an ionic substance. A device for achieving electrolysis require an electrolyte containing an ionic substance, an anode and a cathode of electrode, a diaphragm and a direct electric current supply. In the electrolysis process, water containing NaCl is electrolyzed with a membrane partition, resulting in the production of strongly acidic electrolyzed water(SAcEW) at the anode and strongly alkaline electrolyzed water(SAlEW) at the cathode(Fig.1). The resulting SAcEW contains an available chlorine (AC) such as hypochlorous acid (HClO), which is known to have a strong bactericidal action, and which thus enables the killing of various microorganisms such as viruses and bacteria[1][2].

Electrolyzed water is classified depending on its pH into SAcEW (pH 2.2 to 2.7), slightly acidic electrolyzed water (pH 5.0 to 6.5), and neutral electrolyzed water (pH 6.5 to 7.5), each having different characteristics. SAcEW shows an instantaneous bactericidal activity, but the activity does not persist. In addition, SAcEW stimulates the oral mucosa etc. because of its strong acidity, and is therefore unsuitable for biological applications. On the other hand, in slightly acidic to neutral electrolyzed water, since the activity persists and its pH is physiological levels, it can be applied to various uses. However, production of slightly acidic to neutral electrolyzed water requires electrolysis of hydrochloric acid itself or use of hydrochloric acid as a pH adjuster. Thus, the production is too complicated to perform in domestic homes or long-term care facilities. Then, in order to obtain slightly acidic to neutral electrolyzed water with the conventional system, an additional process, such as mixing the two types of electrolyzed water in different amounts, is necessary. Recently we reported that slightly acidic to neutral electrolyzed water can be produced by automatic mixing controller which consists of solenoid valves[3]. However, the conventional generator is unsuitable for exclusively producing slightly acidic to neutral electrolyzed water.

The present study provides a method of directly producing slightly acidic to neutral electrolyzed water showing excellent persistence of the bactericidal effect in a simple and highly efficiently manner with an electrolyzed water generator having a membrane.

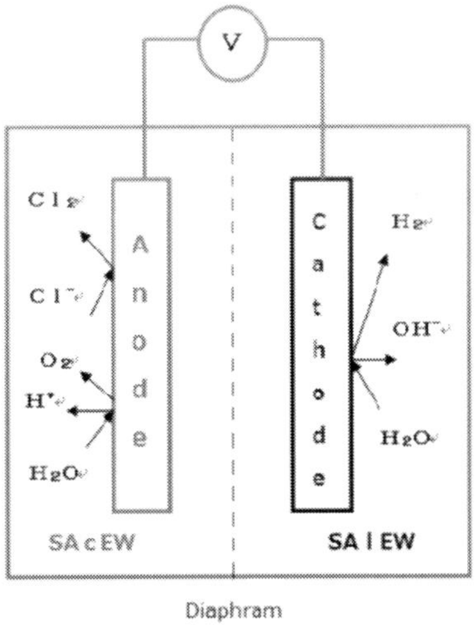

Figure 1. Principle of electrolyzed water production
SAcEW is generated on the anode side and SAIEW is generated on the cathode side.

The chemical reaction is follows,

Anode side

$H_2O \rightarrow 1/2O_2 + 2H^+ + 2e^-$

$2Cl^- \rightarrow Cl_2 + 2e^-$

$Cl_2\,(aq) + H_2O \leftrightarrow HClO + HCl$

Cathode side

$H_2O + 2e^- \rightarrow 1/2H_2 + OH^-$

2. Methods

1) Device

The study provides a vertical-type electrolytic tank with an asymmetric structure having a horizontal membrane, an anode chamber (large chamber) located below the membrane and provided with an anode electrode plate at a position an appropriate distance away from the membrane, and a cathode part (small chamber) disposed above the membrane and provided with a mesh-type cathode electrode plate at a position near the membrane(Fig.2).

Electrolysis is performed by applying a DC 20 voltage between the electrodes in a state in which the anode chamber (large chamber) is filled with saline, but the cathode part (small chamber) contains a small amount of saline, merely immersing the cathode electrode plate therein(Fig.3).

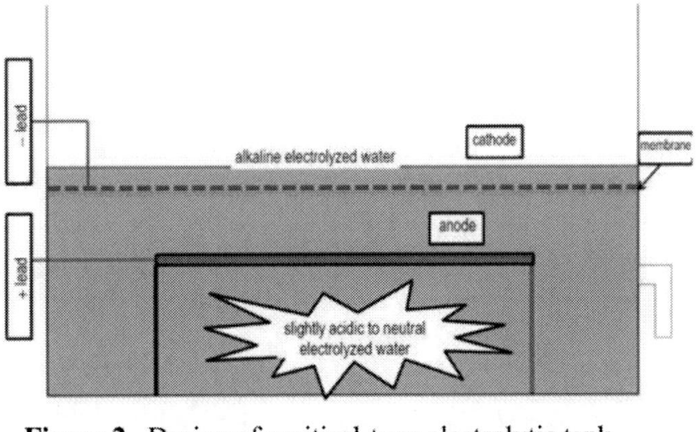

Figure 2. Design of veritical-type electrolytic tank

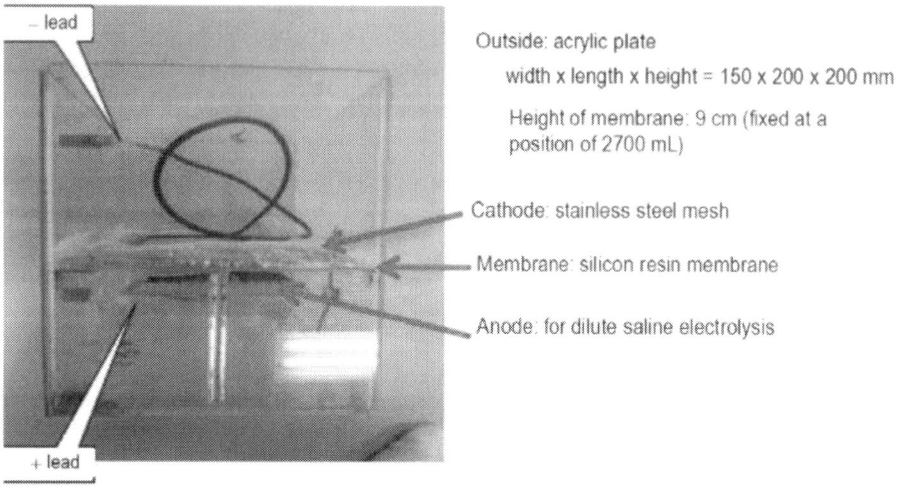

Figure 3. Prototype of veritical -type electrolysis tank electrolyzed water generator

2) Measurement

The values of the pH and AC concentration of the electrolyzed water extracted from the second chamber through the pipe were measured, by changing the time during which power was applied to the pair of the cathode and the anode (electrolysis time). The pH and AC concentration of the extracted electrolyzed water were measured at 0, 2, 4, 6, 8 and 10 minutes after the start of the electrolysis.

The measurement of AC concentration in electrolyzed water were measured with diethyl-p-phenylenediamine by a spectrophotometer.

3) Bactericidal activity

Three strains of bacteria(*E.coli, S.aureus and B.cereus*) were prepared to investigate the bctericidal activity of electrolyzed water after electrolysis for 10 minutes. The bactericidal activity was examined as follows. These three kinds of bacteria were cultivated at 37°C for 24 hours under aerobic-culture in petri dishes. After cultivation, each one colony was incubated with fresh electrolyzed water, also this solutions were added onto the fresh petri dishes and were cultivated for 48 hours. The colony of bacteria in the petri dish was counted and the bactericidal activity was judged.

Results

As a result, a small amount of SAIEW was generated in the upper cathodic small chamber, and a large amount of weakly acidic water generated in the lower anodic large chamber.

A part of this water passed through the mesh-type cathode electrode and the membrane and was mixed with a large amount of weakly acidic water generated in the lower anodic large chamber to produce slightly acidic to neutral bactericidal water in the anodic large chamber.

Though the pH of the solution in the lower tank shifts toward acidic after the start of electrolysis, the solution is mixed with the alkaline electrolyzed water generated in the upper part to become slightly acidic to neutral electrolyzed water 10 minutes later(Table 1).

Table 1. The pH and AC concentration of lower and upper electrolytic tank

	Electrolytic tank lower part		Electrolytic tank upper part	
Min	pH of solution	AC concentration (ppm)	pH of solution	AC concentration (ppm)
0	6.7	0	6.7	0
2	4.1	4.7	—	—
4	3.9	9.8	—	—
6	3.9	15.5	11.4	0
8	3.6	29	11.3	0
10	6.7	39.5	11.7	0

The results of bactericidal activity of electrolyzed water are shown in Table 2. There were many colonies of *E.coli, S. aureus* and *B.cereus* in each petri dish as controls. While there were no colony of bacteria in each petri dish after added electrolyzed water.

Table 2. Bctericidal activity of electrolyzed water

Bacteria	Control (cfu /ml)	SAcEW (cfu/ml)	NEW (cfu /ml)
E. coli	5.2×10^7	0	0
S. aureus	3.7×10^7	0	0
B. cereus	4.8×10^6	0	0

Discussion

Slightly acidic to neutral electrolyzed water has an advantage that its bactericidal activity stably persists for a long time, making it useful for sterilization, killing microorganisms, disinfection, and cleaning, as with strongly acidic electrolyzed water. Though maintenance for membrane replacement is necessary, the present study does not need to discharge water and is therefore highly efficient, without wasting water, and can be suitably applied to intermediate to large systems. In addition, since hydrochloric acid is not used, the electrolyzed water is nontoxic and safe and is also friendly to the environment. Accordingly, it is applicable to electrolyzed water generators producing bactericidal water that can contribute to killing bacteria on equipment and utensils in food factories, sterilization of instruments in medical-related departments, care homes for the elderly, cleaning toilets and the interiors of buildings, and cleaning bathrooms. Furthermore, since bactericidal water can be simply generated and returns to water when discarded, the present study is easy to use for improving public hygiene.

In an electrolyzed water generator having a membrane, equal amounts of acidic water and alkaline water are generated in the respective electrolytic chambers with the same capacities on both sides of the membrane. Consequently, slightly acidic to neutral water cannot be directly produced. A system for mixing appropriate amounts of both types of water generated by electrolysis includes an additional step and is therefore not an efficient method. Alternatively, a system for producing slightly acidic to neutral water by a method using hydrochloric acid as a pH adjuster with a membrane-less system suffers from the problem of handling hydrochloric acid.

The present study is characterized in that the vertical-type electrolytic tank with a horizontal membrane has an asymmetric structure composed of a large anode chamber located below the membrane and a small cathode chamber located above the membrane, and slightly acidic to neutral water with strong bactericidal ability is produced directly in the anode chamber. No agent such as hydrochloric acid is used, and the structure is a simple one that does not need any special tool. In the structure, a small part of the alkaline electrolyzed water generated in the cathode part (small chamber) remains in the chamber. This is not a problem for repeating electrolysis by extracting the slightly acidic to neutral water from the anode chamber and then feeding saline again. Regarding electrolysis of saline using a membrane, one method uses a two-chamber electrolytic tank where the acidic aqueous solution on the anode side and the alkaline aqueous solution on the cathode side are mixed at an appropriate ratio outside of the electrolytic tank to give electrolyzed water having a pH of 3 to 6. In a three-chamber electrolytic apparatus, acidic electrolyzed water, neutral electrolyzed water, and alkaline electrolyzed water are generated in the respective chambers. However, there has been no report of an apparatus that electrolyzes most of the solution into slightly acidic to neutral bactericidal water in an electrolytic tank having a membrane.

Conclusion

We made a vertical-type electrolytic tank with an asymmetric structure having a horizontal membrane and succeeded to produce lightly acidic to neutral water with strong bactericidal ability is produced directly.

Acknowledgment

This research was supported in part by a Grants-in-Aid for Scientific Research from Japan Society for the Promotion of Science.

References S

[1] K. Hotta, K. Kawaguchi, K. Saito, K. Ochi, T. akayama 1994 *Actinomyatologica* 8 51

[2] K. Umimoto, Y. Emori, H. Fujita, K. Jokei 2003 *IEEE-EMBS, APMBE, Proceedings CD-ROM*

[3] K. Umimoto, H. Kawanishi, Y. Tachibana, N. Kawai S. Nagata and J. Yanagida 2009 *Proceedings of IFMBE* 25 306

Journal of Physics: Conference Series **450** (2013) 012036 doi:10.1088/1742-6596/450/1/012036

Advanced Sensors for Accurate, Broadband AC Voltage Metrology

Thomas E. Lipe, Joseph R. Kinard, Donald B. Novotny, and June E. Sims

Quantum Measurement Division
National Institute of Standards and Technology, 100 Bureau Drive, M/S 8172
Gaithersburg, MD 20899-8172, USA[1]

Email: thomas.lipe@nist.gov

Abstract. We report on advances in ac voltage metrology made possible by a new generation of Multijunction Thermal Converters (MJTCs). Although intended for use primarily in high-frequency (1 MHz to 100 MHz) metrology, their exceptional low-frequency qualities, combined with a large dynamic range, makes these MJTCs excellent devices for the frequency range 10 Hz to 100 MHz at voltages from 1 V to 20 V, depending on the design. We anticipate that these devices will form the future basis for ac voltage metrology at the National Institute of Standards and Technology (NIST).

1. Introduction

The most accurate method of ac voltage measurement at frequencies from 10 Hz to 100 MHz is by comparing the heating effects of the unknown ac signal to that of a known dc reference using thermal voltage converters [1]-[4], a technique known as ac-dc transfer. These devices use one or more thermocouples arrayed along a heater structure to sense the heat generated by reference and unknown signals; comparing the output emfs allows us to determine the ac voltage in terms of the dc reference. Many National Metrology Institutes (NMIs) worldwide use planar multijunction thermal converters (MJTCs), fabricated on silicon substrates, as standards for ac-dc difference measurements [5,6].

Unfortunately, silicon has a relatively high dielectric permittivity ($\varepsilon_r \approx 11.8$) resulting in increased capacitive coupling between the heater and thermocouples through the thin silicon nitride membrane and silicon obelisk at high frequencies [7]. The large ac-dc differences (2 % or more at 100 MHz) resulting from this coupling largely restricts the use of silicon-based MJTCs to frequencies of less than 1 MHz. To obtain standards with improved performance at higher frequencies, NIST and other NMIs have developed MJTCs based on both crystalline quartz and fused silica substrates [8]-[10]. Both materials have significantly smaller relative permittivities than silicon ($\varepsilon_r \approx 3.8$) that reduces the high-frequency errors; however, crystalline quartz often exhibits microfractures that make the material too fragile for reliable fabrication, and so we have concentrated our work on electronics grade fused silica.

This paper describes the design and fabrication of fused silica MJTCs, their present use as ac voltage standards in the NIST AC-DC Difference Project, and their future use as both standards and as possible integrated parts of larger-scale sensor arrays.

2. Design of High-Frequency MJTCs

2.1 Membrane

All planar MJTCs share several design features, shown in figure 1. The heater structure is fabricated on a thin membrane, with the hot junctions of the thermocouples in close proximity to the heater. The first silicon-based MJTCs had membranes composed of layers of Si_3N_4 and SiO_2 to compensate the compressive and tensile stresses of each material. The layers were deposited on a silicon wafer using chemical vapor deposition, and then the silicon beneath the membrane removed using KOH etching. After much experimentation, this method reliably resulted in membranes a few micrometers thick. Later generations of silicon-based MJTCs featured pre-deposited, low-stress nitride layers forming the membranes, with the silicon removal accomplished using XeF_2 etching. This method is advantageous when including obelisks beneath the heater to increase the thermal mass and time constant, and reduce low-frequency errors.

The first high-frequency MJTCs fabricated on crystalline quartz used the quartz itself for the membrane; the backside etching was done using a 60:40 mixture of HF and NH_4F at 60 °C to achieve a membrane thickness of about 20 μm. To reduce etching time, the quartz wafers were only 250 μm thick. After etching, the wafers were often so fragile that they broke before any post-etching processing could be done. The few devices that were completely fabricated showed good performance, and served as testbeds for new fabrication techniques.

To increase the yield of the MJTCs, we investigated thicker (500 μm-thick) fused silica wafers back-etched using physical means to form membranes about 15 μm – 25 μm thick. This process is known as micropowder blasting and employs a fine spray of microscopic beads to physically remove material. Control of such a process can be quite good, and we experienced only a few isolated membrane failures due to the backside etching. The resulting membrane is stiff enough to support the heater, but still provide good thermal efficiency.

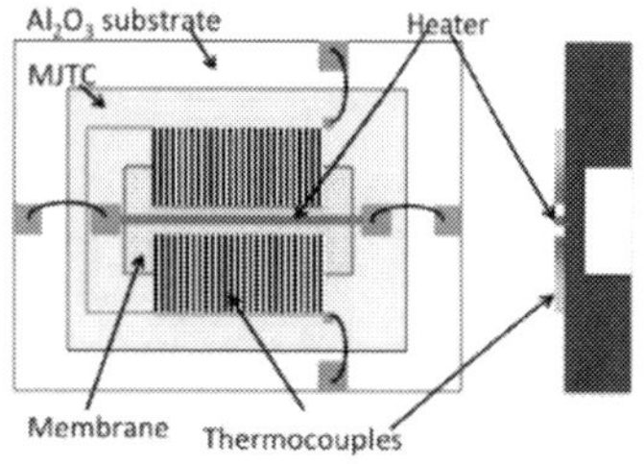

Figure 1. Drawing of a coaxial MJTC showing the primary components.

2.2 Heater

The heater presents possibly the greatest design challenge. The ac-dc difference of a thermal converter depends on both frequency-dependent effects (skin effect and transmission line effects) and on thermoelectric dc effects (Thomson and Joule heating) and any heater design that results in small ac-dc differences will necessitate a compromise between the two. The thermoelectric effects can be minimized using quality resistance material for the heater. We have chosen an alloy of $Ni_{75}Cr_{20}Al_{2.5}Cu_{2.5}$, a well-understood material with a temperature coefficient of resistance of about 10 μΩ/Ω/°C. This material has been used for over a century in the construction of high-precision resistance standards. Although readily sputtered, this material is prone to crazing at deposition thicknesses of more than a few hundred nanometers. We therefore used two heater designs and two heater widths to finely control the resistance. One design features a straight (coaxial) heater, while the other is shaped like a ∩ (bifilar), and both designs are made in 140-μm and 200-μm widths.

Of the frequency-dependent errors, skin effect results in a positive, generally linear error that is inversely dependent on heater resistance. Capacitive errors include distributed coupling between the heater and thermocouples through the membrane, and coupling between the heater bonding pads, and generally make a negative, square-law contribution. Simulations of our design indicate that the frequency-dependent effects should cancel at a heater resistance of about 800 Ω.

2.3. Thermocouples

Our thermocouples are $Cu_{55}Ni_{45}$ and $Ni_{90}Cu_{10}$, with a Seebeck coefficient for the pair of about 65 μV/K. In early devices, problems were encountered with the adhesion of the metals at the thermocouple interfaces. This was caused by material building up at the side of the feature, leading to the "bat ears" shown in figure 2. This effect, combined with inconsistent liftoff during fabrication, led to poor yields.

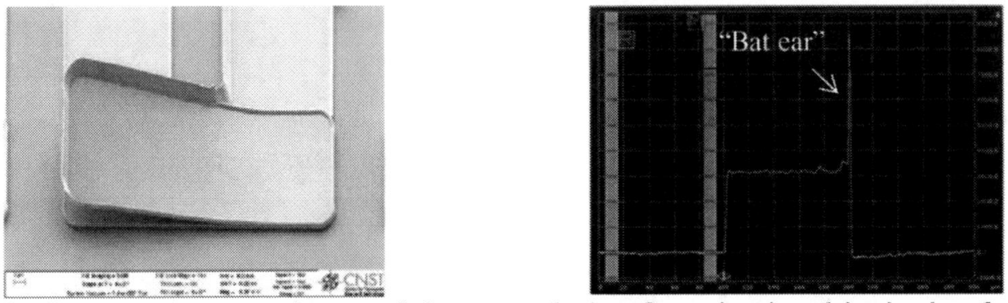

Figure 2. Electron microscope image of thermocouple interface, showing delamination from the material buildup during the metallization (left), and a profilometer image showing a "bat ear" (right). This feature is roughly 500 nm high, compared to the metallization of about 400 nm.

The thermocouple deposition was improved using several techniques. We discovered that the "bat ears" were the result of sputtering the thermocouple metals at a 60° incidence to the vertical. By siting the wafer on a wedge and keeping the platen stationary (instead of rotating), we were able to sputter the metals at normal incidence and prevent the buildup of material at the edges of the metal. The $Ni_{90}Cu_{10}$ tracks are sputtered as a continuous track with the $Cu_{55}Ni_{45}$ legs deposited atop the underlying layer, a structure commonly known as plated thermocouples. The thermocouple junctions are made at the interface between the two metals. Fabricating the thermocouples in this fashion eliminates possible open circuits if the two metals do not form good contacts at the interfaces, and significantly improves the yield. The thermocouple legs are each about 380 nm thick, about 1015 μm in length, and 50 μm wide. With 104 thermocouples in total, these design parameters result in an output resistance of 12 kΩ to 15 kΩ. While it is desirable to reduce the output resistance to prevent common-mode voltage issues when using certain voltmeters, thicker metallization often results in cracking of the metal films.

In the original MJTC designs, the thermocouple banks were rectangular in shape, with the hot junctions equally distant from the heater at both the high and low potential ends, resulting in more leakage current from the heater to the thermocouples at the high potential end. To reduce this effect, more recent MJTCs have wedge-shaped thermocouple banks, designed to present an equipotential gap between the thermocouple banks and heater. Gold contact pads 220 nm thick are deposited at the ends of the heater, and thermocouples, and over the bridge between thermocouple banks to reduce the output resistance.

Figure 3 shows the various designs fabricated and used at NIST.

2.4. Mounting

The MJTCs are mounted on ADS-96R alumina substrates. To test the effect of capacitive coupling from the heater through the silica to the alumina, we tested chips with and without the backside gold etch mask, and fixed to the alumina substrate using both conductive epoxy and nonconductive adhesive. After wire bonding the chips, a protective lid is glued to the substrate.

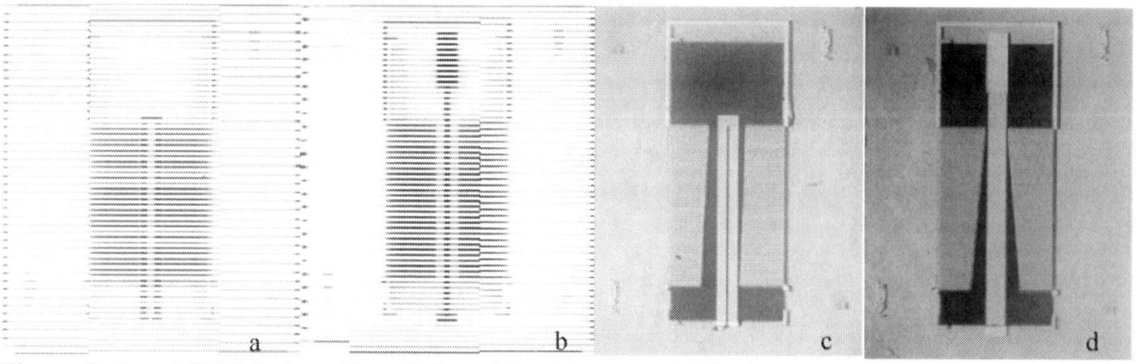

Figure 3. Four MJTC designs. From left to right: (a) narrow bifilar heater, rectangular thermocouple banks, backside gold etch mask stripped, (b) narrow coaxial heater, rectangular thermocouple banks, backside gold etch mask stripped, (c) wide bifilar heater, wedge thermocouple banks, backside gold intact, (d) wide coaxial heater, wedge thermocouple banks, backside gold intact.

The reference plane for ac-dc transfer measurements is usually defined at the center of a tee structure between a reference standard MJTC and a unit under test (UUT). At frequencies exceeding about 1 MHz the length of transmission line between the plane of reference and the MJTC can make a significant contribution to the overall ac-dc difference of the device. The best results were obtained by mounting the MJTCs in a cylindrical enclosure with a tee structure integrated into the front face of the enclosure, minimizing the electrical length between reference plane and MJTC [11].

3. Testing and Results

As the MJTCs were completed, they were mounted as described in Section 2, and measured against the NIST standards for ac-dc and RF-dc difference from 10 Hz to 100 MHz. The applied voltage depended on the resistance of the heater; devices with resistances ranging from 200 Ω to 3.5 kΩ were measured. A summary of the results at frequencies from 1 MHz to 100 MHz follows. The ac-dc differences at lower frequencies are negligible compared to the uncertainties.

3.1. Effect of heater design

To test the effect of heater design on the ac-dc difference, two MJTCs with nearly identical heater resistances were measured against the NIST standards. Both devices have rectangular thermocouple banks, wide heater design, and intact backside gold layer, and are mounted in the same enclosure; the only difference is in the type of heater. The results of the measurements are shown in figure 4a. The positive trend of the coaxial heater can be explained by assuming that a fraction of the heater current couples from the heater to the substrate through the silica membrane and gold etch mask, effectively bypassing the heater. Since this is an ac effect, the thermocouples sense less heat with ac voltage applied, leading by definition to a positive ac-dc difference. The bifilar design would be less prone to this effect because the fields are constrained to a smaller area.

3.2. Effect of backside gold layer

In the early chips, the backside gold layer was left intact during fabrication to facilitate the processing. This layer contributes to the ac-dc difference by creating a ground path for leakage currents from the heater. To determine the magnitude of this effect, two coaxial MJTCs from the same wafer and identical in all respects except that one had the gold layer stripped prior to measurement, were compared to NIST standards. As shown in figure 4b, the gold layer creates a noticeable effect on the performance.

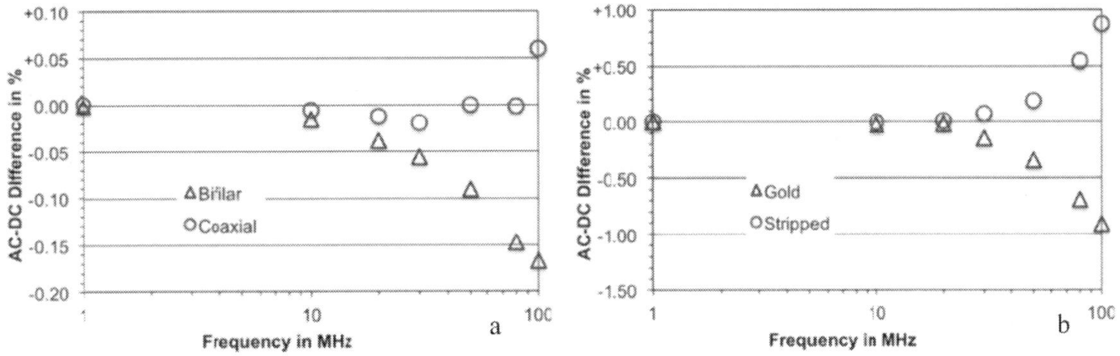

Figure 4. Comparison of heater design on ac-dc difference (a) and effect of the backside gold layer (b). Uncertainties for both graphs are given in Table 1.

3.3. Effect of thermocouple design

The original MJTC designs featured rectangular thermocouples. In order to reduce the leakage current through the membrane at the high potential end of the heater, a wedge-shaped thermocouple design was developed. This design maintains nearly equal potential along the length of the heater, and was applied to both coaxial and bifilar designs. Figure 5a shows the effect of the thermocouple design on the ac-dc difference for both types of heaters. As expected, the effect of the thermocouple design is more pronounced when used with coaxial heaters.

3.4. Voltage and frequency coefficients

The large number of thermocouples, compared to a traditional single junction thermal converter (SJTC), allows the MJTCs to be used effectively with input voltages ranging from less than 20 % to 100 % of rated voltage, as compared to 50 % to 100 % common with SJTCs. In addition, the construction of MJTCs reduced the risk of damage due to over-ranging of the converter. A key demand for the use of MJTCs in a calibration environment is that the variation in ac-dc difference with changing input voltage levels be small. Similarly, these devices are designed to be broadband, with minimal variation in ac-dc difference between 10 Hz and 100 MHz. Figure 5b shows both voltage and frequency coefficients for a typical bifilar MJTC with wedge thermocouples.

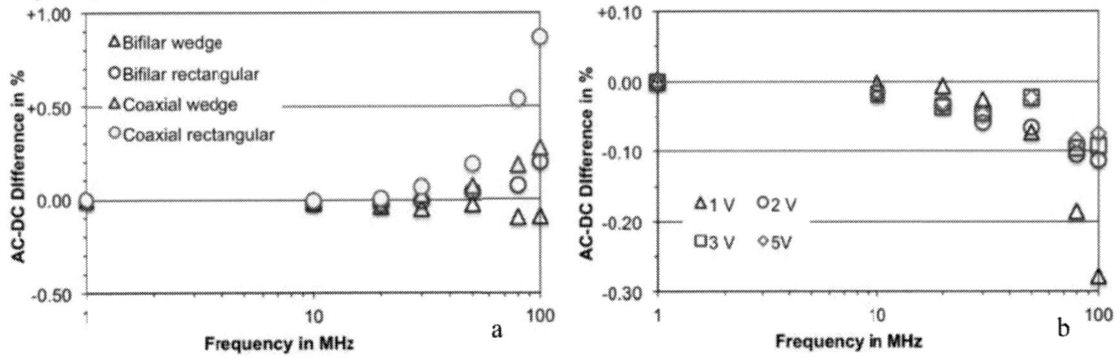

Figure 5. Effect of thermocouple design on various heater designs (a). The MJTCs have roughly identical heater resistances. Voltage and frequency coefficients for a bifilar heater design with wedge thermocouples (b). Uncertainties for both graphs are given in Table 1.

Table 1. Uncertainties for ac-dc differences reported in Figures 4 and 5 ($k = 2$)

1 MHz	10 MHz	20 MHz	30 MHz	50 MHz	80 MHz	100 MHz
0.0018 %	0.015 %	0.025 %	0.035 %	0.080 %	0.150 %	0.180 %

4. Revised Uncertainties

The large dynamic range of the new MJTCs, and their relatively flat frequency response from 10 Hz to 1 MHz will allow us to populate the uncertainty parameter space using significantly fewer scaling steps than are presently used. Preliminary calculations indicate that the uncertainties over the parameter space from 0.5 V through 10 V or possibly 20 V may be reduced by about half through the use of MJTCs. At frequencies above 1 MHz, the small dependence on ac-dc difference with increasing frequencies will allow us to significantly reduce the uncertainties at these points as well. Coupled with the development of an ac standard based on quantum effects [12], MJTCs will enable NIST to meet customers' future requirements for very accurate, broadband ac voltage measurements.

5. Other Applications

MJTCs of similar design have been used in applications such as low-flow gas measurements [13], and pressure sensors. Owing to their small size and fabrication techniques, we envision these devices as components of integrated sensor packages where ac voltages must be accurately known or supplied. For example, a sensor package may require a well-known voltage for measurement of a particular quantity; either an MJTC could accurately sense an external voltage supplied to the package, or an MJTC with integrated micropotentiometer [14] could be used to supply a regulated voltage. In either case, the accuracy of the sensor package would benefit from the integration of an MJTC.

6. Conclusions

MJTCs fabricated on fused silica substrates have ac-dc differences significantly smaller than traditional thermal converters at frequencies up to 100 MHz from 1 V to 10 V. Combined with the small input voltage dependence (negligible at frequencies below 1 MHz) and a large dynamic range, these devices are presently replacing the older designs for ac-dc difference metrology at NIST. It is anticipated that the used of these standards will lead to significant reductions in uncertainty for the NIST calibration services for ac-dc difference and ac voltage. It is also envisioned that similar devices will be integrated into larger sensors or networks to monitor and/or supply accurate ac voltages.

Acknowledgements

This research performed in part at the NIST Center for Nanoscale Science and Technology. Partial funding provided through Calibration Coordination Group (CCG) Project 546.

References

[1] Hermach F 1952 *J. Res. Natl. Bur. Stand.* **48** 121 – 138
[2] Kinard J and Cai T-X 1989 *IEEE Trans. Instrum. Meas.* **38** 360 – 367
[3] Wilkins F 1972 *IEEE Trans. Instrum. Meas.* **21** 334 – 340
[4] Hermach F, Kinard J and Hastings J 1987 *IEEE Trans. Instrum. Meas.* **36** 300 – 306
[5] Klonz M and Weimann T 1989 *IEEE Trans. Instrum. Meas.* **38** 335 – 337
[6] Kinard J, Novotny D, Lipe T and Huang D-X 1997 *IEEE Trans. Instrum. Meas.* **46** 347 – 351
[7] Wunsch T, Manginell R, Solomon O, Kinard J, Lipe, T and Jungling K 2001 *Proc. Of the 18th IEEE Instrumentation and Technology Conference, 2001 (Budapest, Hungary, 21-23 May 2001)* pp 450 – 452
[8] Scarioni L, Klonz M, Laiz H and Kampik M 2003 *IEEE Trans. Instrum. Meas.* **52** 345 – 349
[9] Scarioni L, Lipe T and Kinard J 2009 *IEEE Trans. Instrum. Meas.* **58** 868-871
[10] Lipe T, Kinard J and Novotny D 2012 *Proc. 2012 Conference on Precision Electromagnetic Measurements (CPEM) (Washington, DC USA, 2-6 July 2012)* pp. 434 – 435
[11] Free G, Lipe T, Kinard J and Sims J 2007 *IEEE Trans. Instrum. Meas.* **56** 341 — 345
[12] Lipe T and Kinard J 2012 *IEEE Trans. Instrum. Meas.* **61** 2160 – 2166
[13] Blackburn D and Ortega A 1996 *Proc. 2nd International Workshop on Thermal Investigations of ICs and Microstructures (THERMINIC) (Budapest, Hungary, 25-27 September 1996)*
[14] Kinard J, Huang D-X and Novotny D 1995 *IEEE Trans. Instrum. Meas.* **44** 374 – 376

Journal of Physics: Conference Series **450** (2013) 012037 doi:10.1088/1742-6596/450/1/012037

Investigation of the significance of the 'body effect' on sensitivity to metallic objects in a walk-through metal detector

L A Marsh[1,4], J Makkonen[2], J Vihonen[2], A Visa[2], A Järvi[3], D W Armitage[1] and A J Peyton[1]

[1]School of Electrical and Electronic Engineering, The University of Manchester, Manchester, M13 9PL, UK
[2]Tampere University of Technology, Department of Signal Processing, Korkeakoulunkatu 10, P.O. Box 553, FIN-33110, Tampere, Finland
[3]Rapiscan Systems Oy, Klovinpellontie 3, Torni 2, FIN-02180 Espoo, Finland

E-mail: liam.marsh@manchester.ac.uk

Abstract. An investigation has been carried out to determine the extent to which a walk-through metal detection system is affected by the capacitive and inductive coupling between candidates' bodies and the coil array – known as the 'body effect'. In this experiment both small and large items are investigated to determine ratio of the signal contribution from the candidate compared to the object, and a comparison is made between the response of a small object both with and without the candidate. Also an experiment is presented to demonstrate the inductive / capacitive nature of this signal.

1. Introduction

Walk-through metal detectors (WTMDs) are widely used for personnel screening in security and checkpoint applications. The sensitivity of these detectors is limited by either interference, electronic noise in the receiver circuitry or by spurious inductive or capacitive coupling associated with the presence of the candidate, often known as the *body effect*. A similar limitation is encountered with other metal detection systems such as the *product effect* with on-line conveyor type metal detectors and the *ground effect* with metal detectors for detection of buried objects. This paper investigates the significance of the body effect signal compared to that of metallic targets. There has been little research into this area despite the potential sensitivity limit it presents. It is expected that by better understanding the body effect it should be possible to overcome this restriction.

In this paper a walkthrough metal detector (WTMD) capable of calculating the magnetic polarisability tensor of metallic objects, $\overleftrightarrow{\mathbf{M}}$, is used as a measurement system. The operation of this system is reported elsewhere [1]. Previous studies [2] have shown that the inductive signal from metallic objects can be related to the magnetic field via the magnetic polarisability tensor, $\overleftrightarrow{\mathbf{M}}$, which is represented as a complex, symmetric 3×3 matrix.

[4] To whom any correspondence should be addressed

System measurements, λ, are related to the magnetic polarisability tensor and the incident magnetic field vectors as shown in equation 1. In this expression the vectors $\vec{\mathbf{H}}_{tx}$ and $\vec{\mathbf{H}}_{rx}$ correspond to the values of H-field for the transmit and receive coils respectively. The system inverts the tensor from the computed magnetic field data and measurements. However, this expression does not contain any terms which take into account the effect of the capacitive signal. Consequently any signals of capacitive nature cause errors in the data fitting in the inversion algorithm. In order to improve the data fit, and consequently to perform more accurate tensor inversion it is necessary to account for the capacitive input to the system prior to performing the inversion.

$$\lambda = \overleftrightarrow{\mathbf{M}}\vec{\mathbf{H}}_{tx} \bullet \vec{\mathbf{H}}_{rx} \tag{1}$$

2. Background

Like all electrically conductive objects the human body can store electric charge and display capacitive propoerties. This behaviour is exploited by devices such as touch-screens, however the build-up of static electricity can cause damage to sensitive electronic components and in some electrostatically sensitive processes the operators must also be earthed via a tethered conductor. Typically the human body has a capacitance in the region of 100 pF [3].

In a similar way as eddy currents flow around metallic targets, they are also able to flow around the human body. This leads to the possibility of a distributed inductive response from the candidate as induced eddy currents circulate through conductive and dielectric body tissues. The electrical conductivity of human muscle is approximately 2.5×10^{-1} Sm^{-1} at 10 kHz [4]. Although this is considerably smaller than that of metallic targets e.g. alumimium at $3.58 \times 10^{7} Sm^{-1}$ [5] it is still measurable. As the human body is considerably larger than the pairs of transmit-receive coils throughout the detector the effect is distributed and consequently does not follow a response similar to the case of small metallic targets. Most biological tissues contain water, which is weakly diamagnetic, having a real magnetic susceptibility, χ, of about -9×10^{-6} [6] at room temperature. As a consequence of this it is known that the magnetic properties of water influence the magnetic properties of biological tissues.

An example of an inductive response from a metal object, in this case a model aluminium gun [7] as defined by the *US National Institute for Justice* (NIJ) is shown in figure 1; the x-axis has been normalised such that the object is defined at being in the centre of the detector at t=*0s*. It is possible to compare this signal with the coil geometry used to acquire it, which is shown in figure 3, note the symmetry of the object response, which results from the symmetrical design of the coil array. The coils shown in figure 3 display the fact that the transmit coil (shown in the solid line) contains a single crossover at the point x=*0m* and the receive coil (shown with a dashed line contains two crossover points at x=$\pm$*0.18m*). These three crossover points are shown on figure 1 at the points t=*-0.23*, *0* and *0.2s*.

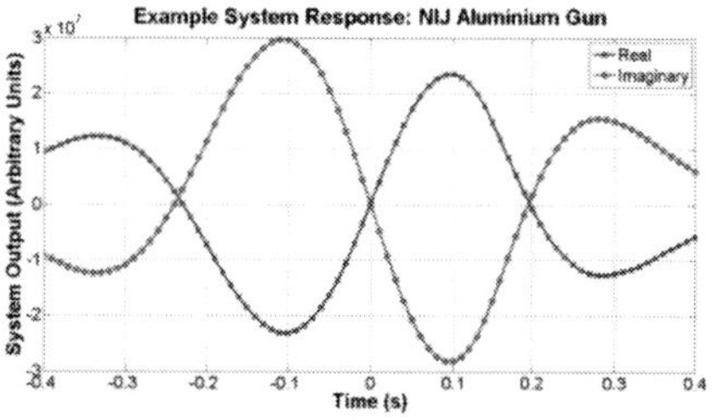

Figure 1. Example inductive signal

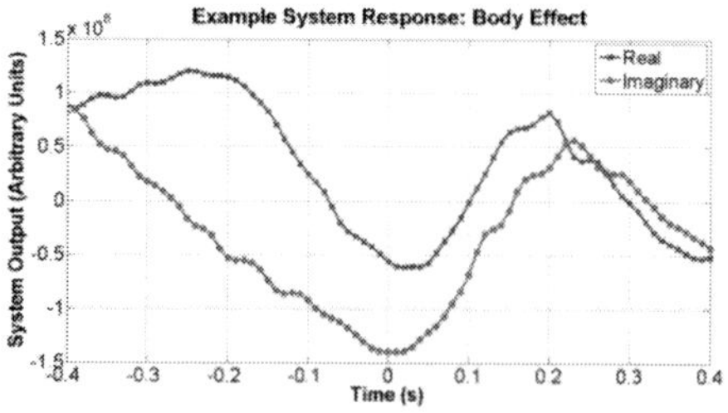

Figure 2. Example body effect signal

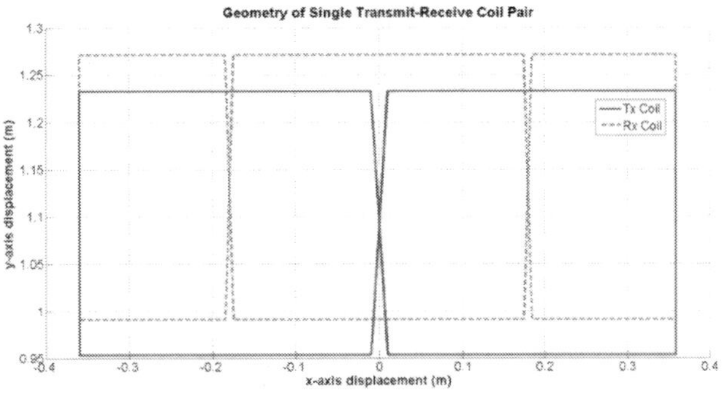

Figure 3. Example tx-rx coil pair

Unlike the high conductivity inductive signal shown in figure 1 a typical body effect signal does not follow a trend. An example of the signal produced by a 'clean scan' – i.e. from the passage of a candidate free from metallic objects is shown in figure 2. The data for this figure was obtained using the same coil pair used for the measurements shown in figure 1. Now however, there is no zero-crossing in the centre of the signal, no symmetry and the real and imaginary components are no longer proportional to each other (i.e. a straight line in the impedance plane), as is the case for the high conductivity inductive measurements.

3. Experimental Setup

Five candidates each made a total of 40 passes through the detector whilst wearing clothing which was free from metallic elements. Ten passes were recorded for when the candidate had no metallic items in their possession, and for the case when they carried steel and aluminium NIJ handguns and an NIJ aluminium knife [7]. On each occasion when the candidate was carrying an item it was placed at the midpoint of the detector with respect to the panel-to-panel displacement, a height of 1.1m with respect to the ground level and all points in the direction of transit through the detector (shown as the x, y and z-axes respectively on figure 4).

In order to determine the nature of the body effect, i.e. if it is predominantly capacitive or inductive, a second investigation has been conducted. This experiment consists of demonstrating the effect that grounding the candidate has on the portal, and the system response to saline of varying conductivity. In the first test a candidate entered the detector and remained inside the portal whilst stretching their right arm outside the sensitive region of the detector. After several seconds the candidate had a grounded wire touched against their right hand for several more seconds; this was then removed. Due to the fact that grounding the candidate is a capacitive action, a change in signal level is expected should the portal be sensitive to the capacitance of the body. The second test involved recording the signal response for a 10 L bucket of saline of varying conductivity from 13.83 mSm^{-1} to 11.18 Sm^{-1}.

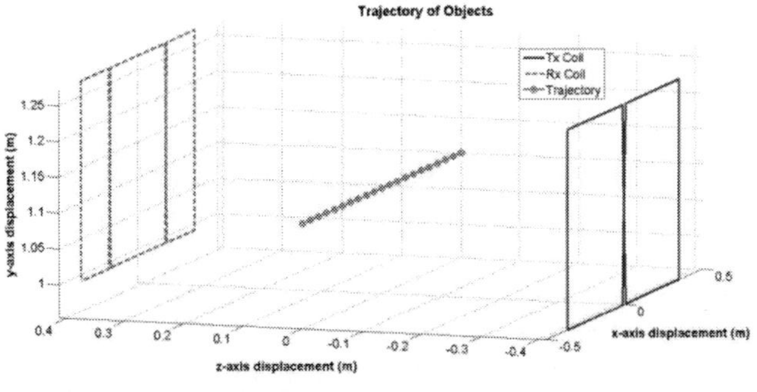

Figure 4. Approximate object trajectory

For each candidate a series of physical measurements were taken to allow for estimation of their body size. In total five parameters were measured – height, the peak width of shoulders, the peak chest size, the waist size and the height of the waist from the ground.

4. Results

The main results from this experiment are shown in figures 5 to 8. In each figure a total of ten walkthrough scans for each of the five candidates is shown. It is possible to see from figures 6 and 8 that the signal from the guns is so strong that it is difficult to see the effect of the body on the measurements. This is reflected in the fact that all of the scans are fully overlaid on one another. Figures 5 and 7 however show that the different candidates give distinguishable responses which are consistent across each of the candidate's walkthroughs. Also inculded in figure 7 are the results for ten scans in the case when the object is not attached to a candidate. To achieve these measurements the object was attached to an insulating pole, and passed through the WTMD.

Although the measurements for the aluminium knife have a far smaller signal than that of the aluminium gun (approximately an order of magnitude difference) it is still possible to see that the two signals share a common phase bias; this is as a result of the response of the aluminium. However, in the case of the knife it is possible to see that this response is superimposed onto that of the body signal to produce the different candidate clusters shown. In the case of the candidate-free scan for the aluminium knife it is possible to see that the response (labelled 'No Candidate'on figure 7) is very linear. This figure clearly shows that the presence of a body along with the object can significantly obscure the target response.

Table 1 shows the Pearson correlation coefficient of the NIJ aluminium gun and aluminium knife for each candidate, averaged across all ten walk-through scans. It also shows the average for the candidate-free scan of the aluminium knife. The aluminium gun measurements show very strong correlation for all candidates with all values in excess of -0.99, thereby verifying that the body signal produced by the candidates has not significantly distorted the response. However, in the case of the aluminium knife the correlation is significantly worse for each candidate, ranging from -0.639 to -0.742. In contrast to these values the coefficient for the 'no candidate' case is in excess of -0.99. These results confirm that the system is capable of measuring a clear, linear response for small targets, however that they are adversely affected by the body effect.

In the case of the clean scans it is also possible to identify clustering of different candidates, however, as would be expected, this is not as obvious as with the aluminium knife. The results show that all five candidates show a distinctly repeatable response when walking through the detector, and with the exception of a single scan for 'candidate 5' the walk-through profiles do not significantly deviate from the main cluster for each candidate.

Table 1. Pearson correllation coefficient for aluminium targets.

Object	Candidate 1	Candidate 2	Candidate 3	Candidate 4	Candidate 5	No Candidate
Knife	-0.7420	-0.6329	-0.7017	-0.6536	-0.6388	-0.9994
Gun	-0.9974	-0.9989	-0.9967	-0.9996	-0.9993	*n/a*

Table 2 shows the recorded dimensions of each of the five candidates. The table shows that candidates 3, 4 and 5 are generally larger than candidates 1 and 2. However, the region of interest (1.1 m from the ground) corresponds approximately with waist-level for all candidates; the order of candidates from largest to smallest waist size is [4; 2; 3; 5; 1]. Analysis of the data in figure 5 shows that there is no clear correlation between waist size, or between the candidates' physical size and the magnitude or phase of the response. This implies that the relationship between the candidate and the clean scan response is not directly based upon physical size of the body.

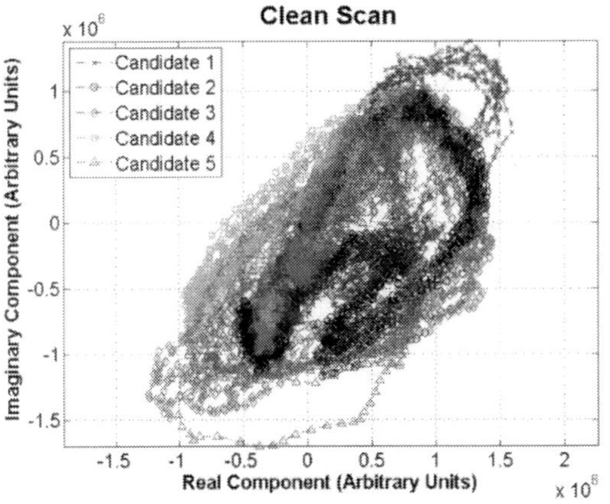

Figure 5. Results of clean scan

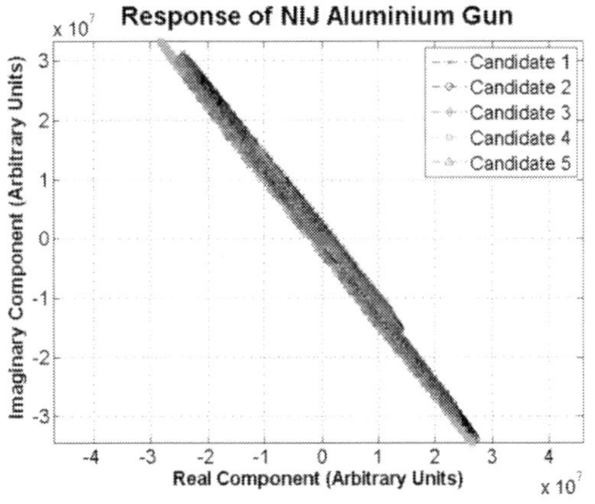

Figure 6. Results for NIJ aluminium gun

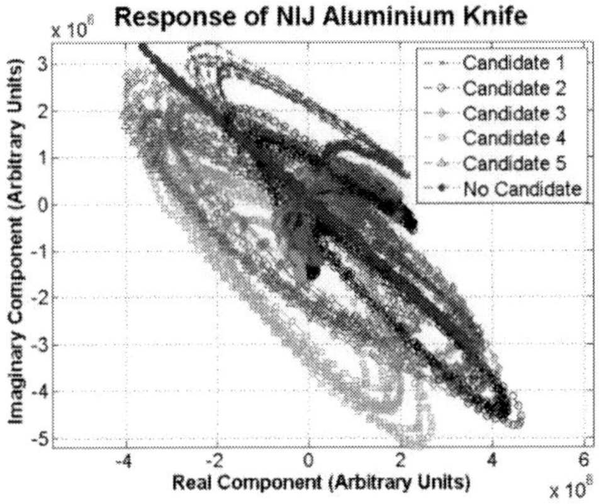

Figure 7. Results for NIJ aluminium knife

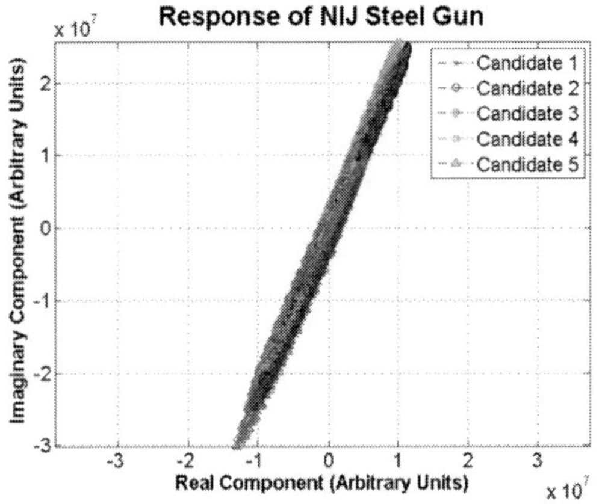

Figure 8. Results for NIJ steel gun

Table 2. Selected dimensions of candidates.

Dimension	Candidate 1	Candidate 2	Candidate 3	Candidate 4	Candidate 5
Height	1.79 m	1.74 m	1.87 m	1.86 m	1.85 m
Shoulders	1.15 m	1.04 m	1.34 m	1.23 m	1.17 m
Chest	0.96 m	0.99 m	1.09 m	1.10 m	1.02 m
Waist	0.86 m	1.06 m	1.02 m	1.10 m	0.94 m
Waist Height	0.98 m	0.95 m	1.07 m	1.05 m	1.00 m

4.1. Investigation of the Nature of the Body Response

Figure 9 shows the results of the tests for capacitive coupling. In figure 9(a) at sample reference $S=0$ the detector is empty. At approximately $S=400$ 'candidate 1' stepped into the detector space; the response jumps up at the point and settles down. At $S=750$ the candidate was grounded, and remained grounded until $S=2350$. The candidate then remained in the portal until $S=2850$. Each of these phases are clearly visible on the figure. The clear change that occurs when the candidate is grounded demonstrates that the body signal contains a significant capacitive component. In figure 9(b) the response is seen to vary almost entirely in the quadrature component as a function of conductivity. The offset of this linear response is also indicative of capacitive coupling between the coils and the saline.

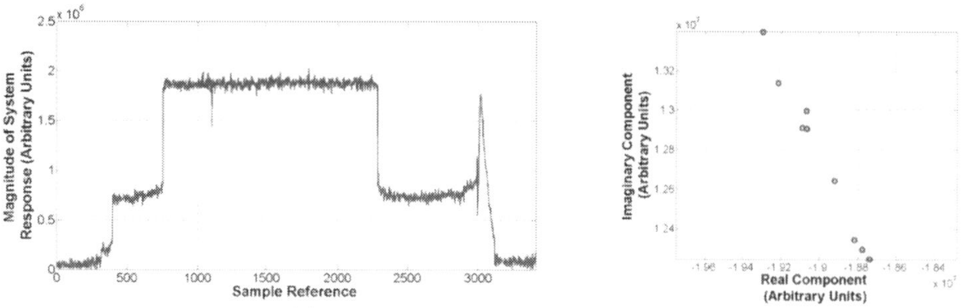

(a) (b)

Figure 9. Tests for capacitive coupling between the candidate and the detector

5. Conclusions and Future Work

The results from this paper show that calibration for the body effect is not required for highly detectable objects such as the NIJ handguns; this is due to the fact that the inductive signal dominates in such cases as figures 6 and 8 show. However, there is a visible body signal in the case of both no metallic targets, and for small metallic targets as in figures 5 and 7 respectively. The experiments presented in this paper have demonstrated that a significant proportion of this body signal occurs as a result of capacitive coupling between the candidate and the coil array. The magnitude of this body effect response compared to that of small targets demonstrates a need for either calibration to account for the body signal, or screening to reduce it.

The results of this experiment demonstrate that the body effect appears to be distinct for each candidate. This is reflected in the fact that each of the five candidates produces a clustered response which is shown for all ten scans. This is an encouraging result as it demonstrates that the body effect is both measurable, and repeatable, which are two requirements for the development of a calibration routine.

The next step for this research, having identified the capacitive element of the body effect, is to establish a measurable, theoretical link which can be used to model it. This would then allow for the inclusion of the body effect into the forward model (equation 1) which would subsequently improve the quality of the measured data.

References

[1] Marsh L A, Ktistis K, Järvi A, Armitage D W and Peyton A J 2013 Three-dimensional object location and inversion of the magnetic polarisability tensor at a single frequency using a walk-through metal detector Meas. Sci. Technol. vol. 24 045102 doi:10.1088/0957-0233/24/4/045102

[2] Norton S J and Won I 2001 Identification of buried unexploded ordnance from broadband electromagnetic induction data IEEE Trans. Geosci. Remote Sens. 39 2253-2261

[3] Electrostatic Discharge Association 2010.ESD Part 5: DeviceSensitivity Testing – Human Body Testing, ESD Association, Rome, NY. pp 1.

[4] Gabriel C, Gabriel S and Corthout E 1996 The dielectric properties of biological tissues: I. Literature survey Phys. Med. Biol. vol 41 pp. 2231-2249

[5] Electrical Conductivity and Resistivity for Aluminum and Aluminum Alloys 1999. Retrieved April 22, 2013 from http://www.ndt-ed.org/GeneralResources/MaterialProperties/ET/ET_matlprop_Aluminum.htm

[6] Arrighini G P ,Maestro M and Moccia R 1968 Magnetic Properties of Polyatomic Molecules: Magnetic Susceptibility of H_2O, NH_3, CH_4, H_2O_2. J. Chem. Phys. 49 (2): 882-889. doi:10.1063/1.1670155

[7] Paulter N and Larson D 2009 NIJ Metal Detector Test Object Report (Gaithersburg, MD: National Institute for Justice)

Acknowledgements

The authors would like to thank *Find A Better Way* and the *Academy of Finland* for their financial support of this research under the programmes *SEMIS* and *MIDAS II (grant no. 251777)* respectively. They are also grateful to Michael O'Toole, Omar Abdel Rehim and John Wilson at the University of Manchester for their participation in this experiment.

Journal of Physics: Conference Series **450** (2013) 012038 doi:10.1088/1742-6596/450/1/012038

Classification of metallic targets using a single frequency component of the magnetic polarisability tensor

J Makkonen[1,4], L A Marsh[2], J Vihonen[1], A Visa[1], A Järvi[3] and A J Peyton[2]

[1] Tampere University of Technology, Department of Signal Processing, Korkeakoulunkatu 10, P.O. Box 553, FIN-33101 Tampere, Finland
[2] School of Electrical and Electronic Engineering, The University of Manchester, Manchester, M13 9PL, UK
[3] Rapiscan Systems Oy, Klovinpellontie 3, Torni 2, FIN-02180 Espoo, Finland

E-mail: jarmo.makkonen@tut.fi

Abstract. A k-nearest neighbour (KNN) classification algorithm has been added to a walk-through metal detection system which is capable of inverting the magnetic polarisability tensor of metallic targets at a frequency of 10 kHz. Pre-computed library data is used to determine the class of the object, e.g. 'knife' or 'mobile phone', and is consequently capable of determining if an object is considered a threat. The results presented show a typical success rate of 95%. An investigation into classification accuracy between different candidates is also presented to determine the significance of the *body effect* on the success of the classification.

1. Introduction

Modern walk-through metal detectors are incredibly sensitive, precision measurement systems. Their widespread use for detection of threats in environments such as airports, prisons and public buildings, combined with the competition between several leading manufacturers has ensured that the current generation of technology is capable of detecting very small items such as handcuff keys and integrated circuits [1]. In the aviation industry metallic threats are principally considered to be knives and guns and are thus considerably larger than the smallest detectable objects. Improved signal-to-noise ratios may continue to be sought by manufacturers in an attempt to increase the detectability of non-magnetic, low conductivity materials such as some stainless steels, however the most significant challenge is no longer to detect the objects, but to determine whether they present a threat or not.

It is reported that between 2000 and 2010 a total of 17 out of an estimated 310 million commercial flights were hijacked worldwide [2]. Using these numbers it is estimated that in this ten year period the chances of a plane being hijacked was in the region of 1 in 18.2 million. This statistically rare threat can be contrasted with the fact that in environments such as airports it is common for people to carry several innocuous items such as mobile phones, jewellery, or keys, and current regulations require that travellers must remove these items prior to screening. This causes a great deal of disruption and inconvenience, and requires a large number of staff to administer. These facts demonstrate that any method capable of determining which items located on a person are threatening, and which are innocuous has the potential to greatly improve the aviation industry.

[4] To whom any correspondence should be addressed.

In this study, a tomographic metal detection system [3] which is capable of inverting the magnetic polarisability tensor, $\overleftrightarrow{\mathbf{M}}$, from a single walk-through scan has been used as the measurement system. A library of tensors produced by this system has been recorded for a variety of different objects. To examine the capacitive and inductive effect of the human body on the measurements, known as the *body effect*, a separate library using two candidates has been collected. A classification algorithm based on the k-nearest neighbour (KNN) approach has been implemented, and the performance of the algorithm is tested for both the general library case, and the body effect case.

Although the method which is reported here is done so with focus on the aviation industry, the methods documented in this paper may be applied to any system which is capable of yielding the magnetic polarisability tensor; including detection systems for UXO/landmines.

2. Method

2.1. Producing the library

The library was constructed by simple walkthroughs using the WTMD described in [3]. This system uses eight pairs of transmit-receive coils, each of which operate at a different frequency. These frequencies range from 8 kHz to 13.8 kHz and are arranged lowest to highest from floor level upwards, with a separation of approximately 700 Hz between neighbouring channels. The system inverts a single tensor, and consequently offers only a single point on the tensor's continuously varying frequency spectrum [5]. Theoretically the inverted tensor is valid at the frequency corresponding to the parallel channel which is level with the object, however to for simplification and as the range of operating frequencies is relatively small the tensors are assumed to correspond to a frequency of 10 kHz – the median frequency value centred about the average height of candidates.

The library was constructed by repeatedly conducting walkthrough measurements. One object was carried through the detector at a time per scan, and its orientation and position on the candidate was varied between scans. The library contains a total of 33 different objects, and was built using data from 1316 walk-through scans. These scans were conducted using three different candidates. The objects belong to 10 different classes which are shown below. Numbers in brackets represent the total number of measurements per class.

- **Belts (40)**
  - *1 belt*
- **Coins (80)**
  - *3 sets of coins*
- **Keys (90)**
  - *3 sets of keys*
- **Knives (424)**
  - *11 knives*
- **MP3 Players (80)**
  - *4 MP3 players*
- **Pocket Mirrors (50)**
  - *1 pocket mirror*
- **Mobile Phones (120)**
  - *4 mobile phones*
- **Wristwatches (40)**
  - *1 wristwatch*
- **Guns and gun parts (352)**
  - *2 guns*
  - *1 set of bullets*
- **Scissors (40)**
  - *1 pair of nail scissors*
  - *1 pair of office scissors*

The data recorded for each walk-through consists of the estimated tensor value, $\overleftrightarrow{\mathbf{M}}$, the residual of the inversion algorithm, r, and the estimated coordinates for the path of the object, **P**. The main library, subsequently referred to as '*Main, all*' contains of all the recorded measurements.

A second library, called *'Reduced'*, was produced which consisted of a single object per class. This was done to balance the probability of classification, as the outcome of KNN algorithms is known to be dependent upon the number of samples per class. The data consisted of measurements from two different locations, in the left trouser pocket and along the central line of the portal at chest height; these are labelled *'Left pocket'* and *'Chest'*. At each position, 10 scans were performed, resulting in 20 measurements for each object. Thus, the total size of this second library was 200 measurements.

Previous testing with a calibration object has indicated that some tensor components deviate by up to 50% when the same object is scanned in different regions of the portal. Typically this variation occurs as the object trajectory moves from one side panel of the detector to the other. In order to demonstrate this effect and to evaluate the performance of the KNN algorithm in such circumstances, the results from the *'Chest'* and '*Left pocket*' libraries described above have been compared.

Tensor reliability can be estimated from the residual value, r, of the inversion algorithm described in [3]. The residual is calculated by taking the L2-norm of the difference between the actual measurements, ρ, and the forward response as a function of the inverted tensor and 3D coordinates, $\hat{\rho}$, and dividing this value by the L2-norm of the inverted measurements, $\hat{\rho}$; this is defined in (1).

$$r = \frac{\sqrt{\sum_{k=1}^{n} \left|\rho_k - \hat{\rho}_k\right|^2}}{\sum_{k=1}^{n} \left|\hat{\rho}_k\right|^2} \quad (1)$$

Experience has shown that residual values greater than 0.50 indicate unreliable tensors, and that anything in excess of 0.35 can be considered to be a poor representation of the object's tensor. As a result of this observation a further two versions of each library have been created. These have been made by selecting all of the samples that have a residual less than 0.5 and 0.35. The libraries are called *'X, r < 0.5'* and *'X, r < 0.35'* respectively, where *X* represents the base library name.

A total of ten objects were used to construct the library for investigation of the body effect, five of which were threat objects and the remaining five were innocuous ones. The threat objects were the aluminium and steel NIJ handguns [6] and three knives - *Opinel Lame Acier 12*, the *Opinel Inox 8* and a large kitchen knife with a stainless steel blade.

Two adult candidates of substantially differing heights were used to conduct this experiment. Candidate A had a height of 1.86 m and candidate B had a height of 1.55 m. Each person performed the same number of walkthroughs, 10 for each object, which resulted in 100 measurements per candidate. The object location was chosen to be central with respect to the distance between the side panels, and at chest height of candidate A. The same height was maintained for candidate B, resulting in keeping the object at approximately neck level.

Figure 1 shows a typical response of how the inverted tensors vary with different candidates. The different colours represent the two candidates, and the different markers represent the three different eigenvalues. The two test objects, made of aluminium and stainless steel, in the case shown represent guns. Due to the clustering of the points for each candidate it is believed that the differences correspond to the body effect rather than noise from the inversion algorithm.

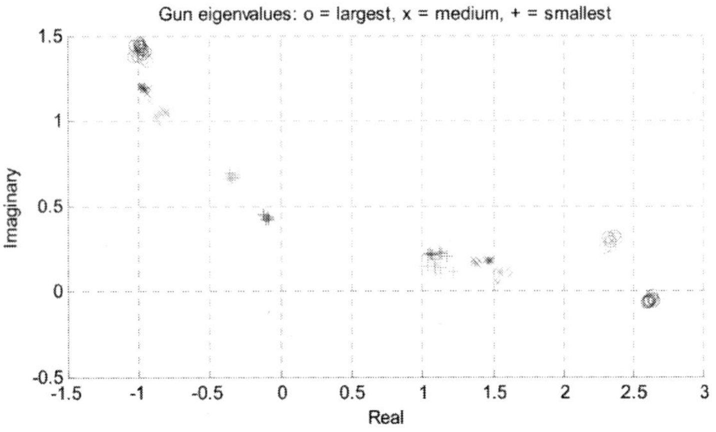

Figure 1. Body effect on eigenvalues

2.2. Implementation of the algorithm

The method is based on the classification algorithm presented in [4]. The magnetic polarisability tensor, $\overleftrightarrow{\mathbf{M}}$, takes the form of a complex, symmetric 3×3 matrix as shown below:

$$\overleftrightarrow{\mathbf{M}}(f) = \begin{bmatrix} M_{11} + jN_{11} & M_{12} + jN_{12} & M_{13} + jN_{13} \\ M_{12} + jN_{12} & M_{22} + jN_{22} & M_{23} + jN_{23} \\ M_{13} + jN_{13} & M_{23} + jN_{23} & M_{33} + jN_{33} \end{bmatrix}$$

The algorithm firstly calculates the eigenvalues of the tensor, $\vec{\lambda}$. These eigenvalues are a rotation-invariant representation of $\overleftrightarrow{\mathbf{M}}$. As $\overleftrightarrow{\mathbf{M}}$ is complex its eigenvalues form a vector containing three complex values. These three values are then sorted in order of increasing magnitude, $\vec{\lambda}_s$. The magnitude is calculated by multiplying each value by its complex conjugate.

Each object is classified by using (2) to compare the Euclidean distance, D_i between the object's sorted eigenvalues and all of the samples in the library, where the subscript i refers to the library index. The vector of distances is sorted in order of smallest to largest values to produce $\vec{\mathbf{S}}$, and the first K values are selected. The classification outcome, ζ, is calculated by taking the statistical mode, Mo(), of the class of each of the K nearest distances as shown in (3); often K is an odd number to reduce the chances of even numbers of neighbours belonging to different, most popular classes.

$$D_i(\vec{\lambda}_a, \vec{\lambda}_b) = \sqrt{(\lambda_{a1} - \lambda_{b1})^2 + (\lambda_{a2} - \lambda_{b2})^2 + (\lambda_{a3} - \lambda_{b3})^2} \tag{2}$$

$$\zeta = \mathrm{Mo}(\mathrm{Class}\{S_1 \quad \dots \quad \dots \quad S_k\}) \tag{3}$$

The results of the classification are evaluated according to two criteria. Firstly, by the accuracy of determining whether an object is considered to be threatening or innocuous, and secondly, by the accuracy of correctly identifying which of the 10 classes the object belongs to. Normalised values for the latter case have also been calculated to correct for differences in class sizes. This is defined as the average recall for all classes.

False negatives, i.e. classifying an object as innocuous when it is really a threat, are not acceptable in the application area. Therefore, the threat recall score should be 100% while still maintaining an acceptable level of overall accuracy. In a security screening application, this means that all threats are spotted while maintaining a low false alarm rate.

2.3. Limitations of the method

The results presented in this paper consider only situations where a single metallic object is within the detection volume of the portal. The classification algorithm would theoretically perform equally well regardless of how many objects are detected, providing that the quality of all tensors remains the same. The case of multi-object classification remains untested for measured data.

Although the algorithm utilises the tensor approximation (which is location invariant), and tensor eigenvalues (which are rotation invariant), the inversion algorithm is sensitive to certain locations within the detector space and also to variations in orientation as the object passes through the detector. In addition to this, it is known that the orthogonal field components are not of equal magnitude, and that as a result the direction with respect to floor-to-ceiling displacement tends to be noisier. Also, the areas at the very top and very bottom of the detector have a reduced concentration of coils, and consequently the tensor quality decreases in these regions. The effect on the classification outcome can be equated to that of cases with large residual values (typically $r > 0.35$).

As a final point it should be noted that KNN algorithms are highly dependent upon the underlying example library. If the library is not exhaustive, samples that are numerically far away from all of the examples tend to be classified as the class that has the most examples.

3. Results and Discussion

The results of the KNN are shown in Table 1. The algorithm was run once for each library with the parameter value K=1. For the library *'Main, all'* the algorithm was run also with parameter values K=3 and K=5. Leave-one-out cross validation was used, unless stated otherwise. The results show that the classifier performance seems to be the best with K=1 when the library *'Main, all'* is used. This is most likely due to the fact that there are only a few samples from some classes and a large number of samples from others. As discussed previously this is a limitation of KNN.

Table 1. Full results of experiment.

Library	K	Accuracy (%), threat/innoc.	Accuracy (%), class	Accuracy (%), class, norm.	Recall (%), threat
Main, all	1	98.9	97.6	96.0	99.4
Main, all	3	98.3	97.1	94.9	99.4
Main, all	5	98.6	96.7	94.1	99.5
Main, R < 0.5	1	99.3	98.9	97.9	99.8
Main, R < 0.35	1	99.5	99.0	97.7	99.7
Chest	1	99.3	97.4	97.4	99.0
Chest, R < 0.5	1	100	99.7	99.8	100
Chest, R < 0.35	1	100	99.4	98.6	100
Left Pocket	1	99.6	96.7	97.2	100
Left Pocket, R < 0.5	1	100	98.1	98.3	100
Left Pocket, R < 0.35	1	100	97.5	97.4	100
Reduced	1	100	98.5	98.5	100
Body Effect, all^	1	100	98.5	97.9	100
Body Effect, all*	1	96.5	88.0	82.9	100
Candidate A	1	100	100	100	100
Candidate B	1	100	97.0	95.7	100

^ based on a 'leave-one-out' classification approach
* based on a classification of Candidate A's measurements using Candidate B's library data and vice versa

Table 1 shows that the classifier appears to execute properly with all of the libraries tested. Recall is 99% or higher for all tested libraries; from a practical point of view this value needs to be higher to prevent security breaches, however this could be improved by improving the complexity of the classification algorithm, and also by biasing the threat objects in the library.

The testing with Candidates A and B shows that the body effect seems to be noticeable in classification results. In the results the recall of threat objects remained at the required value of 100%, however the ability to classify the objects fell substantially. Also, there were a greater number of false positive classifications where innocuous objects were misclassified as threats. In the testing, the performance was reduced when the library and test data was swapped such that the library was for Candidate A's data, whilst Candidate B was being scanned, and vice versa.

It can be seen that removing high residual samples generally improves the performance by a noticeable amount, cutting down misclassifications by 36-55%. However, it can be noticed that in some instances the removal of all samples with a residual higher than 0.35 does not improve performance. This is because there are fewer examples of some object classes such as keys and coins that tend to yield noisy tensor values. This causes bias towards the classes with a higher number of examples, and larger items which are more readily detectable.

The results show that, as expected, the classifier performs better when using only data from a single location, e.g. pocket data, instead of using a mixed set of data.

4. Conclusions and Future Work

The results of this study show that the modified KNN algorithm is capable of classifying targets consistently and with a typical normalised accuracy of over 95%, and a recall value of in excess of 99%. Based on this, it is clear that this improvement shows great promise to the field of inductive metal detection. However, there are a number of improvements that could be made to this algorithm and there is further research which can be conducted which the authors expect would improve the results shown here.

As previously discussed, the KNN approach is a relatively simple algorithm which is subject to several limitations. It is probable that the success of the method reported here could be improved by implementing a more complex algorithm. One way to enhance the method is to use heuristics and the 3D location information of the metallic object that is given by the detector system. For example, if an object was to be classified as a wristwatch, then statistically it is very unlikely that it would be near the floor level. Also, given the variation in tensor components as a function of location within the detector, the location information of the object to be classified could be used to increase the trust in those object samples in the library that have been recorded from the same region of the detector space. An extra level of validation could be added to the algorithm which would take this into account, thereby improving the result.

The results presented in this paper indicate that the detector and classifier are sensitive to substantial changes in the body size of the candidate. Even if samples were to be recorded in the absence of the candidate, e.g. with a robotic arm, there is still a correction that must be made for the body effect. One possible way to reduce this capacitive coupling would be to introduce screening between the coils and the person being measured. Although this paper has shown that the classification remains successful regardless of the contribution of the body effect from the candidate, further study is needed to determine the extent of this effect and how it may be overcome.

Given the strong frequency dependency of the tensor, it is considered that the most significant adaptation which could be made to improve the effectiveness of the classification algorithm would be to use the broadband tensor components rather than those at a fixed frequency; the classification would be fitting a curve rather than a single point. This remains a topic for future research.

5. References

[1] Rapiscan Systems (2012, 13/02/2013) Metor 6S| Rapiscan Systems. Available: http://www.rapiscansystems.com/en/products/item/metor_6s

[2] Bodington, T (2011, 13/02/2013) Is the perceived threat of terrorism greater than the actual threat?, Available: http://www.tsvc.lincoln.ac.uk/papers/submission.php?paper=118

[3] Marsh L A, Ktistis K, Järvi A, Armitage D W and Peyton A J 2013 Three-dimensional object location and inversion of the magnetic polarisability tensor at a single frequency using a walk-through metal detector *Meas. Sci. Technol.* vol. 24 045102 doi:10.1088/0957-0233/24/4/045102

[4] Kauppila J, Ala-Kleemola T, Vihonen J, Jylhä J, Ruotsalainen M, Järvi A and Visa A, 2009 Classification of items in a walk-through metal detector using time series eigenvalues of the polarizability tensor *Proc. SPIE 7303, Detection and Sensing of Mines, Explosive Objects, and Obscured Targets XIV*, pp. 73030P-1~73030P-10

[5] Norton S J and Won I J, 2001 Identification of Buried Unexploded Ordnance From Broadband Electromagnetic Induction Data *IEEE Trans. Geosci. Remote Sens*, vol. 39, pp. 2253-2261

[6] Paulter N and Larson D 2009 NIJ Metal Detector Test Objects Report, National Institute for Justice, Gaithersburg, Maryland

Acknowledgments

This work was supported by the Academy of Finland under the project "Magnetic Induction and Data Analyses for Sophisticated Sensing", grant no. 251777. The authors are also very grateful to Sarah Penny at the University of Manchester for her participation in this experiment.

Defect detection of the weld bead based on electromagnetic sensing

B M Abdullah, A Mason and A Al-Shamma'a
School of Built Environment, Liverpool John Moores University, Henry Cotton Building, 15-21 Webster Street, Liverpool, L3 2ET, UK

E-Mail: b.m.abdullah@ljmu.ac.uk

Abstract. Characterization of flaws of weld bead is imperative for high-quality welding. Methods of weld bead inspection include radiographic, ultrasonic and vision inspection. However, such methods are costly and time consuming. The proposed sensor is light, low-cost and fast. This paper summarizes our work on weld bead monitoring and defect detection using an electromagnetic sensor. Measurements are acquired in the form of S-Parameters, specifically measuring changes in the reflected coefficient S_{11}. The weld bead is scanned using the sensor and any form of weld bead defection such as undercutting and excessive penetration is detected and identified.

1. Introduction

The increasing importance of the integrity of structures has led to the development of new and more sensitive non-destructive evaluation methods for welding technology. Non-destructive testing/evaluation (NDT/NDE) techniques have been used for over 30 years. These techniques are analysis techniques used in science and industry to evaluate the properties of a material, component or system without causing damage to the samples.

Currently, weld bead inspection methods that are commonly used include the following; radiography, ultrasonic, vision, magnetic detection, eddy current, and acoustic measurements. However, these methods have limitations, are bulky, costly and time consuming. Many of these defects associated with weld beads are caused by the welder, because of either a lack of care or a lack of skill, and emphasize the need for adequate training. A simple non-destructive examination technique that is sometimes overlooked is that of a thorough visual examination by a suitably trained and experienced welding inspector. Visual inspection is easy to apply, quick, and relatively inexpensive. Visual testing equipment includes rulers, fillet weld gauges, squares, magnifying glasses, and reference weld samples. Such an examination will identify only some defects and will be time consuming and the need for a fast and reliable method is therefore a necessity.

A radiographic test is a non-destructive method that reveals the presence and nature of discontinuities in the interior of welds. This test makes use of the ability of short wavelength radiations, such as X-rays or gamma rays, to penetrate material that is opaque to ordinary light [1-3]. Ultrasonic testing is a non-destructive method of detecting the presence of internal cracks, inclusions, segregations, porosity, lack of fusion, and similar discontinuities in all types of metals. It can be used as the sole type of inspection, or it can be used with other types of testing. It is often used in conjunction with radiographic testing because it determines the depth of the defect from the test surface. Ultrasonic inspection is widely used in weld production with flexible operation and low cost [4-6]. Main disadvantages are lack of visual record and requirement for high skilful operators for defect recognition. A number of researchers have used indirect monitoring systems to link elements such as welding pool vibrations, superficial temperature distribution and acoustic emissions to size, geometry or welding pool depth [7]. Other researchers showed that it is possible to detect some interference and to assess the welding quality by measuring the sound and optical emissions as infrared band [8-10].

The advance of high speed computers and cameras has increased the use of vision systems in many applications such as weld bead monitoring. Vision systems are particularly useful in hazardous and hard to reach environments since they can be operated remotely. One of the major advantages of using a visual sensing technology to monitor welding operations is the fact that the visual sensor is not touched or interfered with during the welding process and visual images of the weld pool contain more abundant and accurate information about the welding dynamics. B. M. Abdullah developed a vision system for monitoring of arc welding in real-time. The system uses a camera, an illumination source and a combination of filtering and illumination-camera synchronization techniques. Arc light (noise) is totally eliminated and a substantial amount of information is obtained in real-time, e.g. metal transfer and weld pool geometry [11, 12]. Figure 1 (a) shows the block diagram of the main components of the vision system. Meanwhile, figure 1 (b) shows an image of a TIG welding process obtained by the system in real-time. As can be seen the interference from the arc light is totally eliminated and the image is of very high quality.

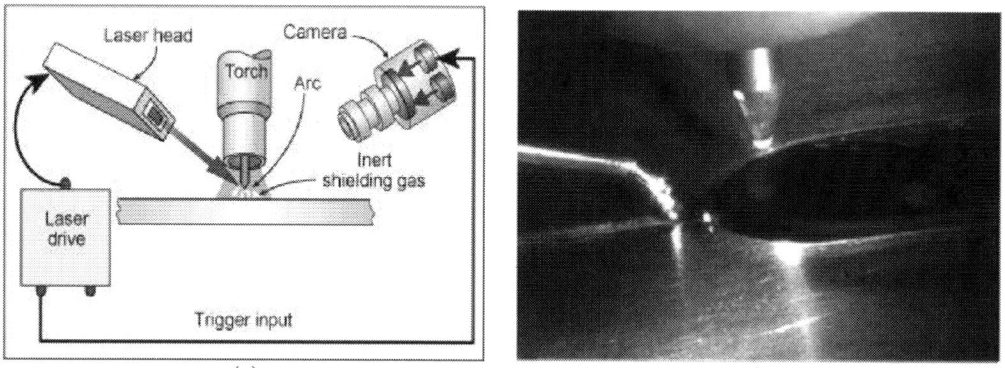

(a) (b)

Figure 1. Block diagram of weld monitoring vision system (b) Image of a TIG welding process obtained in real-time.

2. Theory

Electromagnetic (EM) waves are waves of energy that travel through a vacuum at the speed of light. They consist of two primary components, an electric (E) field and a magnetic (H) field. The electric field and magnetic field oscillate in phase perpendicular to each other and perpendicular to the direction of energy propagation. Analysis using microwaves is non-ionising with a low power output of 1mW (0dBm). The multi-parameter nature of wide band microwave analysis can provide unique signal spectrum signatures. Typically these would be in the form of a reflected signal (reflection coefficient) S_{11}, which are influenced by parameters such as conductivity and permittivity.

Figure 2 shows a block diagram of a typical microwave system. Reflected signal is of paramount importance in this study. A microwave sensing system is constructed of a microwave sensor, an oscillator and a detector. The purpose of the oscillator is to provide a microwave signal to the sensor, while the detector measures the response of the sensor. This functionality is embodied in a typical vector network analyser (VNA) such as the Rohde and Schwarz ZVL 6 which is used throughout this study.

The characteristic impedance of a typical transmission line is highly dependent on key parameters of the transmission line such as, resistance, conductance, capacitance and inductance. From the perspective of a microwave device (i.e. a sensor or antenna) which is exposed to a real world environment, changes which occur in this environment can be measured as they influence the device's characteristic impedance, and therefore its response at varying frequencies.

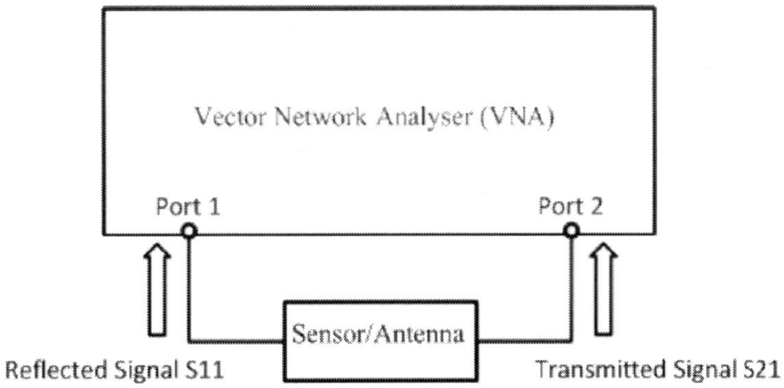

Figure 2. Bock diagram of a typical microwave system

3. Experimental Methodology

For this work a large rectangular steel plate with dimensions of 100cm by 50cm and a thickness of 2.0mm is cut into two pairs of rectangles each 25cm wide and 50cm long. Each pair is welded together as seen in figure 3. Each plate is welded with the intention of producing a weld bead with a variety of defects. Plate A contains less defects than plate B. The welding bead in plate B has more craters as can be visually seen due to improper weld termination. Both plates suffer from spatter due to metal drops expelled from the weld and sticks near the weld bead.

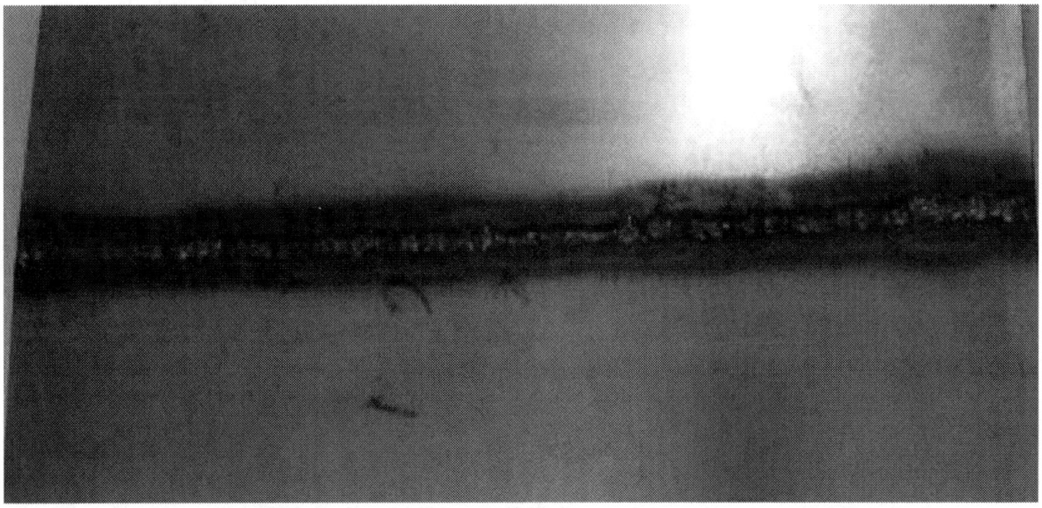

(a)

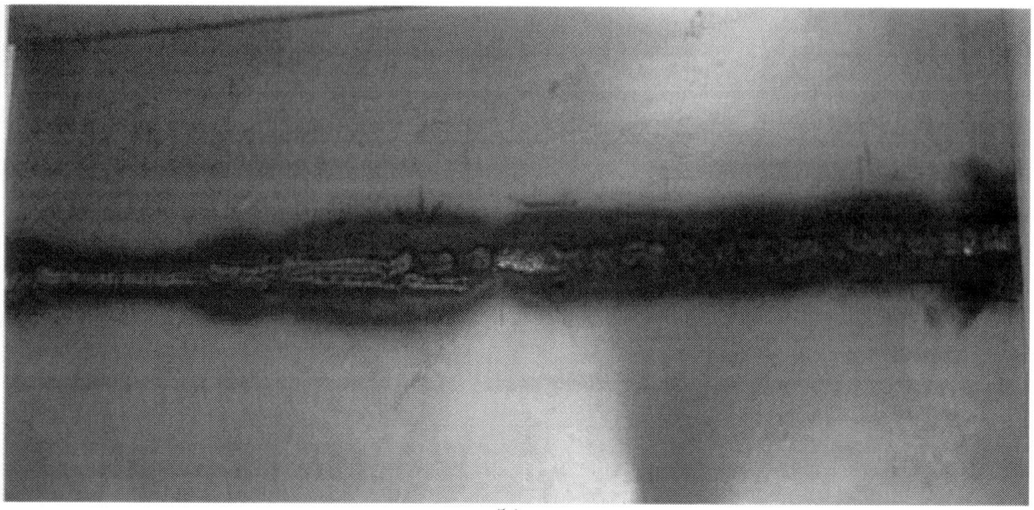

(b)

Figure 3. Two identical steel plates are cut then welded together (a) Plate A, minimum defects (b) Plate B, maximum defects.

A patch antenna operating at microwave frequencies (2.0GHz) is chosen as such devices are flat with a reasonably large contact area. They can be pressed uniformly against the welded piece of metal. A software model is constructed using the Ansys High Frequency Structure Simulator (HFSS) package to ascertain appropriate dimensions for the patch sensor, and this model is shown in figure 4. Once appropriate dimensions are attained, the software model is then constructed on double sided FR4 substrate. Only the top layer is routed, the under layer is left untouched and acts as a ground plane. The patch has a single SMA connector to allow measurements to be taken using a Vector Network Analyzer (VNA) via a coaxial connection. All measurements are acquired in the form of S-Parameters, specifically measuring changes in reflected signature (electromagnetic wave spectrum) back from the antenna when interacting when different media (this is commonly denoted S_{11}).

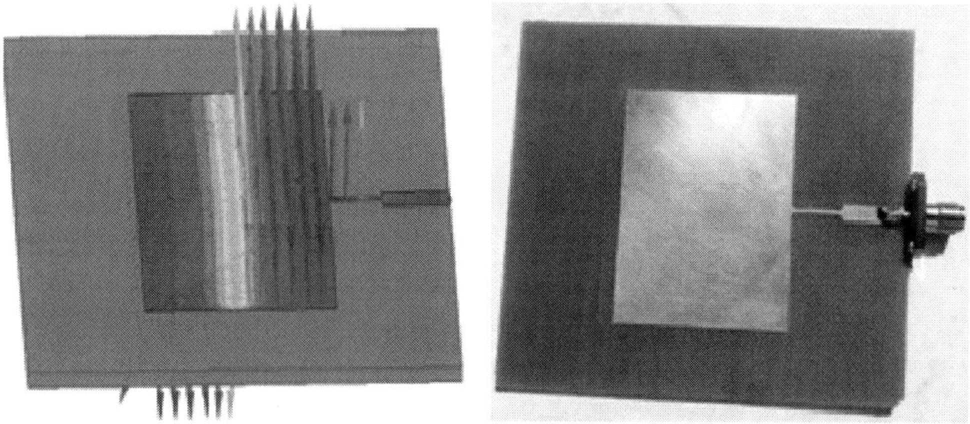

(a) (b)

Figure 4. (a) Ansys HFSS model showing electric field and (b) Constructed patch sensor.

4. Results and Analysis

Figure 5 shows the reflection coefficient measurements for both plates. Each plate is divided into sections and each section is scanned with the sensor. Figure 5 (a) shows the spectrum response when the sensor is used to scan a weld bead on plate A. The different peaks correspond to different defects present in each section of the plate. Meanwhile, figure 5 (b) shows the spectral response when the sensor is used to scan the weld bead on plate B.

As we can see from the results shown, defects elicit a change to the frequency response and each section of the plate where there is a defect produces its own unique signature. Each defective section has a distinctive frequency shift as can be seen. This is a significant result as prediction of the presence and type of defects can be easily determined.

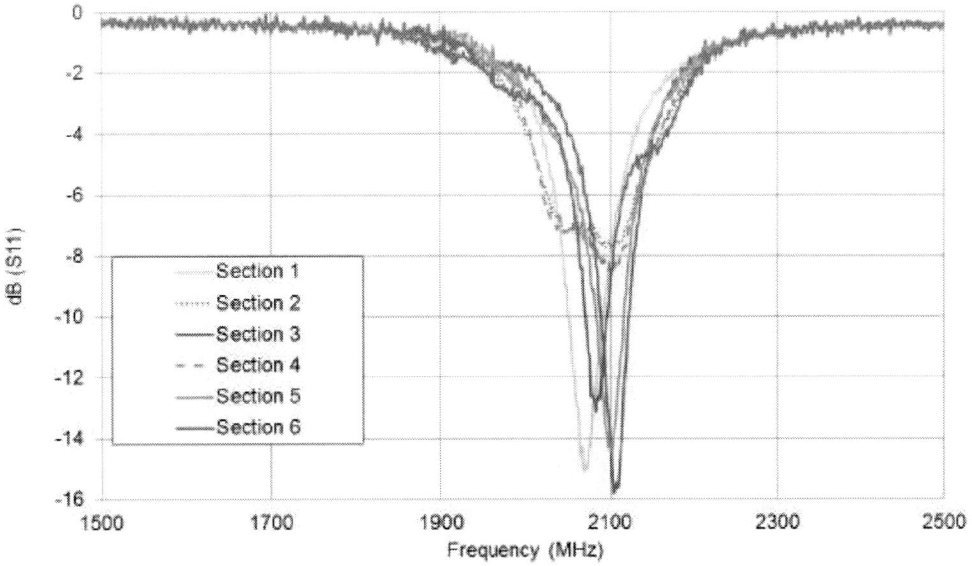

(a) Plate A

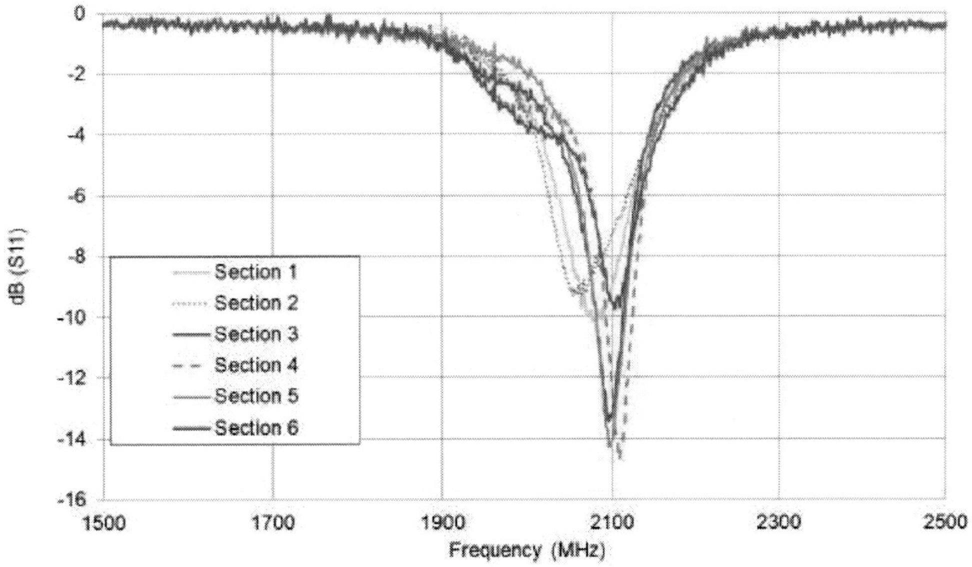

(b) Plate B

Figure 5. Spectral response of the constructed patch sensor highlighting the principle of weld bead defect detection. Peak shifts corresponding to different defects.

5. Conclusion

There are a number of NDT methods available for inspection of weld bead. However, many of these are expensive and time consuming. The results obtained so far using the patch sensor show that weld bead defects can be detected and identified using an electromagnetic sensor. The sensor scans the surface of the weld bead for defects. Defects in the weld bead elicit a change in the sensor spectral response. The results show that the size and type of defects give rise to a unique sensor signature. Each defect has its own signature and hence an appropriate action can be taken to rectify the defective section. Thus the challenge for the subsequent phases of this work is to develop a system which can be applied and be robust enough to operate in the toughest of environments without an operator.

References

[1] C. Hongjun, L. Junyuan, Z. Xiaohua, and D. Zongquan, "Application of visual servoing to an X-ray based welding inspection robot," in *Control and Automation, 2005. ICCA '05. International Conference on*, 2005, pp. 977-982 Vol. 2.

[2] L. Yuan, L. You Fu, W. Qing Lin, X. De, and T. Min, "Measurement and Defect Detection of the Weld Bead Based on Online Vision Inspection," *Instrumentation and Measurement, IEEE Transactions on*, vol. 59, pp. 1841-1849, 2010.

[3] X.-G. Zhang, J.-J. Xu, and Y. Li, "The research of defect recognition for radiographic weld image based on fuzzy neural network," in *Intelligent Control and Automation, 2004. WCICA 2004. Fifth World Congress on*, 2004, pp. 2661-2665 Vol.3.

[4] L. Yuan, W. Qing Lin, L. You Fu, X. De, and T. Min, "On-line Visual Measurement and Inspection of Weld Bead Using Structured Light," in *Instrumentation and Measurement Technology Conference Proceedings, 2008. IMTC 2008. IEEE*, 2008, pp. 2038-2043.

[5] G. J. Schuster, S. R. Doctor, and L. J. Bond, "A system for high-resolution, nondestructive, ultrasonic imaging of weld grains," *Instrumentation and Measurement, IEEE Transactions on*, vol. 53, pp. 1526-1532, 2004.

[6] M. V. Predoi, C. C. Petre, M. Rousseau, and M. de Billy, "Ultrasonic inspection of defects in welded plates and tubes," in *Ultrasonics Symposium, 1997. Proceedings., 1997 IEEE*, 1997, pp. 845-848 vol.1.

[7] S. C. A. Alfaro and F. D. Franco, "Exploring Infrared Sensoring for Real Time Welding Defects Monitoring in GTAW," *Sensors*, vol. 10, pp. 5962-5974, 2010.

[8] J. Tarn and J. Huissoon, "Developing psycho-acoustic experiments in gas metal arc welding," in *Mechatronics and Automation, 2005 IEEE International Conference*, 2005, pp. 1112-1117 Vol. 2.

[9] L. Grad, J. Grum, I. Polajnar, and J. Marko Slabe, "Feasibility study of acoustic signals for on-line monitoring in short circuit gas metal arc welding," *International Journal of Machine Tools and Manufacture*, vol. 44, pp. 555-561, 2004.

[10] H. C. Wikle Iii, S. Kottilingam, R. H. Zee, and B. A. Chin, "Infrared sensing techniques for penetration depth control of the submerged arc welding process," *Journal of Materials Processing Technology*, vol. 113, pp. 228-233, 2001.

[11] B. M. Abdullah, "Monitoring of Welding Using Laser Diodes," in *Semiconductor Laser Diode Technology and Applications* ed, 2012, p. 22.

[12] B. M. Abdullah, J. S. Smith, W. Lucas, J. Lucas, and F. Malek, "Monitoring of TIG welding using laser and diode illumination sources: A comparison study," in *Electronic Design, 2008. ICED 2008. International Conference on*, 2008, pp. 1-4.

A Quasi-Monte Carlo solution for the mutual inductance of misaligned circular coils

L Babić[1] , M Dadić[2],

[1] Končar – Power Plant and Electric Traction Engineering, Fallerovo šetalište 22, Zagreb, Croatia
[2] University of Zagreb, Facultey of Electrical Engineering and Computing, Department of Fundamentals of Electrical Engineering and Measurements, Unska 3, Zagreb, Croatia

[1] lucija.babic@koncar-ket.hr
[2] martin.dadic@fer.hr

Abstract. This paper presents a quasi-Monte Carlo method for calculating the mutual inductance of misaligned circular coils which can be useful in the areas of designing and producing the devices based on the mutual inductance (biomedical applications, sensor modelling, wireless battery chargers, etc.). Validation of the method is presented through the examples. The results obtained highly match previously published data that uses other methods which is shown graphically. Advantages of the quasi-Monte Carlo method are the availability of a deterministic bound for error (unlike the probabilistic in the Monte-Carlo method), a faster convergence rate and the high flexibility.

1. Introduction

Growing needs for developing and designing of new devices based on the mutual inductance with better performances demand new solutions of the calculations. With the powerful and general numerical methods such as finite element method (FEM) and boundary element method (BEM) it is possible to calculate accurately the mutual inductance of almost any 3-D arrangement of conductors. But in some circumstances there is still necessity to address the problem using the other analytical and semi-analytical methods. The quasi-Monte Carlo method based on Van der Corput sequence, presented in this paper, considerably simplifies the mathematical procedures and reduces the computational effort. It also gives a deterministic bound for error. This computation technique can be useful in the area of wireless battery chargers for general purpose and biomedical applications, sensor modeling, coreless printed circuit board transformers, superconducting magnetic levitation, etc.

This paper will deal with the calculation of mutual inductance of coils in perfect alignment, lateral misalignment and angular misalignment. For simplicity, it assumes the case of one-turn coils which are placed in the free-space.

2. Neumann's formula

Neumann's formula for the mutual inductance [1] is expressed by

$$M = \frac{\mu_0}{4\pi} \oint \oint \frac{dl_1 \cdot dl_2}{r_{12}} . \qquad (1)$$

2.1. Perfect alignment case

According to the figure 1, table 1 and [1] mutual inductance is expressed by

$$M_i = \mu_0 \sqrt{ab} \left\{ \left(\frac{2}{k} - k \right) K(k) - \frac{2}{k} E(k) \right\} \qquad (2)$$

where

$$k \equiv \left(\frac{4ab}{(4+b)^2 + d^2} \right)^{1/2} \qquad (3)$$

and K(k) and E(k) are the complete elliptic integrals of the first and second kind.
The bracketed expression from the equation (2) will be denoted by

$$G(r) \equiv \left(\frac{2}{r} - r \right) K(r) - \frac{2}{r} E(r). \qquad (4)$$

Now equation (2) can be written as

$$M_i = \mu_0 \sqrt{ab}\, G(k). \qquad (5)$$

This exact expression will be used as the basis for comparing the mutual inductances in the nonideal cases to follow.

Table 1. Coil and configuration parameters.

Parameter	Meaning
a, b	Coil radii
d	Coil spacing
Δ	Lateral misalignment
α	Angular misalignment
M_i	Mutual inductance – ideal case
M_L	Mutual inductance - Lateral misalignment
M_A	Mutual inductance - Angular misalignment
μ_0	Free-space permeability ($4\pi\times10^{-7}$ H/m)

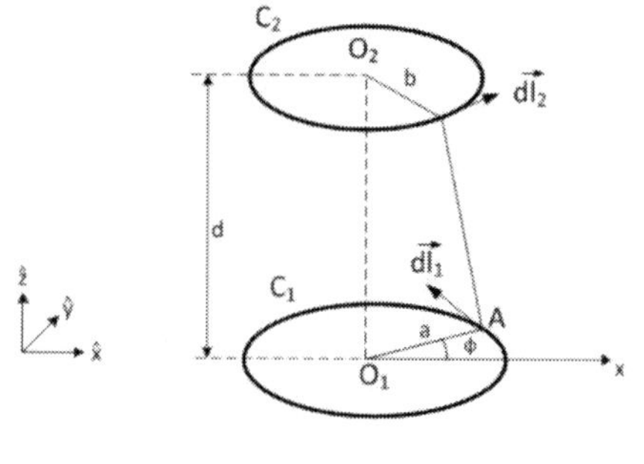

Figure 1. Perfectly aligned coil configuration.

2.2. Lateral misalignment case

Lateral misalignment configuration can be observed in figure 2. By simplifying and adjusting the Neuman's equation, the expression for the mutual inductance in this case can be written as

$$M_L = \frac{\mu_0 ab}{4\pi} \oint \frac{\cos\beta}{\sqrt{ab_L}} G(r)\, d\phi \qquad (6)$$

where

$$b_L \equiv \sqrt{b^2 + \Delta^2 + 2\Delta b \cos\phi}, \quad \tan\beta \equiv \frac{\Delta \sin\phi}{b + \Delta\cos\phi}, \quad r \equiv \left(4a \frac{b_L}{(a+b_L)^2 + d^2} \right)^{1/2}. \qquad (7)$$

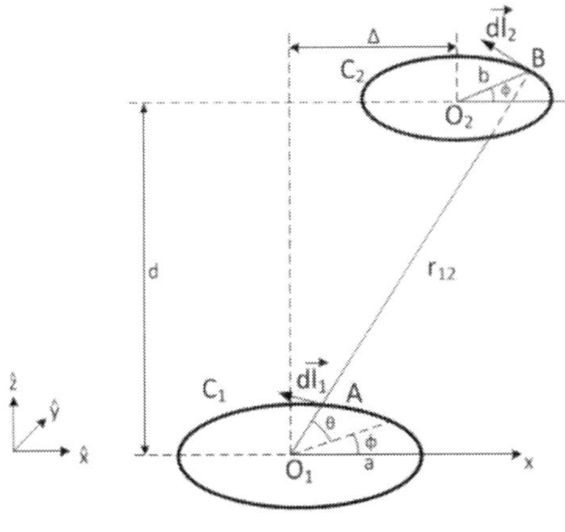

Figure 2. Lateral coil configuration.

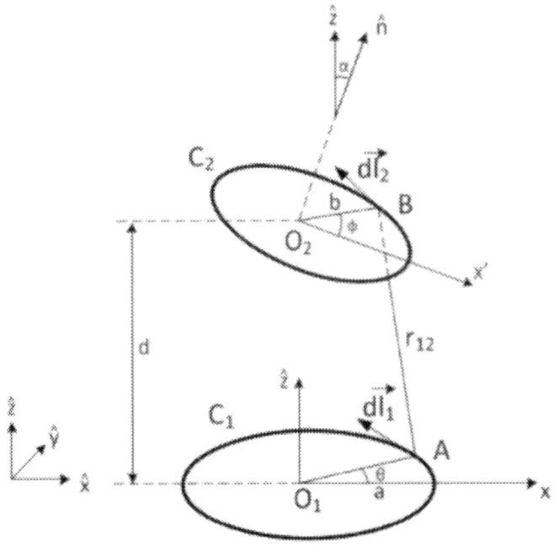

Figure 3. Angular coil configuration.

When $\Delta = 0$, or perfectly aligned, M_L reduces to M_i. When the $\Delta \neq 0$ it is impossible to integrate equation (6) exactly. This paper will introduce a new possible way to calculate the equation (6).

2.3. Angular misalignment case

Coils are in an angular misalignment but the plane perpendicular to coil 1 is passing through the both coil centers as can be seen from the figure 3. Expression for calculating this mutual inductance is derived from the adjustment of equation (1) [1]

$$M_A = \frac{\mu_0\sqrt{ab}}{\pi\sqrt{\cos\alpha}}\int_0^\pi \left(\frac{\cos\lambda}{\cos\phi}\right)^{3/2} G(r)\,d\phi \tag{8}$$

where the $G(r)$ is given in equation (4).

It is worth of mentioning that the λ is independent of θ and for the perfect alignment case, when the $\alpha = 0°$, the mutual inductance M_A is reduced to the M_i.

3. Monte-Carlo integration

Monte-Carlo integration is a powerful technique for numerical integration using random numbers. According to the Monte-Carlo integration, numerical calculation of $\int_0^1 f(x)dx$ can be performed by using the generator of n pseudo numbers (x_i), $(i \in N)$ and by using the numerical method for calculating $\frac{1}{n}\sum_{i=1}^{n} f(x_i)$ where $(x_1, x_2, x_3,\ldots)$ is a sequence of independently random variables distributed on [0,1] interval which gives us the approximation to the integral. [2] Larger values of n will produce more accurate approximations.

Equations (6) and (10) can now be written in a form of Monte-Carlo integration

$$M_L = \frac{\mu_0 ab}{\sqrt{ab_L}}\; G(r)\frac{1}{n}\sum_1^n \cos\beta \tag{9}$$

$$M_A = \frac{\mu_0\sqrt{ab}}{\sqrt{\cos\alpha}}\frac{1}{n}\sum_1^n \left(\frac{\cos\lambda}{\cos\phi}\right)^{3/2} G(r) \tag{10}$$

4. Quasi-Monte Carlo integration using Van der Corput sequence and Error bound

The simple Monte Carlo method can be upgraded by a method that uses a better distributed sequence of numbers unlike the generation of independent uniform random variables. The more these sequences are uniformly distributed and the less is the discrepancy. One of sequences that can be used is Van der Corput sequence. Figure 4 shows the graphical comparison of Van der Corput sequence and the random variables distribution over the [0,1] interval.

Van der Corput sequence

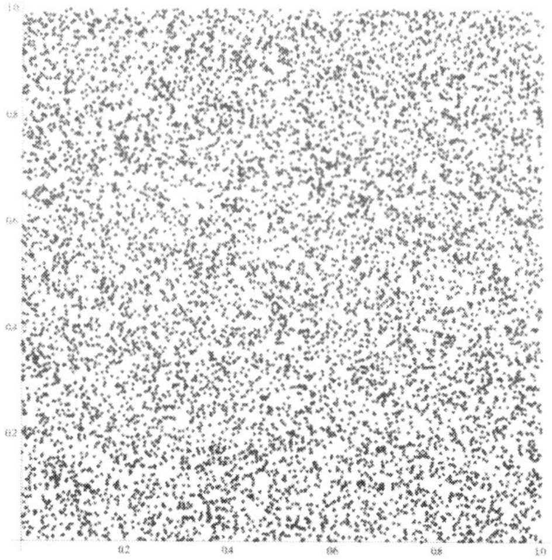

RandomReal (*WolframMathematica*® generator of random numbers)

Figure 4. Comparison of the Van der Corput sequence and RandomReal function from the *WolframMathematica*® distribution on the interval [0,1] with the number of variables n=10 000.

4.1. Van der Corput sequence

It is a low discrepancy sequence over the unit interval whose terms are obtained as follows: to compute the ith point x_i of the Van Der Corput sequence of base b, the integer number i is at first expanded on base b. Then x_n is obtained by reversing the digits after the decimal point. [3]

4.2. Error Bounding and Bounding of the Discrepancy

Other main issue with the generation of independent uniform random variables is that you cannot be drastically sure of the precision of the result because of the probabilistic nature of the error bounding. Thus it is of a great value to be able to determine the error bounding while choosing the different number of variables over the chosen interval.

A useful equation that provides an upper bound to discrepancy $D_N(x)$ for a base 2 Van der Corput sequence is [3]

$$n \cdot D_n \leq \frac{log_{10}(n)}{3 \times log_{10}(2)} + 1 \quad \forall n \geq 1 \tag{11}$$

And so called Koksma-Hlawka inequality will provide an upper bound for the error in quasi-Monte Carlo approximation [3]

$$Error\ bounding \leq V(f) \cdot D_n$$

$$\left| \frac{1}{n} \sum_{i=1}^{n} f(x^i) - \int_0^1 f(t)dt \right| \leq V(f) \cdot D_n \tag{12}$$

with $D_N(f)$ the discrepancy od the function f and $V(f)$ its variation.

The total variation of function f on [0,1] is

$$V(f) = sup \sum_{i=1}^{n-1} \left[\left| f(x^{i+1}) - f(x^i) \right| \right] \tag{13}$$

5. Results

Results presented in this paper are obtained by using the *WolframMathematica®* software and are graphically compared with the results from [1].

5.1. Lateral misalignment example

Dimensions of circular coils and lateral misalignment:

a = 1.0 cm, *b = 1.0 cm*
d = 0.5 cm, *Δ = [0, 0.1, 0.2, 0.3, 0.4, 0.5] cm*
Results obtained with n = 10 000.

Table 2. Lateral misalignment example.

Lateral misalignment Δ (cm)	M_L (H)	Error Bounding percentage (%)
0	1.11261×10^{-8}	0.0542924
0.1	1.0868×10^{-8}	0.0544287
0.2	1.02401×10^{-8}	0.054845
0.3	9.37911×10^{-9}	0.0555645
0.4	8.41474×10^{-9}	0.0566305
0.5	7.43747×10^{-9}	0.0581155

V(f) and the error bounding for this example are

$$V(f_{exL}) = \frac{\mu_0 \cdot a \cdot b}{\sqrt{a \cdot b_L}} \mathrm{G(r)} \cdot V(f) \tag{14}$$

$$Error\ bounding \leq V(f_{exL}) \cdot D_n\ . \tag{15}$$

5.2. Angular misalignment example

Dimensions of circular coils and angular misalignment:

a = 1.0 cm, b = 0.9 cm
d = 0.5 cm, α = [0, 6, 10, 14, 16, 20] °
Results obtained with n = 10 000.

Table 3. Angular misalignment example.

Angular misalignment α (°)	M_A (H)	Error Bounding percentage (%)
0	9.85367×10^{-9}	0.0542924
6	9.9207×10^{-9}	0.0645385
10	1.00368×10^{-8}	0.072094
14	1.02015×10^{-8}	0.0800614
16	1.02962×10^{-8}	0.0840651
20	1.04848×10^{-8}	0.0915331

V(f) and error bounding for this example are

$$V(f_{exA}) = \frac{\mu_0 \sqrt{a \cdot b}}{\sqrt{\cos \alpha}} \cdot V(f) \tag{16}$$

$$Error\ bounding \leq V(f_{exA}) \cdot D_n\ . \tag{17}$$

5.3. *Graphical comparison*

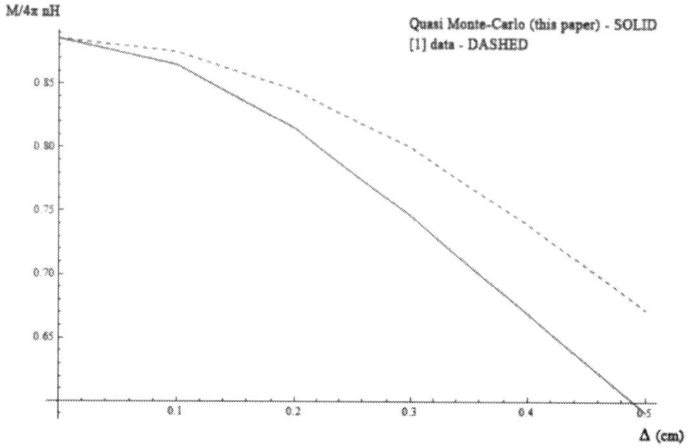

Figure 5. Comparison of mutual inductance under lateral misalignments M_L (normalized by 4π nH) with the data from [1].

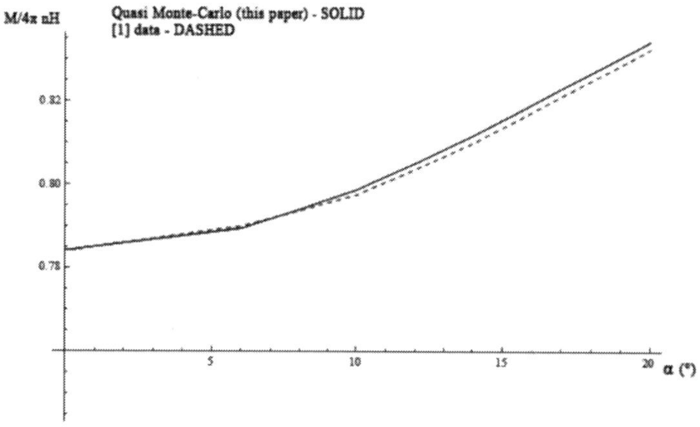

Figure 6. Comparison of mutual inductance under angular misalignments M_A (normalized by 4π nH) with the data from [1].

6. Conclusion

This paper presents a new method for the calculation of the mutual inductance of the circular coils. This method offers an alternative to FEM and BEM. It is characterized by the speed of the calculation as well as the possibility to calculate the specific equation within the determined percentage of error bounding. It can be useful in the process of designing and developing new devices when mutual inductance of coils is set and their lateral or angular misalignment should be calculated.

References

[1] Soma M, Galbraith D C and White R L 1987 Radio-frequency coils in implantable devices: misalignment analysis and design procedure IEEE Trans. Biomed. Eng. **34** 276-282

[2] Sarapa D N 1992 *Teorija vjerojatnosti* (Zagreb: Školska knjiga)

[3] Legrand X, Xémard A, Fleury G, Auriol P and Nucci C A 2008 A quasi-Monte Carlo integration method applied to the computation of the Pollaczek integral *IEEE Trans Pow. Del.* **23** 1527-34

Magnetic sensing for microstructural assessment of power station steels: Magnetic Barkhausen noise and minor loop measurements

JW Wilson[1*], N Karimian[1], W Yin[1], J Liu[2], CL Davis[2], AJ Peyton[1]
[1]School of Electrical and Electronic Engineering, University of Manchester, Manchester M60 1QD, UK
[2]School of Metallurgy and Materials, University of Birmingham, Edgbaston, Birmingham B15 2TT, UK

Abstract. There are currently no techniques available to monitor the microstructural condition of power station steel components in-service (at elevated temperatures). Electromagnetic (EM) inspection methods have the potential to provide a solution to this problem. Tests have been carried out on power generation steel (P9 and T22) samples with different microstructural states using major and minor B-H loop measurements and correlations established between EM properties and material properties such as Vickers hardness. These correlations will be used to develop a field deployable tool for the quantification of degradation in power station steels.

1. Introduction

Current procedures for the assessment of the condition of components in power stations involve site inspections during costly shut-down periods and inspection of steel components often involves lengthy procedures such as replica metallography [1, 2]. The use of electromagnetic (EM) sensors for inspection has the potential to provide information on microstructural changes in steel by exploiting the link between the microstructure and magnetic domain structure of the material. EM inspection [3-5] has the advantage that it can be performed in-situ, at elevated temperatures, with minimal surface preparation.

In this paper, the results of tests carried out on EM characterisation of power station steels are provided. Section 2 details the equipment constructed for the tests and gives an overview of the steel samples. The experimental results are provided in section 3, including; Magnetic Barkhausen Noise (MBN) with major *B-H* loop excitation; the derivation of incremental permeability curves using minor *B-H* loop excitation and MBN measurement with minor loop excitation. The paper concludes with a discussion and comparison of the results.

2. Measurement System and Sample Summary

2.1. Measurement system

A schematic of the measurement system developed for the tests is shown in figure 1. A low frequency time varying signal is fed to two power amplifiers, which supply current to two excitation coils wrapped around a silicon-steel core. The cylindrical sample is fitted into a slot in the core, to maximise coupling between core and sample. The axial applied field (*H*) is measured using a Quantum Well Hall sensor, developed at the University of Manchester. The flux density of the induced field (*B*) is measured using a 20-turn encircling coil connected to an instrumentation amplifier. For MBN

* John.Wilson@manchester.ac.uk

measurements, the 20-turn coil is replaced with a 6000-turn encircling coil and the low frequency component of the signal is rejected through the addition of a passive 5 kHz high-pass filter.

For the major loops, 1 Hz sinusoidal excitation is used and 9 cycles are recorded and averaged. A 10 Hz sinusoidal excitation is used to generate the minor loops, with two types of minor loop being recorded; 1) Deviations from the main *B-H* loop. In this case, the sample is taken through several major loop cycles before the applied field is held constant at a pre-determined *H* value and several minor loop cycles are recorded; 2) Deviations from the initial magnetisation curve. The sample is demagnetised by the application of 10 Hz sinusoidal excitation, gradually reducing in amplitude. The applied field is then increased to a pre-determined *H* value and several minor loop cycles recorded. For both types of minor loop, up to 90 cycles are acquired and averaged, to reduce noise.

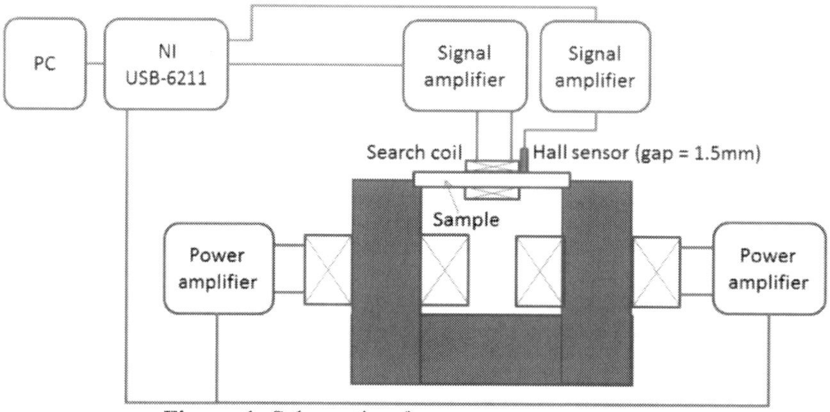

Figure 1. Schematic of measurement apparatus

The 1 Hz sinusoidal excitation is used to generate major loop MBN profiles, with the signal from the MBN pickup coil and the applied axial field from the Hall sensor recorded simultaneously. The signal from the coil is then high-pass filtered at a frequency of 5 kHz, rectified, and a moving average technique used to generate the MBN profile, which is then plotted against H. A similar process is used for the minor loop MBN readings, with the minor loop generated as outlined above.

2.2. *Test Samples*

Two sample sets have been studied, consisting of three P9 samples and three T22 samples. The P9 samples were machined from material which was in service for approximately eleven years at 520°C. Samples were heat treated to simulate service entry microstructure by normalising and tempering, as normalised samples were also assessed. Heat treatments, composition details and Vickers Hardness numbers (HV) are given in Table 1. Cylindrical rods with a diameter of 4.92±0.03 mm and a length of 49.59±0.54 mm were removed from the samples at each stage to create taken-from-service, tempered (and normalised), and normalised samples for both materials.

Table 1. Sample composition (in wt%), heat treatments and Vickers harness number (HV)

	P9					T22				
	Cr	**Mo**	**C**	**Si**	**Mn**	**Cr**	**Mo**	**C**	**Si**	**Mn**
%	8.4	0.97	0.12	0.52	0.44	1.90-2.60	0.87-1.13	0.05-0.15	0.5 (max)	0.30-0.60
	P9-TEMP	**P9-TFS**	**P9-NORM**	**T22-TEMP**	**T22-TFS**	**T22-NORM**
	Tempered at 760°C for 1h	Taken from service	Normalised at 950°C for 1h	Tempered at 760°C for 1h	Taken from service	Normalised at 950°C for 1h
HV	212	158	401	203	129	316

Complete metallographic tests were carried out at each stage; micrographs for the tests have been presented previously [3, 4]. The microstructure of the as-normalised P9 consists of predominantly martensite mixed with some bainite. Subsequent tempering produces a simulated service entry microstructure, i.e. tempered martensite / bainite. After long service exposure, the microstructure showed equiaxed ferrite with large carbides distributed within ferrite grains or on grain boundaries.

The as-normalised T22 steel shows a mixed microstructure of bainite and some proeutectoid ferrite. No carbides are present in the ferrite, but plate-like carbides can be seen within the bainite region. After tempering, many carbides can be observed along prior austenite grain boundaries, on ferrite boundaries or within bainite regions. The microstructure of T22 after service exposure consists of equiaxed ferrite and a great many carbides outlining the ferrite grain boundaries or finely dispersed within the ferrite grains.

The HV values follow the same trend for each sample set; lowest for the taken-from service sample and highest for the normalised sample. As P9-NORM is predominantly martensitic, it exhibits the greatest HV value. As bainite typically has a lower hardness than as-quenched martensite, but higher hardness than ferrite, the bainite / proeutectoid ferrite microstructure of T22-NORM shows a reduction in HV in comparison to P9-NORM. The decrease in hardness for service exposure is reflected by a decrease in HV for both sample sets.

3. Measurement and experimental results

3.1. Major B-H Loop and MBN Measurements

Figure 2 shows major *B-H* loops for both sample sets and corresponding MBN profiles (only the profile corresponding to half of the magnetisation cycle is shown for clarity). Examination of the coercivity (H_C), remanence (B_R) and MBN peak (MBN_{PK}) values in comparison to HV shows that H_C exhibits the strongest correlation with hardness, though the decrease in H_C for service exposure is relatively small. It can be seen from figure 2 that the MBN peak position also follows the trend in H_C.

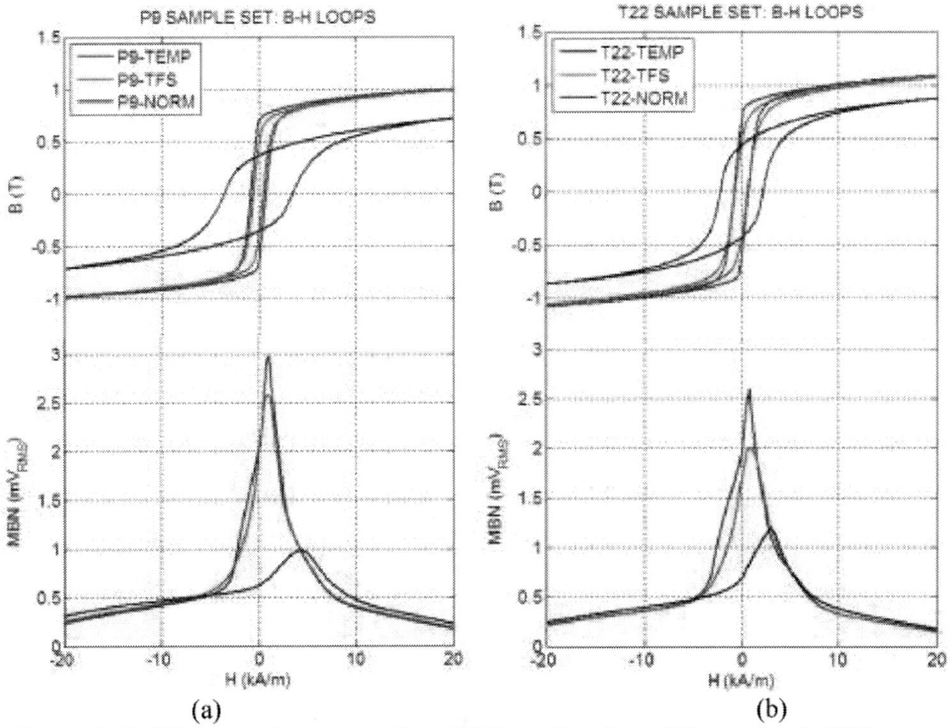

Figure 2. *B-H* loops and corresponding MBN profiles for a) P9 samples, b) T22 samples

3.2. Minor Loop Measurements

The evolution of the minor loop as deviations from the initial magnetisation curve is shown in figure 3. The origin of the first minor loop corresponds to the demagnetised state ($B=0$, $H=0$). In this state, for a small applied field (H), the magnetisation (M) of the material can be described by Raleigh Law [6]:

$$M = \chi_0 H + \alpha_R \mu_0 H^2 \quad (1)$$

Where χ_0 is initial susceptibility, describing the reversible part of magnetisation, the Rayleigh constant α_R describes the irreversible Barkhausen jumps and μ_0 is the permeability of free space. Thus, in this region, magnetisation is a combination of reversible and irreversible components, resulting in a loop enclosing a relatively large area, as show in figure 3 (bottom left). As the initial magnetisation curve approaches saturation, domain walls are swept away by field pressure and the dominant magnetisation mechanism is the progressive alignment of the field against anisotropy [6]. Thus, only reversible components remain, resulting in a loop with a much smaller area, with a smaller ΔB for a given ΔH, as shown in figure 3 (bottom right).

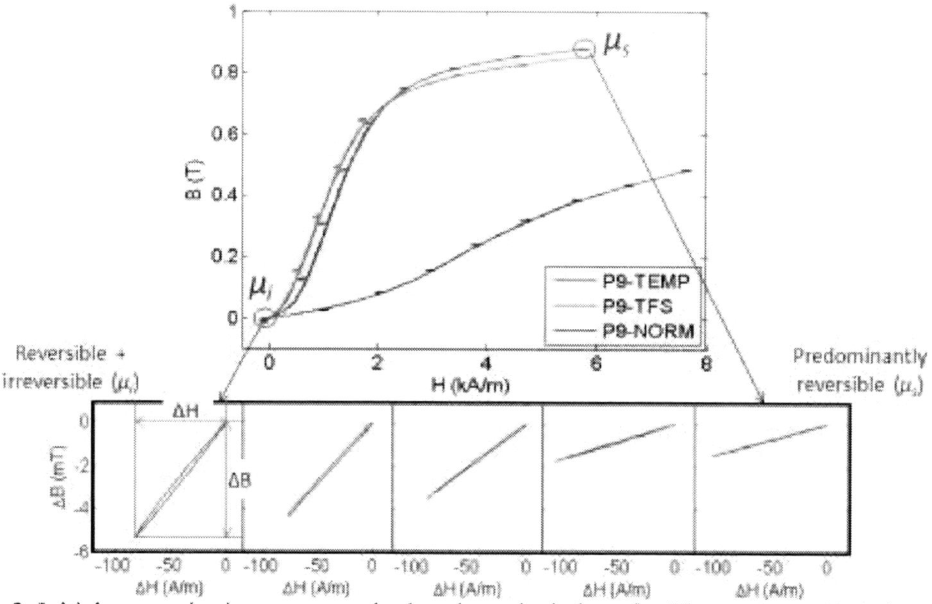

Figure 3. Initial magnetisation curves and minor loop deviations for P9 sample set (*B* and *H* offsets removed from minor loops for comparison)

Incremental permeability (μ_Δ) is calculated as the ratio between the change in flux density (ΔB) and the change in the applied field (ΔH) scaled with respect to the permeability of free space (μ_0); $\mu_\Delta = \Delta B/\Delta H \mu_0$. Figure 4a shows the resultant incremental permeability curves for minor loop deviations from the initial magnetisation curve. It can be seen from the plot that the maximum permeability values for the initial magnetisation curves correspond to the initial permeability (μ_i) reading; the point at which domains have the greatest degree of freedom to overcome pinning sites, resulting in the greatest change in *B* for a given applied field. There is a sharp decrease in μ_Δ with increasing *H* and some convergence in μ_Δ values for the three samples as saturation is approached and contributions from domain wall pinning sites are reduced, giving way to reversible domain rotation effects. P9-NORM exhibits a much smaller variation in μ_Δ for increasing *H*, as the high dislocation density of the predominantly martensitic sample results in heavy domain wall pinning and irreversible magnetisation effects are minimised. Figure 4b shows the μ_Δ curves for minor loop deviations from the major *B-H* loop. It can be seen from the plot that the maximum μ_Δ value ($\mu_{\Delta\text{-}max}$, table 2) corresponds to the coercive field. As with the initial curve results, this is the point at which $B=0$ and domain walls have the greatest degree of freedom to move.

Figure 4c shows the P9-TEMP curves from figures 4a and 4b together for comparison. Although the curves have different values at $H=0$, they converge at H_C as the initial magnetisation curve and major *B-H* loop converge. It is apparent from the plot that the random domain structure of the demagnetised sample corresponding to the μ_i reading results in a greater variation in *B* for a given *H* field than the systematic reorganisation of domains at H_C.

As operating in the Raleigh region results in the highest permeability values and it has been found that minor loop measurements in this region are highly sensitive to lattice defects [7], the effect of varying the amplitude of the minor loop used to calculate μ_i has been investigated. Figure 5 shows plots of μ_i for P9 and T22 for a variation in minor loop amplitude. It can be seen from the plots that as the minor loop amplitude increases, so does μ_i. At low minor loop amplitudes, reversible magnetisation dominates ($\chi_0 H$, equation (1)) as the minor loop amplitude increases, the irreversible component ($\alpha_R \mu_0 H^2$ equation (1)) is introduced and the gradient of the minor loop increases, as a greater ΔB is generated for a given change in H. Polynomial fitting has been employed to extrapolate values for μ_i if the minor loop amplitude could be made to equal zero; see $\boldsymbol{\mu}_{i\text{-}H0}$, table 2.

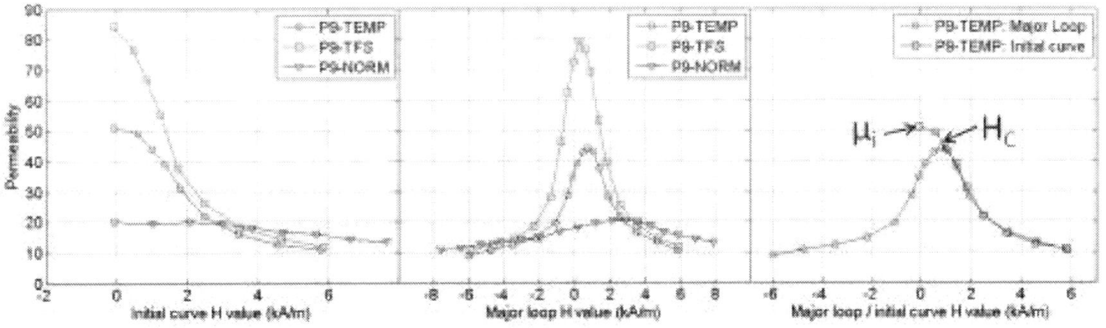

(a) (b) (c)

Figure 4. a) Incremental permeability curves for initial magnetisation, b) μ_Δ curves for major *B-H* loop, c) μ_Δ curves for initial magnetisation and major *B-H* loop for sample P9-TEMP

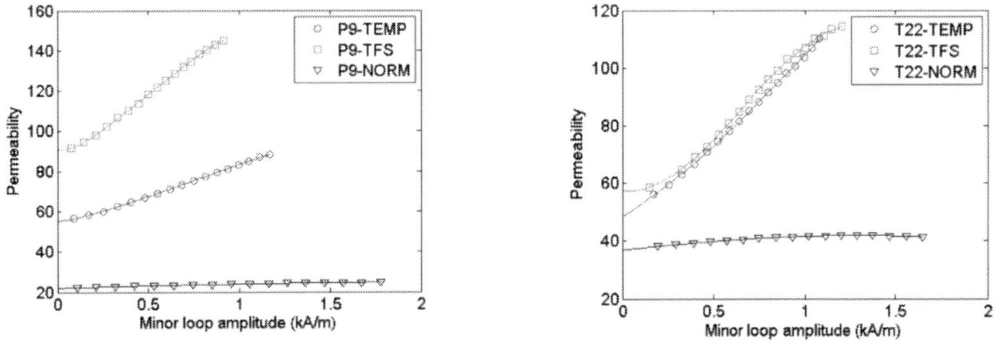

(a) (b)

Figure 5. Initial permeability (μ_i) values derived from a minor loop amplitude sweep for; a) P9 sample set b) T22 sample set

3.3. Minor Loop Magnetic Barkhausen Noise

The change in MBN_{RMS} values derived from a minor loop amplitude sweep for initial permeability is shown in figure 6a. It can be seen from the plot that at higher minor loop amplitudes, MBN_{RMS} follows a similar trend to the permeability values derived from the minor loop, as shown in figure 5a; with the MBN value for the P9 ex-service sample increasing rapidly and reaching the highest amplitude, increasing less rapidly for the tempered sample and exhibiting very little change for the normalised sample. However, at lower amplitudes the plots for the three samples converge, only showing a significant increase in amplitude at around 0.3 kA/m for P9-TFS and 0.5 kA/m for P9-TEMP. This indicates that only at these higher applied fields do the domain walls gain enough energy to overcome particular pinning sites in the material.

The plots for the T22 samples (figure 6b) exhibit a similar trend. However in contrast with the results for P9, T22-TFS and T22-TEMP start to cross over, reflecting the fact that tempered samples give the highest level of MBN for major loop excitation (see figure 2). It is also notable that the point at which the samples exhibit a significant increase in amplitude is indicative of the trend in H_C.

In order to provide a single minor loop MBN reading for each sample ($MBN_{\mu i}$, table 2), it was decided to choose the readings at the point where the minor loop amplitude reaches the coercive field value for P9-TEMP and T22-TEMP for the P9 and T22 sample sets respectively. This point was chosen because the tempered samples represent the service entry microstructure of the two steels, therefore the H_C values represent a fundamental magnetic property of these steels.

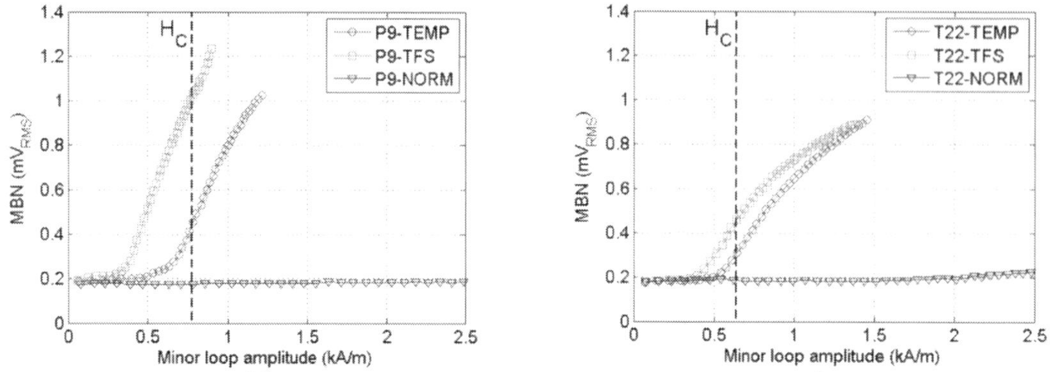

(a) (b)

Figure 6. MBN_{RMS} values derived from initial permeability (μ_i) minor loop amplitude sweep for; a) P9 sample set b) T22 sample set

4. Discussion and Conclusions

Table 2 shows the various signal features collected from the tests in this paper. All the minor loop features ($\mu_{i\text{-}H0}$, $\mu_{A\text{-}max}$, $MBN_{\mu i}$) have an inverse relationship with HV, increasing from normalised to tempered to taken-from-service. Although H_C is not directly proportional to HV, it does follow the same trend, though the change in H_C for service exposure is much smaller than for the minor loop parameters, indicating that the minor loop features may be more sensitive to changes in microstructure due to service exposure.

Table 2. Major loop, permeability and MBN values

	P9-TEMP	P9-TFS	P9-NORM	T22-TEMP	T22-TFS	T22-NORM
HV	212	158	401	203	129	316
B_R (T)	0.68	0.44	0.36	0.63	0.46	0.43
MBN_{PK} (mV_{rms})	2.98	2.58	0.98	2.59	1.10	1.18
H_C (kA/m)	0.78	0.51	3.51	0.72	0.64	2.15
$\mu_{i\text{-}H0}$	55.2	90.9	22.0	48.5	57.4	36.9
$\mu_{A\text{-}max}$	44.3	79.6	20.3	38.6	51.5	35.2
$MBN_{\mu i}$ (mV_{rms})	0.46	1.03	0.18	0.31	0.46	0.19

This work shows that there are clear relationships between EM features and microstructural changes in power station steels. Although major *B-H* loop features are useful, this type of measurement is difficult to achieve on open samples (i.e. pipes and tubes). MBN and permeability readings derived from minor loops are easier on open samples, require less power, and by looking at the change in readings rather than absolute values, may not require an accurate *H* field measurement; as *H* is heavily dependent on geometry. The next step of this work will be to exploit the correlations established using closed magnetic loop tests to develop a tool for the inspection of pipes and tubes in power stations.

References

[1] Maharaj C, Dear JP, and Morris A 2009 *Strain* **45(4)** 316-331
[2] Sposito G, Ward C, Cawley P, Nagy PB and Scruby C 2010 *NDT & E Int.* **43(7)** 555-567
[3] Liu J, Hao XJ, Zhou L, Strangwood M, Davis CL, Peyton AJ, 2012 *Scripta Materialia* **66** 367–370
[4] Yin W, Karimian N, Liu J, Hao XJ, Zhou L, Peyton AJ, Strangwood M and Davis CL 2012 *NDT & E Int.* **51** 135-141
[5] Wilson JW, Tian GY, Moorthy V and Shaw BA 2009 *IEEE Trans. Magn.* **45** 177-183
[6] Bertotti G 1998 Hysteresis in magnetism: for physicists, materials scientists, and engineers (San Diego:Academic Press)
[7] Takahashi S, Kobayashi S, Kikuchi H, and Kamada Y 2006 *J. Appl. Phys.* **100** 113-908

Magnetic sensing for microstructural assessment of power station steels: Differential permeability and magnetic hysteresis

N Karimian[1*], JW Wilson[1], W Yin[1], J Liu[2], CL Davis[2], AJ Peyton[1]
[1]School of Electrical and Electronic Engineering, University of Manchester, M60 1QD
[2]School of Metallurgy and Materials, University of Birmingham, Edgbaston, B15 2TT

Abstract. Failure of power station steel components can have severe economic impacts and also present significant risks to life and the environment. Currently components are inspected during costly shut-downs as no in-situ technique exists to monitor changes in microstructure of in-service steel components. Electromagnetic inspection has the potential to provide information on microstructure changes in power station steels in-situ. In this paper, tests have been carried out on pipe and tube samples in different microstructural conditions, using a lab-based closed magnetic circuit and impedance measurement systems. EM properties have been identified with correlations to material properties, which can quantify degradation in-situ and at elevated temperatures.

1. Introduction

Measurement of the electromagnetic (EM) properties of materials can be vital; as such properties are related to the materials' microstructure, condition-related parameters and consequently performance. For instance, the permeability and conductivity of porous Cu and Fe foams, manufactured by the lost carbonate sintering process, have proven to be related to their porosity [1]. In steels, ferrite fraction variations [2, 3], creep damage and precipitation [4] cause changes to EM properties. Factors such as residual stress [5], temperature and case hardening [6] also have an affect on the steels' EM properties. Such EM properties can be obtained through inspection of the material using purpose built EM sensors.

Power generation steel components such as boiler tubes and steam line pipes are exposed to high temperature and pressure during their lifetime, and therefore experience carbide coarsening and occurrence of creep cavitation [7], possibly leading to component failure; an unplanned shutdown period can cost approximately ~$2M per day per power station. These components are currently inspected during costly shut down periods and no in-situ technique exists to directly monitor the changes in microstructure of such in-service components at an elevated temperature.

Magnetic induction techniques have been used extensively in measuring the EM properties of materials at relatively low frequencies (i.e. <10 M Hz). In doing so, various sensor configurations have been used, for example, cylindrical air-cored sensor [8,9], H-shaped ferrite-core coils and printed circuit board (PCB) based planar coils [10]. The ability to monitor any microstructural changes in steel components through their EM property variation has great potential and may become very useful in the future. In this paper, a lab-based closed magnetic circuit measurement system and a cylindrical air-cored sensor were designed to measure the EM properties of 50 mm long cylindrical power generation steel (P9 and T22) samples in different microstructural states. The obtained EM properties from both sets of systems are then compared with each other. Analytical and numerical methods (finite elements (FE)) have been used to calculate the sensor response, through which the conductivity and

* Noushin.Karimian@postgrad.manchester.ac.uk

permeability of the samples were inferred. In addition, the relative incremental permeability of P9 and T22 steel samples have been obtained through the measurement of minor loop deviations from the initial magnetisation curve, the results of which are connected to the permeability values obtained from the spectroscopy of these samples. From the two magnetic hysteresis and incremental permeability measurements, correlations have been established between the electromagnetic properties and material properties such as hardness. Such correlations may allow one to quantify degradation in power station steels both in-situ and at elevated temperatures; and could, therefore, lead to considerable savings while allowing frequent and detailed component inspection.

2. Methodology

2.1. Physical Principle

Any microstructural variation in steel may lead to changes in its EM properties, e.g. permeability and conductivity. EM sensors function on the basis of detecting and identifying variations in these quantities measured from samples. By measuring the response of such EM sensors over a range of frequencies, the permeability and conductivity can then be inferred.

The effect of eddy currents in the sensor response is very weak at low frequencies, and any contribution to the inductance change is mainly from the magnetisation of the sample. Therefore the inductance measured at low frequencies is related to the sample permeability. However, the effect of eddy currents in the sample becomes stronger with increasing frequency; and therefore variations in inductance are gradually dominated by the effect of conductivity or polarisation delay of the sample. The electric and magnetic properties of the samples can be determined by complete analysis of the inductance-frequency spectra.

2.2. Experimental Procedure and Sensor FE Simulation

Two groups of power generation steels were studied, the first group being P9 steel (8.40Cr-0.97Mo-0.12C-0.52Si-0.44Mn wt%) and the second group being T22 (1.90-2.60Cr-0.87-1.13Mo-0.05–0.15C-0.50Si-0.30–0.60Mn wt%), both removed from service at 520 °C for approximately 11 years. Selected samples were heat treated to simulate the service entry microstructure by normalising at 950 °C or 940 °C for 1 hour and then tempering at 760 °C for 1 hour or 720 °C for 1.5 hours, for the P9 and T22 steels respectively. In addition to the tempered and ex-service samples, as normalised samples were also examined. Cylindrical rods 50 mm long and 4.95 mm in diameter for each condition were prepared for the EM sensor measurements.

Samples were measured by inserting them into an air-cored cylindrical sensor, which consists of two identical coils arranged with their axes aligned, one exciting and the other receiving. The coils have a length of 9.8 mm, inner diameter of 6.9 mm and outer diameter of 8.3 mm with a separation of 10.4 mm. Measurements of trans-impedance were taken with an impedance analyser (Solartron 1260) at frequencies from 10 Hz to 1 MHz. Analytical and numerical methods have also been employed to calculate the sensor response of these samples and presented in [9]. In addition to the analytical methods, FE models have been built in Maxwell (ANSYS, Inc.) to model the actual measurement arrangement, taking full account of the sensor and sample interaction, in order to determine the EM properties such as relative permeability and conductivity, of these samples by fitting with the experimental data. The geometry, setup and the FE model for the sensor are shown in Fig 1.

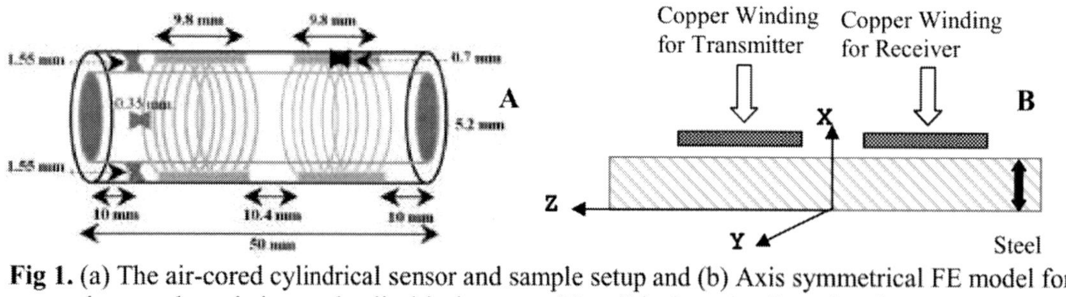

Fig 1. (a) The air-cored cylindrical sensor and sample setup and (b) Axis symmetrical FE model for the sample and air-cored cylindrical sensor - Note Z is the axis of rotational symmetry.

3. Measurement and Experimental Results

3.1. Metallographic Tests

The microstructure for the P9 and T22 grade samples have been analysed and presented previously [8, 9,11]. However, for completeness, the microstructure of the P9 and T22 samples with different conditions are shown in Fig 2. Referring to Fig 2(a), the microstructure of the as-normalised P9 consists of predominantly martensite mixed with some bainite. Subsequent tempering produces a simulated service entry microstructure, i.e. tempered martensite / bainite as shown in Fig 2(b). After long service exposure, the microstructure showed equiaxed ferrite with large carbides distributed within ferrite grains or on grain boundaries as shown in Fig 2(c). The as-normalised T22 steel shows a mixed microstructure of bainite and some proeutectoid ferrite, as shown in Fig 2(d). After tempering, many carbides can be observed along prior austenite grain boundaries, on ferrite boundaries or within bainite regions, as shown in Fig 2(e). The microstructure of T22 after the service exposure consists of equiaxed ferrite and a great many carbides outlining the ferrite grain boundaries or finely dispersed within the ferrite grains, as shown in Fig 2(f).

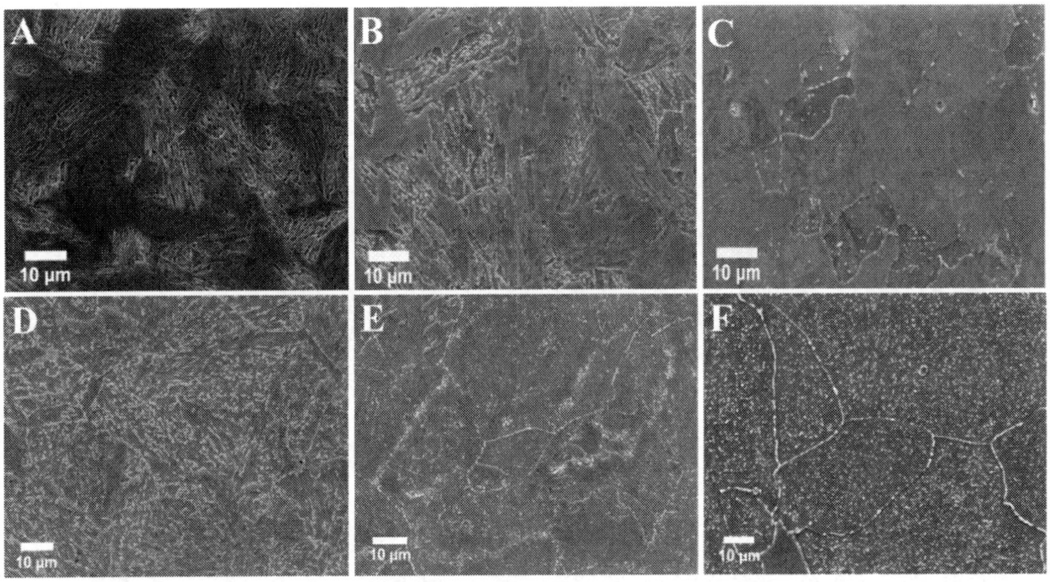

Fig 2. Microstructure of the P9 samples in different conditions: (a) as normalised (b) as tempered and (c) ex-service and microstructure of T22 samples in different conditions: (d) as normalised (e as tempered and (f) ex-service

3.2. Electromagnetic Sensor Measurements

The measured mutual inductance spectra for P9 and T22 with different conditions are shown in Fig 3.

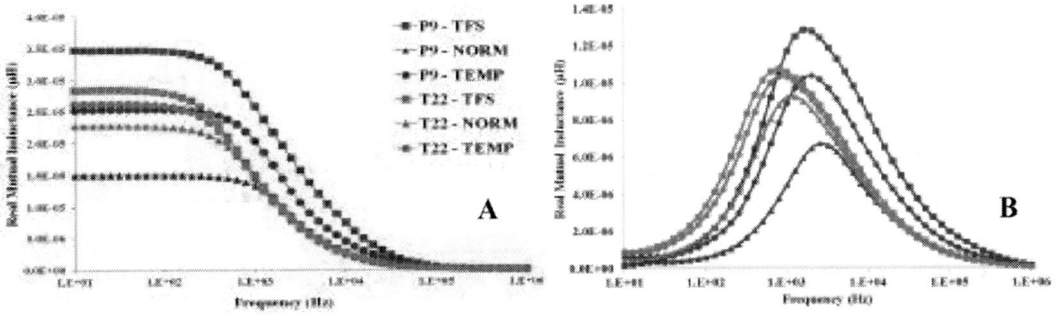

Fig 3. (a) Real and (b) Imaginary Inductance spectra for P9 and T22 samples of different conditions (TFS: Taken from Service, NORM: Normalised, TEMP: Tempered) measured using cylindrical air-cored sensor

Journal of Physics: Conference Series 450 (2013) 012042 doi:10.1088/1742-6596/450/1/012042

The real inductance graphs, shown in Fig 3 (a), have a characteristic shape, with a flat section for frequencies below approximately 100 Hz; a roll off over the intermediate frequencies and finally a flat high frequency response. The flat low frequency response is determined by the differential permeability of the sample, suggesting that the permeability is constant up to a frequency of at least 100 Hz. The roll-off over the intermediate frequency response results from eddy currents circulating coaxially in the sample; here the electromagnetic skin depth is reducing with frequency and eventually becomes much smaller than the radius of the sample. At high frequencies, the skin depth effectively approaches zero and the real mutual inductance approaches a constant slightly negative value determined only by the dimensions of the coil / sample geometry. The graphs of the imaginary component in Fig 3(b) show a characteristic curve with maximum responses which coincide with maximum energy transfer eddy current heating from the transmitter coil to the sample.

3.3. B-H Measurements

Another aspect of obtaining the magnetic properties of a sample is to measure its magnetic hysteresis. The B-H curves for the three P9 and T22 steel samples in different conditions have been measured (Fig 4). The B-H curves demonstrate the micro-magnetic properties of steels including coercivity, remanence, saturated magnetisation etc. These properties and the relative permeability can be affected by different microstructural features. Therefore, the B-H measurement and EM sensor can be complementary techniques.

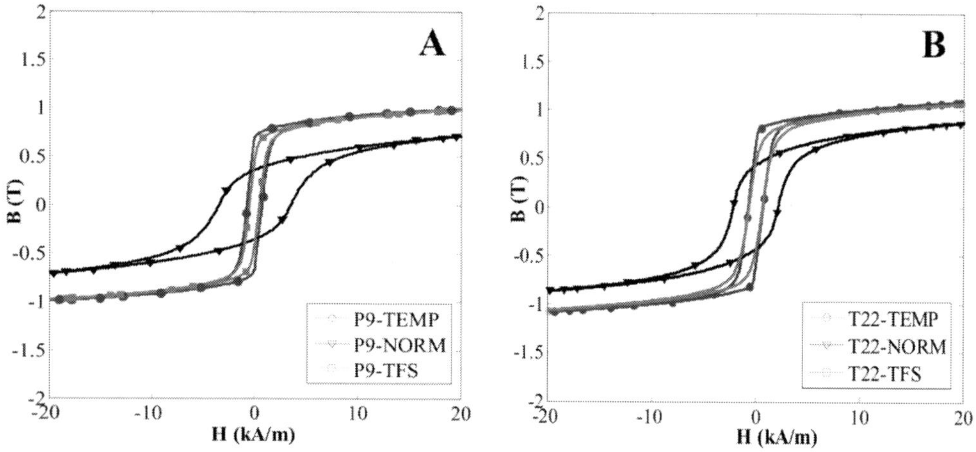

Fig 4. B-H measurements for (a) P9 and (b) T22 samples with different conditions (TEMP: Tempered, NORM: Normalised, TFS: Taken from Service)

As shown in Fig 4, the B-H loops for both set of samples (P9 and T22) follow the same trend; the ex-service sample results in the lowest coercivity (H_C), with a small increase in H_C for the tempered samples and a large increase in H_C for the normalised samples (Table 1). The coercivity values reflect the magnetic hardness of the samples, which in turn is indicative of material hardness. As the P9 normalised sample is predominantly martensitic with a Vickers hardness value of HV401, it exhibits the greatest H_C value. As bainite typically has a lower hardness than as-quenched martensite, but higher hardness than ferrite, the bainite / proeutectoid ferrite microstructure of T22 normalised sample (HV316) shows a reduction in H_C in comparison to the P9 normalised sample. The difference between the ex-service and tempered samples is less pronounced, with a decrease in hardness after the in-service exposure being reflected by a small decrease in H_C for both sample sets; with HV212 and HV203 for P9 and T22 TEMP and also HV158 and HV129 for P9 and T22 TFS respectively.

Fig 5 shows the evolution of the minor loop as deviations from the initial magnetisation curve. The reason for such measurements is to obtain the relative incremental permeability of these power station steels through their magnetic hysteresis, the results of which are connected to the permeability values obtained from the spectroscopy of these samples. The origin of the first minor loop corresponds to the demagnetised state, where B and H are equal to zero. As the initial magnetisation curve approaches saturation, domain walls are swept away by field pressure. The dominant magnetisation mechanism

Journal of Physics: Conference Series 450 (2013) 012042 doi:10.1088/1742-6596/450/1/012042

can be described as the progressive alignment of the field against anisotropy, as the domains rotate from their magnetic easy axes towards the direction of the applied field [12]. Consequently, only the reversible components remain, resulting in a closed loop, with a smaller variation in Fig 5(b) for a given applied field, as shown in the graph (bottom right). As illustrated in Fig 5, the incremental permeability (μ_Δ) can be obtained from the minor loops. Such a quantity can be calculated as the ratio of the variation in flux density (ΔB) and the corresponding change in the applied field (ΔH), scaled with respect to the permeability of free space (μ_0); presented in equation 1.

$$\mu_\Delta = \frac{1}{\mu_0} \times \frac{\Delta B}{\Delta H} \qquad (1)$$

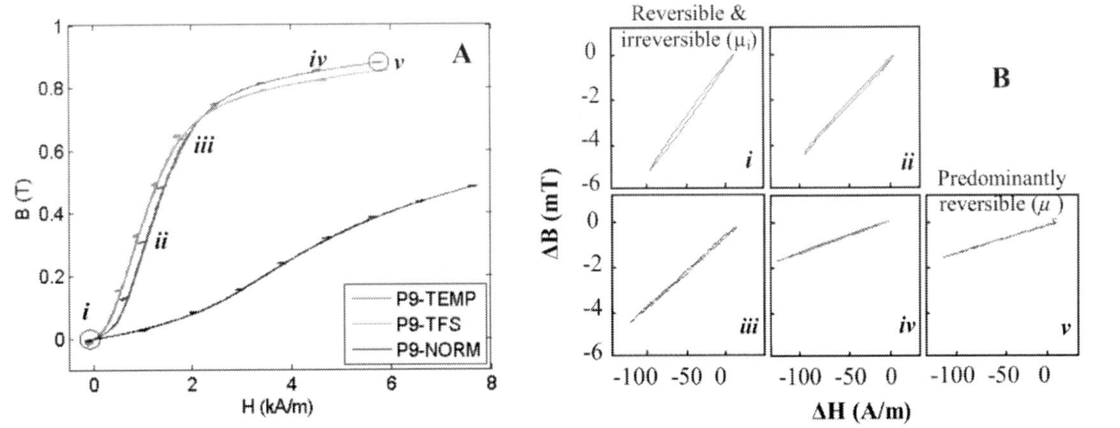

Fig 5. (a) Initial magnetisation curves and (b) minor loop deviations for P9 Tempered sample - B and H offsets removed from minor loops for comparison

3.4. Permeability and Conductivity Evaluation

Finite element models were set up to simulate the sensor response. The simulated results are compared to the measured results to fit the conductivity and permeability in a least squared sense. The comparison of the simulated results with the measured results for the cylindrical sensor has been covered in [9]. It is evident that the measurement and simulation results are in good agreement. Permeability and conductivity values for the P9 and T22 samples can therefore be inferred, a list of which is provided in Table 1. This table presents the permeability values for the power station steels (P9 and T22) for both impedance analyser measurement (using the cylindrical air-cored sensor) and also their corresponding coercivity and remanence values obtained from the magnetic hysteresis loop measurements. It is interesting to see how the values for P9 and T22 normalised, tempered and ex-serviced samples increase from small to large values respectively. It is also evident that when a series of samples all have basically similar metallurgical microstructure, i.e. the tempered P9 and T22 samples, which are all forms of tempered martensite, then the variation in permeability is very small. Thus, the permeability values for P9 and T22 tempered samples (P9-TEMP and T22-TEMP) fall into a narrow range of 52-53.9 for the cylindrical sensor. However, when there is a microstructural change, there is a big change in permeability. As a result, the samples with un-tempered martensitic microstructures (P9-NORM and T22-NORM) have permeability of 28.7 and 45.1 for the cylindrical sensor. On the other hand, samples with a predominantly ferritic structure (P9-TFS and T22-TFS) have permeability of 83 and 60.4 for the cylindrical sensor. Although the permeability values are not precisely the same for the two sets of measurements, similar trends for the samples can be observed.

As presented in Table 1, the samples with ex-service heat treatment have the lowest coercivity (H_C) values, with a minor increase in H_C for the tempered samples, followed by a considerable increase in H_C for the normalised samples. As the P9 normalised sample is predominantly martensitic (with an associated high dislocation density), higher fields are required for demagnetisation, and therefore it exhibits the greatest H_C value. The lower (mechanical) hardness bainite / proeutectoid ferrite microstructure of the T22 normalised sample shows a reduction in H_C in comparison to the P9 normalised sample. On the other hand, the difference between the ex-service and tempered samples is less pronounced, with the decrease in hardness following in-service exposure being reflected by a small decrease in H_C for both sample sets.

Table 1. Summary of relative permeability and conductivity for different P9 and T22 steel samples

Samples and conditions		μ_R obtained from Cylindrical Sensor	Conductivity (MS/m)	μ_i obtained from BH curves	Coercivity (A/m)	Remanence (T)
P9-(MTG L3101A-1)						
P9-TEMP	Normalised and Tempered at 760°C for 1h	**52**	**1.82**	**55.2**	**0.778**	**1.057**
P9-NORM	Normalised at 950°C for 1h	**28.7**	**1.69**	**22.0**	**3.514**	**0.559**
P9-TFS	Taken from service	**83**	**1.83**	**90.9**	**0.512**	**0.684**
T22 - (MTF C33-3)						
T22-TFS	Taken from service	**60.4**	**3.87**	**57.4**	**0.644**	**0.712**
T22-TEMP	Tempered at 720°C for 1.5h	**53.9**	**3.53**	**48.5**	**0.718**	**0.988**
T22-NORM	Normalised at 940°C for 1h	**45.1**	**2.99**	**36.9**	**2.154**	**0.679**

It is also clear from Table 1 that the coercivity has an inverse relationship with permeability (obtained from impedance measurements), as the samples with the lowest coercivity values have the highest permeability values. The coercivity values reflect the magnetic hardness of the samples, which in turn is indicative of material/physical hardness, therefore both the permeability and coercivity values are linked to magnetic and material hardness.

4. Conclusion

In this paper, a cylindrical air-cored sensor has been designed to measure the EM properties of 50 mm long, 4.95 mm diameter power generation steel (P9 and T22) samples taken from pipes and tubes exposed to different heat treatments. Numerical methods (finite elements) were used to calculate the sensor response, and as a result the conductivity and permeability of these samples were inferred. Magnetic hysteresis curves for the power station steel samples were also measured, from which relative incremental permeability values were obtained. The incremental permeability values measured through minor loop deviations from the initial magnetisation curve were compared to the permeability values obtained from the spectroscopy of these samples, and their corresponding EM properties have been identified with strong correlations with material properties such as hardness. Correlations between EM properties and changes in the microstructure (normalised only, normalised and tempered and ex-service) were established and will be used in future to develop a field deployable tool for in-situ inspection.

References

[1] Ma X, Peyton. AJ, Zhao YY 2006 *NDT&E Int.* **39** 562–568
[2] Haldane RJ, Yin W, Strangwood M, Peyton AJ, Davis CL 2006 *Scripta Mater.* **54** 1761–1765
[3] Yin W, Peyton AJ, Strangwood M, Davis CL 2007 *Mater. Sci.* **42** 6854–6861
[4] Sposito G, Ward C, Cawley P, Nagy PB and Scruby C 2010 *NDT & E Int.* **43(7)** 555-567
[5] Gou R, Zhang Y, Xu X, Sun L, Yang Y 2011 *NDT&E Int.* **44** 387–393
[6] Wilson JW, Tian GY, Moorthy V, Shaw BA 2009 *IEEE Trans. Magn.* **45** 177-183
[7] Ennis P, Czyrska-Filemonowicz A 2003 *Sadhana Acad. Proc. Eng. Sci.* **28** 709–730
[8] Yin W, Karimian N, Liu J, Peyton AJ, Davis CL 2012 *I2MTC 2012, IEEE Int.* 1603 – 1606
[9] Yin W, Karimian N, Liu J, Hao XJ, Peyton AJ and Davis CL 2012 *NDT & E Int.* **51** 135-141
[10] Tumanski S 2007 *Meas. Sci. Technol.* **18(3)** R31-R46
[11] Liu J, Hao XJ, Zhou L, Strangwood M, Davis CL, Peyton AJ 2012 *Scripta Materialia.* **66** 367-370
[12] Bertotti G 1998 Hysteresis in magnetism: for physicists, materials scientists, and engineers (San Diego:Academic Press)

Evaluation of Uncertainty in AC Power Calculation with Asynchronously Sampled Data

D Lindenthaler[1] **and H Zangl**[2]

Institute of Electrical Measurement and Measurement Signal Processing, Graz University of Technology, 8010 Graz, AT

E-mail: [1]`lindenthaler@tugraz.at`, [2]`hubert.zangl@tugraz.at`

Abstract. This article presents a comparison of several standard based methods for determination of the effective and the apparent AC power from sampled voltage and current signals. The comparison is based on the accuracy achieved in three different measurement scenarios, which is evaluated by means of Monte-Carlo-simulations. The comparison shows that the accuracy of the power estimation from asynchronously sampled data can be as good as the power estimation from synchronously sampled data. This allows for a reduction of the effort for synchronization between voltage and current channels while the overall performance remains equal.

1. Introduction

Measuring power is a standard task in electrical engineering. High accuracy of power estimation is of major interest in the in context of energy efficiency. For example, the optimization of a power factor correction device [1] with the intention to improve the efficiency will require a validation by measurement. Unfortunately, the power loss is not directly accessible by electrical measurement as rather the small difference between large quantities (input and output power) needs to be determined. This results in high demands on the accuracy of the measurement approach. By reaching physical limits of the efficiency factor, the improvements get smaller and the power measurements needs to be more and more accurate. For instance, an efficiency improvement from 98 % to 99 % reduced the loss by a factor of 50 % (and thus the generated heat!), whereas it can not be validated if the uncertainty of the power estimation is in the range of 1 % as this relates to the input and output power respectively.

It should be noted that power estimation has achieved very high accuracy for the sinusoidal case at mains frequencies. However, modern inverter technology often leads to signals that drastically deviate from the sinusoidal case and high order harmonics and also non-periodic components are commonly observable. Consequently, highly accurate methods for this situations are of major interest. Due to the advantages of the digitalization modern measurement systems operate in the discrete time domain and the arising question is, how accurate the various methods for the calculation are.

1.1. Problem description and objective

In order to estimate the electrical AC power in the digital domain it is necessary to sample the current and voltage and process them appropriate to approximate the continuous-time integral. When the sampling frequency is an integer multiple of the signal frequency, the power determination can accurately be performed in the (discrete) frequency domain. However, when the sampling frequency is not an integer multiple of the signal frequency, spectral leakage occurs, which reduces the accuracy. Things get even worse when the current and voltage signals are sampled asynchronously, as the spectral line obtained from the Discrete Fourier transform (DFT) are no longer coherent.

Generally, voltage and current measurement require individual measurement devices. These devices can share the same clock generator, be synchronized, or run completely independent. The more they have in common, the more restricted is their general applicability as synchronization signals need to be transmitted besides the measurement data, which consumes additional hardware, wires and/or bandwidth. On the other side, an algorithm for computing the AC power from two devices has lower complexity when the quantities share common timings.

In context of asynchronous or synchronous samples, we investigate three different scenarios for the measurement setup:

- The two analog-to-digital converter (ADC) for the current and the voltage have a common clock generator (or accurate synchronization by means of a stable Phase Locked Loop) and thus have the same sampling frequency. This scenario will be denoted as OC (one clock generator).
- The second scenario is that the two ADCs run independently with two different clock generators and thus have slightly different sampling frequencies and the samples are not coherent in time. However, time stamps are available. The abbreviation of this scenario is TC (two clock generators).
- The last measurement-setup is that the two ADCs run independently but the clock generators are synchronized before every measuring interval and thus they have different sampling frequencies but no initial time-lag between the current and voltage. This scenario will be denoted as TC-S (two clock generators with synchronization).

The approximation of the continuous-time integral from samples can be done by different interpolation methods or by evaluating the integral in the frequency-domain. The focus of this paper is to compare these methods concerning their accuracy in the three different measurement scenarios.

The organization of this paper is as follows. Section 2 explains the basic theory for measuring electrical power in the digital domain. Section 3 deals with the uncertainty of the various methods and scenarios. The conclusion is given in section 4.

2. Digital calculation of electrical power

This section deals with the theoretic background on the calculation of the AC power in the discrete time and frequency domain. The underlying time-continuous case is conform with the IEEE standard 1459 – 2010.

The evaluation of the AC power from the power grid and a linear load is considered, and thus it is legitim to assume voltage $u(t)$ and current $i(t)$ to be a composition of sinusoidal waves:

$$u(t) = \sum_{i=1}^{N_\mathrm{w}} \hat{U}_i \cdot \sin(\omega \cdot i \cdot t + \varphi_0) \qquad \text{and} \qquad i(t) = \sum_{i=1}^{N_\mathrm{w}} \hat{I}_i \cdot \sin(\omega \cdot i \cdot t + \varphi_0 + \varphi_i), \tag{1}$$

where N_w is the number of assumed harmonic waves. The cycle time T_S of these signals is defined as $T_S = \frac{2\pi}{\omega}$ and the effective power P and the apparent power S are then calculated as

$$P = \frac{1}{N_S T_S} \int_{t_0}^{t_0+N_S T_S} u(t)\, i(t)\, \mathrm{d}t \qquad \text{and} \qquad S = \frac{1}{N_S T_S} \sqrt{\int_{t_0}^{t_0+N_S T_S} u(t)^2\, \mathrm{d}t \cdot \int_{t_0}^{t_0+N_S T_S} i(t)^2\, \mathrm{d}t}, \qquad (2)$$

where N_S is the number of signal-periods.

The continuous time signals are reconstructed from the discrete time samples $u[n]$ and $i[n]$ which are in general not coherent in time and have different sampling-periods T_u and T_i. The sampled signals are denoted as

$$u[n] = u(nT_u) \quad \forall n \in \{0, 1, \dots N-1\} \qquad \text{and} \qquad i[n] = i(nT_i + t_s) \quad \forall n \in \{0, 1, \dots N-1\},$$

where t_s is the time-shift due to the unsynchronized ADCs and N is the number of available samples. In the case of continuous running ADCs it is sufficient to limit t_s to be in the range $0 \le t_s < T_i$ since it is always possible to arrange the samples such that the time-shift is in the given range.

2.1. *Evaluation in the time domain*

The approximation of the continuous-time signals for evaluating (2) are obtained by interpolating the discrete time signals $u[n]$ and $i[n]$ with different interpolation functions $f_{\text{int}}(x)$:

$$u_{\text{int}}(t) = \sum_{n=0}^{N-1} u[n] \cdot f_{\text{int}}\left(\frac{t - nT_u}{T_u}\right) \qquad \text{and} \qquad i_{\text{int}}(t) = \sum_{k=0}^{N-1} i[k] \cdot f_{\text{int}}\left(\frac{t - kT_i - t_s}{T_i}\right).$$

The effective power P in (2) and the apparent power S from (2) are then evaluated as

$$P_{\text{int}} = \frac{1}{N_S T_S} \cdot \sum_{n=0}^{N-1} \sum_{k=0}^{N-1} u[n] \cdot I_{\text{int}}(n, k, T_u, T_i, 0, t_s) \cdot i[k],$$

$$S_{\text{int}} = \frac{1}{N_S T_S} \sqrt{\sum_{n=0}^{N-1} \sum_{k=0}^{N-1} u[n]\, I_{\text{int}}(n, k, T_u, T_u, 0, 0)\, u[k] \cdot i[n]\, I_{\text{int}}(n, k, T_i, T_i, t_i, t_i)\, i[k]}$$

where $I_{\text{int}}(n, k, T_n, T_k, t_n, t_k)$ is

$$I_{\text{int}}(n, k, T_n, T_k, t_n, t_k) = \int_{t_0}^{t_0+N_S T_S} f_{\text{int}}\left(\frac{t - nT_n - t_n}{T_n}\right) f_{\text{int}}\left(\frac{t - kT_k - t_k}{T_k}\right) \mathrm{d}t$$

Due to the convenient framework and versatileness of splines, cardinal splines of different orders are compared with the sinc function. A detailed comparison of them and further interpolation functions are given in [2]. Since the number of samples is finite and the interpolation functions have in general an infinite support, the effect of truncating the interpolation function is considered. A further aspect of truncation is the computational complexity of P and S since all integrals $I_{\text{int}}(n, k, T_n, T_k, t_n, t_k)$ without a common support are zero a priori.

The cardinal splines $\eta_\alpha(x)$ with order $\alpha \in \{0, 1, 2, 3, 5\}$ are calculated according to [3]. The truncated versions of them $\eta_\alpha^S(x)$ and the truncated sinc function $\text{sinc}^S(x)$ have a supported on $(-S, S)$. The cardinal spline of zeroth and first order $\eta_0(x)$ and $\eta_1(x)$ (nearest-neighbor and linear interpolation) have inherently finite support and are not truncated.

2.2. *Evaluation in the frequency domain*
Since the DFT represents a scaled version of the Discrete Fourier Series under the assumption that the sampled signal is periodic, it is necessary to select a window such that it contains an integer number of signal cycles. Actually, this is only possible when the sampling frequency is an integer multiple of the signal frequency, otherwise it is an approximation. The signal-lengths N_u and N_i of the truncated voltage and current signals for the evaluation with the DFT are chosen such that the truncated versions include almost a full number of signal-periods.

The DFT of $u[n]$ and $i[n]$, denoted as $U[k]$ and $I[k]$ are defined as:

$$U[k] = \frac{1}{N_u} \sum_{n=0}^{N_u-1} u[n] \cdot e^{-j\frac{2\pi k}{N_u} n} \qquad I[k] = \frac{1}{N_i} e^{j\frac{2\pi t_\Delta k}{N_i T_i}} \sum_{n=0}^{N_i-1} i[n] \cdot e^{-j\frac{2\pi k}{N_i} n}$$

where k is in the interval $\{0, 1, \ldots, N_u - 1\}$ and $\{0, 1, \ldots, N_i - 1\}$ respectively.

The effective power P and the apparent power S of the underlying signal model (1) is then calculated as

$$P = 2 \cdot \sum_{l=1}^{N_w} \mathrm{Re}\Big(U\Big[l\frac{\omega}{2\pi} N_u T_u\Big] \cdot \overline{I}\Big[l\frac{\omega}{2\pi} N_i T_i\Big]\Big) \qquad S = 2 \cdot \sum_{l=1}^{N_w} \Big|U\Big[l\frac{\omega}{2\pi} N_u T_u\Big]\Big| \cdot \Big|I\Big[l\frac{\omega}{2\pi} N_i T_i\Big]\Big|,$$

where the summation is over the indices corresponding to the harmonics and $\overline{(.)}$ denotes the complex conjugation. In the following, the AC power calculation using the DFT is denoted by DFT.

2.3. *Simple power calculation in the time domain*
The simplified calculation is based on the assumption that the sampling frequency $\frac{1}{T}$ of voltage and current is equal, the signals are coherent in time and the sampled signal is periodic. In this case the AC power is calculated as

$$T = T_i = T_u \qquad \tilde{N} = \left\lfloor \frac{N_s \cdot T_s}{T} \right\rfloor \qquad P = \frac{1}{\tilde{N}} \sum_{n=0}^{\tilde{N}-1} u[n]\, i[n] \qquad S = \frac{1}{\tilde{N}} \sqrt{\sum_{n=0}^{\tilde{N}-1} u[n]^2 \cdot \sum_{n=0}^{\tilde{N}-1} i[n]^2},$$

where $\lfloor . \rfloor$ denotes rounding towards $-\infty$. In the remainder of this article this method will be denoted by SPC.

3. Uncertainty evaluation
Due to the enormous amount of parameters for the current and the voltage the uncertainty is evaluated by means of Monte-Carlo-simulation. The voltage is generated at random with the baisc conditions from the standard IEC 60038.

3.1. *Simulation setup*
The common parameters for all three measurement setups are:

- The fundamental frequency f may vary between 47 and 52 Hz but is assumed to be constant for each sampling interval and the phase φ_0 is random between 0 and 2π.
- The number of harmonics N_w is defined to be $N_w = 25$ The total harmonic distortion (THD) is smaller than 8 % and the $\hat{U}_i$ are bounded according to IEC 60038.
- The number of samples N is fixed to 4100 which corresponds to a sampling interval of approximately 1 s. Sampling a longer time can result in a higher accuracy but on the other side the time resolution of the calculated AC power decreases.

Additionally a load is generated as a RLC-combination at random with values from 50 to about 2500 VA.

With the voltage $u[n]$, which is randomly generated for each Monte-Carlo-simulation, the current is calculated by means of complex impedances. Depending on the measurement scenarios OC, TC or TC-S the measured current $i_{\mathrm{m}}[n]$ is calculated.

In all three scenarios the measured voltage $u_{\mathrm{m}}[n]$ and current $i_{\mathrm{m}}[n]$ are independently corrupted by Additive White Gaussian Noise (AWGN) $\nu_u[n]$ and $\nu_i[n]$, with a standard derivation of $\sigma_u = 1\,\mathrm{mV}$ and $\sigma_i = 30\,\mu\mathrm{A}$.

The effective power P and the apparent power S is calculated with the measured signals $u_{\mathrm{m}}[n]$ and $i_{\mathrm{m}}[n]$ and the methods explained in section 2. The true values P_{true} and S_{true} are calculated by means of complex AC-analysis.

All methods for calculating the AC power need the the fundamental frequency of the voltage and current signal. Therefor, the fundamental frequency is calculated by counting the average number of samples between the zero-crossings in $u_{\mathrm{m}}[n]$. This method is abbreviated by ZC. Additionally, the frequency is estimated by finding the position of the first maximum in the estimated autocorrelation-function (abbr. ACF) by means of Newton's method.

3.2. Simulation results

The following figures show the average of all Monte-Carlo-iterations relative absolute errors of the effective- and apparent power P and S, which are calculated as

$$e_{\mathrm{P}} = \mathbb{M}\left[\left|1 - \frac{P_{\mathrm{Meth.}}}{P_{\mathrm{true}}}\right|\right] \qquad e_{\mathrm{S}} = \mathbb{M}\left[\left|1 - \frac{S_{\mathrm{Meth.}}}{S_{\mathrm{true}}}\right|\right].$$

$\mathbb{M}[.]$ denotes the mean operator and the number of Monte-Carlo-iterations is chosen to be 100. A larger number of iterations did neither significantly influence the error-mean nor the error-variance. The abscissa show the different methods for calculating the AC-power. Beside the mean absolute error the ordinate shows the quantile for 84.1 % and 15.9 %. Thus, 68.2 % of the absolute errors lie in the interval denoted by the quantiles.

Figure 1(a) shows the absolute error of the effective power e_P in the case of a setup with one clock generator (OC). The frequency estimation with ZC (denoted by •) outperforms the ACF method (denoted by $*$). The accuracy limit of each method is achieved if the frequency is known exactly and the measurement noise is absent. This case gives a potential estimation and is denoted by x.

The accuracy of the setup with two independently running ADCs (TC) is shown in figure 1(b). The interpolation methods with the measured samples perform slightly worse than in the case OC, but the possible potential of the configuration (perfect frequency estimation and no measurement noise) is noticeable worse.

The case TC-S, is illustrated in figure 1(c). The accuracy in this case is similar to the accuracy in the case OC and slightly better than in the case TC. A possible reason for this is that the uncertainty of interpolation between two samples is maximal in the middle of them and is zero at the sample points. Since the sample frequencies differ with a maximum of 100 ppm and the sampling period is one second, the worst possible time delay between the current and the voltage is 0.41 samples.

The error of the apparent power e_S in figure 1(d) is independent of an possible time-shift t_{s} between the voltage and current signal. Therefor the absolute error is the same in all three scenarios and just depends on the method of calculation.

4. Conclusion

It is not advisable to use a truncated sinc or cardinal spline of order 1 as interpolation function, since the sinc function has the worst performance in all scenarios for a given support S and the

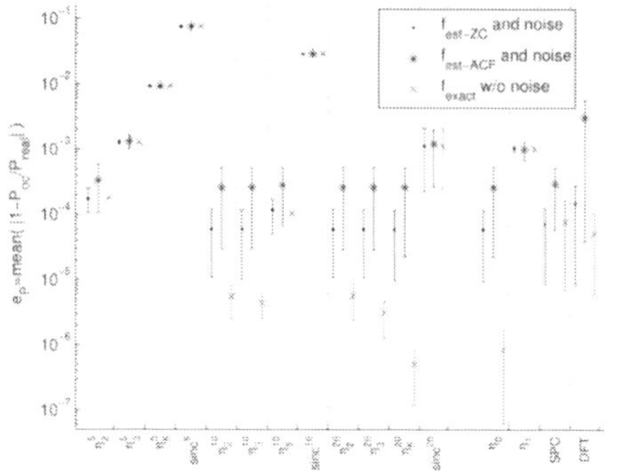

(a) e_P in the scenario of using a common clock generator.

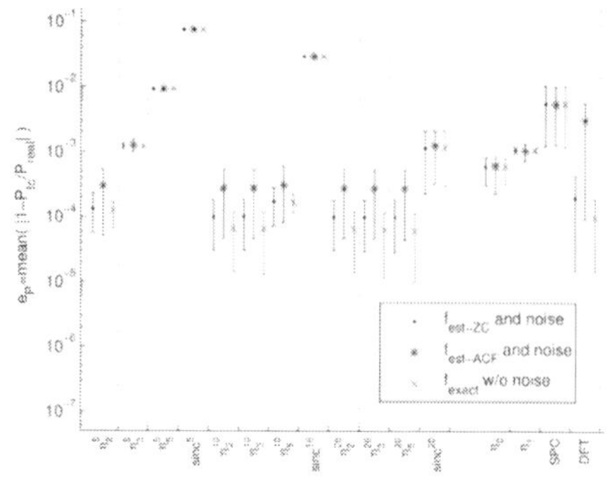

(b) e_P in the scenario of using two independently free running ADCs.

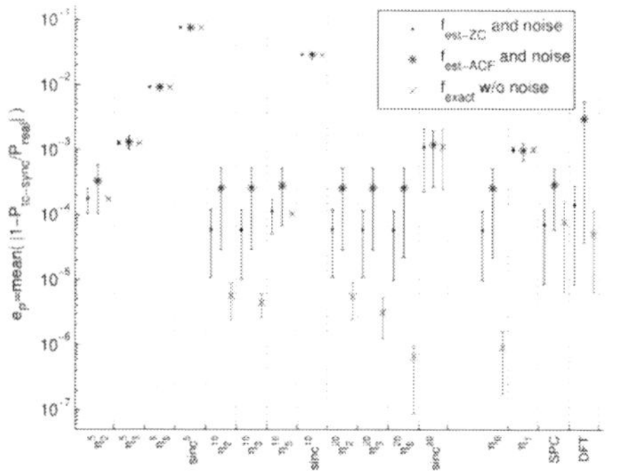

(c) e_P in the scenario of using two synchronized free running ADCs.

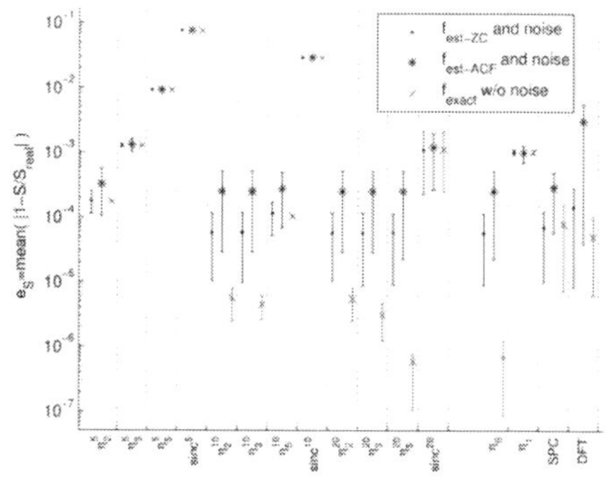

(d) Error of the apparent power e_S. The plot is equal for all three scenarios.

Figure 1. Relative absolute error of the AC power in the three scenarios OC, TC or TC-S.

cardinal spline of order 0 outperforms always the cardinal spline of order 1.

The evaluation in the frequency domain is an alternative to interpolation in the time domain for all three scenarios because the computational costs are very low. A drawback is the high sensitivity concerning the estimate of the fundamental frequency.

The method SPC has a competitive accuracy in calculating the effective power P in the case of one common clock generator or synchronized signals, but performs poor in the case of free running ADCs.

The overall conclusion is that the measurement setup does hardly influence the accuracy if a cardinal spline with an order greater than 1 or the DFT is used. A benefit through the additional effort of the scenarios OC or TC-S will make sense if the frequency is measured more accurate and the measurement noise is low or if the SPC method is used.

References

[1] Stupar A, Friedli T, Minibock J and Kolar J W 2012 *Power Electronics, IEEE Transactions on* **27** 1732–44 ISSN 0885-8993 URL `http://dx.doi.org/10.1109/TPEL.2011.2166406`

[2] Thévenaz P, Blu T and Unser M 2000 *Medical Imaging, IEEE Transactions on* **19** 739–58 ISSN 0278-0062 URL `http://dx.doi.org/10.1109/42.875199`

[3] Aldroubi A, Unser M and Eden M 1992 *Signal Processing* **28** 127–38 ISSN 0165-1684 URL `http://dx.doi.org/10.1016/0165-1684(92)90030-Z`

Pervasive Sensing: Addressing the Heterogeneity Problem

Michael J. O'Grady, Olga Murdoch, Barnard Kroon, David Lillis, Dominic Carr, Rem W. Collier and Gregory M. P. O'Hare

CLARITY: Centre for Sensor Web technologies, University College Dublin, Belfield, Dublin 4, Ireland.

E-mail: michael.j.ogrady@ucd.ie

Abstract. Pervasive sensing is characterized by heterogeneity across a number of dimensions. This raises significant problems for those designing, implementing and deploying sensor networks, irrespective of application domain. Such problems include for example, issues of data provenance and integrity, security, and privacy amongst others. Thus engineering a network that is *fit-for-purpose* represents a significant challenge. In this paper, the issue of heterogeneity is explored from the perspective of those who seek to harness a pervasive sensing element in their applications. A initial solution is proposed based on the middleware construct.

1. Introduction

Historically, sensor networks are tightly coupled to particular application domains, for example, environmental monitoring. In the case of a scientific experiment, a dedicated network would be designed such that the data required for the experiment in question would be collected within defined quality-of-service parameters. Pervasive sensing radically challenges the conventional perception of sensor networks in many dimensions. Specifically, a sensing infrastructure may be envisaged as multi-functional, fulfilling the requirements of a diverse end-user constituency. Conceptually, a standard security alarm system for a home might also serve as a basis for an Ambient Assisted Living (AAL) service. However, ensuring that the needs of both services are sufficiently addressed is a complex endeavour.

Mote platforms that encapsulate a range of sensing modalities, many in a *Plug-and-Play* manner, are increasingly available. The SunSPOT [1] and Tyndall [2] platforms are archetypical of platforms that incorporate a variety of physical sensors, and are augmented with sufficient computational resources on the platform to engage in relatively sophisticated pre-processing. Indeed, this development is a logical continuation of the process of moving intelligence to the fringe nodes of networks [3]. However, the inherent heterogeneity that arises from the variety of hardware and software platforms poses significant problems for those aspiring to design, implement and deploy sensor networks. In this paper, this issue of heterogeneity is explored more detail, and an initial solution based on the middleware paradigm is illustrated.

2. Heterogeneity in Wireless Sensor Networks

Heterogeneity amongst Wireless Sensor Networks (WSNs) is a significant barrier to integrating various types of sensor platforms into a common application. Utilising a number of disparate sensor motes and platforms may require familiarity with multiple implementation programming languages, operating systems (OSs) and deployment mechanisms. Table 1 outlines a number of WSN products in terms of the languages and OSs they support.

Table 1. Sensor Motes Comparison

Namef	Programming Language	Operating System
EcoSpire	C	Enix
Namimote	C, C++	RTOS
Shimmer	nesC	TinyOS (default)
SunSPOT	Java ME	Java VM
TelosB	nesC	TinyOS
Tyndall	nesC, C	TinyOS, Contiki
Waspmote	C++	Embedded

In this table, the phrase "Operating System" is used loosely, as the Java Virtual Machine supported by SunSPOT motes and the Waspmote's approach of running compiled C++ code directly do not technically constitute OSs. However, the table indicates the range of technologies that an application developer must master in order to combine the capabilities of these platforms. More detailed comparisons of commonly-used WSN technologies can be found in [4] and [5].

This problem is exacerbated further if the definition of a "sensor" is expanded beyond these traditional WSNs to other data-producing entities. For instance, modern mobile smartphones typically include a range of sensors including accelerometers, GPS capabilities, digital compasses, etc. Additionally, both raw data and more complex information is available from internet sources. In many cases, it may be desirable to use this data to augment that gleaned from WSNs deployed directly by an application developer in a physical environment.

2.1. Standardization

The vast amount of different sensors generate a diverse set of issues when trying to integrate the heterogenuous data types and communication models into a sensor network. The Open GeoSpatial Consortium (OGC) have proposed the Sensor Web Enablement (SWE) Initiative to help alleviate these issues, especially when exposing these sensor networks on the internet [6]. The SWE proposes a standard data model, in the form of the Observations and Measurements XML Schema, to encode observations and measurements from sensors. Additionally a Sensor Model Language (SensorML) is also given, and this allows for describing the deployed sensor, including such things as the location of the sensor and its various tasking capabilities.

SWE also describes a number of service protocols for accessing the sensors, and these include the Sensor Observation Service (SOS), Sensor Planning Service (SPS), Sensor Alert Service (SAS) and Web Notification Services (WNS). SOS and SPS provide access mechanisms for requesting, filtering and retrieving sensor observations and information. The SAS provides mechanisms for subscribing and publishing sensor alerts, and WNS provides mechanisms for asynchronous access to SAS and SPS web services.

2.2. Ontological issues

The OGC SWE Initiative is one effort to move sensors to wider audience; however the SWE standards are quite verbose and challenging to implement [7]. Another more generic effort to enable wider interaction and better understanding between different systems is the emerging semantic web. There have been also efforts to semantically annotate the SWE standards and encode them using the semantic notation [8].

Other attempts at creating formal Ontologies to represent sensors and their data have also emerged. These include the Sensor Data Ontology (SDO) [9], OntoSensor [10], Ontonym [11] and the CSIRO Sensor Ontology [12].

All these diverse attempts at creating sensor Ontologies have motivated the World Wide Web Consortium (W3C) to create a working group with the express purpose of creating a standardized sensor ontology, namely the W3C Semantic Sensor Network (SSN) Ontology [13]. This uses the CSIRO Sensor Ontology as a base, incorporating elements from various other existing Ontologies. It should be noted that the W3C SSN uses entities defined in the Dolce Ultralight (DUL) Ontology, and has defined all sensors to be physical entities. This means that any sensors that are not also physical entities cannot be described using the W3C SSN without causing contradictions and inconsistencies in the resulting Ontology.

3. Towards an Adaptive Middleware for Heterogeneous Sensors

In pervasive systems, "context" comprises any information that characterises the situation of any person, place, or object that is considered relevant to the interaction between a user and an application. Context data may be gathered from many sources including physical sensors deployed in the environment (e.g. sensing light or sound levels) and non-physical sources such as calendar information and data accessible via the World Wide Web (WWW). As illustrated in Figure 1 and Table 2, the current WWW encompasses diverse documents and services that produce and/or maintain real-time and static information relating to a wide range of users, places, topics, and concepts. With the movement towards the vision of the Internet of Things (IoT) and Web of Things (WoT), the availability of web-accessible data providing resources is expected to increase dramatically in the coming years.

The term *cyber sensor* describes software that accesses data from non-physical sources. Specifically, a cyber sensor programmatically monitors a non-physical environment, in contrast with physical sensors that can only monitor the physical environment in which they are situated.

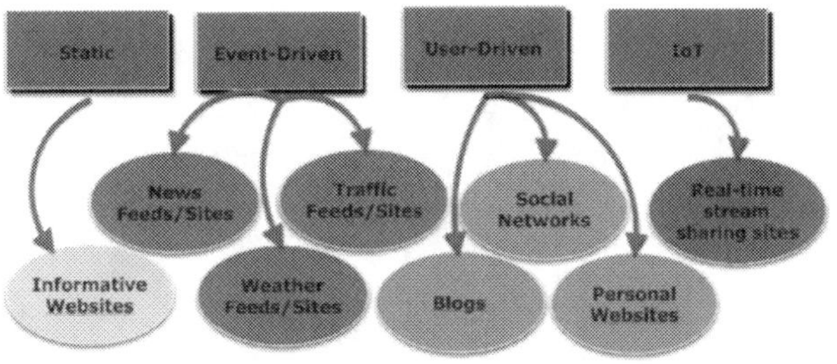

Figure 1. Taxonomy of Web-based Sources of Context

The fusion of web-based sources of context data with that emitted from physical sensors will bridge the divide between the physical and cyber worlds, enabling a wealth of web-based and immersive complex context-driven applications. As a simple example, in the context of a building subject to real-time fluctuating energy prices, physical sensors monitoring electricity usage can be combined with cyber sensors that capture real-time price information as part of a smart energy management system.

Development of systems driven by WSNs is complicated by the dynamic and uncertain nature of the physical devices and the data they produce. Programming with data from cyber sensors poses alternative challenges to those inherent with physical sensors. Many web-based environments produce data that is manually entered by users (e.g. social networks, news websites). Manually entered data is mostly unstructured, prone to human error and can often be

Table 2. Classification of Web-based Context Sources

Type	Definition	Example
Static	Knowledge or fact	Documents, Web1.0
User Driven	Data availability relies on user updates	Blogging, social networks
Event Driven	Scheduled or manual reporting	Weather site/API, newsfeed
IoT	Live stream from remote networks/devices	COSM, Thing Speak
Service	Provision of information via data processing	Klout, Peer Index

subjective and dependent on the provider's perspective. These factors not only impact on how reliable the data is but also make managing and interpreting the data a difficult task. Situation identification and uncertainty management techniques in pervasive computing typically aim to manage uncertain data from physical sensors. For systems comprising both physical and cyber sensor data to become a reality it is imperative that these techniques begin to take into account diverse sources of data.

Each non-physical environment has different access policies and restrictions that impact how often and how much data can be accessed. Application Programming Interfaces (APIs) most often have a limit on the number of requests that can be made per minute/hour. Especially for applications that require near real-time data, fusing related context from multiple sources can be problematic when considering the different rates at which data will reach the system.

Cyber sensors are more complex than physical sensors in terms of location and time. Physical sensors have a single location at any given point in time and as such data recorded at a particular time is related to a specific location. Web-based sources often provide information relating to many geographical locations and the information may be (near) real time or historical/futuristic to some degree. For example, web sites providing weather information provide data relating to many locations and while the data provided can be considered real time the information doesn't always relate to 'now', i.e., weather forecasts and historical weather data. These differences offer alternative perspectives on data other than that available through use of physical sensors. While such diversity of location and time offers a richer contextual basis for applications it also makes the task of managing and incorporating these diverse data sources more difficult.

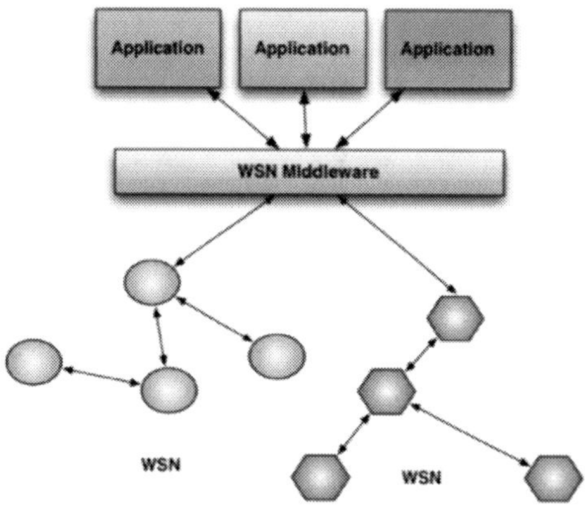

Figure 2. Wireless Sensor Network middleware

In order to address the challenges of pervasive sensing outlined above, a sensor middleware solution is proposed. This acts as an intermediary between an array of sensors that capture data on one hand and applications that make use of this sensor data on the other. By addressing the low-level challenges, it empowers the developer to rapidly develop complex real-time applications. The role of a sensor middleware system in providing access to WSNs is illustrated in Figure 2.

In this configuration, application developers are not required to become familiar with the specific implementation details of each of the sensor types with which they wish to interact. This is a particularly important concern when heterogeneous sensors support a variety of operating systems, programming languages and deployment strategies. Instead, the middleware provides a uniform interface whereby applications can gain access to sensor data and event information in addition to being able to re-task individual sensor networks. The incorporation of cyber sensors adds substantially to the complexity of such a system.

4. SIXTH: A case study

The SIXTH sensor middleware has been created to address the challenges outlined in the previous section. In addition to the basic WSN middleware model shown in Figure 2, SIXTH also provides support for other heterogeneous sensing resources that can provide context, particularly cyber sensors. The key features of the SIXTH middleware is as follows:

- **Uniform adaptor interface:** Integrating a particular sensor type requires an *adaptor* to be written for it. This allows the sensor's data to be used by any SIXTH application in a consistent way. An example of the structure of an adaptor created for a SunSpot WSN is shown in Figure 3. Here, sensor data generated by the WSN is routed through the adaptor to the middleware, which can then distribute this data to appropriate applications. The applications need not be concerned with the mechanism by which this data was sourced. As adaptors can be created for both physical and cyber sensors, this is the core enabler for heterogeneous sensor support.

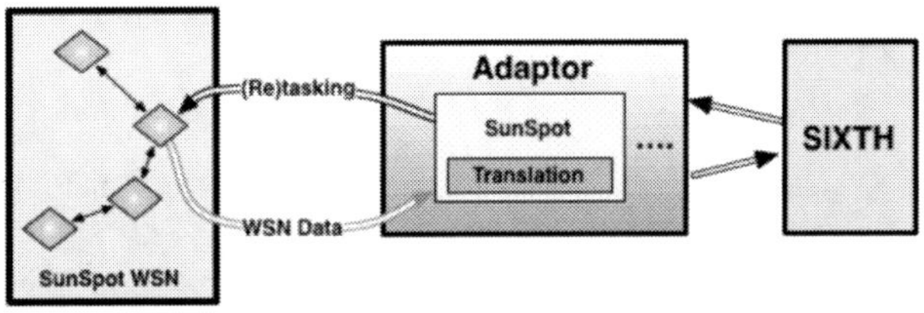

Figure 3. Anatomy of a SIXTH Sensor Adaptor

- **Retasking capabilities:** The interaction between applications and sensors is not one-way. In addition to gaining access to sensor data, it is also important for certain applications to be capable of retasking the sensors that they use. An example of this is when a change to the sampling frequency of the sensor is required. Again, the application can create a retasking message in a uniform way, which is passed to the sensor by the middleware using the adaptor. Here, the adaptor's role is to translate the SIXTH retasking message into the native instructions required by the sensor itself.
- **Distribution and Discovery:** SIXTH can run on a variety of computer types including servers, desktop computers, laptops and mobile devices (currently Android devices are supported). Because each individual device may have direct access to a different set of sensors, it is important that the sharing of sensor data between devices be supported. Direct access to a sensor or sensor network may come through software hosting of cyber sensors, bluetooth connections, USB connections to base-stations of WSNs and in the case of mobile devices a range of sensors is typically built in to the device itself. In addition to the possibility of distributed deployment of SIXTH, a discovery service is also provided whereby SIXTH deployments can be identified and their capabilities shared.

- **Querying:** In a system where multiple applications are utilising a range of sensor types, it is necessary to avoid flooding the communication channels with unwanted sensor data. For this reason, a querying system is available whereby applications can describe the type of data in which they are interested either programmatically or using a specialised query language. Only sensor data matching the query is forwarded to the interested application.
- **Data Management:** In some instances, sensor data may require aggregation or contextualisation using data from other sources. SIXTH provides a framework to support the management of data through the system prior to consumption by applications.
- **Event Notifications:** As sensor data not is all an application may require, SIXTH also notifies interested components about events that affect the middleware. This includes details of new sensors or sensor networks being added to the system, removal or failure of existing sensor capabilities and any other events in which an application may be interested.
- **Standards-based:** SIXTH uses a range of existing standards. The Java programming language facilitates the use of the Open Services Gateway Initiative framework (OSGi) to allow for the dynamic addition and removal of components at runtime. RESTful transfer of messages and data is inbuilt and JmDNS is used for service registration and discovery.

5. Conclusion

Heterogeneity increases complexity; thus engineers will find it increasingly difficult to design sensor networks that serve multiple purposes while remaining fit-for-purpose. In this paper, some of the key sources of heterogeneity were discussed. The middleware construct offers one viable approach for mitigating the complexity inherent in heterogeneous sensor networks.

Acknowledgments

The authors would like to acknowledge the support of Science Foundation Ireland (SFI) under the grant 07/CE/I1147.

References

[1] Smith R 2007 *Information Processing in Sensor Networks, 2007. IPSN 2007. 6th International Symposium on* pp 565–566
[2] O'Flynn B, Lynch A, Aherne K, Hare P, Barton J, Harte S, O'Mathuna C, Diamond D and Regan F 2006 *Adjunct Proceedings of the 4th International Conference on Pervasive Computing* pp 21–26
[3] O'Grady M J, O'Hare G M P, Chen J and Phelan D 2009 *Information Systems Frontiers* **11**(1) 61–73 ISSN 1387-3326
[4] Johnson M, Healy M, van de Ven P, Hayes M, Nelson J, Newe T and Lewis E 2009 *IEEE Sensors* pp 1439–1442 ISSN 1930-0395
[5] Madan V and Reddy S N R 2012 *VSRD International Journal of Electrical, Electronics and Communication Engineering (VSRD-IJEECE)* **2** 50–55
[6] Botts M, Percivall G, Reed C and Davidson J 2008 *GeoSensor Networks* (*Lecture Notes in Computer Science* vol 4540) ed Nittel S, Labrinidis A and Stefanidis A (Springer Berlin / Heidelberg) pp 175–190
[7] Rouached M, Baccar S and Abid M 2012 *Services (SERVICES), 2012 IEEE Eighth World Congress on* pp 65–72
[8] Broring A, Foerster T and Jirka S 2010 *Pervasive Computing and Communications Workshops (PERCOM Workshops), 2010 8th IEEE International Conference on* pp 732–737
[9] Eid M, Liscano R and El Saddik A 2007 *Computational Intelligence for Measurement Systems and Applications, 2007. CIMSA 2007. IEEE International Conference on* pp 59–62
[10] Goodwin C An ontology-based sensor network prototype environment
[11] Stevenson G, Knox S, Dobson S and Nixon P 2009 *Proceedings of the 1st Workshop on Context, Information and Ontologies* CIAO '09 (New York, NY, USA: ACM) pp 9:1–9:8
[12] Michael Compton Holger Neuhaus K T and Tran K N 2009 *2nd International Workshop on Semantic Sensor Networks, CEUR-WS* pp 33–48
[13] Semantic sensor network xg final report `http://www.w3.org/2005/Incubator/ssn/XGR-ssn/` last visited 2013/05/13

Enhancing performance of wireless sensor networks in glacial environments using wake-up receivers

Vana Jelicic[1], Daniel Cesarini[2] and Vedran Bilas[1]

[1] University of Zagreb, Faculty of Electrical Engineering and Computing, Unska 3, 10000 Zagreb, Croatia

[2] University of Pisa, Department of Information Engineering, Largo Lucio Lazzarino 1, 56122 Pisa, Italy

E-mail: `{vana.jelicic, vedran.bilas}@fer.hr`, `daniel.cesarini@iet.unipi.it`

Abstract. Development of radio telemetry enabled long-term monitoring of hard-to-reach and harsh environments. This paper compares two WSN deployment projects for gathering sensor data in glacier monitoring application — GlacsWeb and PermaSense, in terms of system design and wireless communication. We discuss the potential benefits of energy-efficient event detection using wake-up receivers together with duty-cycled communication. We show that adding a WURx would increase the average power consumption of Dozer protocol for 10%, but it would reduce the delay from 2 minutes to several milliseconds. Besides for event detection, WURx could be used for synchronizing the beginning of the TDMA communication, which would eliminate the need for clock drift compensation, making the protocol simpler and lighter.

1. Introduction

Environmental monitoring was one of the first telemetry applications to attract attention, due to the need of unattended and unobtrusive measurements interesting to a wider community (e.g. weather and climate monitoring and forecast), as well as field scientists (biologists, zoologists, geologists, etc.). Monitoring hard-to-reach and harsh environments, like glaciers, has posed a special challenge, since the measuring equipment has to survive extreme conditions (low temperatures, wind, ice...) and there is a high maintenance cost in case of a failure. Automatic Weather Stations (AWSs) are measurement units developed and deployed in various locations in the past 30 years. They provide a single point of measurement, usually for temperature, humidity, wind speed, etc. To collect the data, the location has to be visited, or the data can be communicated wirelessly (e.g. via WiFi or satellite) [1].

Although field scientists benefit significantly in understanding many phenomena from data collected by AWSs, they are also interested in obtaining dense spatial distributed measurements [2]. Technology enabling that requirement is called Wireless Sensor Networks (WSNs) [3]. WSNs are suitable for long-term monitoring of various phenomena on a large area (temperature, humidity, snow height, etc.). To reduce high energy consumption of the transceivers, but ensure reliable communication at the same time, various low power medium access control (MAC) protocols for WSNs have been developed [4]. Their main characteristic is duty-cycling of the radio, that reduces consumed energy, but introduces latency. Many phenomena in glacial environments are quite slow, but there are some events where precise timing is interesting for scientists (start of snowing, subglacial movement, etc.) or safety related, e.g.

Journal of Physics: Conference Series **450** (2013) 012045 doi:10.1088/1742-6596/450/1/012045

rock falls or avalanches. A promising solution for eliminating latency and keeping low energy consumption of the communication unit are wake-up receivers (WURx) [5].

By now, only two projects had WSN deployments in glacial environments. They gather spatially distributed sensor data at a low frequency (<1 Hz). In order to detect interesting events in WSNs, we propose to use WURxs, that would enhance the performance of by-now used TDMA (time division multiple access) protocols, and communicate the event to the user in real-time, enabling timely reaction in case of safety-critical events. Another benefit introduced by WURx in MAC protocols is synchronization of the nodes for beginning of TDMA communication. The remainder of the paper is organized as follows. Section 2 presents requirements of WSN monitoring in glacial environments, with comparison of two real-life deployments and their TDMA communication protocols (Dozer and GWMAC). Section 3 presents the benefits of using the WURx technology in low-power TDMA MAC protocols, for event detection in glacial environments, and for TDMA synchronization. Section 4 concludes the paper.

2. Monitoring glacial environments with WSNs

Environmental sensor networks [6] have evolved from passive logging systems that require manual downloading, into intelligent sensor networks that comprise sensor nodes which communicate their sensor data to a network server. The first widely famous environmental WSN application was in 2002, and it monitored bird habitats. Besides habitat monitoring, WSNs are often developed and deployed in agriculture [7]. Harsh environments like glacial areas pose even bigger challenges for WSN development and deployment because of more extreme weather conditions and more difficulties (i.e. costs) to reach the equipment for maintenance. In this section, we provide an overview of glacial monitoring with WSNs, focusing on two most famous research projects using WSNs in this field: PermaSense and GlacsWeb. We also survey the low-power MAC protocols they developed (Dozer and GWMAC).

2.1. PermaSense

The PermaSense project investigates the influence of climate change on permafrost. The main scientific research goal is to understand the heat transport in frozen rock walls and its influence on the stability, as well as large scale mass movements (rock glaciers) [8]. From 2006 by now, there were 2 deployments in Swiss Alps, at 3500 m a.s.l. Although there were various modifications of the system over time, the main principle has remained the same: sensor nodes placed on the surface, and the probes with sensors (temperature and crack meter) placed in the ground. The WSN comprises TinyNodes (with TI MSP430 MCU and Semtech XE1205 transceiver) as sensor nodes and an embedded computer, running the Linux OS, as base station.

2.1.1. Dozer incorporates a MAC layer (TDMA), topology control, and a routing protocol [9]. It builds a data gathering tree on top of the underlying network topology and provides nodes with precise wakeup schedules for all communication only relying on local synchronization (without a global schedule). Nodes are organized in parent-child relationships. Parents schedule precise rendezvous times for all communication with their children. While in theory wakeup times can be calculated perfectly at both parent and children, clock drift has to be considered in real-world applications. In Dozer, the receiver node is responsible for clock drift compensation and worst-case guard times are used to guarantee a prior wake up of the receiver before the sender starts its transmission. Furthermore, Dozer addresses the problem of temporary network partition and energy efficient tree adaptation in case of local link failures. Despite these additional considerations, Dozer attains low radio duty cycles in both single-hop and multi-hop networks. The Semtech XE1205 radio transceiver operates at 868 MHz using 0 dBm transmission power and a bandwidth of 75 kbps. The obtained range is about 300 m. Using a sampling period of two minutes, Dozer achieved an average duty cycle of less than 0.2% on all nodes.

2.2. *GlacsWeb*
GlacsWeb project uses a WSN to understand subglacial processes, and relationship between glacier dynamics and climate change [10]. They had several deployments from 2004 to now (in Norway and in Iceland), with different improvements of the system design. The latest deployment in Iceland is a heterogeneous sensor network consisting of: probes in the ice and till layers, geophones embedded within the ice and a differential GPS network in order to produce precise measurements of movement distance. The nodes deployed into the glacier are based on the Energy-Micro ARM Cortex-M3 MCU and Radiometrix RPM1 transceivers. Geophones, on the other hand, communicate the data over the air from their probes to the base station using the TI CC1120 transceiver. Sensors used on the probes and/or geophones are: temperature, pressure, strain, accelerometer, compass, gyroscope and conductivity. The base station is a platform with TI MSP MCU and a Beagle Bone Linux computer [11].

2.2.1. GWMAC is a centralised TDMA-based protocol, designed for unreliable networks where contention is completely eliminated and control packets are minimised [12]. Since in GlacsWeb latency is secondary, GWMAC reduces the duty cycle of the nodes to almost zero, providing only one small communication window (few minutes) per day. Even during the communication window, nodes have their transceivers turned off for the major part and only turn them on either during time slots in which they are expected to receive data or slots uniquely assigned to them for transmission. A centralized algorithm is used to synchronise the entire network at boot phase and, additionally, each time a command packet is received. In GlacsWeb deployment, the base station is connected to some of the nodes in the ice (called anchor nodes) via a serial cable. The anchor nodes communicate wirelessly with other nodes deployed even deeper under the surface. In order to get the nodes synchronised, the base station first synchronises its own RTC with the average of the closest RTCs of the anchor nodes. The remaining nodes are synchronised through the diffusion of any message packet initiated by the anchor nodes. When a node receives a message packet, it can uniquely determine its clock by considering the time stamp value embedded within the packet, the time of flight and time of processing.

3. Enhancing communication with wake-up receivers
Extremely low duty cycles of Dozer and GWMAC are implemented because the data delivery latency is not considered to be important, i.e. it is important to deliver the data, but it doesn't matter if it occurs immediately or the day after. In case of GlacsWeb, glacial and sub-glacial movements were monitored. Instead of moving continuously throughout the year, the movement of a glacier occurs through a series of slip events ("stick-slip" motions), but it is a rather slow process and the glaciologist are satisfied to receive the information rarely (once a day). PermaSense project studied heat transport in frozen rock walls and its influence on large scale mass movements and stability. For now the benefit of the WSN system was to gather the data from a wider area and use it for offline analysis. They used an average duty cycle (D) of 0.2%. There are, though, some events that require prompt reaction to avoid catastrophes (e.g. avalanches or rock falls), that can't be detected with such low duty cycles. Wake-up receivers are a suitable solution to optimize delay, since they are continuously monitoring the communication channel and wake up the node upon message reception.

In 2007, when both PermaSense and GlacsWeb projects published their low-power MAC solutions (Dozer and GWMAC, respectively), WURxs were only at their beginning. Just in 2009 and after, there have been some significant attempts of designing WURx prototypes [13]. In this section, we will analyze the benefits of a wake-up receiver in glacial monitoring WSNs on a case study of enhancing Dozer behavior with AS3932 WURx.

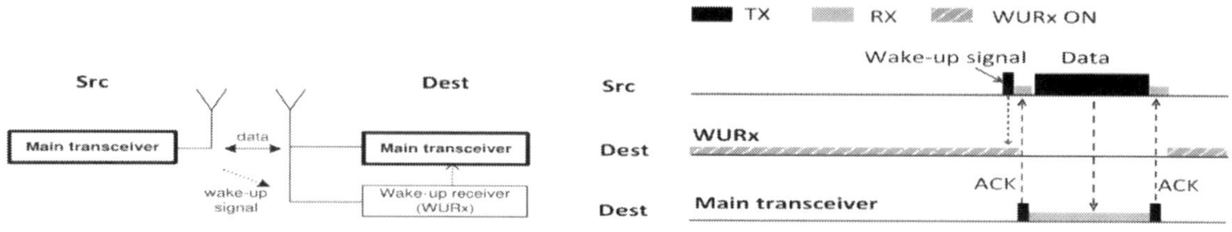

Figure 1. Implementing wake-up receivers on wireless sensor nodes

Table 1. Characteristics of low-power MAC protocols (DOZER, GWMAC) and AS3932 WURx

Name	Frequency	D [%]	Rate [kbps]	P [μW]
Dozer	868 MHz	0.2	75	82
GWMAC	173 MHz	~ 0	5	386
AS3932	125 kHz	100	5.5	~ 8

3.1. Wake-up receivers
A wake-up receiver (WURx) is a simple, very low-power receiver added to a transceiver, that continuously monitors the communication channel and wakes up the transceiver upon message reception to receive the data (Fig.1 left). Fig.1 (right) shows communication between two nodes when employing a WURx. The main radio is turned off most of the time, both for the source and the destination node. When a node wants to communicate, it sends a wake-up signal, usually containing the address of the destination node to awake only the desired neighbor.

WURxs have been a hot topic in the last couple of years, but are still under development. Since it is a new component with quiescent power consumption added to the radio circuitry, it is important to reduce its power consumption as much as possible. As a consequence, the downside of WURxs is the low reception range and low throughput [5]. Low throughput is not a problem, since for wake-up signal only a short message is used. Low reception range could be solved by placing nodes in smaller distances. One of the very few commercially available is a low-power, low-frequency wake-up receiver chip with addressing capability AS3932 [14]. Gamm *et al.* [15] designed a low-power WURx circuitry around it. The main transceiver produces a 125 kHz wake-up signal OOK (On Off Keying) modulated on an 868 MHz carrier. They report communication range of 45 m when transmitting with main high-consuming transceiver with +11 dBm and over 15 m when using 0 dBm output power.

3.2. Benefits of WURx combined with Dozer protocol
Table 1 compares the characteristics of AS3932 WURx to Dozer and GWMAC protocols. For Dozer and GWMAC, power consumptions are average values simulated from the tests and current measurements on the nodes [9, 12]. Average power consumption of the WURx depends on the number of received wake-up signals. But, since the duration of the wake-up signal is very short, the power consumption of the idle state dominates in practical applications. Unfortunately, the Dozer and GWMAC have not been implemented on the same platform, making it difficult to compare their performance. Current consumption of TinyNodes and GWnodes while in sleep, receive (RX) and transmit (TX) state, as well as of AS3932 WURx in idle and while receiving, are shown in Table 2. Hereafter, we discuss the benefits that the WURx would introduce in Dozer protocol, in terms of asynchronous and synchronous communication.

Table 2. Current consumptions of a TinyNode, GWnode and AS3932 WURx

	GWnode	TinyNode	AS3932
RX	18 mA	13 mA	12 μA
TX	35 mA	16 mA	-
sleep/idle	1 μA	6 μA	2.7 μA

Table 3. Time duration: communication period in Dozer, data message (t_{data}) and wake-up signal (t_{wup})

Dozer period	t_{data}	t_{wup}
2 min	5 ms	1 ms

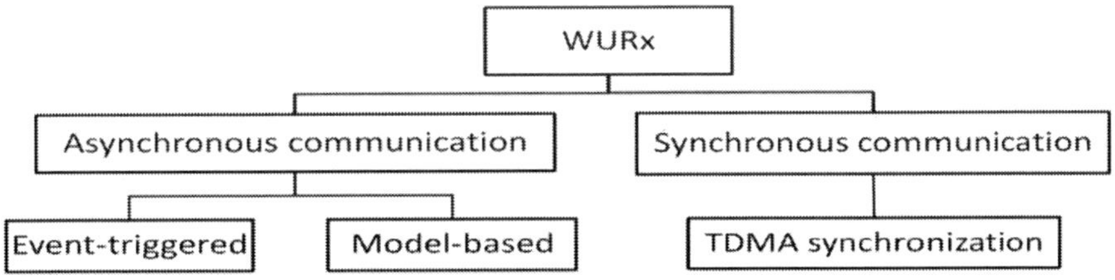

Figure 2. Benefits of using a WURx in a WSN with TDMA protocol

3.2.1. Asynchronous communication: Table 3 shows the times in Dozer communication: period and data message. Since the time necessary to transmit the data is about 5 ms, we can approximate the maximal delay in the communication to be 2 minutes. If we use the AS3932 WURx in addition to the main nodes transceiver, we can reduce that delay to orders of milliseconds. Since the detected event can be important for the whole network, it is not necessary to include addressing in the wake-up signal, but to wake all the nodes within the communication range, instead. That means that the wake-up signal has to be at least 550 μs long (by the specifications). We will suppose it is 1 ms long, as a worst case. Let's suppose that each node detects maximum 1 event per hour. An interesting event can be a crack in the rock, or a sudden rise in snow height. It can be recognized by superating a certain threshold or by differing significantly from a model of the monitored phenomena, executed on the MCU. Further on, let's suppose that a node receives in average 5 wake-up signals in an hour, like in Fig.1. Adding the AS3932 WURx to the TinyNode, and having 1 wake-up signal to send and 5 wake-up signals to receive each hour, causes 10% additional power consumption to the node with only Dozer protocol.

3.2.2. Synchronous communication: Another benefit of adding a WURx in low-power MAC protocols like Dozer is for synchronizing timeslots for TDMA communication. A similar solution was proposed for a single-hop star wireless body area network, where authors used their own WURx design and showed about 14 times lower communication consumption compared to their low power TDMA protocol [16]. Precise clock compensation is a complex and energy-consuming task in WSNs, especially in harsh conditions like glacial environment, due to large temperature variations causing clock drifts. For approximating the benefits of using AS3932 in Dozer, it would be necessary to implement the whole system. Namely, the Dozer protocol would have to be modified in a way that it doesn't perform the synchronization. Instead, the synchronization would be performed by WURx, and the Dozer would only be used for the communication handshake. In that way, the protocol would be simpler and less energy-consuming.

4. Conclusion
In this paper, we propose using a WURx to enhance the existing low-power TDMA MAC protocols, in two aspects, as showed in Fig.2: (i) for asynchronous communication upon detecting an interesting event or discrepancy between sensed and the modeled phenomenon; (ii) to

synchronize the start of TDMA communication. Those two challenges (event detection and synchronization) were also recognized as future work challenges in Dozer [9], but at that time WURxs were not that known technology as they are today. Comparison of characteristics of AS392 WURx and Dozer MAC protocol shows that there would be 10% of power consumption increment when adding the WURx, but it would reduce the latency of the system from 2 minutes to several milliseconds. Latency avoidance is important for detection of safety-related events in glacial environments, and timely reaction useful for avoiding catastrophes like rock falls or avalanches. In case of using WURx for synchronizing the TDMA communication, reliable and robust data gathering producing continuous data logs already implemented with Dozer would be preserved. In addition, there wouldn't be the necessity to compensate the nodes clock drift for synchronization, and the communication could also be event-triggered. It is expected that in the next few years there will be more WURxs available on the market, with even better characteristics than the AS3932, enabling a promising enhancement for power saving and delay reduction in wireless asynchronous systems.

Acknowledgment
This work has been supported by the European Community Seventh Framework Programme under grant No. 285939 (ACROSS).

References
[1] Abbate S, Avvenuti M, Carturan L and Cesarini D 2013 Deploying a communicating automatic weather station on an alpine glacier *3rd Internat. Workshop on Sensor Networks for Intelligence Gathering and Monitoring (SNIGM)* ed Springer

[2] Cesarini D, Jelicic V, Avvenuti M and Bilas V 2013 From automatic weather stations to sensor networks in long-term glacier monitoring *IOP Sensors & Their Applications XVII (accepted for publication)*

[3] Akyildiz I F, Su W, Sankarasubramaniam Y and Cayirci E 2002 *Computer Networks* **38** 393–422

[4] Langendoen K 2008 *Medium Access Control in Wireless Networks* **II: Practice and Standards**

[5] Demirkol I, Ersoy C and Onur E 2009 Wake-up receivers for wireless sensor networks: Benefits and challenges *IEEE Wireless Comm.* pp 88–96

[6] Hart J K and Martinez K 2006 *Earth-Science Reviews* 177–191

[7] Jelicic V, Razov T, Oletic D, Kuri M and Bilas V 2011 Maslinet: A wireless sensor network based environmental monitoring system *Proc. 34th Internat. Convention on Information and Comm. Technology, Electronics and Microelectronics (MIPRO)* (Opatija, Croatia) pp 172–177

[8] Beutel J, Gruber S, Gubler S, Hasler A, Keller M, Lim R, Thiele L, Tschudin C and Ycel M 2009 The PermaSense remote monitoring infrastructure

[9] Burri N, von Rickenbach P and Wattenhofer R 2007 Dozer: Ultra-low power data gathering in sensor networks *IPSN* (Cambridge, Massachussets, USA)

[10] Martinez K, Basford P J, Jager D D and Hart J K 2012 A wireless sensor network system deployment for detecting stick slip motion in glaciers *IET Internat. conf. on Wireless Sensor Systems 2012* URL `http://eprints.soton.ac.uk/339176/`

[11] Martinez K, Basford P J, Jager D D and Hart J K 2013 Poster abstract: Using a heterogeneous sensor network to monitor glacial movement URL `http://eprints.soton.ac.uk/348427/`

[12] Elsaify A, Padhy P, Martinez K and Zou G 2007 GWMAC – A TDMA based MAC protocol for a glacial sensor network *PE-WASUN* (Chania, Crete Island, Greece)

[13] Jelicic V, Magno M, Brunelli D, Bilas V and Benini L 2012 Analytic comparison of wake-up receivers for wsns and benefits over the wake-on radio scheme *Proc. 7th ACM workshop on Performance monitoring and measurement of heterogeneous wireless and wired networks (PM2HW2N)* (Paphos, Cyprus) pp 99–106

[14] Austriamicrosystems 2010 *AS3932 3D Low Frequency Wakeup Receiver*

[15] Gamm G U, Sippel M, Kostic M and Reindl L M 2010 Low power wake-up receiver for wireless sensor nodes *ISSNIP* pp 121–126

[16] Marinkovic S J and Popovici E M 2011 Power efficient networking using a novel wake-up radio *5th Internat. Conf. on Pervasive Computing Technologies for Healthcare (PervasiveHealth)*

Calibrations of phase and ratio errors of current and voltage channels of energy meter

P Mlejnek[1], P Kaspar[2]
[1]Czech Technical University in Prague, University Centre for Energy Efficient Buildings, Nam. Sitna 3105, 272 01 Kladno, CZ
[2]Czech Technical University in Prague, Faculty of Electrical Engineering, Technicka 2, 166 27 Prague 6, CZ

E-mail: pavel.mlejnek@uceeb.cz

Abstract. This paper deals with measurement of phase and ratio errors of current and voltage channels of a new produced energy meter. This fully digitally controlled energy meter combines the classical static energy meter with power quality analyzer. The calibration of phase and ratio errors in wide frequency range is then necessary. Paper shows the results of error measurement, introduces the mathematical approximations and describes the calibration constants. It allows error compensation and power calculation of particular harmonics. The electric power of the higher harmonics can be interesting information of distributed electric energy quality.

1. Introduction

Nowadays electric energy is distributed to most of households, offices and industrial buildings around the globe. Precise measurement of consumed energy is very important for consumers and distributors. Electromechanical induction watt-hour meters with rotating disc are now replaced by static energy meters. Static energy meters are mostly based on digital processing of sampled instantaneous voltage and current waveforms.

The heart of the static energy meter is a microcontroller (MCU) that provides timing, sampling, calculations and communication with superior systems. Voltage channel is usually very simple because there are not so high dynamic range requirements. Most of energy meters assume to be connected to distribution network with constant voltage level (e.g. 230V) with tolerances in order of ten percent (e.g. ±15%) or the energy meters are connected to voltage transformer with standard ratio. Therefore voltage channels mostly contain only voltage divider, buffer and low pass filter (LPF). Following analog-digital converter (ADC) is the last part of the measurement chain that provides digital data.

On the other hand current channel is rather complicated. It is due to the high dynamic range of measured signal, needs galvanic insulation and DC immunity and limited power dissipation. Mostly the range is from ones of miliamperes (e.g. 10 mA) up to tens of amperes (e.g. 100 A). All these requirements fulfils only sensors that are based on measurement of the induced magnetic field like Hall probe, current transformer (CT) and Rogowski coils. More details about sensors in energy meters can be found in our previous work [1]. Current sensor is, like in voltage channel, followed by a buffer, amplifier, LPF and ADC.

2. Current transformers in energy meter

In this work we focus only on current transformers used for sensing the current in energy meters. There are special requirements defined in International and European Standards (e.g. EN 62053-21 or EN 50470-3). Except the others the immunity to dc current is also defined there. It is defined as a magnitude of half wave rectified current that causes power measurement error greater than predefined limit (according to class of the energy meter). The reason of the requirement is that some loads can be supplied through a diode to reduce power or saturate current sensors, cheat the electricity meter and steal energy.

Special type so-called DC tolerant CTs have to be used in this application. These CTs use low permeability core with large saturating intensity of magnetic field. So they can work even with large dc magnetic field induced by dc current without saturating the core. Two types of the transformers matching these demands: linear CTs and dual core CTs. Dual core CTs are cheaper than the linear ones but there are problems due the different phase errors with and without presence of dc current [2].

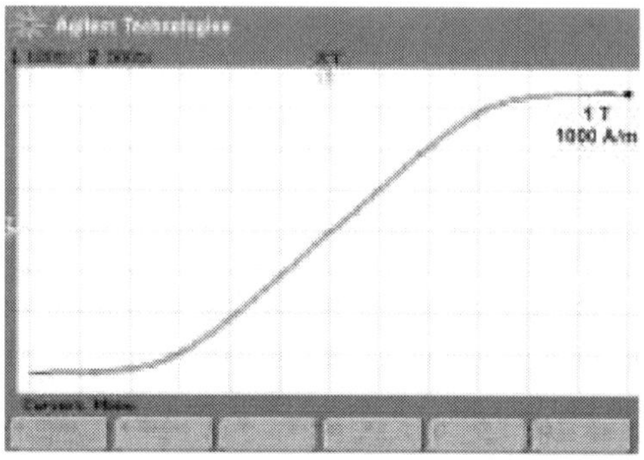

Figure 1. Example of BH loop of linear CT

In our work we deal only with linear CT (see figure 1) with almost constant permeability and phase error φ_I - about 4 deg in wide range of measured current. We use VAC 4626-X101 current transformers with amorphous core that can work even with 100 A peak value [3].

$$\varphi_I = \varphi_{secondary} - \varphi_{primary} \quad \text{(deg)} \tag{1}$$

3. Concept of modern energy meter with power quality analyzer

The basic concept of energy meter has been presented in section 1. As stated above voltage and current are sampled by ADCs with frequency f_S and discrete samples are processed by MCU. The electric power P is defined as a product of corresponding samples of voltage u_i and current i_i (N samples per period) and consumed energy E is sum of the power in time.

$$P = \frac{1}{N}\sum_{i=1}^{N} u_i i_i \quad \text{(W; V, A)} \tag{2}$$

$$E = \frac{1}{3600 \cdot f_S} \cdot \sum_{0}^{t} P \quad \text{(Wh; W)} \tag{3}$$

This calculation assumes that there is no phase shift between voltage and current caused by the sensors. But as was stated above linear CT has constant phase error about 4 deg but only at frequency 50 Hz. Therefore the current samples have to be shifted by this phase.

In present time there are many appliances that consume no harmonic current moreover some power plant (e.g. solar) produce dc current and use DC/AC converter that can be source of higher harmonics. These facts lead to the necessity of measurement of particular harmonics component and analyses the electric power of higher harmonics.

Due to large performance of modern MCUs complicated calculation can be done. It is no problem to use FFT to compute spectrum in real time, get the amplitudes and phases of particular harmonics and use this information to analyze the power quality (PQ) of electric energy. Moreover it allows power calculation of higher harmonics.
Previously described functionality along with communication ability allows using these electricity meters in Smart Grids that is required by Directive of the European Parliament and Council Directive 2009/72/EC.

4. Frequency dependence of sensors

Previously described method for electric power calculation needs constant phase error in wide frequency range. In fact phase shift of CT as well as following electronic circuits (amplifier, LPF ...) are frequency dependent. It is necessary to calibrate whole measurement channel. Similar calibrations have to be done in voltage channel. Moreover the amplitude errors of both channels also have to be calibrated.
We prepare PCB with analog part of the energy meter consisting of the three voltage and current channels where all the errors can be measured and calibrated. Measurement system contains arbitrary generator HP 33120A for generating variable frequency voltage. Voltage is amplified by Krohn Hite 7500 amplifier and with the matching transformer MT-56R is converted to testing current. This current and voltage from amplifier are connected to current and voltage channel of the energy meter as well as to Multi-Channel Precision Power Analyzer LMG500. Outputs of analog parts of the energy meter are connected to sensor inputs of the second channel of the analyzer. Raw data from the analyzer are transferred to PC where the LabView program computes spectrum, phase shifts and ratio errors. Same program controls the generator. Multitone signal is used for testing current to speed up calibration procedure. Figure 2 shows the result of measurement of the errors for 30 frequencies in range 10 Hz – 10 kHz (points).

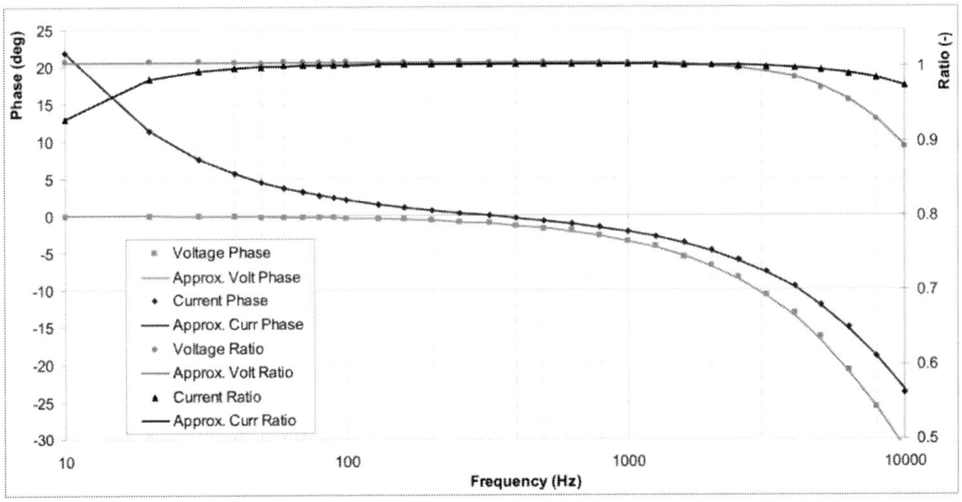

Figure 2. Phase and ratio errors of voltage and current channel (points – real measured data, line – mathematical approximations)

All the errors have to be digitally compensated. Compensating by lookup table is very easy solution and does not need high performance of MCU but it needs huge memory (with 10 Hz frequency bin and 16bit data number it needs 1000×4×2 = 8kB per channel of energy meter). Better solution is to describe the dependence by mathematical formulas with few parameters. Channels contain only LPFs and high pass filters (HPF) represented by CTs and errors can be described by following formulas.

$$\varphi_U = -\arctan(\omega K_1) \tag{4}$$

$$\varphi_I = \arctan\left(\frac{1}{\omega K_2}\right) - \arctan(\omega K_3) \tag{5}$$

$$\varepsilon_U = \left(1 + (\omega K_4)^2\right)^{-\frac{1}{2}} + K_5 \tag{6}$$

$$\varepsilon_I = (\omega K_6) \cdot \left(1 + (\omega K_6)^2\right)^{-\frac{1}{2}} + \left(1 + (\omega K_7)^2\right)^{-\frac{1}{2}} + K_8 \tag{7}$$

where φ_U, φ_I is phase error of voltage and current channel respectively
ε_U, ε_I is ratio error of voltage and current channel respectively
ω is angular frequency = $2\pi f$
K_1 - K_8 are calibration constants given from calibration procedure

Calibration constants are obtained from curve fitting. Figure 2 (lines) shows the approximation curves for measured data. The calibration constants are stored in the table 1. Calibration process during energy meter production has to be as fast and easy as possible and thus it is impossible to measure in details all the errors. Most of calibrating generators are able to generate only base harmonic (50 Hz) and few (3 or 5) higher harmonic. Therefore the optimal calibrating frequencies have to be chosen. Figure 3 shows the approximation curves fitted from only 3 points 50, 500 and 5000 Hz (nominal frequency, 10th and 100th harmonics).

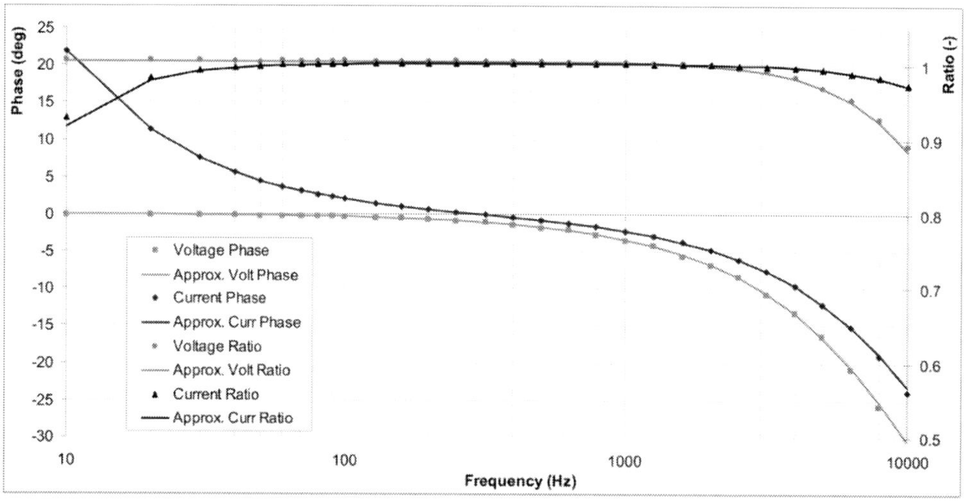

Figure 2. Phase and ratio errors of voltage and current channel (line - curve fitting from 3 points – 50, 500 and 5000 Hz)

Next figure shows the differences of real measured data and fitted curves. It can be seen that even only three points calibration provides satisfying results. Phase error is less than 0.5 degree and ratio error is less then 0.5% in frequency range from 30 Hz up to 5 kHz. If there is need to fit precisely at lower and higher frequencies the next two points should be included to calibration.

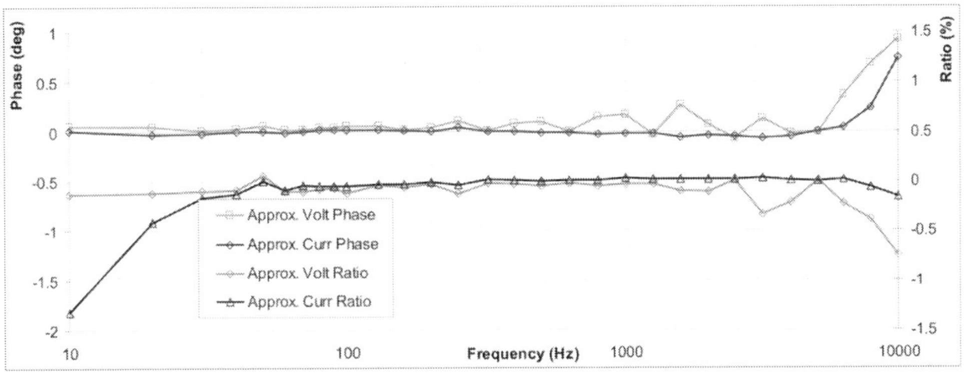

Figure 3. Differences of measured data and fitted curve (phase in degree, ratio in %)

Table 1. Calibration constants from 3 and 30 points calibration

Const.	3 points calibration (50, 500 and 5000 Hz)	30 points calibration (10 Hz – 10 kHz)	Error type
K_1	9.40E-6	9.70E-6	Voltage phase
K_2	3.96E-2	3.96E-2	Current phase
K_3	6.81E-6	6.90E-6	
K_4	8.59E-6	8.30E-6	Voltage ratio
K_5	5.14E-3	6.09E-3	
K_6	3.60E-2	3.94E-2	Current ratio
K_7	4.02E-6	3.94E-6	
K_8	-9.97E-1	-9.97E-1	

5. Conclusion

Increasing number of alternative energy sources, inferential application of DC/AC converters and increasing number of nonlinear loads lead to necessity of power quality analysis and consider electric power of higher harmonics. All these tasks can be solved by new generation of fully digital energy meters. The phase and ratio error of voltage and current channels have to be corrected in whole frequency range. In this work we focus on calibration in range from 10 Hz up to 10 kHz. The dependencies can be approximated by mathematical formulas derived from LPF and HPF. It is necessary to know only 8 constant per channel of energy meter to make corrections. Only the three frequencies are enough to find a calibration constant that allows compensation of phase error under 0.5 degree and ratio error under 0.5% in frequency range from 30 Hz up to 5 kHz. Final conclusion can be done only with final PCB of real energy meter.

6. References

[1] Mlejnek P, Kaspar P 2007 Current Sensors Suitable for Using in Electricity Meters *The 2nd International Conference on Sensing Technology* ed Massey University (Palmerston North) pp 519–522 ISBN 978-0-473-12432-8

[2] Mlejnek P, Kaspar P 2009 Drawback of Using Double Core Current Transformers in Static Watt-Hour Meters *Sensor Letters* **7** 394-396 ISSN 1546-198X

[3] Vacuumschmelze GmbH & Co. KG 2011 T60404-E4626-X101 Specification *online at*: http://www.vacuumschmelze.com/index.php?id=397

Acknowledgments

This work has been supported by the European Union, OP RDI project No. CZ.1.05/3.1.00/13.0283 – Intelligent Buildings.

Journal of Physics: Conference Series **450** (2013) 012047 doi:10.1088/1742-6596/450/1/012047

Active induction balance method for metal detector sensing head utilizing transmitter-bucking and dual current source

D Ambruš, D Vasić and V Bilas

University of Zagreb, Faculty of Electrical Engineering and Computing
FER-ZESOI, Unska 3, HR-10000 Zagreb, Croatia

E-mail: davorin.ambrus@fer.hr, darko.vasic@fer.hr, vedran.bilas@fer.hr

Abstract. A central problem in a design of frequency domain electromagnetic induction sensors used in landmine detection is an effective suppression of a direct inductive coupling between the transmitter and the receiver coil (induction balance, IB). In sensing heads based on the transmitter-bucking configuration, IB is achieved by using two concentric transmitter coils with opposing exciter fields in order to create a central magnetic cavity for the receiver coil. This design has numerous advantages over other IB methods in terms of detection sensitivity, spatial resolution, sensor dimensions and suitability for model-based measurements. However, very careful design and precise sensing head geometry are required if a single excitation source is used for driving both transmitter coils. In this paper we analyze the IB sensitivity to small perturbations of geometrical properties of coils. We propose a sensor design with dual current source and active induction balance scheme which overcomes the limitations of geometry-based balancing and potentially provides more efficient compensation of soil effects.

1. Introduction

Electromagnetic induction (EMI) sensors have been used for the detection of metallic parts of landmines since the World War II and are nowadays considered a rather mature technology [1]. However, in spite of recent developments in other landmine detection techniques, EMI sensors still remain an area of active research, [2]. Due to their sensitivity, robustness and high-speed operation, they are practically the only sensors that are currently used in humanitarian demining for close-in detection in the field [1], [2].

Time-domain (TD) or pulse induction (PI) EMI sensors transmit a primary magnetic field of pulse waveform and detect a weak secondary field induced by eddy-currents in the metallic object after the transmitter has been shut-off. Since the receiver responses corresponding to the primary and the secondary field are separated in time, TD sensors are inherently balanced which greatly simplifies sensor design. However, their fundamental limitation is the available excitation spectrum band (i.e. the equivalent time frames that eventually cannot capture 'very early' or 'very late' time data) [2].

Frequency-domain (FD) EMI sensors provide better control over the power and frequency content of the excitation spectrum. Also, they generally claim higher sensitivity and improved SNR (signal-to-noise) response in metal detector applications when compared to TD sensors [2]. On the other hand, in order to explore their benefits an efficient method of primary field suppression (induction balance, IB) must be provided. There are several IB methods that are normally used for design of FD EMI sensors: physical separation of transmit (TX) and receive (RX) coils [3], RX coils in gradiometer configuration, overlapping TX and RX coils (OO and DD types) [4], orthogonal coil arrangement, etc.

Journal of Physics: Conference Series **450** (2013) 012047 doi:10.1088/1742-6596/450/1/012047

All of the above IB implementations have some apparent drawbacks when it comes to design of sensing head for a handheld landmine detector with additional model-based metal characterization features. Such sensor needs to have simple and compact geometry, high detection sensitivity, high spatial resolution and pinpointing accuracy. It also must provide good invertibility of the measured data so that the parameters of a model can be reliably computed. Based on these requirements, a sensing head configuration based on the transmitter-bucking approach is proposed [3].

2. Sensing head design

The initial parameter for the design of a landmine detector sensing head is the diameter of its main transmitter coil D_1 which can be used as a rough estimate of the sensor's maximum ground penetration depth [2]. Having in mind the practical requirements of landmine detection, we set D_1 at 30 cm. Other geometrical properties of the sensing head (coil diameters and number of turns) are obtained from D_1 and the given mathematical model of the chosen coil configuration.

2.1. Transmitter-bucking configuration

The transmitter stage consists of two concentric coplanar coils (TX1 and TX2) driven by AC current of opposite direction so as to create a central region in which the receiver coil senses zero primary field, figure 1. If a circular current loop approximation is used for both TX1 and TX2 coils, the expression for vertical component of magnetic field B_Z as a function of radial distance r from the coil centre and observed at a plane coplanar with the sensing head (i.e. at z=0) can be written as [3]:

$$B_Z(r) = \frac{\mu_0 IN}{2R}\left[1 + \sum_{n=1}^{\infty}\left[\frac{(2n-1)!!}{(2n)!!}\right]^2 (2n+1)\left(\frac{r}{R}\right)^{2n}\right] \qquad \text{for } r \leq R, \text{ inside loop} \qquad (2.1)$$

$$B_Z(r) = -\frac{\mu_0 IN}{2R}\left[\frac{1}{2}\left(\frac{R}{r}\right)^2 + \sum_{n=1}^{\infty}\left[\frac{(2n+1)!!}{(2n+2)!!}\right]^2 (2n+2)\left(\frac{R}{r}\right)^{2n+2}\right] \qquad \text{for } r \geq R, \text{ outside loop} \qquad (2.2)$$

I is the transmitter coil current, μ_0 is the magnetic permeability of vacuum (μ_r=1 for free space), N is the number of turns and R is the radius of the transmitter coil. The expressions are obtained by differentiating the Taylor series representation of magnetic scalar potential and assuming azymuthal symmetry [2]. The voltage induced in the RX coil is proportional to the magnetic flux, i.e. the surface integral of magnetic field components from TX1 coil ($B_z^{\,1}$) and TX2 coil ($B_z^{\,2}$) over the circular area defined by the receiver coil radius r, figure 2. The magnetic cavity is created under condition (2.3).

$$\int_0^r B_Z^1(r)\,2\pi r dr = \int_0^r B_Z^2(r)\,2\pi r dr \qquad (2.3)$$

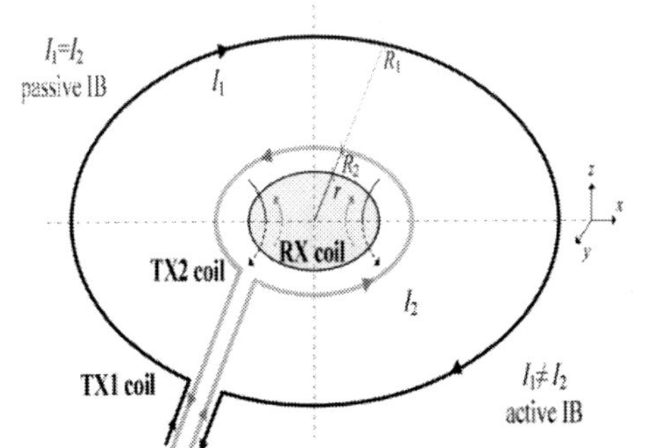

Figure 1. Block diagram of the sensing head in transmitter-bucking configuration.

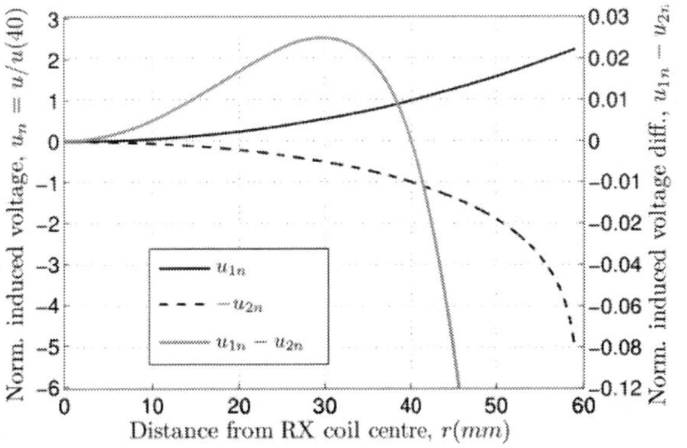

Figure 2. Normalized voltages induced in RX coil in response to TX1 and TX2 coils. IB is obtained for RX coil with radius r=40 mm.

Substituting (2.1) into (2.3) for TX1 and TX2 coils yields an expression (2.4), which describes the mutual dependence of all geometrical parameters of a sensing head (coils radii and number of turns).

$$\frac{N_1}{R_1}\sum_{n=0}^{\infty}\left[\frac{(2n-1)!!}{(2n)!!}\right]^2\frac{(2n+1)}{(2n+2)}\left(\frac{r}{R_1}\right)^{2n}=\frac{N_2}{R_2}\sum_{n=0}^{\infty}\left[\frac{(2n-1)!!}{(2n)!!}\right]^2\frac{(2n+1)}{(2n+2)}\left(\frac{r}{R_2}\right)^{2n} \quad (2.4)$$

Based on (2.4) and initially selected sensing head size, we set the following coil parameters: R_1=15cm, N_1=12 (TX1 coil), R_2=6cm, N_2=4 (TX2 coil) and r =4cm (RX coil).

2.2. Induction balance sensitivity analysis.

In order to analyze the influence of small variations of coil's geometrical parameters on sensor balancing, we define the induction balance sensitivity S_x^u to the parameter x, (2.5)-(2.7).

$$\frac{\Delta u}{u_{IB}}=\frac{\Delta u_1}{u_1}-\frac{\Delta u_2}{u_2} \quad (2.5)$$

$$u_{IB}(r,R,N)=u_1(r,R_1,N_1)=u_2(r,R_2,N_2)=\frac{kN_1}{R}r^2\sum_{n=0}^{\infty}\left[\frac{(2n-1)!!}{(2n)!!}\right]^2\frac{(2n+1)}{(2n+2)}\left(\frac{r}{R}\right)^{2n} \quad (2.6)$$

$$S_x^u=\frac{\Delta u}{u_{IB}}\left(\frac{\Delta x}{x}\right)^{-1}=\left(\frac{\Delta u_1}{u_1}-\frac{\Delta u_2}{u_2}\right)\left(\frac{\Delta x}{x}\right)^{-1}=S_x^{u_1}-S_x^{u_2} \quad (2.7)$$

In (2.5)-(2.7), u_1 and u_2 are voltages induced by TX1 and TX2 coils (respectively), u is the total voltage induced in RX coil by both TX1 and TX2 coils (u_1-u_2), u_{IB} is the voltage induced by a single transmitter coil for a balanced sensor, and k is the coefficient dependent on transmitter current, frequency and RX coil properties. From (2.7), IB sensitivity can be calculated as a difference of equivalent sensitivities of voltages induced by TX1 coil ($S_x^{u_1}$) and TX2 coil ($S_x^{u_1}$), respectively.

The induction balance sensitivity to small variations of TX1 coil radius R_1 can be found by differentiating the expression for $u_1(R_1)$, (2.6), yielding equivalent sensitivity of u_1, (2.8).

$$S_{R_1}^{u_1}=\frac{\partial u_1}{\partial R_1}\frac{R_1}{u_1}=-1-\frac{1}{u_1}\frac{kN}{R_1}r^2\sum_{n=0}^{\infty}\left[\frac{(2n-1)!!}{(2n)!!}\right]^2\frac{(2n+1)}{(2n+2)}2n\left(\frac{r}{R_1}\right)^{2n} \quad (2.8)$$

As expected, the obtained sensitivity to ΔR_1 is nonlinear and contains the expression part that is almost identical to u_1, except for the *2n* term that multiplies each component of the series. Rewriting the Taylor series components in form of two vectors of length $m+1$ ($\boldsymbol{u_1}$ and $\boldsymbol{w}$) and neglecting the higher-order terms (under the assumption of the Taylor series sum convergence), (2.9)-(2-10), we obtain the approximate expression for $S_{R_1}^{u_1}$, (2.11). Since $S_{R_1}^{u_2}=0$, IB sensitivity $S_{R_1}^{u}$ is equal to $S_{R_1}^{u_1}$.

$$\boldsymbol{u_1}=\left[0 \quad \cdots \quad \left(\frac{(2m-1)!!}{(2m)!!}\right)^2\frac{(2m+1)}{(2m+2)}\left(\frac{r}{R_1}\right)^{2m}\right] \quad (2.9)$$

$$\boldsymbol{w}=[0 \quad 2 \quad 4 \quad \cdots \quad 2m], \boldsymbol{i}\text{ - the unit vector of length } m+1 \quad (2.10)$$

$$S_{R_1}^{u}=S_{R_1}^{u_1}-S_{R_1}^{u_2}\approx-\left(1+\frac{\boldsymbol{u_1}\cdot\boldsymbol{w}}{\boldsymbol{u_1}\cdot\boldsymbol{i}}\right)-0=-\left(1+\frac{\boldsymbol{u_1}\cdot\boldsymbol{w}}{\boldsymbol{u_1}\cdot\boldsymbol{i}}\right) \quad (2.11)$$

It can be easily shown that the same analysis applies to the case of IB sensitivity to small variations of TX2 coil radius R_2. If R_1 is substituted with R_2 in (2.8) and (2.9), and the corresponding vector notation $\boldsymbol{u_2}$ is introduced, the approximate expression for the IB sensitivity $S_{R_2}^{u}$ is given by (2.12).

$$S_{R_2}^{u}=S_{R_2}^{u_1}-S_{R_2}^{u_2}\approx 0-(-1)\left(1+\frac{\boldsymbol{u_2}\cdot\boldsymbol{w}}{\boldsymbol{u_2}\cdot\boldsymbol{i}}\right)=1+\frac{\boldsymbol{u_2}\cdot\boldsymbol{w}}{\boldsymbol{u_2}\cdot\boldsymbol{i}} \quad (2.12)$$

The above procedure can also be applied to the calculation of IB sensitivity to small variations of RX coil radius r. By differentiating $u_1(r)$ and $u_2(r)$, and calculating equivalent voltage sensitivities as in (2.8)-(2.11), an expression for S_r^u can be derived. In (2.13), $\boldsymbol{w} = [0\ \ 2\ \ 4\cdots(2m+2)]$, length $m+1$.

$$S_r^u = S_r^{u_1} - S_r^{u_2} \approx \frac{\boldsymbol{u_1} \cdot \boldsymbol{w}}{\boldsymbol{u_1} \cdot \boldsymbol{i}} - \frac{\boldsymbol{u_2} \cdot \boldsymbol{w}}{\boldsymbol{u_2} \cdot \boldsymbol{i}} \tag{2.13}$$

Finally, IB sensitivity to x may also be expressed in absolute terms, (2.14). The voltage imbalance Δu additionally depends on the number of turns of RX coil N_r and the operating frequency ω.

$$\Delta u = S_x^u \frac{\Delta x}{x} u_{IB}, \qquad u_{IB} \sim N_r \omega \tag{2.14}$$

3. Active induction balance technique

The induction balance method analyzed so far is of passive type, based only on geometrical properties of the transmitter-bucking coil configuration, (2.4). By that, it is assumed that TX coils are fed with the excitation current of the same amplitude and of opposite phase. If TX1 and TX2 coil currents are controlled separately (I_1 and I_2), an active IB scheme is achievable under the condition (3.1).

$$\frac{I_1 N_1}{R_1} \sum_{n=0}^{\infty} \left[\frac{(2n-1)!!}{(2n)!!}\right]^2 \frac{(2n+1)}{(2n+2)} \left(\frac{r}{R_1}\right)^{2n} = \frac{I_2 N_2}{R_2} \sum_{n=0}^{\infty} \left[\frac{(2n-1)!!}{(2n)!!}\right]^2 \frac{(2n+1)}{(2n+2)} \left(\frac{r}{R_2}\right)^{2n} \tag{3.1}$$

3.1. *Effects of active induction balance*

In order to investigate the effectiveness of the active induction balance (AIB) approach, a simple sensitivity analysis is performed. The corresponding sensitivities of voltages induced in the receiver coil to small perturbations of I_1 and I_2 can be easily calculated from (3.1), based on considerations given in chapter 2.2, yielding (3.2) and (3.3).

$$S_{I_1}^u = S_{I_1}^{u_1} - S_{I_1}^{u_2} = 1 - 0 = 1 \tag{3.2}$$

$$S_{I_2}^u = S_{I_2}^{u_1} - S_{I_2}^{u_2} = 0 - 1 = -1 \tag{3.3}$$

From the above expressions, it is evident that the obtained sensitivities are linear and that small sensor imbalances caused by geometry imperfections or by coils of finite length can be effectively compensated by applying a correction signal to either I_1or I_2, i.e. ΔI. In principle, it is better to use I_2 as a balancing source so that the far field distribution of the main transmitter (i.e. coil with a higher magnetic moment, TX1) remains virtually unaffected.

Another important advantage of the AIB approach is that a dynamic IB method can be introduced to automatically compensate for imbalances caused by the effects of non-cooperative (mainly magnetic) soils. Although the transmitter-bucking configuration is clearly not the best choice when it comes to sensing head sensitivity to soil effects [4], other coil configurations lack the practical opportunity to implement AIB. Sensors with passive IB generally cope with the soil problem on the receiver side, by appropriate signal processing techniques. However, these methods usually suffer from the reduction of dynamic range and sensitivity of metal detection [2].

3.2. *Current source design*

Proper design of the excitation source circuit is crucial for the implementation of AIB technique. For a landmine detector application, a current source is a preferred solution over a voltage source. In the former case, the magnitude of the excitation current (and thus the transmitted magnetic field) is kept constant and basically unaffected by the coil impedance, soil properties, as well as the lift-off and orientation of the sensing head (when the sensor is operated over magnetic soil) [4].

AIB can be achieved with a number of different current source configurations. The simplest one is to use two separate current sources; one supplying the nominal excitation current to TX1 and the other supplying the same current with a superimposed compensation signal (shifted by 0° or 180°) to TX2 .

The problem with this approach is that current sources of high accuracy (<1%) and resolution (<0.1%) are needed. The phase synchronization of their output signals might also become a challenging task.

The alternative current source configuration is shown in figure 7. With the proposed approach, the main current source is used for driving both TX1 and TX2 coils (I_m), while the secondary current source supplies only the small balancing current to TX2 coil (I_b). In this way, excitation and balancing functions are separated, which loosens the initial requirements on both current sources. The reference of the balancing source is fed from the current feedback circuit of the main source, so that I_m/I_b remains independent of I_m. For multi-frequency operation, this feature keeps the set balancing point constant for all components of the excitation current spectrum.

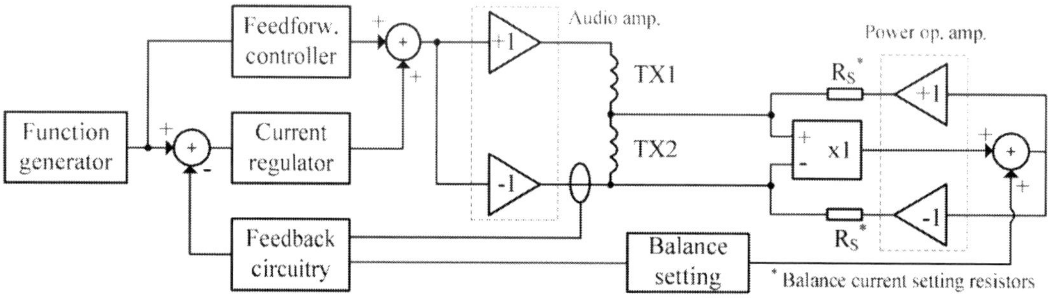

Figure 7. Block diagram of the proposed dual current source.

4. Results and discussion

Induction balance sensitivities to small perturbations of TX1 coil radius R_1, TX2 coil radius R_2 and RX coil radius r are calculated from equations (2.11), (2.12) and (2.13), respectively. For a selected sensing head geometry (R_1=15cm, R_2=6cm, r=4cm, N_1=12, N_2=4, N_r=180), the obtained sensitivities are $S_{R_1}^u$ = -1.057, $S_{R_2}^u$ = -1.555 and S_r^u = 0.498. The variations of the induced voltage imbalance Δu (in both relative and absolute terms) with respect to variations of R_1, R_2 and r (for three characteristic values of the respective parameter change) are given in table 1.

From the results given in table 1 it can be concluded that the IB sensitivities of the proposed coil configuration to ΔR_1, ΔR_2 and Δr are generally low. This can be also explained qualitatively, by observing the curves of spatial distribution of induced voltages corresponding to TX1 and TX2 coils near the sensor's centre, which are rather flat, figure 2. However, if combined effects from ΔR_1, ΔR_2 and Δr are taken into account and IB is observed in absolute terms, where N_r and ω bring a significant influence, the total sensor imbalance can become quite large, as shown in table 1.

In general, very careful coil design and high precision in the production of a sensing head are needed for practical application to landmine detection.

Table 1. Induction balance sensitivities to small variations of R_1, R_2 and r for a given sensor geometry.

TX1 coil radius		TX2 coil radius		RX coil radius		Induced voltage (excitation current, I=1A)		
							f=1kHz	f=100kHz
$\Delta R_1/R_1$ (%)	ΔR_1 (mm)	$\Delta R_2/R_2$ (%)	ΔR_2 (mm)	$\Delta r/r$ (%)	Δr (mm)	$\Delta u/u_{IB}$ (%)	Δu, (mV)	Δu, (V)
1	1.5	0	0.0	0	0.0	-1.057	3.1	0.310
2	3.0	0	0.0	0	0.0	-2.114	6.2	0.621
3	4.5	0	0.0	0	0.0	-3.171	9.3	0.931
0	0.0	1	0.6	0	0.0	-1.555	4.6	0.457
0	0.0	2	1.2	0	0.0	-3.110	9.1	0.913
0	0.0	3	1.8	0	0.0	-7.775	22.8	2.284
0	0.0	0	0.0	1	0.4	0.498	1.5	0.146
0	0.0	0	0.0	2	0.8	0.996	2.9	0.293
0	0.0	0	0.0	5	2.0	2.490	7.3	0.731

The dual current source, figure 7, and the analyzed sensing head configuration were implemented as laboratory prototypes, figure 8. The main current source is designed of the two bridge-mode audio amplifiers and the current feedback and control circuitry (I_n=1 A, f_H=100 kHz). The secondary current source is built from two power operational amplifiers with current control stage (I_n=100 mA). Laboratory function generator (Voltcraft VG-506) was used as a reference voltage source in the experiments. Voltage induced in RX coil, Δu, was fed to unity-gain instrumentation amplifier and measured with an oscilloscope. As a proof of AIB concept, Δu was measured at different frequencies (absolute value and normalized to u_{IB}) for a balanced sensor with passive IB (I_1=I_2) and with AIB, figure 9. The excitation current magnitude was set to 0.5A. Active balance was adjusted separately for each frequency. The results clearly indicate that the imbalances of the passive IB can be effectively compensated with AIB. Additional experiments are needed in order to fully characterise the method.

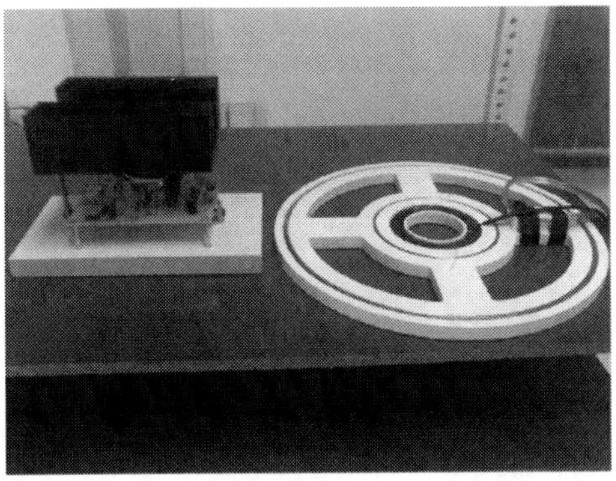

Figure 8. Implementation of laboratory prototypes (sensing head, current source).

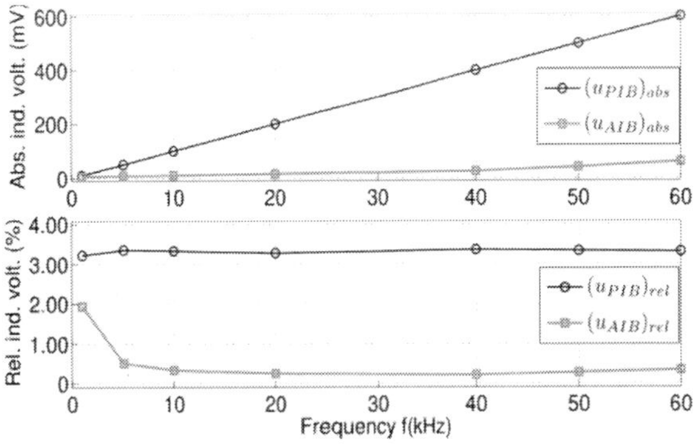

Figure 9. Magnitude of voltage induced in RX coil for passive IB and AIB at different frequencies.

5. Conclusions

For a design of novel, frequency-domain EMI landmine detector with additional model-based metal characterization features, we propose a sensing head configuration based on the transmitter-bucking approach. For this configuration, we analyzed the induction balance problem in terms of balance sensitivity to small perturbations of geometrical properties of the sensing head. The obtained results suggest that although the overall IB sensitivities are quite low, the total sensor imbalances expressed in absolute terms can become significant, therefore complicating the sensor design. In order to overcome this limitation and to explore new methods of coping with soil effects on IB, we developed and experimentally verified a prototype sensor with dual current source and active induction balance scheme. Future work will focus on further characterization of the method and new techniques for automatic compensation of imbalances caused by the effects of non-cooperative soils.

Acknowledgement

This work has been supported by the European Community Seventh Framework Programme under grant No. 285939 (ACROSS).

References

[1] Ishikawa J, Furuta K 2006 *Anti-personnel Landmine Detection for Humanitarian Demining* (London: Springer-Verlag London Limited)

[2] Bruschini C 2006 *A multidisciplinary Analysis of Frequency Domain Metal Detectors for Humanitarian Demining* (PhD Thesis:Vrije Universiteit Brussel)

[3] Won I J, Keiswetter D A, Hanson D R, Novikova E and Hall T M 1997 GEM-3 A Monostatic Broadband Electromagnetic Induction Sensor *J. Env. Eng. Geophys.* **2** 53-64

[4] Druyts P, Das Y, Craeye C, Acheroy M 2009 Modelling the Response of Electromagnetic Induction Sensors to Inhomogenous Magnetic Soils With Arbitrary Relief *IEEE Trans. Geosci. Rem. Sens.* **47** 8 2627-2638.

A matched Bow-tie antenna at 433MHz for use in underwater wireless sensor networks

A A Abdou, A Shaw[1], A Mason, A Al-Shamma'a, J Cullen, S Wylie and M. Diallo

BEST Research Institute, School of the Built Environment, Liverpool John Moores University, Liverpool, L3 2ET, United Kingdom
Email: a.shaw@ljmu.ac.uk

Abstract. Electromagnetic (EM) wave propagation underwater is been disregarded because of attenuation at high frequencies, however the theory predicts that propagation is possible at some useful distance in the lower Industrial, Scientific and Medical (ISM) band. Common transceivers rely on narrowband antennas and matching circuit. The aim of this paper is to design a broadband 433MHz bow-tie antenna and experiment it in air and water without a matching circuit. This antenna could be attached to wireless transceivers and form a Wireless Sensor Network for deployment in various underwater applications. The bow-tie antennas were designed, simulated and constructed in laboratory. Experiments were setup carefully by using a completely isolated transmitter from electronics to avoid airborne transmission. The 433MHz. bow-tie proved its suitability for use in Underwater.

1. Introduction

Electromagnetic (EM) communication has been disregarded due to attenuation at high frequencies. With a propagation velocity up to 3×10^7 m/s, EM waves propagation offer many advantages over conventional acoustic techniques, including bandwidth, data rate and better transmission across boundaries. Attenuation has been the biggest obstacle in radio propagation underwater. However in [1] it was demonstrated both, experimentally and theoretically that EM wave propagation is possible through the water column at useful distances in the lower unlicensed Industrial, Scientific and Medical bands. In the context of deploying Wireless Sensor Network in unusual places for electromagnetic wave propagation, further experimental research in [2] demonstrated communication underwater with off the shelf radio transceivers. It is been concluded that the main factors enabling the increase in communication range were the antenna and the operating frequency. In [3], experiments involving narrowband antennas built for the 433MHz ISM band, this frequency of operation was targeted because it presents equilibrium in attenuation and data rate transmission in ISM bands, experiments demonstrated that type and size of antenna also have an impact on underwater communication systems. Simulation of a monopole antenna in a water environment in [4] illustrated how narrowband antenna designed for air presents a shift in the frequency of operation with respect to the return loss due to the change of medium air/water making the antenna almost not transmitting in its new environment (water) at the frequency designed for. Loop antennas were used in [5] and [6] and other than being directional antennas, they strongly responds to magnetic fields making them more destined for seawater.

Return Loss (RL) also known as S11(dB)is a good measure of how much power is reflected back (logarithmic scale) from the terminal [7]. It is known that a good antenna must be at least -10dB to allow 90% transfer of energy. Therefore this will be the targeted RL for the underwater antenna.

Impedance matching is an important aspect in any antenna design and the use of a matching circuit is common for air designed antenna. As the antenna impedance is different when the antenna is in air or water the impedance measurements must be taken with the antenna in the medium it will operate to design the appropriate matching circuit. Furthermore antennas utilising a matching circuit may suffer from complete signal loss if any changes in temperature and salinity occur in the water due to the high dielectric property of water.

Another technique for choosing an underwater antenna is to use a broadband antenna design that can accommodate the change in such a heavy dielectric medium. With dipoles being one of the simplest but widely used types of antenna, the bow-tie is one form that potentially has the correct properties [7] to be used as an underwater antenna. The aim of this paper is to validate the use of this antenna design to form an underwater wireless sensor network.

2. Electromagnetic wave propagation underwater

2.1. Permittivity of water

Permittivity of a material is a characteristic which describes how it affects any electric field set up in it. When the material is lossy (dissipates energy), the relative permittivity is a complex quantity [8] given by Equation (1).

$$\varepsilon r = \varepsilon' r - j\varepsilon'' r \tag{1}$$

The reel part is related to the energy stored within the medium. Because water is a polar molecule and rotates when exposed to an alternating electric field, the imaginary part is then associated to the dissipation of the energy due to collisions during that rotation and to the loss due to free charge conduction. At 433MHz, the relative complex permittivity of water is $80.17 - j1.92$ [9]. These values correspondent to permittivity of water with salinity S=0ppt and temperature T=20°C. Meissner and Wentz introduced in [10] a model as in Equation (2) with double Debye relaxation wavelength that they consider as a necessary parameter to provide an accurate fit for the dielectric constant over a wider frequency range than the single Debye model does.

$$\varepsilon(T,S) = \frac{\varepsilon s(T,S) - \varepsilon 1(T,S)}{1 + i\frac{v}{v1}(T,S)} + \frac{\varepsilon 1(T,S) - \varepsilon\infty(T,S)}{1 + i\frac{v}{v2}(T,S)} + \varepsilon\infty(T,S) - i\frac{\sigma(T,S)}{(2\pi\varepsilon 0).v} \tag{2}$$

With $i = \sqrt{-1}$, v the radiation frequency in GHz, $\varepsilon_s(T,S)$ the static (zero frequency) dielectric constant, ε_∞ the dielectric constant at infinite frequencies, ε_1 the intermediate frequency dielectric constant, ε_0 the vacuum electric permittivity = 8.854×10^{-12}, $v_1(T,S)$ and $v_2(T,S)$ the first and second Debye relaxation frequencies, T the temperature in °C, S the salinity in ppt (parts per thousands) .

Using Equation (2), for operating frequency at ISM band 433MHz the permittivity of water with salinity concentration S=0.2345ppt (tap water used for the tank experiment), water with S=0.29011ppt (measured value in Liverpool Stanley Canal) and water with S=0ppt (≈ distilled water), are calculated and displayed in Table 1.

Table 1 Real and Imaginary parts of permittivity at 433MHz ISM freq in different water environment

ISM Frequency	**Permittivity** of water with S=0.29011ppt (≈canal water), T=15°C		**Permittivity** of water with S=0.2345ppt (≈tap water in tank), T=15°C		**Permittivity** of water with S=0ppt (≈distilled water), T=15°C	
	Real	Imaginary	Real	Imaginary	Real	Imaginary
433MHz	81.930	4.177	81.945	3.81	82.010	2.27

Those values are used to calculate the effective conductivity (σeff) and the attenuation (α).

2.2. Attenuation of signal in water

The complex propagation constant γ describes the behaviour of an EM wave [11] is defined in Equation (3) with α attenuation constant and β phase constant:

$$\gamma = \alpha + j\beta \quad (3)$$

The main factor that effects EM propagation underwater is the attenuation. Attenuation refers to the decrease in the magnitude of a signal as it propagates through a medium (e.g. air, rock, water). Attenuation may be due to the spreading of energy as it propagates away from its source as well as due to the medium itself absorbing the energy of the signal. Attenuation of a plane wave in a lossy medium can be calculated from the homogeneous Helmholtz equation which leads to Equation (4).

$$\alpha = \omega\sqrt{\mu\varepsilon'}\sqrt{0.5\sqrt{1+(\sigma_{eff}/\omega\varepsilon')^2}-1} \quad (4)$$

With α the attenuation in Nepers/m, ω the frequency in radians, μ the permeability of the medium in N/A^2, ε′ the permittivity of the medium in F/m as $\varepsilon' = \varepsilon_0\varepsilon'_r$ and σ_{eff}, the effective conductivity in S/m as $\sigma_{eff} = \omega \times \varepsilon_0\varepsilon''_r$

Using permittivity values from Table 1 in Equation (4), attenuation (α) is then calculated for operating frequency at 433MHz ISM band in the water with salinity concentration S=0.2345ppt (tap water used for the tank experiment), water with S=0.29011ppt (measured value in Liverpool Stanley Canal) and water with S=0ppt (≈ distilled water), and displayed in Table 2.

Table 2 Attenuation of signal at 433MHz ISM frequency in different water environment

ISM Frequency	**Attenuation (dB/m)** of water with S=0.29011ppt (≈canal water), T=15°C	**Attenuation (dB/m)** of water with S=0.2345ppt (≈tap water in tank), T=15°C	**Attenuation (dB/m)** of water with S=0ppt (≈distilled water), T=15°C
433MHz	18.178	16.599	8.465

According to table 2 Attenuation in water with salinity equals or smaller than canals water and freshwaters salinities are not high in the 433 MHz ISM band and seems encouraging for building a wireless sensor network using commodity nodes. As the important remaining factor to enable communication in water is the antenna, a potential type "bow-tie" is designed, constructed and tested.

3. Research approach

At first, using High Frequency Structure Simulator, a bow-tie antenna is designed and iterated to resonate at 433MHz in air with at least a return loss of -10dB. A standard 1.6mm thickness Printed Circuit Board made of 0.0356 thick copper sheet and FR4 glass-reinforced epoxy laminate sheet was used for the antenna which is modelled in a $0.30m^3$ box of air. Then a bound body of water with conductivity 0.02 S/m (approximated value of tap water) and absorbent boundary condition was used to model the antenna in a water environment.

Then, the antenna is sketched in Eagle PCB software and transferred into the Computer Numerical Control routing machine Bungard CCD2 for prototypes in the sensors laboratory.

Journal of Physics: Conference Series **450** (2013) 012048 doi:10.1088/1742-6596/450/1/012048

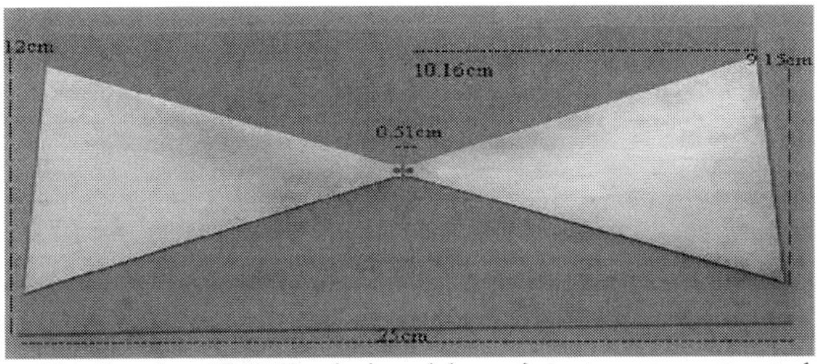

Figure 1 Angle-view of the Bow-tie antenna constructed in laboratory

Finally, the 433MHz designed bow-tie antenna constructed as shown in Figure 1 is tested in a plastic tank filled with 1200L of tap water with conductivity of 0.035S/m and temperature of 15°C. The antenna is fully waterproofed with glue and connected with an RG58 cable to a MS2024A Anritsu Network Analyzer through a plastic waterproofed container and a hose (Figure 2). This setup is used to collect the S11 antenna parameters. For collecting S21 parameters, the experimental setup in Figure 3 involves on the transmitting side an electronic circuit (voltage controlled oscillator, a variable resistor, a voltage regulator and a pack of batteries) in a plastic waterproofed container connected to the bow-tie antenna and on the receiver side a vertically moveable waterproofed bow-tie and plastic container connected to a 8594E HP Spectrum Analyzer through an RG58 cable in hose, supported by a PVC tube. The transmitter is completely submerged with no RF cable floating in air/water. This setup eliminates airborne transmission.

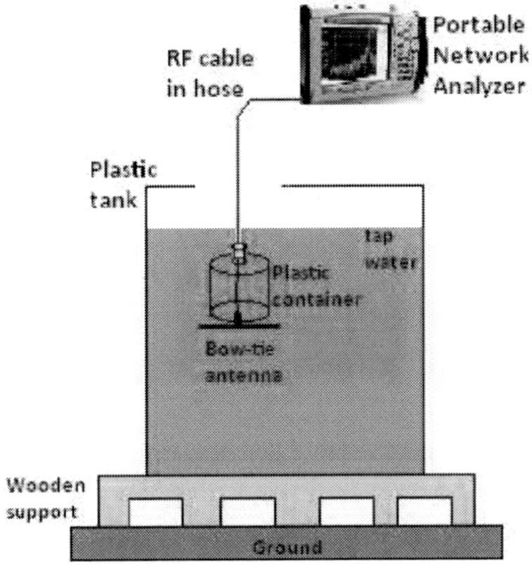

Figure 2 Experiment setup in tank for S11

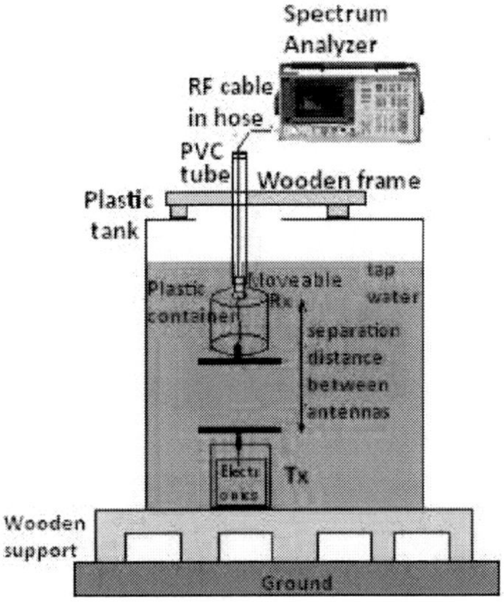

Figure 3 Experiment setup in tank for S21

4. Results

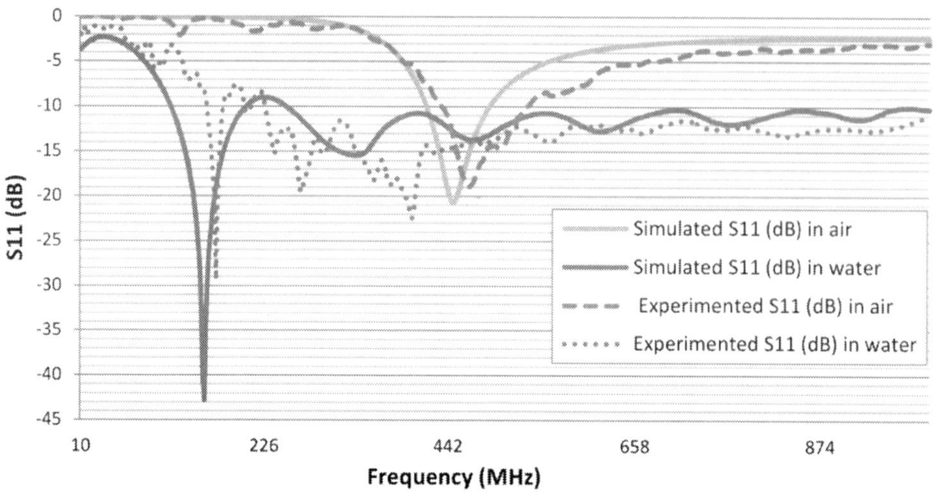

Figure 4: Simulated and experimented S11(dB) of a bow-tie antenna in air and in water

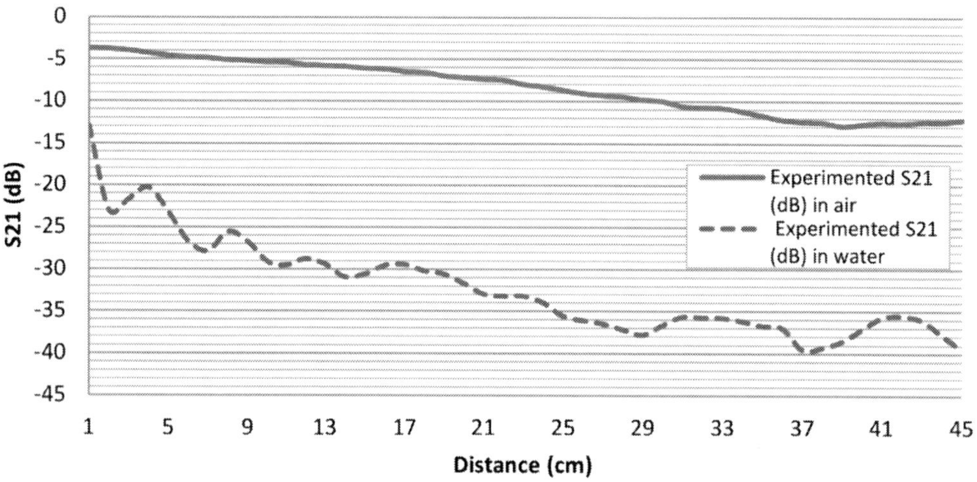

Figure 5: Preliminary experimental S21 (dB) of two bow-tie antennas in air and in water

5. Discussion

The HFSS simulation and experimental RL results are compared for the 433MHz designed bow-tie antenna in air and water in Figure 4. In air, the simulated antenna exhibits a RL of -16dB at 433MHz which means that more than 95% of the power is transmitted. In water, the simulation presents a sharp valley at low frequencies of 154MHz with a high value -43dB of RL and bandwidth of 90MHz. More importantly, is the behaviour of that antenna from 226MHz up to 1GHz shown as a ripple region. In this region the antenna becomes broadband with at least -10dB RL. Simulation results in air and water envisaged the bow-tie antenna to be a good candidate for development of an underwater WSN antenna. Experimental results in air show a slight shift of the valley however at 433MHz the RL is -10dB and lies within the targeted RL, this is also in good agreement with the simulation. In water, the experimental curve follows the overall shape of the simulated curve and the ripple region in figure 4 shows the broadband aspect of the antenna and therefore agreement with the simulation and suitability to be used as an underwater antenna.

Most of underwater RF experiments in the literature review utilise on the transmitting side a signal generator to create the signal to the antenna through a RG58 RF cable. Trials with this type of setup have shown that the RF cable acts as an antenna and by consequent results in erroneous data received from airborne transmission. In this paper for collecting S21 parameters, the transmitter is completely submerged with no RF cable floating in air/water to eliminate airborne transmission.

Figure 5 displays S21 preliminary results for the bow-tie antenna in air and in water over a distance of 45cm.The graph shows that in air the antenna signal strength has dropped 10dB over 45cm separation distance. However it seems that this is the maximum loss the antenna will suffer from in air. In water, at 1cm the signal loses 10dB compared to the air environment. This value added to the power of the VCO (6dB at 433MHz) is in agreement with the theoretical attenuation value in table 2 (16.6dB). Two regions can be distinguished a near-field region up to 30cm separation distance and a far-field region onwards. The near field region includes the totality of the signal drop. In the far field the signal lies in the -40dB whereas most common transceivers have a sensitivity of -100dBm.

6. Conclusion

In this paper the steps that can permit a bow-tie antenna to work in water without a matching circuit has been presented from design to practice. Preliminary through water transmission experiment is carried on to support the use of the constructed 433MHz bow-tie in a two way communication sensor system. Further underwater trials are planned in a canal to determine the maximum separation distance and the path loss at 433MHz through water.

7. References

[1] A. A. Abdou, A. Shaw, J. Cullen, S. Wylie, A. Mason, and A. Al-Shamma, "A Review of Underwater EM Wave Propagation to Investigate the Development of a Through Water WSN," in *6th Built Environment And Natural Environment*, 2011, pp. 1–6.

[2] A. A. Abdou, A. Shaw, A. Mason, A. Al-Shamma, J. Cullen, and S. Wylie, "Electromagnetic (EM) Wave Propagation for the Development of an Underwater Wireless Sensor Network (WSN)," in *Sensors, 2011 IEEE*, 2011, no. 1, pp. 1571 – 1574.

[3] A. A. Abdou, A. Shaw, A. Mason, A. Al-Shamma, S. Wylie, and J. Cullen, "Wireless Sensor Network for Underwater Communication," in *Wireless Sensor Systems*, 2012, pp. 3–8.

[4] A. A. Abdou, A. Shaw, A. Mason, A. Al-Shamma, S. Wylie, and J. Cullen, "Antennas Experiment , Design and Simulation for Underwater EM Wave Propagation at ISM Frequency Bands.," in *Built Environment and Natural Environment*, 2012, pp. 1–6.

[5] J. H. Goh, A. Shaw, and A. I. Al-Shammaa, "Underwater wireless communication system," *Journal of Physics: Conference Series*, vol. 178, p. 012029, Jul. 2009.

[6] A. Shaw, A. I. Al-Shammaa, S. R. Wylie, and D. Toal, "Experimental Investigations of Electromagnetic Wave Propagation in Seawater," in *2006 European Microwave Conference*, 2006, no. September, pp. 572–575.

[7] Y. Huang and K. Boyle, *Antennas from theory to practice*. John Wiley & Sons, Ltd, 2008.

[8] R. W. P. King and G. S. Smith, *Antennas in Matter*. Cambridge, MA: MIT Press, 1981.

IOP Publishing
doi:10.1088/1742-6596/450/1/012048

[9] D. Lide, *CRC Handbook of Chemistry and Physics*, 2010th ed. CRC Press/ Taylor and Francis, 2010.

[10] T. Meissner and F. J. Wentz, "The complex dielectric constant of pure and sea water from microwave satellite observations," *IEEE Transactions on Geoscience and Remote Sensing*, vol. 42, no. 9, pp. 1836–1849, Sep. 2004.

[11] D. K. Cheng, *Field and Wave Electromagnetics*, Second. Addisson Wesley Publishing Company, 1989, p. 367.

Journal of Physics: Conference Series **450** (2013) 012049 doi:10.1088/1742-6596/450/1/012049

Drive system alignment calibration of a microgravity drop tower of novel design

J Trunins[1], **B P Osborne**[2] **and A Augousti**[1]
[1] Kingston University, Penrhyn Road, Kingston, KT1 2EE, UK
[2] The University of New South Wales, High St Kensington, NSW 2052

Abstract. We report here the calibration of the drive system of a new scientific facility for production of microgravity, operating on a novel design of electromagnetically driven platform. The construction achieves the design specification of alignment of the guide rails to better than 0.254mm across the entire guide rail height of 8m, despite a small lean to the right (within tolerance) and it was noted that this alignment is improved by the presence of the trolley that carries the platform.

1. Facility design

A microgravity drop tower has been developed at Kingston University London (KUL) [1], based on novel electromagnetic linear actuation. Microgravity research is important in a range of applications, including fundamental studies on the behaviour of liquids injected through nozzles for understanding fuel flow and transfer in orbit [2], as well as broader applications such as growth of crystals [2] for pharmaceutical and other industrial applications.

1.1 Outline of drop tower design

The KUL drop tower facility is 8m in height (figure 1), and is scheduled for final commission at the end of July 2013. According to its design, programmable velocity profiles will be available, and up to 2.2s of microgravity will be achievable. The payload is situated inside housing with a drag shield, to which the actuating motors are attached, and the whole is housed within a safety surrounding. The remainder of this paper is concerned with the calibration of the alignment of the drive system of this novel facility.

1.2 Drop tower drive subsystem design

The drive system consists of several primary components, namely the base support beam, the motor installed on the trolley and the magnetic track, the guide rails and the position fixing L-brackets (figure 2). The base support beam is an 8.4 m long I-beam. This beam is manufactured by contractors with pre-cut holes for magnetic tracks and the installation of fixing brackets. The magnetic tracks are fixed on the I-beam in line with no spacing in between. The physical attachment of the trolley itself helps to keeps the motors aligned. It also has the guide rail rollers installed with the right spacing from the motors. The position of the motors has been designed so that the offset from the centre line of the magnetic track is kept within 0.254 mm. This guideline is set by the providers of the motor and magnetic track. The guide rails, which are installed along both sides of the magnetic track on the I-beam, are used to keep the position of the motor relative to the magnetic track within the required limits.

Figure 1. Drop Tower. Left – drop tower CAD model. Right – drop tower assembly

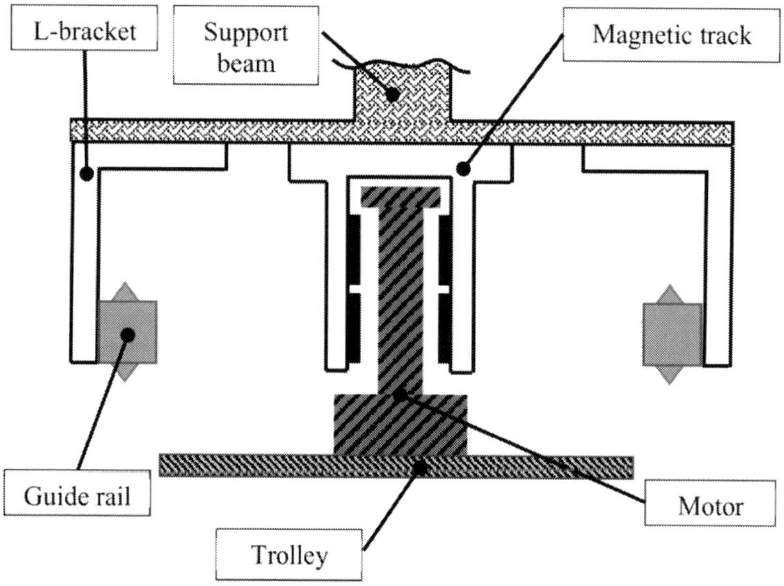

Figure 2. Drive system design. Arrows show the main components of the system

2. Drive subsystem assembly

The first step of the assembly procedure was to install the magnetic track. The magnetic track was attached to the I-beam with untightened bolts. The long line gauge and brackets were used to align the track of the magnets and after that the bolts were tightened. The bottom magnetic tracks were not installed in order to allow the motors to be slid in later after the evaluation of the system had been completed.

The next step was to attach the L-brackets, followed by installation of the guide rails. Using a pre-manufactured block gauge and the digital vernier callipers the rails were fixed at the right position.
After this milestone was achieved the motors could be installed on the trolley. But prior to this installation the alignment check of the guide rails was carried out using a laser distance gauge, as described in the following section.

3. Alignment check

The digital alignment check was done using the Baumer laser digital sensor (OADM 20i4460/S14C) [3] and National Instruments NI USB-6210 digital acquisition platform (DAq) [4]. The DAq has a USB interface which simplified the initial setup. Signal processing was performed using National Instruments LabVIEW SignalExpress software. Microsoft Excel was used for the signal analyses at a later stage.
The sensor was powered by a voltage supply of 10.00 ± 0.02V. The sensor range was 30 to 130 mm [3] but for our measurements a distance of around 80 mm was selected. The resolution of the sensor is 0.05 to 0.07 mm, which is well below our requirements of0.01 inch = 0.254 mm. The linearity error is ± 0.2 mm, again within the design requirements. The signal was acquired continuously at a frequency of 50Hz.

3.1 Sensor installation

The size of the sensor and the allowable minimum measurement distance did not permit the sensor beam to be positioned normal to the magnetic track surface. For that reason the sensor was positioned at a precise angle of 45° to the surface. A range of tests were carried out and different sensor installations were explored. The first of these involved attaching the sensor to a single guide rail, while another was used to attach the sensor with the trolley installed.
Due to the high level of electromagnetic noise the DAq was enclosed inside a steel casing. This confirmed that the noise level was visibly reduced.

3.2 Alignment check procedures

Several procedures were employed to confirm the alignment. The initial step was to check the random electronic noise within the system. For this several static measurements at a constant position were taken. The confirmed the noise level was 0.1098 mm.
Following this static measurements were taken at several places on both sides of the magnetic track. Table 1 shows the results obtained.

Table 1. Static position measurements along the I-beam

Reading place (m)	Average reading (mm), Left ±0.11	Average reading (mm), Right ±0.11
1	80.51	80.54
3	80.52	80.53
5	80.53	80.52
7	80.54	80.51
8	80.55	80.50

The reading location was measured with respect to the ground. The average reading is the average value of a set of continuous measurements taken over duration of 2-5 s. These measurements were taken with a simple attachment to the guide rail.
The final test included the dynamic measurements, performed when the trolley was moved with a velocity that was approximately constant. These latter measurements were usually performed over a maximum period of 10 s at an average velocity of 0.8 ms^{-1}.

4. Results

Table 1 summarises the static point readings taken at different positions along the tower. The evident outcome is that the alignment does shift to the right. The measurements were assessed also for the introduction of measurement uncertainties which might arise following movement of the sensor while taking the measurements. Table 2 gives the values of this uncertainty, and confirms that the measurement uncertainty does exist but is well below the noise level of the sensor.

Table 2. Uncertainties for static readings

Read place (m)	Total distance (mm ±0.11)	Difference (mm × 10^3 ± 0.11)
1	113.051	0.89
3	113.052	0.05
5	113.054	- 1.61
7	113.049	2.81
8	113.054	- 2.14
Average of distance:	113.052	

To confirm these measurements and also to include the alignment along the whole length of the rails the dynamic measurements were undertaken. Figure 3 shows a graph of initial data readings from a typical run. Here it is possible to see the spikes which arise from gaps between the magnetic tracks. This is due to the fact that the signal beam passing over the gap gets trapped, which results in a non-reading value returned. In this data, at our selected frequency of data acquisition, this corresponds to corruption of a set of 5-10 readings.

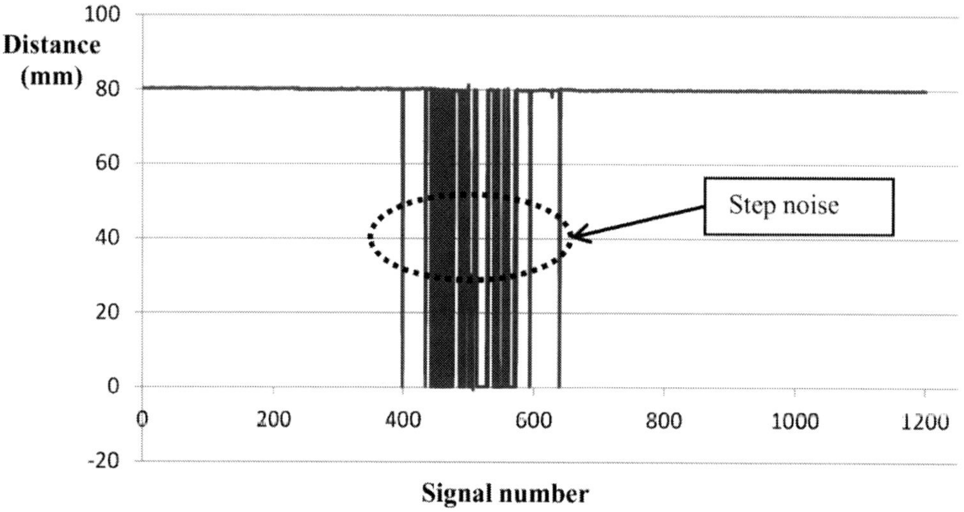

Figure 3. Initial signal readings. The arrow shows the location of corrupted data

These values were identified in the data set and were substituted with previously received non-corrupted readings. Additional smoothing was achieved by performing a boxcar average over 7 values as shown in figure 4. It is important to note here that the quantity of the selected readings is set so as not to affect the overall measurement quality. From the design perspective the L-brackets were positioned with a centre to centre spacing of 125 mm. As the beam does not buckle it was only important to make sure that at least two reading were taken to characterise this spacing, which results in needing to take measurements at least every 62.5 mm. With our measurement rate (50 Hz) and velocity (0.8 ms^{-1}) this corresponds to 7.8 measurements.

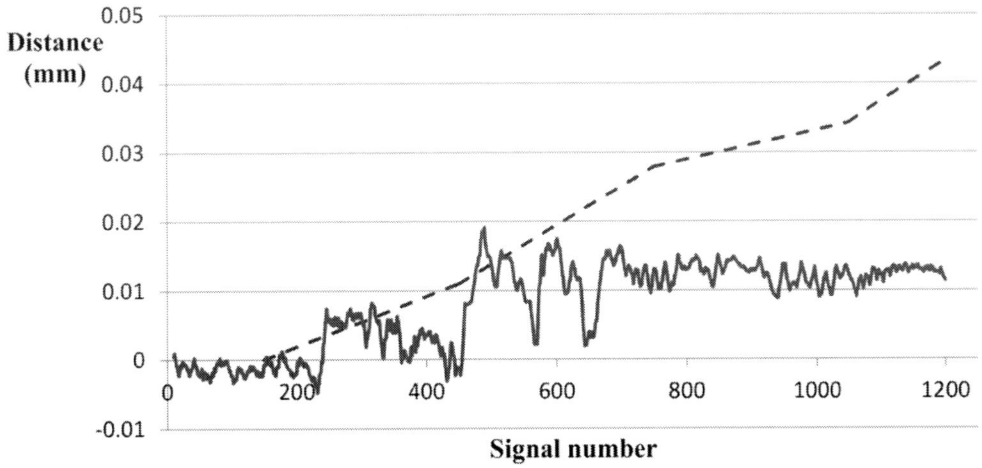

Figure 3. Processed signal readings. Solid line – dynamic test, Dashed line static test

The graph shows the values with respect to the base value which was used from static readings. The dashed line represents the static readings that were taken before. The difference here is explained by the presence of the trolley which is connected to both guide rails and provides additional alignment stability when the system is disturbed. Nonetheless the presence of an offset to one side has been confirmed, but within the tolerance limits of the design as both tests have verified.

5. Conclusion of the work

The installation of 8 meters of magnetic tracks and two linear motors were installed with a precision of 0.01inch (0.254 mm). Prior to that the guide rails were installed as motors guides to allow for this linearity to be achieved. A laser distance sensor with a DAq was used to check this alignment and confirm any required corrections. Several types of tests were used to determine the alignment achieved: the first was a static single point reading, and the second was a dynamic sliding test with the trolley installed. The results show that the alignments of the rails are within the tolerance requirements, despite a slight measured offset of the track. The maximum offset from the permissible tolerance was determined to be 17% and 5% respectively for the left and right rails respectively. It was noted that the presence of the trolley provides additional alignment due to self-correction achieved via averaging the misalignment of both sides of the guide rails. When this confirmation was achieved the motors were successfully installed and now await the first test run.

References

[1] Osborne B and Welch C 2012 Short duration reduced gravity drop tower design and development *Journal of the British Interplanetary Society* **65** 71-76

[2] Anderson S 1992 *Microgravity Honours Thesis* Mechanical Engineering The University of Queensland

[3] Baumer [Internet] OADM 20I4460/S14C [updated 2013 Apr 9; cited 2013 May 13] Available from: *www.baumer.com*

[4] National Instruments [Internet] NI USB-6210 [updated 2012; cited 2013 May 13] Available from: *www.uk.ni.com*

From single point of measurement to distributed sensing in long-term glacier monitoring

Daniel Cesarini[1], Vana Jelicic[2], Vedran Bilas[2] and Marco Avvenuti[1]

[1] University of Pisa, Department of Information Engineering, Largo Lucio Lazzarino 1, 56122 Pisa, Italy

[2] University of Zagreb, Faculty of Electrical Engineering and Computing, Unska 3, 10000 Zagreb, Croatia

E-mail: daniel.cesarini@iet.unipi.it, vana.jelicic@fer.hr

Abstract. Glacial environment monitoring is a key task in understanding natural phenomena related to global warming. For the last 30 years, Automatic Weather Stations (AWSs) have been spreading among the meteorological and geophysical community, and are on the way to become a *de facto* standard to perform long-lasting unattended data acquisitions in single localized points of interest. Sensor Networks (SNs), on the other hand, promise the possibility to perform measurements with a higher spatial density and lower cost. Designing and developing a SN for glacial environment face particular challenges for embedded electronics and sensor systems, which is why SNs are still under research and development in this field. This paper surveys the AWSs and SNs for glacial monitoring applications and compares their characteristics.

1. Introduction

Studying the evolution of glaciers became an essential task to follow global climate changes and its local effect [1]. Monitoring the state of glaciers (and in general of glacial environment) is important also for safety reasons. It enables early warning generation in case of an avalanche, or permafrost cracks and falls from mountains. Scientists today benefit from a broad range of technologies to better understand the phenomena in glacial environments. These means can be categorized based on whether they sense physical quantities remotely or locally.

Remote measurements can be carried out using satellites, planes or using ground-based devices, to take images in the visible light range (photogrammetry), or other spectral frequencies like infra-red (thermal imaging). Thermal imaging enables obtaining a temperature distribution image over the whole terrain or snow [2]. Laser or radar ranging allows to reconstruct 3D surfaces of the terrain, and compare two subsequent images to detect changes or movements. Local measurements, on the other hand, are carried out using sensing devices with output that varies based on proximal physical quantities. The sensors (temperature, water level, deformations, movements, etc.) are attached to devices able to store data (dataloggers).

Automatic Weather Stations (AWSs) are specialized measuring units, aimed at the monitoring of weather-related parameters. They are built around bigger dataloggers, that enable multiple sensors readings. In particular, AWSs deployed in glacial regions allow to investigate the micro-meteorological variables and processes related to glacier behavior caused by global climate changes. Establishing remote communication with the AWS is necessary to enable real-time data availability and timely detection of damage or malfunctions.

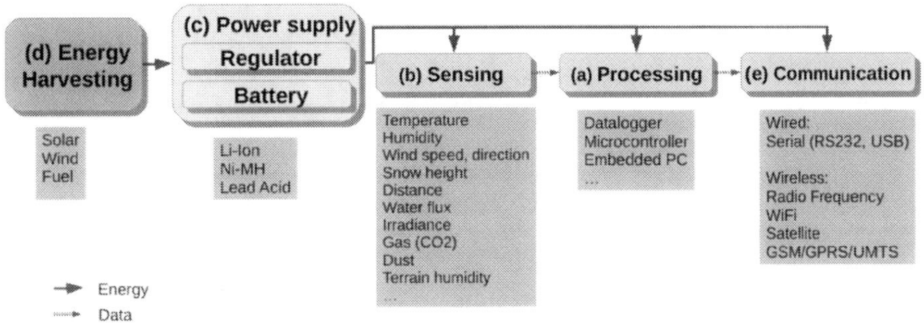

Figure 1. Scheme of typical components of an Automatic Weather Station

As the cost of hardware is decreasing, it becomes feasible to develop distributed measurements over glaciers. Deploying a network of spatially distributed sensing systems on different places of the same glacier, or in different glaciers, allows to perform real-time streamflow monitoring or modeling, enriched by dense environment data [3]. Sensor Networks (SNs), composed of multiple distributed wired or wirelessly communicating sensing devices, represent a possible enhancement to AWSs in terms of spatial resolution and costs. Open problems with SNs are related to their relatively complex setup phase and commercial availability.

A thorough study of existing literature showed the lack of a systematic overview of measuring devices used in glacial monitoring, particularly with respect to Automatic Weather Stations. These devices have been a common tool within a highly skilled community interested for measuring climate-related physical events and quantities. In this paper, we present an overview of AWSs, as they represent a standardized environmental monitoring solution. In addition, we analyse key opportunities and issues related to SNs, that could offer a higher spatial resolution of the AWSs measurements. Section 2 presents AWSs in glacial monitoring, from the hardware and software point of view. Section 3 discusses possibilities of using SNs to enhance the available solutions based on AWSs. Section 4 concludes the paper.

2. Automatic Weather Stations for glacier monitoring

A typical AWS is composed of a datalogger, sensors, a power supply, energy harvesting unit (optional), and a communication unit (optional). Fig. 1 shows a schematic view of an AWS, with the energy and data flow. Possible sensors, communication devices, harvesting and battery technologies are presented in the grey boxes in the figure. Energy harvesting unit enables replenishing the battery with energy from the environment (solar, wind, etc.), providing longer autonomy of the AWS. Communication unit enables remote access to the collected data and notifications in case of malfunctions of the AWS. Both enhancing autonomy and real-time data access are useful for geologists, since usually a glacier station can be reached only during short time slots during the year.

Programming the behavior of battery-powered acquisition systems, in terms of sampling and communication rates, impacts on storage, energy and transmission. A higher acquisition rate causes higher energy consumed by the sensors, requires more storage resources and more data transmission. Wireless data communication also causes high energy consumption and, if using GSM or satellite connections, non-negligible financial costs. This motivates researchers to adapt sampling rate and use compression algorithms to obtain energy awareness in the system, enabling better usage of energy and financial budget.

A Glacial Automatic Weather Station (GAWS) can be considered as standardised within the scientific community, that accepted and adopted the basic composition of GAWS introduced

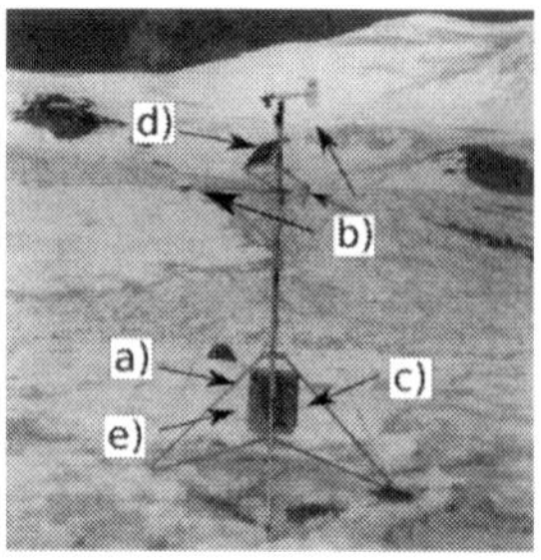

Figure 2.
a) Datalogger: MCU, EPROM, ADC
b) Sensors: temperature, wind, humidity, snow height, irradiance
c) Power supply: battery (Li-Ion, lead acid), charge regulator
d) Energy harvesting: photovoltaic panel
e) Communication device: satellite

Producer	Product	Temperature range	Sleep, Active Power	Enclosure
Gemini	TGP-4020 [5]	-40°C ... +85°C	0.5mW Average	Waterproof
Campbell Scientific	CR1000 [6]	-25°C ... +50°C	6mW, 15mW	Simple
Vaisala	QML201 [7]	-50°C ... +60°C	< 60mW Active	Simple
Delta-T Devices	DL2e [8]	-20°C ... +60°C	Not documented	Simple
Aanderaa	3660s [9]	-40°C ... +60°C	0.5mW, 150mW	Simple
MSR Electronics	MSR145 [10]	-20°C ... +60°C	Not documented	Waterproof
LSI-Lastem	e-log [11]	-20°C ... +60°C	2.4mW, 240mW	Simple
IMAU	i-AWS4Polar [12]	Not documented	Not documented	Waterproof
SensorScope	DS3 [13]	Not documented	Not documented	Waterproof

Table 1. Specifications of most popular dataloggers

in the 1990s by IMAU [4]. A typical GAWS is composed of 5 main subsystems: datalogger, energy harvester (solar panel), communication device (satellite, RF, WiFi), sensors, battery. The sensors of a GAWS are: snow level, wind speed and direction, temperature and humidity, direct and indirect irradiance. In Fig. 2 a typical GAWS is presented. Since the datalogger is the main component of an AWS, we will deepen the analysis of its features and compare the most used models.

2.1. Dataloggers

Table 1 reports about commercially available complete systems for the development and deployment of AWSs. The comparison should give a quick overview of systems and producers, to speed up the process of choosing a particular one. From the specifications of reported loggers and producers, it is extremely difficult to understand which producer is using which software architecture or programming language. The only producer that is documenting hardware and software of dataloggers is Campbell Scientific (CS). Their CR1000 datalogger (as well as other dataloggers from the same family) executes an operating system (OS). The OS is not directly accessible, but they allow to code scripts in CRBASIC (a Basic dialect) to manage the acquisition, processing and communication policies. Other producers don't enable programming of their products. The absence of fine grained programmability in the majority of products is an advantage for the primary customers (field scientists). Although most of the dataloggers are nowadays based on programmable microcontrollers, often the producers offer only preconfigured solutions, without the possibility of programming the system independently. So, for researchers interested in experimenting with different sensing, processing and communication policies, the possibilities offered by CS remain the only viable solution.

2.2. Software and data management

Programming the datalogger of an AWS often only requires setting some variables in a graphical interface on a PC connected to the datalogger, or to regulate physical handles on the housing

Institute	Location	Producer	Hardware	Time
Universities (ITA) [14]	Italian Alps	Campbell Scientific	GAWS+Sat	2004-present
IMAU (NED) [15]	Worldwide [12]	Campbell Scientific	GAWS	1995-present
Universities (ITA) [16]	Italian Alps	LSI-Lastem	GAWS+GSM	2007-present
University (CHI) [7]	Himalaya	Vaisala	AWS	2000-2010

Table 2. Some AWS deployments (with datalogger producers). Acronyms: Italy (ITA), China (CHI), Netherlands (NED).

of the logger, that set the sensors acquisition rate. Measurements taken on dataloggers are typically stored as simple log files, containing the raw sampled values. In some cases, the data is preprocessed before being logged, performing some kind of aggregation (e.g. averaging, min, max) on it, in order to reduce the amount of stored samples. Stored data are collected manually during field visits to the AWS or, if the system is equipped with some transmission device, like RF or satellite, data can be remotely sent to a computer (telemetry).

The World Glacier Monitoring Service (WGMS) is a unique place to store standardised observations on changes of mass, length, volume, area of glaciers over time [1]. It is a kind of a database, but its consistency and meaning are maintained by people. It would be a great advantage to automate at least some of the steps needed to populate the WGMS, e.g. through systems like Global Sensor Network (GSN), that ease the job of managing sensors data and process them using the notion of virtual sensor. In that way, it becomes possible to have real-time elaboration and storage of sensor data [2].

2.3. Deployments

Table 2 presents a selection of AWS deployments in glacial environments. The first three are maintained until today, while the last one has been implemented using mobile AWS (transferred from place to place every few days), to perform measurements in different places on Himalaya. Automatic Weather Stations represent a stable technology and, as we can see from Table 2, they have been used for the last 20 years in different parts of the world and by different organizations. However, recent glaciologist conferences show that there is a growing interest in spatially distributed measurements in such harsh environments [12]. In the following section, we will discuss main challenges in designing and deploying Sensor Networks that need to be reliable, easy to use and robust, just as AWSs are today.

3. From Automatic Weather Stations to Sensor Networks

Automatic Weather Stations, built around dataloggers, are a well known and widely used measurement systems for monitoring glacial environments. They are robust and reliable, but as a drawback, they represent a single point of measurement and a single point of failure.

During the last decade, in the scientific community a new requirement emerged: enhancing single points of measurements with spatially distributed measurements. Small rugged dataloggers, like the TinyTag2 [5] or the MSR145 [10], have been used for years by scientists to acquire temperatures and humidities during long periods. They are much cheaper than AWSs. The data from the sensors is stored and collected manually afterwards. Distributed Thermal Sensors (DTS), on the other hand, allow to monitor temperature along fiber-optical lines, with a spatial resolution of about 1 m. Some experiments even use DTS to measure soil moisture [17].

A technology enabling distributed measurements and communicating data to the user in an automated fashion is Sensor Networks [18]. The architecture of a sensor node comprises the

[1] World Glacier Monitoring Service - http://www.wgms.ch/about.html
[2] Global Sensor Network - http://www.swiss-experiment.ch/index.php/GSN:Home

Project	Locations	Nodes	Number of nodes	Time
SensorScope [13]	Switzerland	TinyNode	~ 15	2007
PermaSense [23]	Switzerland	TinyNode	~ 25	2006-present
GlacsWeb [24]	Norway, Iceland	GWNode	~ 20	2004-present

Table 3. Some WSN deployments on glaciers

AWS	Centralized	Single point of failure	Established	Easy to use
WSN	Distributed	Redundancy	New	Complex

Table 4. Key characteristics of AWSs and WSNs

same components like an AWS (Fig. 2), only in smaller scale. The sensors are connected to a microcontroller, that controls the acquisition, storage, processing and communication of the acquired data. The microcontrollers are often programmed using standard languages (C/C++), thus it is possible to develop new behavioural paradigms by computer scientists or electronic engineers.

In wired sensor networks, nodes are connected with wires to the base station, like in [19]. An interesting commercial solution is Thermistor Chains RBR [20] — cables with termistors attached, that can be used in water, cement, ice, etc. In a Wireless Sensor Network (WSN), sensor nodes communicate wirelessly, usually containing a radio in GHz or MHz frequency band, using WiFi, ZigBee or similar standardized protocols.

WSNs have become a hot topic in the last 10 years. The biggest challenge in WSNs is to provide enough energy to the node to enable autonomous operation [21]. Another challenge is to obtain reliable wireless communication in harsh environments (with water, ice, etc.), often requiring using lower communication frequencies (MHz or lower). Those are key reasons that WSN applications in glacial environment are still under development.

There are few successful deployments of WSNs in glacial environment. Table 3 shows most popular ones. Comparing with Table 2, we notice that some WSN deployments lasted almost as long as some AWS deployments. It is necessary to emphasize, though, that WSN deployments were part of multidisciplinary research projects and experienced operation outages (of the whole network or only of particular nodes), requiring often maintenance and modifications during the years. On the other hand, AWSs are based on mature and commercially available technologies and have had a more stable operation, although they were deployed directly by field scientists. PermaSense and GlacsWeb deployments are still active and under development. In [22], we compare those two projects and introduce possibilities of performance enhancements.

WSNs represent an interesting opportunity to enhance measurements of AWSs. Nowadays, programming and electronic skills are still needed to configure, deploy and maintain WSNs successfully. In order to become an established technology, they need to be improved in terms of simplicity of usage and robustness.

4. Conclusion

In this paper we survey the technologies for sensing in glacial environments, in order to help: a) glaciologists — to choose equipment for their application and deploy it with minimal effort and engineering skills; b) engineers — to understand the field requirements and design standardized solutions that would be easy to maintain, interoperable and scalable, in terms of software and hardware. We describe the characteristics of Automatic Weather Stations, and discuss their benefits and drawbacks. In addition, we discuss the possibilities to use sensor nodes (wireless or wired) to obtain dense spatial resolution of the measurements. The wireless sensor networks present a promising advantage for spatially distributed measurements in remote areas, but, as summarized in Table 4, they still need to be improved in terms of robustness and simplicity of usage.

Acknowledgment

This work has been funded by the Italian MIUR Project PRIN 2010-11: "Response of morphoclimatic system dynamics to global changes and related geomorphological hazards", and by the European Community Seventh Framework Programme under grant No. 285939 (ACROSS). Authors would like to thank Luca Carturan for glaciology-related discussions.

References

[1] Oerlemans J 2005 *Science* **308** 675–677

[2] Light J, Parthasarathy S and McIver W 2012 Monitoring winter ice conditions using thermal imaging cameras equipped with infrared microbolometer sensors *Procedia CS* vol 10 pp 1158–1165

[3] Abbate S, Avvenuti M, Carturan L and Cesarini D 2013 Deploying a communicating automatic weather station on an alpine glacier *3rd Internat. Workshop on Sensor Networks for Intelligence Gathering and Monitoring (SNIGM)* ed Springer

[4] Box J E, Anderson P S and van den Broeke M R 2004 Lessons to be learned: Extended abstracts *Automatic Weather Stations on Glaciers* (Univ. Utrecht, The Netherlands)

[5] GeminiDataloggers Tinytag plus 2 logger URL `http://www.geminidataloggers.com/`

[6] CampbellScientific Campbell scientific cr1000 datalogger URL `http://www.campbellsci.com/`

[7] Vaisala Qml201 datalogger URL `http://www.vaisala.com/`

[8] Delta-T Devices Dl2e - data logger URL `http://www.delta-t.co.uk/`

[9] AAnderaa Datalogger 3660s datasheet URL `http://www.aadi.no/Aanderaa/`

[10] MSR-Electronics Msr 145 datalogger URL `http://www.msr.ch/`

[11] LSI-Lastem URL `http://www.lsi-lastem.it/`

[12] 2011 *Workshop on the use of automatic measuring systems on glaciers.* (Univ. Utrecht, The Netherlands)

[13] Ingelrest F, Barrenetxea G, Schaefer G, Vetterli M, Couach O and Parlange M 2010 *ACM Trans. Sen. Netw.* **6** 17:1–17:32 ISSN 1550-4859 URL `http://doi.acm.org/10.1145/1689239.1689247`

[14] Carturan L, Cazorzi F and Fontana G D 2012 *Journal of Glaciology* **58** 467–486

[15] Shepherd A, Ivins E R, Geruo A, Barletta V R, Bentley M J, Bettadpur S, Briggs K H, Bromwich D H, Forsberg R, Galin N *et al.* 2012 *Science* **338** 1183–1189

[16] Diolaiuti G, Senese A, Mihalcea C, Verza G, Mosconi B and Smiraglia C 2011 AWS measurements on glaciers in the Italian Alps *Workshop on the use of automatic measuring systems on glaciers.* pp 31–35

[17] Steele-Dunne S C, Rutten M M, Krzeminska D M, Hausner M, Tyler S W, Selker J, Bogaard T A and van de Giesen N C 2010 *Water Resources Research* **46** URL `http://dx.doi.org/10.1029/2009WR008272`

[18] Hart J K and Martinez K 2006 *Earth-Science Reviews* **78** 177–191

[19] Mohamed N, Jawhar I, Al-Jaroodi J and Zhang L 2010 Monitoring underwater pipelines using sensor networks *12th IEEE Internat. Conf. on High Performance Computing and Communications (HPCC)* pp 346–353

[20] RBR Xr thermistor chains URL `http://www.rbr-global.com/`

[21] Abbate S, Avvenuti M, Cesarini D and Vecchio A 2012 Estimation of energy consumption for TinyOS 2.x-based applications *Procedia CS* vol 10 pp 1166–1171

[22] Jelicic V, Cesarini D and Bilas V 2013 Enhancing performance of wireless sensor networks in glacial environments using wake-up receivers *IOP Sensors & Their Applications XVII (accepted for publication)* (Dubrovnik, Croatia)

[23] Beutel J, Gruber S, Gubler S, Hasler A, Keller M, Lim R, Thiele L, Tschudin C and Ycel M 2009 The PermaSense remote monitoring infrastructure

[24] Martinez K, Basford P J, Jager D D and Hart J K 2012 A wireless sensor network system deployment for detecting stick slip motion in glaciers *IET Internat. conf. on Wireless Sensor Systems 2012* URL `http://eprints.soton.ac.uk/339176/`

Channel equalization for indoor lighting communications networks

S Hadjiloucas[1], A Hashem[1], VM Becerra[1], and RKH Galvão[2]
[1]*School of Systems Engineering, The University of Reading, RG6 6AY, UK*
[2]*Divisão de Engenharia Eletrônica, Instituto Tecnológico de Aeronáutica, São José dos Campos, SP, 12228-900 Brazil*

E-mail: s.hadjiloucas@reading.ac.uk

Abstract. We consider indoors communications networks using modulated LEDs to transmit the information packets. A generic indoor channel equalization formulation is proposed assuming the existence of both line of sight and diffuse emitters. The proposed approach is of relevance to emergent indoors distributed sensing modalities for which various lighting based network communications protocols are considered.

1. Introduction

With the wide proliferation of RFIDs, smart-phones and other portable communication devices, as well as distributed sensing modalities proposed by sensors manufacturers for pervasive sensing in medical/healthcare applications (e.g, hospitals), in domestic electronics e.g. for smart homes as well as in environmental sensing e.g. food/agriculture applications within controlled environments (greenhouses, plant growth rooms super market shelves etc) it is realised that current radio communication links, have limited bandwidth and are unlikely to be able to cope with the large datasets generated in the long run. Although terahertz links have been suggested for indoors communications [1,2], a simpler solution for setting up multi-hop local indoor communications networks is to use the existing lighting infrastructure in a building and modulate indoor lighting appropriately to enable non-directed communications in an ad-hoc manner. LEDs are well suited to such applications as they are cheap, robust, more efficient from an energy consumption perspective and can be very quickly modulated in the time domain to transmit data packets at high data rates. They are also rapidly becoming a preferred lighting solution by many architect firms. LED-based indoor communications can be conveniently achieved assuming direct line-of-sight or diffuse links.

Optical wireless systems use simple baseband modulation schemes such as on-off keying (OOK), or pulse position modulation (PPM) [3, 4]. Alternatives that are superior from a power efficiency perspective include digital pulse-interval modulation (DPIM) [5], dual header pulse-interval modulation (DHPIM) [6] and differential pulse-position modulation (DPPM) [7]. Differential amplitude pulse-position modulation (DAPPM) is a new hybrid modulation technique that has attracted interest recently [8] as it incorporates aspects from both pulse-amplitude modulation (PAM) as well as DPPM. Demodulation of the encoded signals is commonly performed using hard-threshold decision (HTD), maximum-likelihood sequence detection (MLSD) or zero-forcing decision-feedback

[1] To whom any correspondence should be addressed.

equalization (ZF-DFE) modalities assuming intensity demodulation is performed using a direct detection technique [8].

Line-of-sight links can assume a simple exponentially decaying channel model whereas diffuse links require a ceiling-bounce channel model [9, 10]. In the time domain, the output $y(t)$ from the communications channel is given by:

$$y(t) = Rh(t) * x(t) + n(t) \tag{1}$$

where $h(t)$ is the channel impulse response, R is the detector responsivity, $n(t)$ is white Gaussian noise due to the lighting in the room and $*$ denotes the convolution operator. A ceiling-bounce model such as the one assumed by Carruthers and Kahn [11] can be used for the channel model:

$$h(t,a) = L_0 \frac{6a^6}{(t+a)^7} u(t) \tag{2}$$

where $u(t)$ is the unit step function and a depends on the room size, and the transmitter and receiver position. The parameter L_0 is given from:

$$L_0 = \frac{rA_r}{3\pi L^2} \tag{3}$$

where L is the height of the ceiling, r is the reflectivity of the ceiling above the transmitter and receiver and c is the speed of light. The parameter $a = 2L/c$ is related to the rms delay spread D_{rms} through the following expression [8]:

$$D_{rms} = \frac{a}{12}\sqrt{\frac{13}{11}} \tag{4}$$

In the following sections, a generic channel equalization scheme with amplitude, slew rate and power constrains is developed for the above channel.

2. Optimization of the input waveform

The theory of matched filters [12] could be adopted to customize the excitation waveform in order to either maximize the efficiency of energy transfer from the transmitter to the receiver and improve the signal-to-noise ratio of the data in frequency regions where the channel identification used for the equalization process is ill-conditioned. The problem of maximizing the power transfer can be stated as follows. Given the Laplace transform $H(\omega)$ of $h(t)$, find the Laplace transform $U(\omega)$ of $u(t)$ that maximizes the following function:

$$J = \int_0^{\infty} | H(\omega)U(\omega) |^2 \, d\omega \tag{5}$$

subject to the following restrictions related to the capabilities of the emitter:

$$| U(\omega) | \leq M(\omega), \forall \omega > 0 \tag{6}$$

where $M(\omega)$ is an upper bound for the spectral profile of the input pulse at each frequency and the restriction on the overall power of the generator is given from:

$$\int_0^\infty |U(\omega)|^2\, d\omega \le P_{\max} \tag{7}$$

where $M(\omega)$ is the maximum rms value that the generator is able to deliver at frequency ω and P_{max} is the maximum overall power output of the emitter. This problem can be re-stated by introducing a variable $Z(\omega) = |U(\omega)|^2$ so that the objective function becomes:

$$J = \int_0^\infty |H(\omega)|^2\, Z(\omega) d\omega \tag{8}$$

subject to:

$$Z(\omega) \le M^2(\omega), \forall \omega > 0 \tag{9a}$$

$$\int_0^\infty Z(\omega) d\omega \le P_{\max} \tag{9b}$$

Under the hypothesis that H is known, Eq. (8) is a linear functional of $Z(\omega)$, which must be minimized subject to the linear constraints expressed in (9a) and (9b). This optimization problem could be solved by using variational calculus techniques. Alternatively, by discretizing the frequency axis and considering discrete frequencies up to a certain maximum ω_{max}, the problem becomes a finite-dimensional Linear Program, which can be solved in an efficient manner by techniques such as the Simplex method.

A different problem would be to shape the pulse in order to obtain a specified pulse shape at the output of the receiver. This problem can be formulated as follows. Let $Y_d(\omega)$ be the frequency profile of the desired output pulse. Given $H(\omega)$, find $U(\omega)$ that minimizes the following function:

$$J' = \int_0^\infty |Y_d(\omega) - H(\omega)U(\omega)|^2\, d\omega \tag{10}$$

subject to the generator restrictions:

$$|U(\omega)| \le M(\omega), \forall \omega > 0 \tag{11a}$$

$$\int_0^\infty |U(\omega)|^2\, d\omega \le P_{\max} \tag{11b}$$

This problem is more complex than the previous one, because the cost function J' cannot be transformed in a linear function by a change of variables, as it was done previously. In this case, the problem must be solved by numerical nonlinear programming techniques. The formulation can be regarded as a pre-equalization process. In fact, instead of transmitting the desired pulse shape, a different pulse shape is realized in order to account for the distortion caused by the transmission through the devices and channel under consideration. It is worth noticing that, if no restrictions are placed on the generator capabilities, then the solution to this problem is:

$$U(\omega) = \frac{Y_d(\omega)}{H(\omega)} \tag{12}$$

which may not be a feasible solution in practice.

A different approach consists of formulating the pre-equalization as an optimal control problem in the time domain. Assuming that y_d[k] is the desired pulse shape at the output of the device under consideration, the cost function can be written as:

$$J' = \sum_{k=0}^{K} (y_d[k] - y[k])^2 \tag{13}$$

subject to the dynamics of the device:

$$\begin{aligned} x[k+1] &= Ax[k] + Bu[k] \\ y[k] &= Cx[k] + Du[k] \end{aligned} \tag{14}$$

and also the time-domain restrictions of the generator:

$$-u_{\max} \le u(k) \le u_{\max},\ k = 0,1,\ldots,K \qquad \text{(amplitude)} \tag{15a}$$

$$-\Delta u_{\max} \le \Delta u(k) \le \Delta u_{\max},\ k = 0,1,\ldots,K \qquad \text{(slew rate)} \tag{15b}$$

$$\sum_{k=0}^{K} (u[k])^2 \le P_{\max} \qquad \text{(power)} \tag{15c}$$

where $\Delta u[k] = u[k] - u[k-1]$. Equations (13), (14), (15a), (15b) and (15c) define a linear-quadratic control problem with input constraints. This problem can be solved numerically. Moreover, if the power restriction (15c) is found to be of small importance and is eliminated from the optimization problem, then efficient quadratic programming methods can be used to find the solution, because the remaining input restrictions (15a and 15b) are both linear. An alternative consists of transforming the hard input constraints into soft input constraints by adding input-related terms to the cost function (13):

$$J'' = \sum_{k=0}^{K} (y_d[k] - y[k])^2 + \rho(u[k])^2 + \eta(\Delta u[k])^2 \tag{16}$$

where ρ and η are design weights. An exact solution to this problem can be obtained by the use of a Ricatti equation. If it is found to violate the input restrictions, then weight ρ or η can be increased in order to force a reduction in the input amplitude and rate of change respectively. This process can be iterated until a satisfactory solution is found. Examples from the application of the above formulation to indoors DAPPM encoded communications channels will be discussed at the conference.

References

[1] Piesiewicz R, Jemai J, Koch M, and Kurner T, THz channel characterization for future wireless gigabit indoor communication systems, in *Proc. SPIE—Terahertz and Gigahertz Electronics and Photonics IV*, 2005, vol. 5727, pp. 166–176.

[2] Galvão R K H, Hadjiloucas S, Izhac A, Becerra V M and Bowen J W, Wiener system subspace identification for mobile wireless mm-wave networks, *IEEE Transactions Vehicular Technology* vol. 56, No 4, 1935-1948, 2007.

[3] Gfeller F R and Bapst U H, Wireless in-house data communication via diffuse infrared radiation, *Proc. IEEE,* vol. 67, no. 11, pp. 1474–1486, 1979.

[4] Barry J. R., *Wireless Infrared Communications*, KluwerAcademic Publishers, Norwell, Mass, USA, 1994.

[5] Z. Ghassemlooy, A. R. Hayes, N. L. Seed, and E. D. Kaluarachchi, Digital pulse interval modulation for optical communications, *IEEE Commun. Mag.*, vol. 36, no. 12, pp. 95–99, 1998.

[6] Aldibbiat N M, Ghassemlooy Z, and McLaughlin R, Dual header pulse interval modulation for dispersive indoor optical wireless communication systems, *IEE Proceedings Circuits, Devices and Systems*, vol. 149, no. 3, pp. 187–192, 2002.

[7] Shiu D -S and Kahn J M, Differential pulse-position modulation for power-efficient optical communication, *IEEE Trans. Commun.*, vol. 47, no. 8, pp. 1201–1210, 1999.

[8] Sethakaset U, and Gulliver T A, Differential Amplitude Pulse-Position Modulation for Indoor Wireless Optical Communications *EURASIP Journal on Wireless Communications and Networking* vol 1, pp. 3–11, 2005.

[9] Barry J R, Kahn J M, Krause W J, Lee E A, and Messerschmitt D G, Simulation of multipath impulse response for indoor wireless optical channels, *IEEE J. Select. Areas Commun.*, vol. 11, no. 3, pp. 367–379, 1993.

[10] Fernando X.N., and Balendran B., Adaptive Denoising and Equalization of InfraredWireless CDMA System, *EURASIP Journal on Wireless Communications and Networking* vol 1, pp. 20–29, 2005.

[11] Carruthers J B and Kahn, J M, Modeling of nondirected wireless infrared channels, *IEEE Trans. Commun.,* vol. 45, no. 10, pp. 1260–1268, 1997.

[12] Taub H, and Schilling D L, *Principles of Communication Systems*, 2nd ed. McGraw-Hill, New York, 1986.

Analytical modelling of soil effects on electromagnetic induction sensor for humanitarian demining

D Vasić, D Ambruš and V Bilas

University of Zagreb, Faculty of Electrical Engineering and Computing
FER-ZESOI, Unska 3, HR-10000 Zagreb, Croatia

E-mail: darko.vasic@fer.hr

Abstract. Accurate compensation of the soil effect is essential for a new generation of sensitive classification-based electromagnetic induction landmine detectors. We present an analytical model for evaluation of the soil effect suitable for straightforward numerical implementation. The modelled soil consists of arbitrary number of conductive and magnetic layers. The solution region is truncated leading to the solution in form of a series rather than infinite integrals. Frequency-dependent permeability is inherent to the model, and time domain analysis can be made using DFT. In order to illustrate the model usage, we evaluate performances of three metal detector designs.

1. Introduction

Electromagnetic induction (EMI) sensors in the form of metal detectors for humanitarian demining are the most important devices for close-in landmine detection well suited to the field applications due to their robustness and sensitivity achievable by a skilled operator [1]. The research of EMI sensors for landmine detection is directed towards reducing the number of false alarms and increasing the speed either by combining the metal detector with other established methods such as ground penetrating radar (GPR) or by employing more advanced mine detection EMI techniques based on mine or clutter classification of metallic objects according to their material, shape and dimensions [1], [2].

Degradation of metal detector performances in certain soil types has been known since World War II, but the research on the soil effects was somewhat neglected by the mine detection community until recently [3]. The studies of the soil effects employed models of homogenous half-space soil based on the infinite integral formulation of the soil response, usually followed by the approximations for low or negligible electrical conductivity and magnetic permeability [3], [4]. Although useful for understanding of the soil effects on metal detectors in general, e.g. dominant effect of frequency-dependent magnetic permeability, such models may be insufficient for new generation of metal detectors based on the classification algorithms that require accurate soil effect compensation [2].

In this paper we present an analytical frequency-domain model for calculation of the magnetic field and induced voltage for a sensor head located above arbitrary number of conductive and magnetic soil layers. The solution region is truncated leading to the solution in form of a series rather than infinite integrals [5], [6]. This allows easier and faster implementation and error control [6]. The frequency dependence of permeability is inherent to the model and time domain analysis is possible using DFT. The equations are presented in a form suitable for straightforward stable numerical implementation. Using the model, we evaluate three typical configurations of metal detectors with respect to ratio of small metallic target response to the soil response [2], [4].

2. Analytical model

2.1. Vector magnetic potential and magnetic field

We derived the analytical model of the magnetic potential of a single-turn transmitter coil positioned above a half-space of arbitrary number of conductive and magnetically permeable layers, as depicted in figure 1. In contrast to the usual Dodd and Deeds integral formulation [5], we truncated the solution region and, as a consequence, formulated the solution of the vector magnetic potential equation as a series [6]. The problem domain should be at least ten times larger than the characteristic length of the sensor head: transmitter coil diameter or transmitter-receiver distance, whichever is greater.

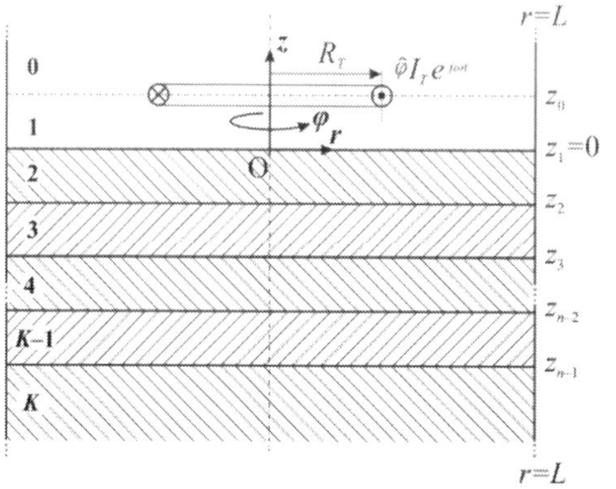

Figure 1. Geometry used for modelling of the soil effects.

The transmitter single-turn coil of radius R_T at height z_0 from the ground is driven by a sinusoidal current i_T of constant amplitude I_T and frequency $\omega = 2\pi f$. The ground is assumed to be composed of $K-1$ linear, isotropic, and homogeneous layers. Electrical conductivity of k-th layer ($k = 2,...K$) is σ_k, its relative magnetic permeability is μ_{rk}, and its thickness is $\Delta z_k = z_{k-1} - z_k$, except that of the K-th layer, where $z_k < z_{k-1} < 0$ as in figure 1. The K-th layer is the region $z < z_{K-1}$ and its thickness is infinite. The problem domain is truncated at $r = L$, whereas it is unbounded in both z directions. In the cylindrical coordinate system, the excitation current has φ component only, and so will the vector magnetic potential **A** too. Taking the axial symmetry into the account and using the quasistatic approximation, the governing equation for the magnetic potential is

$$\frac{\partial^2 A_k(r,z)}{\partial r^2} + \frac{1}{r}\frac{\partial A_k(r,z)}{\partial r} + \frac{\partial^2 A_k(r,z)}{\partial z^2} - \frac{A_k(r,z)}{r^2} - j\omega\mu_0\mu_{rk}\sigma_k A_k(r,z) = 0, \tag{1}$$

where A_k is the φ component of **A** in the k-th region. The method of separation of variables reduces (1) into two equations (ordinary differential equation in z and first-order Bessel equation in r).

The magnetic potential for each region must satisfy the homogenous Dirichlet condition

$$A_k(r = L, z) = 0. \tag{2}$$

Because of (2), the eigenvalues are discrete and the solution for each region can be written in terms of the series truncated to N elements

$$A_k(r,z) = \sum_{i=1}^{N}\left[C_{k,i}\exp(-\alpha_i z) + D_{k,i}\exp(\alpha_i z)\right]J_1(\alpha_i r), \quad k = 0 \text{ or } k = 1, \tag{3}$$

$$A_k(r,z) = \sum_{i=1}^{N}\left[C_{k,i}\exp(-\alpha_{k,i} z) + D_{k,i}\exp(\alpha_{k,i} z)\right]J_1(\alpha_i r), \quad k = 2,\ldots K, \tag{4}$$

where $C_{k,i}$ and $D_{k,i}$ are constants to be found out and J_1 is the first-order Bessel function of the first kind. The eigenvalues α_i for regions 0 and 1 (air) are given as solutions of the equation

$$J_1(\alpha_i L) = 0,\ \alpha_1 < \alpha_i < \alpha_N,\ i = 1, \ldots N. \tag{5}$$

The zeros of the Bessel function can be readily determined using the prepared tables or a numerical root finding algorithm. The eigenvalues for the regions $k = 2,\ldots K$ are

$$\alpha_{k,i} = \sqrt{\alpha_i^2 + j\omega\mu_0\mu_{r,k}\sigma_k},\quad k = 2,\ldots K. \tag{6}$$

Constants $D_{0,i}$ and $C_{K,i}$ are 0 because of the divergence of the respective exponential functions for $z \to \pm\infty$. The remaining constants have to be calculated from the standard interface conditions for the magnetic field. Using the orthogonal property of the Bessel functions, one can write N systems (for each i) of $2K$ linear equations with $2K$ unknowns $[C_{0,i}\ C_{1,i}\ D_{1,i}\ C_{2,i}\ D_{2,i} \ldots C_{K-1,i}\ D_{K-1,i}\ D_{K,i}]^{\mathrm{T}}$. We will report only the potential for $k = 1$, since it is the one required for evaluation of detector performances.

Let $\mathbf{F}_k$ be the characteristic matrix of the k-th region:

$$\mathbf{F}_{k,i} = \begin{bmatrix} 1 & -\dfrac{\mu_{rk}}{\alpha_{k,i}}\tanh(-\alpha_{k,i}\Delta z_k) \\ -\dfrac{\alpha_{k,i}}{\mu_{rk}}\tanh(-\alpha_{k,i}\Delta z_k) & 1 \end{bmatrix},\quad i = 1,\ldots N,\ k = 2,\ldots K-1. \tag{7}$$

It can be shown by solving the linear system that if

$$\mathbf{G}_i = \begin{bmatrix} g_{1,i} \\ g_{2,i} \end{bmatrix} = \begin{bmatrix} 1 & -1/\alpha_i \\ 1 & 1/\alpha_i \end{bmatrix} \left(\prod_{k=2}^{K-1} \mathbf{F}_{k,i} \right) \begin{bmatrix} 1 \\ \alpha_K/\mu_{rK} \end{bmatrix},\quad i = 1,\ldots N, \tag{8}$$

then the relation between the constants for the region $k = 1$ is

$$C_{1,i} = \frac{g_{1,i}}{g_{2,i}} D_{1,i},\quad i = 1,\ldots N. \tag{9}$$

Furthermore, it can be shown that $D_{1,i}$ does not depend on the soil properties and that it is the same as in the case when there is no soil in the vicinity of the coil (coil in the air), i.e.

$$D_{1,i} = \frac{\mu_0 R_T I_T}{L^2} \frac{J_1(\alpha_i R_T)}{\alpha_i J_0^2(\alpha_i L)} \exp(-\alpha_i z_0),\quad i = 1,\ldots N. \tag{10}$$

Finally, combining (3) with (7)–(10), one can write for the potential for $k = 1$ ($0 \le z \le z_0$)

$$A_1 = A_{1,air} + A_{1,soil} = \sum_{i=1}^{N} D_{1,i} \exp(\alpha_i z) J_1(\alpha_i r) + \sum_{i=1}^{N} \frac{g_{1,i}}{g_{2,i}} D_{1,i} \exp(-\alpha_i z) J_1(\alpha_i r). \tag{11}$$

The magnetic potential is sum of the magnetic potential of the coil in air and the soil response. The only soil-dependent part in (11) is ratio $g_{1,i}/g_{2,i}$. Due to the axial symmetry, the magnetic field has only r and z components, and these can be calculated from (11) (calculating $\nabla \times \mathbf{A}$)

$$B_{1,r} = -\frac{\partial A_1}{\partial z} = -\sum_{i=1}^{N} \alpha_i D_{1,i} \exp(\alpha_i z) J_1(\alpha_i r) + \sum_{i=1}^{N} \alpha_i \frac{g_{1,i}}{g_{2,i}} D_{1,i} \exp(-\alpha_i z) J_1(\alpha_i r), \tag{12}$$

$$B_{1,z} = \frac{A_1}{r} + \frac{\partial A_1}{\partial r} = \sum_{i=1}^{N} \alpha_i D_{1,i} \exp(\alpha_i z) J_0(\alpha_i r) + \sum_{i=1}^{N} \alpha_i \frac{g_{1,i}}{g_{2,i}} D_{1,i} \exp(-\alpha_i z) J_0(\alpha_i r). \tag{13}$$

2.2. Induced voltage

Voltage U_R induced in a receiver coil of radius R_R placed at location (r_R, z_R) can be calculated using

$$U_R = j\omega \oint_{C_{RX}} \mathbf{A} \cdot d\vec{l} = j\omega \int_0^{2\pi} A\left(r(\theta), z_R\right) \frac{R_R^2 + R_R r_R \cos(\theta)}{r(\theta)} d\theta, \tag{14}$$

where $r^2(\theta) = R_R^2 + 2R_R r_R \cos(\theta) + r_R^2$. Integral in (14) can easily be calculated using standard numerical integration techniques. The two components of U_R are direct-coupling component, $U_{R,air}$, and soil-related component, $U_{R,soil}$.

2.3. Sensor head sensitivity

A useful concept for analysing and comparing different types of sensor heads is the sensitivity defined for a location in space as

$$S = \frac{\mathbf{H}_T \mathbf{H}_R}{I_T I_R}, \tag{15}$$

where $\mathbf{H}_T$ is the magnetic field produced by the transmitter coil in air energized with current I_T, and $\mathbf{H}_R$ is the magnetic field that the receiver coil in air would produce if it was energized with the current I_R [4], [7]. In the first approximation, if a small and spherical metallic object is introduced in the vicinity of the transmitter-receive pair, the change in the receiver's induced voltage is

$$U_{R,obj} = MS, \tag{16}$$

where M depends only on the object's properties and excitation frequency. The sensitivity defined with (15) does not depend on the metallic object or on the excitation frequency [7].

2.4. Numerical implementation

The discrete eigenvalues allow easier numerical implementation and accuracy control. The convergence of the series is discussed in [6]. The required number of series elements primarily increases with the size of the solution region. For all calculations in this paper, we placed the solution region boundary at $L = 5$ m and we truncated series (11) to $N = 20000$. Using these settings, the numerical accuracy of the magnetic potential was at least 5 significant digits at the regions' boundaries and more than 8 digits otherwise. The model was implemented in MATLAB. It takes about 9 ms to calculate the potential at one point in space for a fixed geometry and $N = 20000$ using Intel Quad CPU and 8 GB of RAM. The execution time increases linearly with N.

3. Results

Using the numerical implementation of the model presented in section 2, we analysed performances of the three typical sensor head designs, schematically depicted in figure 2. The circular sensor head is the most basic design normally used with pulsed excitation in order to avoid excessive direct coupling component in the received voltage. Here, we use it with the harmonic excitation as a standard for comparison against other two types of sensor heads from figure 2. In the OO balanced sensor, transmitter and receiver overlap precisely so that the induced voltage is minimal or preferably zero in air. Finally, the transmitter-bucking configuration employs two transmitter coils with the radii and excitation currents precisely calculated so that the induced voltage is zero. The latter allows easier and more precise control of the minimal induced voltage.

Figure 3 depicts sensitivity isolines in r-z plane ($z < 0$) for all three investigated sensor heads positioned at z = 5 cm above the ground. The circular head sensitivity and the transmitter-bucking head sensitivity are symmetrical around $r = 0$ and they are highest at a given depth along the line. The OO balanced head sensitivity is highest at a given depth along the line $r = 14$ cm that is different from the centre of the coils' overlapping region (at $r = 11.4$ cm). Along these maximum lines, the circular and OO balanced sensor heads have similar sensitivities because of the same transmitter design. The transmitter-bucking design has the sensitivity smaller for 5 to 10 times compared to the circular head. This means that for the same spherical target, the transmitter-bucking sensor head will produce 5 to 10 times smaller induced voltage depending on the target depth. It should be noted, however, that the transmitter-bucking head has the smallest receiver coil.

Figure 4 depicts the relationship between the sensor head position above the ground (lift-off) and the soil-related component of the induced voltage at the excitation frequency of 100 kHz and for the soil conductivity of 0.01 S/m and relative permeability of 1.01. The inductive balance design (OO balanced and transmitter-bucking heads) significantly reduces the soil effect.

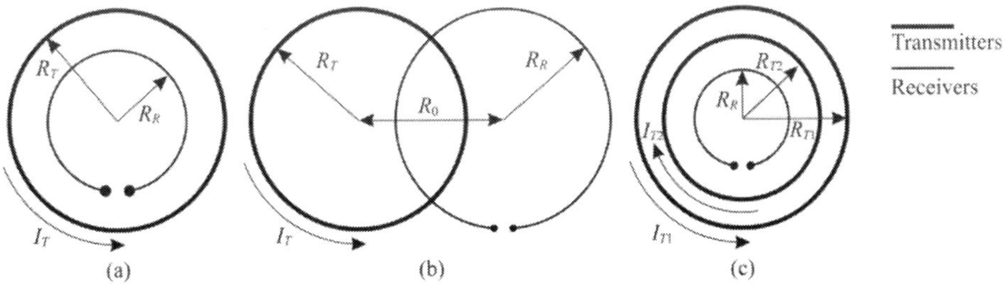

Figure 2. (a) Concentric sensor head (R_T= 15 cm, R_R= 10 cm); **(b)** OO balanced sensor head (R_T= R_R= 15 cm, R_0= 22.74 cm); **(c)** transmitter-bucking sensor head (R_{T1}= 15 cm, R_{T2}= 6 cm, R_R= 4 cm, I_{T1} = $3I_{T2}$).

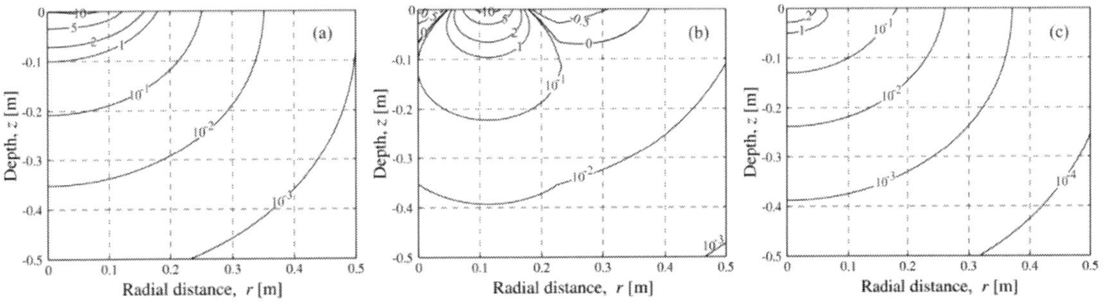

Figure 3. Sensitivity isolines [m^{-2}] for **(a)** circular sensor head, **(b)** OO balanced sensor head, **(c)** transmitter-bucking sensor head. Sensor heads are positioned at z = 5 cm.

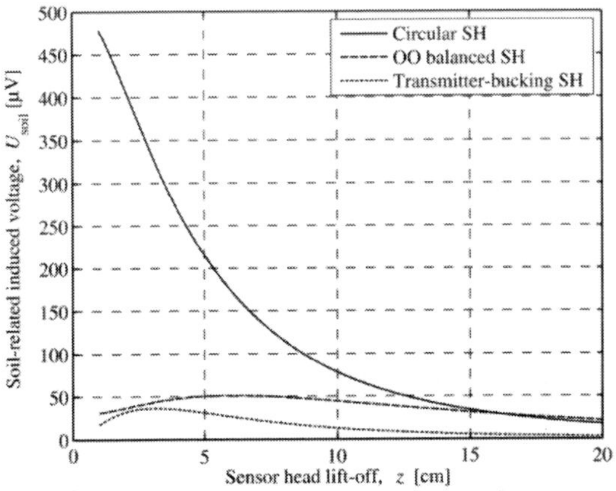

Figure 4. Soil-related (excitation frequency 100 kHz, half-space soil model, σ = 0.1 S/m, μ_r= 1.01) component of the receiver voltage vs. sensor head lift-off.

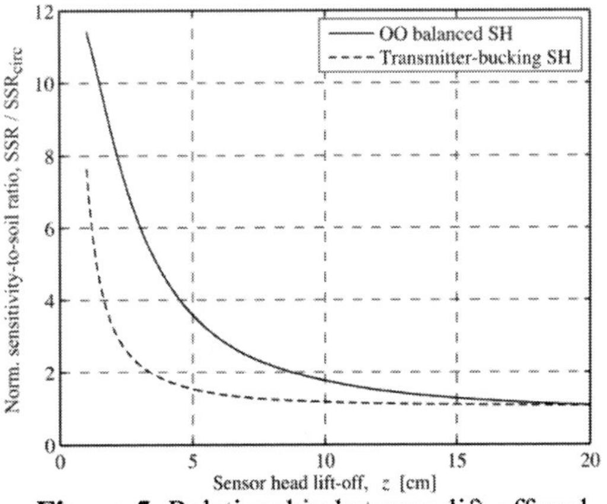

Figure 5. Relationship between lift-off and normalised SSR according to (18) along the lines of the highest sensitivity (constant r). The soil responses are from figure 4.

In order to compare performances of the OO balanced and transmitter-bucking sensor heads in detecting small metallic objects in the presence of soil ($\sigma = 0.01$ S/m and $\mu_r = 1.01$), we introduce the sensitivity-to-soil signal ratio (*SSR*)

$$SSR = \frac{U_{R,obj}}{U_{R,soil}} = \frac{MS}{U_{R,soil}}, \tag{17}$$

and to avoid the effect of the metallic object properties, we normalize it with respect to *SSR* of the circular sensor head

$$\frac{SSR}{SSR_{circ}} = \frac{S}{S_{circ}} \frac{U_{R,soil}^{(circ)}}{U_{R,soil}}. \tag{18}$$

The normalized *SSR* vs. lift-off is shown in figure 5. For a given metallic object, the ratio of the target and soil components of the induced voltage for OO balanced and transmitter-bucking sensor heads is larger (up to 10 times) than the ratio for the circular sensor head. The ratio is the largest for the OO balanced sensor, but the transmitter-bucking head has smaller dimensions, easier balancing control and smaller absolute values of the direct coupling and soil-related components of the induced voltage.

4. Conclusions

We presented an analytical model for evaluation of the soil effects on EMI landmine detector in a form suitable for straightforward numerical implementation. The modelled soil consists of arbitrary number of homogeneous, conductive and permeable layers. The solution region is truncated leading to discrete eigenvalues and the solution in form of a series rather than infinite integral. This allows better convergence control and faster numerical implementation. In order to illustrate the usage of the model, we calculated sensitivities of three typical metal detector designs, evaluated the soil effect and the ratio of the metal target signal and soil response. The model is applicable to soil effect correction procedures for new generation of sensitive classification-based EMI landmine detectors. Future work will focus on extending the model to accommodate any orientation and shape of transmitter coils using the second-order vector potential.

Acknowledgement

This research has been partly supported by the Croatian Ministry of Science, Education and Sport through grant agreement 036-0361621-1625, and by the Seventh Framework Programme of the European Commission through grant agreement no. 285939 FP7-REGPOT-2011-1 (ACROSS).

References

[1] Ishikawa J, Furuta K 2006 *Anti-personnel Landmine Detection for Humanitarian Demining* (London: Springer-Verlag London Limited)
[2] Bruschini C 2006 *A Multidisciplinary Analysis of Frequency Domain Metal Detectors for Humanitarian Demining* (PhD Thesis: Vrije Universiteit Brussel)
[3] Das Y 2006 Effects of Soil Electromagnetic Properties on Metal Detector *IEEE Trans. Geosci. Rem. Sens.* **44** 6 1444–1453
[4] Druyts P, Das Y, Craeye C, Acheroy M 2009 Modelling the Response of Electromagnetic Induction Sensors to Inhomogenous Magnetic Soils With Arbitrary Relief *IEEE Trans. Geosci. Rem. Sens.* **47** 8 2627-2638
[5] Dodd C V, Deeds W E 1968 Analytical Solutions to Eddy-current Probe-coil Problems *J. Appl. Phys.* **39** 6 2829–2838
[6] Theodoulidis T P, Kriezis E 2005 Series Expansions in Eddy Current Nondestructive Evaluation Models, *J. Mater. Process. Technol.* **161** 343–347
[7] Silvester P P, Omeragić Dž 1996 Sensitivity Maps for Metal Detector Design, *IEEE Trans. Geosci. Rem. Sens.* **34** 3 788–792

Emotional recognition from the speech signal for a virtual education agent

A Tickle[1], S Raghu[1] and M Elshaw[2]
[1]Department of Aviation, Aerospace, Electrical and Electronic Engineering, Faculty of Computing and Engineering, Coventry University, Priory Street, Coventry, UK
[2]Department of Computing, Faculty of Computing and Engineering, Coventry University, Priory Street, Coventry, UK
Contact Email: andrew.tickle@coventry.ac.uk

Abstract. This paper explores the extraction of features from the speech wave to perform intelligent emotion recognition. A feature extract tool (openSmile) was used to obtain a baseline set of 998 acoustic features from a set of emotional speech recordings from a microphone. The initial features were reduced to the most important ones so recognition of emotions using a supervised neural network could be performed. Given that the future use of virtual education agents lies with making the agents more interactive, developing agents with the capability to recognise and adapt to the emotional state of humans is an important step.

1. Introduction

This paper describes a neural network approach to speech emotion recognition from a set of acoustic features of a speech waveform. Emotions would enable a communication virtual agent to both maintain a user's positive emotional state and allow it to judge and refine its current dialogue strategy. One way to determine the emotional state of the user in this context is through an interpretation of the speech signal collected using a microphone. This approach outlined in this paper focuses on using auditory sensors rather than visual ones as they are not as intrusive and rely less on environmental conditions.

In order to use virtual agents in the role of educator, it is necessary to provide them with the capability to sense the emotional state of the human user [1, 2] so they can offer the appropriate and natural response. By recognising and expressing emotions, the interactive educational agent could act in an enthusiastic manner towards the teaching material and also show empathy towards student development. If the interactive agent has an appealing and believable personality, this would have the effect of making the interaction more enjoyable for the students [6]. It is felt that students will relate to virtual agents if they can connect with their personality and their emotional representation.

There has seen a growth in interest in emotional speech recognition using intelligent approaches to aid agent-person interaction. For example, Zhang et al. used back propagation neural network to recognise emotions such as anger, clam, happy, sad and surprise [9]. Traista and Elshaw (2012) [7] developed a self-organising neural network approach to perform emotional speech recognition that made use of a emergent unsupervised approach. Vogt et al. (2008) [8] developed EmoVoice a real-time emotion recognition system based on a state vector machine within a virtual agent named Greta. Despite the research carried out into emotional speech recognition, there is little agreement on how to perform feature selection, and so this paper will offer an indication of an approach to do so.

2. Methodology

As can be seen from Figure 1, the methodology for developing an emotional speech recognition system involves extracting features from the speech signal collected by a microphone, reducing the feature set to a more manageable level, and then recognising the emotions using an intelligent technique. In the remainder of this paper we will provide a more detailed description of how this was achieved.

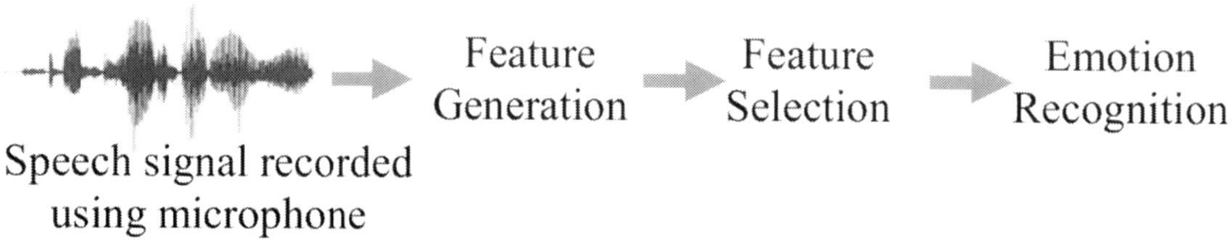

Figure 1. The components of the emotion recognition system.

2.1 Speech corpus

The "Berlin Database of Emotional Speech" is used to train and test the algorithm in this paper [3]. The Berlin Database consist of 535 speech sample, which contain German utterances related to emotions such as anger, disgust, fear, joy, sadness, surprise and neutral, acted by five males and five females. The Berlin database was chosen to be used because of the quality of its recording and its popular use in emotion recognition based research work. Some of the utterance translated into English in the Berlin Corpus include: 'The tablecloth is lying on the fridge', 'She will hand it in on Wednesday' and ' In seven hours it will be'. Of the 535 speech sample, features extraction was carried out on 105 samples, consisting of 15 occurrences of each emotion. To consider how the trained model performed on samples from a person not in the data set, two samples of each emotion were recorded in English by a male volunteer.

2.2 Feature generation

To extract the features from the speech samples the data mining tools OpenSMILE was employed. OpenSMILE toolkit is a flexible feature extractor for speech processing and machine learning applications, which is fully written in C++ [4]. In this research, OpenSMILE was configured to extract a set of 988 acoustic features. The feature set that was extracted from the microphone speech signal is specified by the configuration file "emobase.conf" in the toolbox. The features included the following low-level descriptors (LLD): Intensity, Loudness, 12 MFCC, Pitch (F0), Probability of voicing, F0 envelope, 8 LSF (Line Spectral Frequencies) and Zero-Crossing Rate. Delta regression coefficients were computed from these LLD, and the following functionals were applied to the LLD and the delta coefficients: Max/Min. value and respective relative position within input, range, arithmetic. mean, 2 linear regression coefficients and linear and quadratic error, standard deviation, skewness, kurtosis, quartile 1-3, and 3 inter-quartile ranges. Figure 2 displays the output from OpenSmile when creating the features from the emotional speech signals.

2.3 Feature selection

The performance of the learning algorithm increases according to the quality of the input data. Among the 988 acoustic features extracted using the OpenSMILE tool, several of these features had overlapping information and a few features contained irrelevant information. In order to improve the processing speed of the system to a practical standard, it was necessary to severally reduce the features set to the most relevant ones. A view of the features in the 'explore' tool of WEKA (http://www.cs.waikato.ac.nz/ml/weka/) provides a basic understanding of how each feature effects the emotion the samples describe. Graphs of each feature can be viewed under the pre-process tab.

For the process of feature selection, filtering and wrapping algorithms were employed. Using the WEKA machine learning tool, "info-gain attribute" algorithm was used to rank the features in the order of importance according to their respective weights. Similarly, using the "classify Subset Eval" algorithms the best-first list of features was generated. A hybrid set of features selected by considering both these algorithms was then used as the final training set for the multilayer perceptron.

The results of "classifySubset Eval" algorithm provided 11 features as the ones with highest importance. However, the performance of these 11 features alone was very poor, hence a bigger list of features were needed and so these features were combined with the 63 features available by rank by using the "info-gain attribute" ranker algorithm of WEKA. The combined feature set contained a total of 71 features.

Figure 2. Screenshot of the OpenSMILE command console during feature extraction.

2.4 Multi-layer perception neural network

It was decided to use a neural network for emotion recognition as this learning technique has proved successful in the past. Multilayer perceptron is an important class of neural networks. As shown by Figure 3, it commonly consists of neurons that constitute the input layer, one or more hidden layers and an output layer of computational nodes. The learning rule typically used for the multilayer neural network is the back-propagation rule that allows the network to learn to classify. This rule creates the output of the network compares this with the required output and by propagating the error back through the network alters the weights to reduce the error [5]. During the forward pass the synaptic weights are combined with the input vector to produce on the output layer the classification. During the backward pass, the classification produced by the system is subtracted from the actual classification to get the error signal. The error signal is passed backwards through the network to adjust the weights.

3. Results

For the samples from the Berlin database after initial training was performed, a 10-fold cross validation evaluation was performed. This system's performance is measured based on the number and percentage of correctly recognised emotions. The number of instances correctly classified for the Berlin Database is 88 out of 105, producing an average accuracy of 83.8%. Considering the confusion matrix for samples from the Berlin Emotional Database in Table 1, accuracies differ significantly between the emotions, with some being easier to identify than others. For example, the multilayer neural network seems to struggle to differentiate between emotions anger and disgust. A few inaccuracies also occur in the neutral emotion class as well.

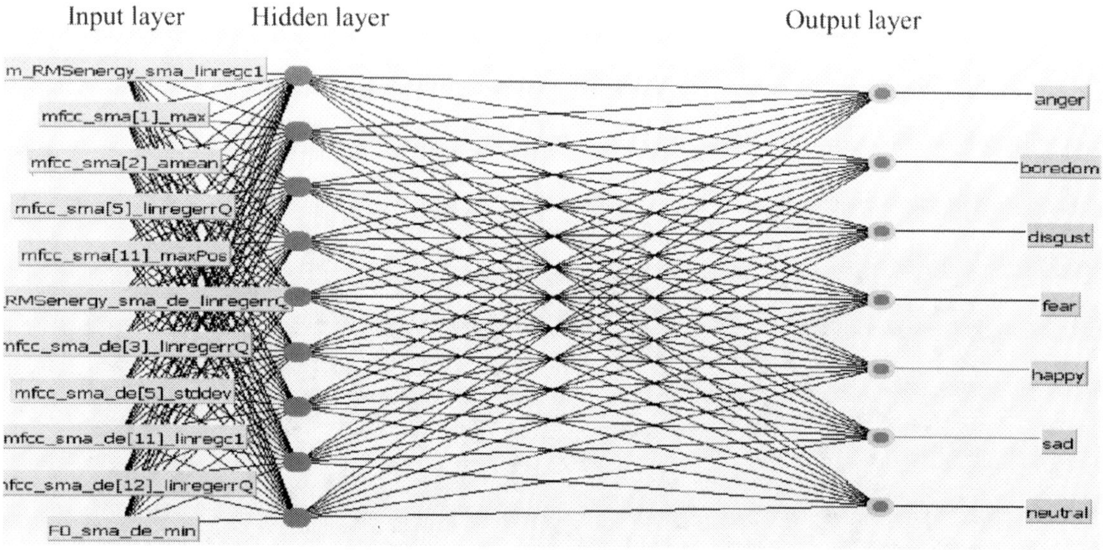

Figure 3. Example multi-layer perceptron model from WEKA showing the 11 features provided from the wrapping algorithms

The performance of the system was also tested by using the speech samples recorded from an English speaking male volunteer, as a preliminary study. These samples had 2 occurrences of each emotion. The average accuracy in this test was much poorer at 42.8%. The fall in accuracy can be explained by the lack of microphone quality, as a generic laptop microphone was used. The noise interference is a major reason for the fall in accuracy. As seen in Table 2, from the confusion matrix it can be observed that the error occurrence is random and has no conclusive patterns.

Anger	Boredom	Disgust	Fear	Happy	Sad	Neutral
12	0	2	0	0	0	1
0	14	0	0	0	1	0
3	1	10	1	0	0	0
0	0	0	13	0	1	1
0	0	0	0	14	0	1
0	1	0	0	0	14	0
0	1	0	0	2	1	11

Table 1. Confusion matrix – performance Results of Berlin Database

Anger	Boredom	Disgust	Fear	Happy	Sad	Neutral
1	0	0	0	1	0	0
0	0	1	0	0	1	0
0	1	0	1	0	0	0
0	0	0	1	0	1	0
0	0	0	0	1	0	1
0	1	0	0	0	1	0
0	0	0	0	0	0	2

Table 2. Confusion matrix – performance on samples taken from an English male volunteer

4. Discussion

Implementing an emotion recognition system for a virtual educational agent would be a huge step forward. Although, the approach presented in this report achieves good performance, further work would help reduce the feature set even further. The number of features used is a primary factor concerning the processing time taken by the system to complete its operations. In order to be able to implement this approach in hardware components, it would require a huge amount of processing memory.

Although part of the reduction in performance on the speech from the male volunteer collected as a preliminary study might be the attribute to the training set being in German and this data being recorded in English, it is not the real reason as both these languages represent emotion in speech in the same manner. Considering the closeness of the origin of these languages, it is anticipated that this emotion recognition system is language independent in this scenario. The poor performance is likely because of the low quality of the microphone and the noise level in the environment during the recording. If virtual educational agents are to be commonplace, there is a need to develop approaches to overcome noise in the environment without making the microphone sensor too costly.

5. Conclusions

The project presents a system for emotion recognition from speech whose accuracy is similar to present works using the Berlin Emotional Database. The report provides descriptions of feature extraction from the Berlin Database samples, various ways the features have been selected for the final set of 80 features and the multilayer perceptron model using which the results have been produced. The approach by which the features have been reduced offers new opportunities for emotion recognition. The system offers a real indication that intelligent autonomous emotion recognition system could be incorporated into social robots to enable them to interact with human users in a natural way.

Although the performance of the multi-layer perceptron neural network proved successful, there is an opportunity to achieve more reliable performance. This will be achieved by performing classification of the emotion using multi-classifiers and then combine the outcomes with a voting based system. There is also the opportunity to incorporate other sensor reading to give the emotional state of the student using the virtual education system. For example, the system could use a camera to explore the facial expressions of the student and a heart rate sensor to establish if emotions influence heart rate. As shown in Figure 4, it would be possible to combine top-down features such as the words and facial expression recognised with the lower level feature such as those gained from the speech signal. By being able to recognise emotions using the auditory input from the users, there are many areas as well as virtual educational agents, such as social assistive robots and telephone call centres that would benefit from the research outlined in this paper.

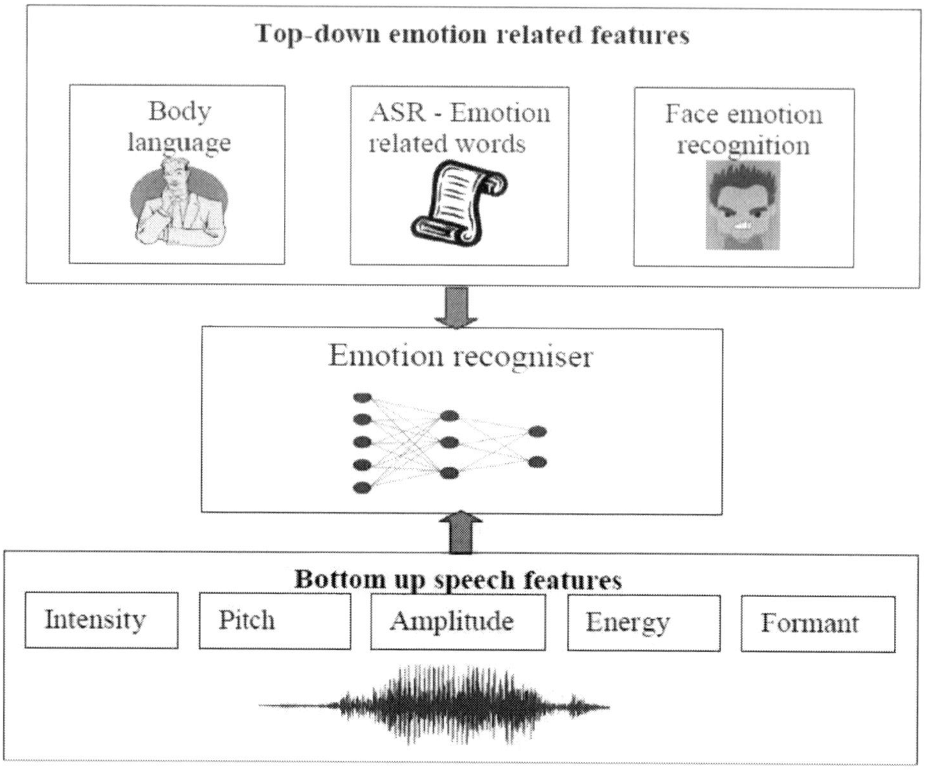

Figure 4. Top-down and bottom-up emotion recognition from multimodal sensors.

6. References

[1] Creed, C and Beal, R. 2005 Using emotion simulation to influence user attitudes and behaviour. *Proceedings of the 2005 Workshop on the role of emotion in HCI*

[2] Cerezo, E., Baldassarri, S., and Seron, F. 2007 Interactive agents for multimodal emotional user interation, *IADIS International Conference Interfaces and Human Computer Interaction*, pp. 35-42.

[3] Burkhardt, F., Paeschke, A., Rolfes, M., Sendlmeier, W., and Weiss, B. 2005 A Database of German Emotional Speech. *Proceedings Interspeech 2005, Lisbon, Portugal*

[4] Eyden, F., Wollmer, M., and Schuller, B. 2010 Opensmile: the munich versatile and fast open-source audio feature extractor, *MM '10 Proceedings of the international conference on Multimedia*, pp. 1459-1462.

[5] Haykin, S 2005 Neural Networks: A comprehensive foundation, 2nd edition. Essex,Pearson Education.

[6] Hertzum, M., Andersen, H., Andersen, V. And Hansen, C. 2002 Trust in information sources: seeking information from people, documents, and virtual agents, *Interacting with Computers*, 14(5), pp. 575-599.

[7] Traista, A., and Elshaw, M., 2012 A hybrid neural emotion recogniser for human-robotic agent interaction, *Proceeding of 13th Engineering Applications for Neural Networks..*

[8] Vogt, T., André, E., and Bee., N. 2008 EmoVoice - A framework for online recognition of emotions from voice. *In Proceedings of Workshop on Perception and Interactive Technologies for Speech-Based Systems, Springer, Kloster Irsee.*

Use of an Infrared Thermometer with Laser Targeting in Morphological Scene Change Detection for Fire Detection

Andrew J. Tickle, Harjap Singh and Josef E. Grindley
Department of Aviation, Aerospace, Electrical and Electronic Engineering, Faculty of Computing and Engineering, Coventry University, Priory Street, Coventry, UK
Contact Email: andrew.tickle@coventry.ac.uk

Abstract. Morphological Scene Change Detection (MSCD) is a process typically tasked at detecting relevant changes in a guarded environment for security applications. This can be implemented on a Field Programmable Gate Array (FPGA) by a combination of binary differences based around exclusive-OR (XOR) gates, mathematical morphology and a crucial threshold setting. This is a robust technique and can be applied many areas from leak detection to movement tracking, and further augmented to perform additional functions such as watermarking and facial detection. Fire is a severe problem, and in areas where traditional fire alarm systems are not installed or feasible, it may not be detected until it is too late. Shown here is a way of adapting the traditional Morphological Scene Change Detector (MSCD) with a temperature sensor so if both the temperature sensor and scene change detector are triggered, there is a high likelihood of fire present. Such a system would allow integration into autonomous mobile robots so that not only security patrols could be undertaken, but also fire detection.

1. Introduction

This continues on from the work presented in [1 – 4] inspired by [5 – 6], for various applications from intruder detection to the detection of leaky pipes. The hope being that this technology will be eventually incorporated into an autonomous mobile robot that could perform patrol duties as in [4]. Morphological Scene Change Detection (MSCD) previously developed was for the detection of intruders [3] and then the result was passed through a skin-filter in order to detect human facial features [4]. This version would now allow people to detect fire where it may not be feasible to place an ordinary smoke detector such as in building sites where the system is not yet in place, or on agricultural land to protect crops from fire during warm periods. Another application is to provide extra help for the elderly or disabled, a patrol mentioned in [7] could operate inside or outside a building and could detect fires that are not yet significant enough in magnitude to trigger a conventional fire alarm system, the aim being to stop criminals from praying on the venerable. As with previous work, this technology will be used to augment existing safeguards rather than replacing them and can be easily moved from location to location via the robot or a mobile platform.

Video surveillance systems with detection tasks usually include a module aimed at detecting relevant changes in a guarded environment [3, 5 – 6]. If the input to the Morphological Scene Change Detector (MSCD) module is pre-processed, it can improve the accuracy of the result; in this case a threshold is applied to the incoming greyscale image in order to remove unwanted intensity information to assist in extracting certain information from the scene. A similar process was carried out in [4] where the skin tone of the intruder was determined to assist in facial detection. By extracting features here that result from a change due to a fire burning, this can be related to either a high intensity from the brightness of the flame or as the environment is consumed by the flames. The latter will result in the environment undergoing a shape change as the material burns, allowing for it to be picked up by original algorithm. As in the original system [3], the output of the MSCD circuitry should undergo a noise reduction filter to assist in removing any unwanted data, helping to improve the accuracy of the counting used to trigger the alarm system and compensate for any camera movement during the process. This will be done again using a combination of mathematical morphology [1].

Figure 1 shows the basic results of a simple situation, showing the reference image, the current image, the XOR difference between the two and the mathematical morphology to assist in cleaning away the noise from camera mismatch. The threshold setting is a crucial step because different threshold values provide extremely different results, in this situation, different thresholds will be experimented with in order to produce the optimum level with which to detect the flames when burning various materials. Using video for fire detection is not new [8 – 9]; various techniques have been used previously used in order to detect flames whilst others focus on the smoke that they emit. In this situation however we will be augmenting the system with an infrared thermometer so that both visual information and temperature can be used as triggering mechanisms.

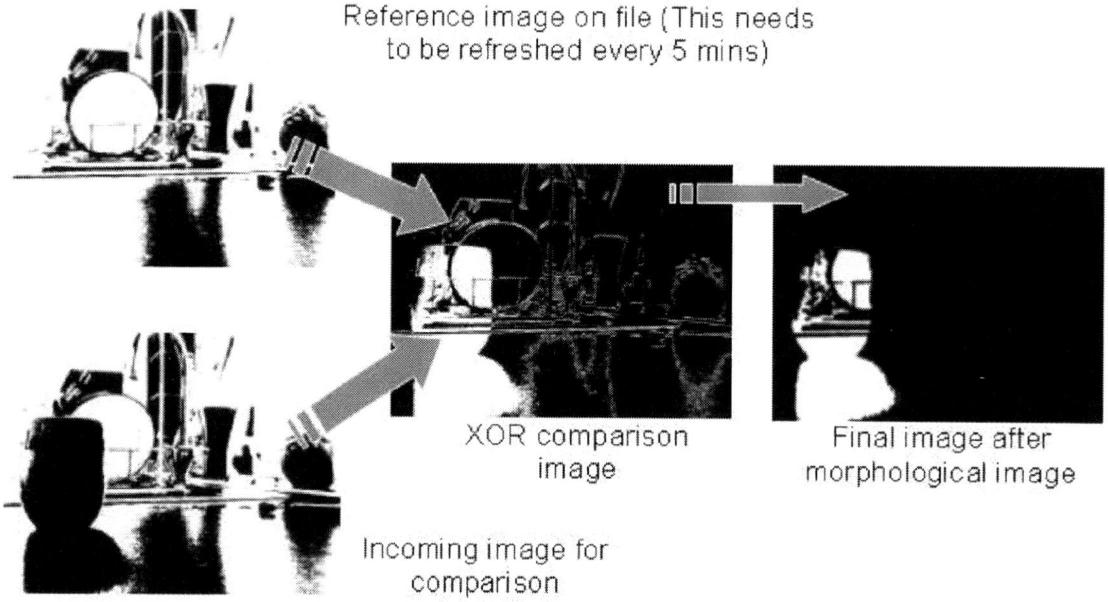

Figure 1. The basic process behind Morphological Scene Change Detection

Fire is one the disasters that occurs in everyday life and causes losses in both terms of materials and life. In fire fighting, fires are identified according to one or more fire classes. Each class (Figure 2) designates the fuel involved in the fire, and thus the most appropriate extinguishing agent. The classifications allow selection of extinguishing agents along lines of effectiveness at putting the type of fire out, as well as avoiding unwanted side-effects. For example, non-conductive extinguishing agents are rated for electrical fires, so to avoid electrocuting the firefighter. Figure 2 shows a table of the different classes according to the American standard. "Ordinary combustible" fires are the most common type of fire, these occur when a solid, organic material such as wood, cloth, rubber, or some plastics become heated to their ignition point. At this point the material undergoes combustion and will continue burning as long as the four components of the fire tetrahedron (heat, fuel, oxygen, and the sustaining chemical reaction) are available. In this piece of work the main fuel sources will be plastic, polystyrene, wood, paper, kerosene and vegetable oil, the latter two being classes B and K [10].

The remainder of this paper is structured thusly: Section II provides an overview of the adaptations to the previously designed systems; Section III discusses the simulation results and the triggering levels, whilst Section IV concludes the paper and discusses the future of this research.

CLASSES OF FIRES	TYPES OF FIRES	PICTURE SYMBOL
A	Wood, paper, cloth, trash & other ordinary materials.	
B	Gasoline, oil, paint and other flammable liquids.	
C	May be used on fires involving live electrical equipment without danger to the operator.	
D	Combustible metals and combustible metal alloys.	
K	Cooking media (Vegetable or Animal Oils and Fats)	

Figure 2. The different classes of fire according to the American standards [10]

2. System Design

The system required very few modifications in order to make it into a flame detection system; these changes are shown in Figure 3 and 4. At the Register Transfer Level (RTL), very little modifications were required and these changes are highlighted in red. The red connection shows how the temperature sensor is added to the system; this is achieved by using a 10 bit unsigned data line that is connected to the "trigger_analysis" block, also highlighted in red. The remainder of this level remains the same as in the original system from [3]. Figure 4 shows the updated counter block at the gate level; here the temperature sensor data is connected to a comparator module. The comparator compares the incoming temperature against a value contained within the value tag set by the user, if the incoming value exceeds this, the output of the comparator is high, otherwise it remains low. The output from this comparator is then sent into an AND gate with the output from the image data trigger comparator, if both are high then the fire alarm mechanism is activated and sent out via a binary port.

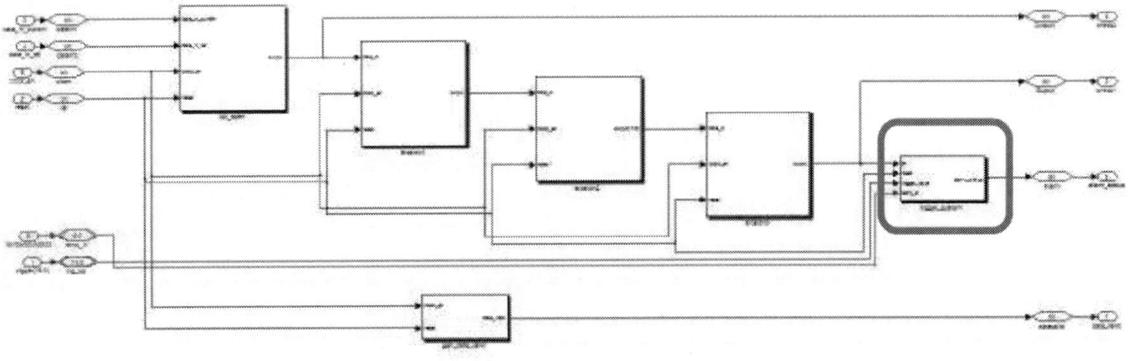

Figure 3. The new RTL level for the system

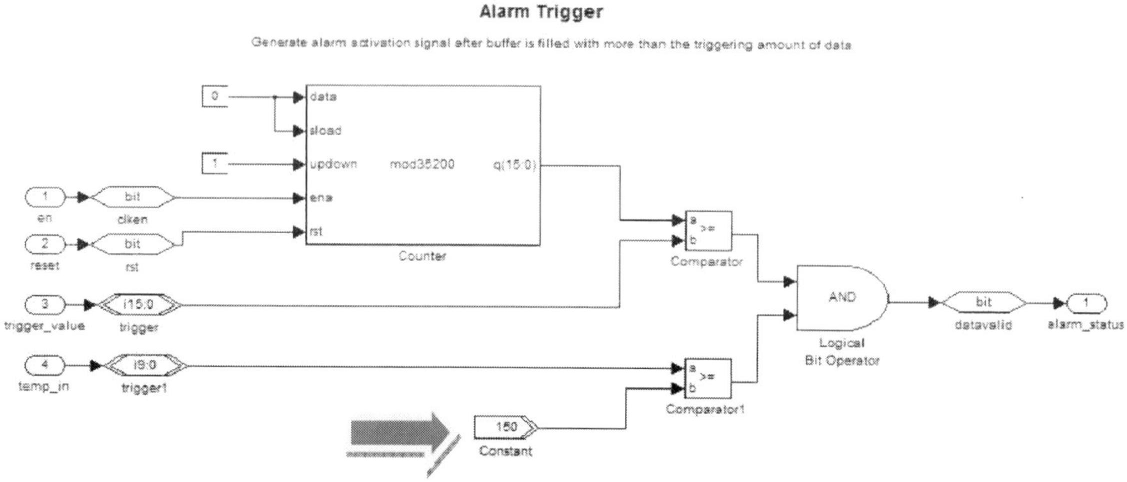

Figure 4. The new counter module for the system

3. Simulation Results

When undertaking the simulations, there were three cases to consider:

(i) From the reference image to the fire image (Figure 8)

(ii) Successive frames of fire images (Figure 5)

(iii) From the reference image before the fire to the image after the fire (Figure 6)

The first task was to set an appropriate thresholding level and as such different 8-bit thresholds were applied after the greyscale image had been input. The ranges considered ranged from 100 to 240 and these differences are shown on some example test data in Figure 7. As can be seen from figure (a), at the lower levels there is still a lot of noise in the image from camera mismatch which would interfere with the triggering mechanism and increase the number of false alarms. On the other hand, shown in figure 7b), when the level is set too high it begins to erode the image down too much and so for when the flame is not as obvious as in Figure 5, the flame data would be completely removed. The final threshold value to be used was selected to be 185.

When measuring the temperature data, it was found to range greatly from 90°C for paper at the start of its combustion to 340°C towards the height of its burning. The kerosene recorded the highest temperature using the sensor at 356°C at a distance of 5m whilst the body recorded a temperature of 21 – 35°C at various distances. Human body temperature was used as a control as the burning experiments took place outside and since the temperature varied from one session to the next, a control was required. As a result of these findings, the threshold for the temperature sensor was set to 150°C as nothing detected during the tests reached this temperature apart from the fire. If the environment does contain items above this setting, the temperature can be changed by modifying the value tag highlighted in Figure 4.

Figure 8 shows a complete series of tests and resultant images where an object is shown in its pre-burnt state and compared to when it is burning. This image was selected as the actual flame data is small and will provide a measure as to how accurate the system is. After the threshold and XOR gate was applied, the noise reduction filter was able to screen all of the unwanted information away and left just the flame data which is shown in Figure 8d). When all the data was analysed from the 3 different test cases for the 6 different combustibles, the most suitable lower threshold level was found to be 0.1% pixel change in the image. The images in these tests were 648 x 484 so that means that the trigger is set to 188 pixels and can be easily calculated for other image sizes. This can of course be set higher if the user is looking for a much more substantial sized flame.

Figure 5. Comparison between two successive frames of fire image (kerosene)

Figure 6.Comparison between before the fire and afterwards (plastic)

Figure 7. Different 8-bit greyscale threshold levels, (a) 100; (b) 240;

Figure 8. Successful detection of flame via image and temperature sensing, (a) Original reference image; (b) Current image with flame and synchronous temperature data input; (c) XOR result prior to mathematical morphology processing; (d) Final image

4. Conclusions

This paper has shown that it is possible to augment the Morphological Scene Change Detector to be able to detect the movement of a flame and additionally verify it by the use of a infrared temperature sensor. The system is accurate and can not only detect large areas of flame but relatively small ones too, which would be advantageous for if a fire has just started. The issue about pixel compensation for animals and wildlife outside is solved by the fact that the temperature sensor is present, this prevents small occurrences moving across the scene that may be a cat, dog or bird from triggering the system as it would not be warm enough to trigger the temperature sensor. Future work will now consist of combining this system will all the other MSCD based systems to form the core of the autonomous mobile robot software so that its construct can begin.

Acknowledgements

Dr Tickle would like to thank Assistant Lecturer Mr Kennedy Iroanusi for his assistance in covering some of his teaching duties in order to enable him to undertake the above given research.

References

[1] Tickle, A. J., Smith, J. S. and Wu, Q. H. (2007) Development of Morphological Operators for Field Programmable Gate Arrays, *Journal of Physics: Conferences Series 76(1)*, 012028

[2] Tickle, A. J., Smith, J. S. and Wu, Q. H. (2007) Feasibility of a Multifunctional Morphological System for Use on Field Programmable Gate Arrays, *Journal of Physics: Conferences Series 76(1)*, 012055

[3] Tickle, A. J., Smith, J. S. and Wu, Q. H. (2008) Feasibility of a Portable Morphological Scene Change Detection Security System for Field Programmable Gate Arrays, *Proc. SPIE 6978*, 69780V

[4] Tickle, A. J., Smith, J. S. and Wu, Q. H. (2008) A Grayscale Skin and Facial Detection Mechanism for Use in Conjunction With Security System Technology Via Graphical Block Methodologies on Field Programmable Gate Arrays, *Proc. SPIE 6978,* 69780Q

[5] Stringa, E. (2005) Morphological Scene Change Detection Notes, *Institute for Systems, Informatic, and Safety*, pp 1 – 12

[6] Stringa, E. and Regazzoni, C. S. (2000) Real Time Video-Shot Detection for Scene Surveillance Applications, *IEEE Trans. Image Processing (9),* pp 69 – 79

[7] Tickle, A. J., Meng, Y. and Smith, J. S. (2010) Simulation of a Dead Reckoning Embedded System Security Patrol Robot for Deployment inside Structures and Buildings, *Proc. SPIE 7833*

[8] Kang, N., Tung, T. X, and Kim, J. M. (2013) Efficient Video-Equipped Fire Detection Approach for Automatic Fire Alarm Systems, *Optical Engineering 52(1)*, 017002

[9] Budi, T. A. and Suwardi, I. S. (2011) Fire Alarm System Based-on Video Processing, *2011 International Conference on Electrical Engineering and Informatics*

[10] http://njfireequip.com/equipment-services/fire-extinguishers/know-your-fire-extinguisher-abcs/ (accessed May 2013)

Journal of Physics: Conference Series **450** (2013) 012055 doi:10.1088/1742-6596/450/1/012055

Wearable sensor network for health monitoring: the case of Parkinson disease

M Pastorino[1], MT Arredondo[1], J Cancela[1] and S Guillen[2]

[1] Life Supporting Technologies, Universidad Politécnica de Madrid, Madrid 28040 Spain
[2] Tecnologías para la salud y el Bienestar - Universidad Politécnica de Valencia, Valencia 46022, Spain

E-mail: mpastorino@lst.tfo.upm.es

Abstract. The aim of this paper is to show how wearable sensors can be useful in health solutions, improving the continuous monitoring and management of patients. This paper is focused on the available solution for motion analysis, providing a description of human motion features which can be measured through the use of wearable sensors. Moreover, this paper presents an example of wearable solution used for the objective assessment of Parkinson's disease symptoms. Results indicate that wearable sensors are useful for the objective evaluation of motor fluctuation and clinicians can benefit from these tools in order to adjust and personalise the treatment.

1. Introduction

The importance of bio-medical engineering and wearable solutions for healthcare is growing during the last decades thanks to the improvement and the availability of many devices and technological solutions, as well as their cost reduction. As a consequence, the interest in applying and combining those technologies to the monitoring and treatment of several kinds of diseases has increased.

This paper addresses the problem of continuous monitoring of patients using a wearable sensors network to improve the following and management of them, with a particular attention to those affected by Parkinson's disease (PD).

1.1. Parkinson disease

Parkinson's Disease (PD) is a neurodegenerative disorder of the central nervous system that affects motor skills and speech [1]. The primary biochemical abnormality in PD is a deficiency of dopamine due to degeneration of neurons in the substantia nigra pars compact [2]. The characteristic motor features of the disease include bradykinesia (i.e. slowness of movement), tremor, rigidity (i.e. resistance to externally imposed movements), flexed posture, postural instability and freezing of gait. Furthermore, PD is usually characterised by the loss of normal prosody of the speech [3].

According to the World Health Organisation (WHO]) there are more than six million people worldwide affected by PD. The syndrome typically appears around the age of 60. It affects Europeans and North Americans more often than Asians or Africans and it is more common in men than in women. PD affects about 2% of the population over the age of 65 years, figure that is expected to double by 2020. For those reasons, PD poses a significant public health burden, which is likely to increase in the coming years [4].

Technology in general and specifically wearable sensors' solutions might be an affordable alternative for PD's patients' treatment and management [5]. The development and integration of specific sensors for the remote health status monitoring, are been able to provide reliable data for the qualitative and quantitative assessment of people suffering from neurodegenerative diseases [6]. This new scenario is expecting to provide in the future a remarkable improvement in patients' management as well as a substantial cutting-off of the economic burden generated by the disease. New technologies allow monitoring the evolution of the disease through the employment of a wide range of wearable and user-friendly micro-sensors. Moreover, the last advances in data processing and data mining algorithms is bound to provide more accurate information about the diverse aspects of PD evolution [7].

1.2. Sensors for motion analysis
Over the past decades various technologies, methodologies and systems have been proposed for the monitoring and the assessment of the Parkinson's disease. A significant number of studies investigated the characteristic parameters of the different symptoms of PD. Others focused on the evaluation and quantification of the patients' motor status by the use of computerized motion tests (e.g. handwriting, inserting pegs, and games). The main goal of these studies was the objective assessment of PD parameters through the motion analysis, using ICT technologies. Sensors play an important role in this steps, considering that the technology associated to them must be able to assure reliable data without interfere with the actions performed by the patients during the motion analysis test.
Table 1 describes some features of the human motion, as well as the characteristics which can be measured through the use of wearable sensors, with a specific focus on PD motion analysis.

Table 1. Parkinson's disease – wearable sensors for human motion related measurements.

Features	Characteristics	Sensor
Gait	Speed of Locomotion Variability of the gait Rigidity of legs	Accelerometer
Posture	Trunk inclination	Gyroscope
Leg movement	Speed Motion sensor Motion sensor Motion sensor Length of Step Step Frequency Stride	Accelerometer
Hand Movement	Speed Angle Amplitude	Accelerometer + Gyroscope
Tremor	Amplitude Frequency Duration Asymmetry	Accelerometer
Fall	Fall Detection	Accelerometer
Freezing of Gait	Leg movement analysis	Accelerometer
Levodopa induced Dyskinesia	Duration Angle amplitud Intensity	Accelerometer + Gyroscope Inertial sensors
Bradykinesia	Speed Duration Frequency	Accelerometer + Gyroscope
Aphasia	Pitch	Microphone

The accuracy of measurements of the parameters above described depends on several technical features that are often in conflict with other needs such as usability, wearability, technical feasibility and the social acceptance of the devices used by the subjects. In Table 2 a description of these desirable properties along with their conflicts is presented.

Table 2. Wearable sensors desirable properties & conflicts

Properties	Conflict
Small sensor	The size of sensor is definitely an important factor, especially for portability and mobility matters. However, small sensors may not have enough room for long-lasting battery or storage capacity.
Smart sensor	Sensors possessing many characteristics are often bigger in size, expensive and consume more power
Sensor storage capacity	Due to a limit in storage capacity, sensors have to upload data frequently to the data personal server. So it is important to employ a good wireless communication technology that does not drain excessive power from the sensors.
Sensor processing capability	Because sensors do not often have large processing capability, they may not be able to process all data before the upload to the personal server. This means that large amount of raw data should be stored and eventually sent. Therefore it is important to have an efficient communication channel.
Sensor communication range	Whilst sensors are only able to communicate over short range, it is crucial to define a specific radius of action.

2. Systems for Parkinson's disease monitoring

Most of the research work carried out in the field of PD monitoring focuses on the assessment of the motor status of PD's patients, focused on the development of a wearable system able to objectively quantify the severity of the motor disturbances using motion sensors [8]. An important number of these studies are based on the study of various parameters of motor behaviors [9], while others are focused on the identification of the symptoms fluctuation along the day, combining the objective evaluation of the motor symptoms, through the data gathered from the selected sensors, usually accelerometers and gyroscope. Some research groups are also committed to use electromyogram (EMG) or voice analysis [10]. Even though many advances have been done in the last years, it must be said that there is still a lack of an all-inclusive system able to provide reliable assessment of the status of PD patients being at the same time economically affordable.

2.1. PERFORM system

PERFORM is a project partially funded by the European Commission under the Seventh Framework Program, aiming at providing an innovative and reliable tool that is able to monitor and evaluate motor neurodegenerative disease patients, such as PD patients. The PERFORM project is based on the development of an intelligent closed loop system that seamlessly integrates a wide range of wearable micro-sensors constantly monitoring motor signals of the patients. Data acquired are pre-processed by advanced knowledge processing methods, integrated by fusion algorithms to allow health professionals to remotely monitor the overall status of the patients, adjust medication schedules and personalize treatment [11–15]. Personalization of treatment occurs through PERFORM's capability to keep track of the timing and doses of the medication and meals that the patient is taking.

PERFORM moitoring System

The wearable device used to recording the motor signals consists of a tri-axial wearable accelerometers' network (Figure 1) used to record the accelerations of the movements at each patient limb, one accelerometer and gyroscope (on the belt) used to record body movement accelerations and angular rate, and a data logger (also located on the belt) that receives and stores all recorded signals in a SD card (Figure 2).Sensors allow the system detecting and quantifying a wide range of symptoms and measures of Parkinson's disease patient i.e. tremor, bradykinesia, dyskinesias and freezing of gait. All sensors transmit data using Zigbee protocol to a logger device, with 62.5 Hz sampling rate (16 milliseconds between samples).

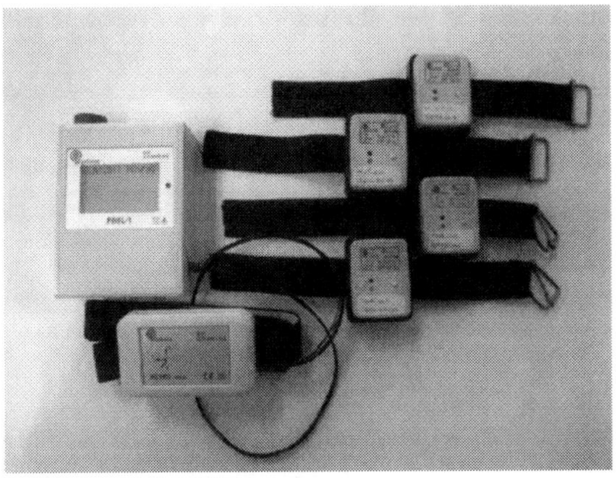

Figure 1. PERFORM wearable sensors network system

Figure 2. PERFORM system set up on patients

Special attention was paid in order to ensure the monitoring system usability and an easy placement for the patient and the caregivers. The sensor size is not bigger than a small matchbox. Sensors on the arms and legs are attached on specially designed elastic Velcro bands, which allow fixation to any wrist or ankle size. The sensors are placed inside an elastic pocket on the band, which secures it firmly on the patient body avoiding motion artifact due to cloth movement.

The sensor on the trunk is placed within a zipped elastic pocket on a vest. The vest is also equipped with Velcro straps to firmly adjust the sensor on the patient chest. The selected design allows the easy wearing and attachment/detachment of sensors.

Once the data has been stored in the SD card, the patient needs to connect the logger to a PC where the patient unit, called Local Base Unit (LBU) is responsible for the identification and quantification of the patient symptoms and the recording of other useful information for the evaluation of the patient status.

PERFORM symptom detection

The LBU processes patient signals gathered from the wearable sensors and detects the targeted patient symptoms (Tremor, Levodopa Induced Dyskinesia, Akinesia). Symptoms strictly related to the activity of the patients (Freezing f Gait, Gait and Bradykinesia) are detected using the results of the activity recognizer algorithm. For each symptom, a dedicated sub module processes the relevant signals (Figure3), detects the symptom episode and quantifies it into a severity scale from 0 to 4, according to the UPDRS [16] scale for PD's patients. Other features such as duration, frequently and amplitude might also be provided for further clinician review and system evaluation.

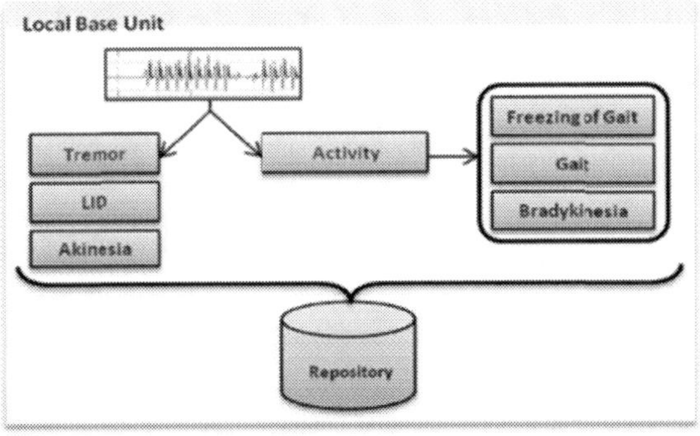

Figure 3. PERFORM LBU schema

PERFORM System Evaluation

PERFORM followed an iterative design process. Three different phases where designed involving 92 PD patients and 20 health subjects. On each phase of the design raw signals were collected using the wearable sensor network.

- **Phase I:** A preliminary version of the system was tested with 20 healthy subjects. Also this initial phase was used to test the technical performance of the platform in terms of data transmission and communication, identify bugs and redesign the system.
- **Phase II:** The second phase was carried out in the hospital under the supervision of the doctors. The goal of this phase was to validate not only the technical performance of the system but also the clinical compliance of the system. A total number of participants of 36 PD patients were involved.
- **Phase III:** It consisted in the evaluation of final PERFORM prototype. In this last phase the final version of the system was tested in everyday practice at patients' home. The total number of participants was 44 PD patients plus 12 patients with Parkinsonisms (Parkinsonism is any condition that causes a combination of the movement abnormalities seen in Parkinson's disease such as tremor, slow movement or muscle stiffness).

3. Results

PERFORM project has released promising results in patients monitoring and status assessment. The system has already been tested in hospitals in Navarra (Spain) in Ioannina (Greece) and in in Modena (Italy).

Obtained results suggest that the wearable sensor solution adopted is a valid solution able to detect PD symptoms based on motor signals, giving a high accuracy level of the overall status of the. Due to the short-term nature of the clinic trials that have been carried out it is difficult to determine the exact future impact on patient management; however it is possible to at least provide a preliminary quantification of the system performances. PERFORM has been designed to assess in an objective way the motor symptoms and the general status of the patients in order to:
determine the activity of the subject
provide a quantification of symptoms severity based on the clinical scales and present such an information to the physician through remote communication

The validation has proved that the PERFORM system is able to provide a very reasonable assessment of the daily activity of the patients using data classification techniques based on accelerometers and gyroscopes. PERFORM system is able to detect and quantify Parkinson's disease symptoms with the following results:

- LID severity classification accuracy: 93.73% [17]
- Bradykinesia severity accuracy: 86% [15]
- Tremor severity accuracy: 87 % [18]

As well as it has been developed a special module for the assessment of the gait parameters i.e. step frequency, velocity, arm swing frequency and entropy of the gait signal [14] and for the detection of the Akiniesia, used as an excellent discriminator of the ON/OFF status, reaching a good correspondence (88.2 +/- 3.7 %) between patient's diary and ON/OFF phases identified by the system.

4. Conclusion

This paper has presented how the continuous monitoring of patients can be achieved using wearable sensors network to improve the following and management of such subjects. Moreover, an example of wearable sensor network for patients affected by Parkinson's Disease has been provided. Clinical assessment is usually done through the analysis of patients' diaries and short punctual tests during the medical examination, this practice cannot give an overall and objective evaluation of the patient's status, as a result, treatment is not adjust following objective data, causing a causing that the treatment modifications are not undertaken in time.

On the other hand, wearable sensors solutions are able to provide quantitative and objective information about the motor complications. Data gathered from sensors, will be used in order to give to clinicians an overall and objective clinical status about each patient, so treatment can be adjust for any case and treatment modification can be applied according to changes in the symptoms fluctuation.

5. Acknoledgment

Authors thank the PERFORM consortium for their contribution to this work, especially the University Clinic of Navarra (Spain), Dr Alexandros T. Tzallas and Dr. Markos G. Tsipouras of the University of Ioannina (Greece), the Nuovo Ospedale Civile di Modena (Italy), the University of Bologna (Italy) and the Technical University of Madrid (UPM - Spain).

References

1. Jankovic J. Parkinson's disease: clinical features and diagnosis 2008. Journal of neurology, neurosurgery, and psychiatry .;79(4):368–76.

2. DeLong MR, Wichmann T. 2007 Circuits and circuit disorders of the basal ganglia. Archives of neurology ;64(1):20–4.

3. Darkins AW, Fromkin VA, Benson DF. 1988 A characterization of the prosodic loss in Parkinson's disease. Brain and Language 315–27.

4. Noyes K, Liu H, Li Y, Holloway R, Dick AW 2006. Economic burden associated with Parkinson's disease on elderly Medicare beneficiaries. Movement disorders official journal of the Movement Disorder Society, 21(3):362–72

5. Lawo M, Herzog O 2011. Wearable computing for medical applications: Personal health system for Parkinson's disease patients.8th International Conference Expo on Emerging Technologies for a Smarter World. IEEE; 2011. p. 1–5.

6. Synnott J, Chen L, Nugent CD, Moore G 2011. WiiPD An approach for the objective home assessment of Parkinson's disease . In: 2011 Annual International Conference of the IEEE Engineering in Medicine and Biology Society. IEEE; 2011. p. 2388–91.

7. Mera TO, Heldman DA, Espay AJ, Payne M, Giuffrida JP 2011. Feasibility of home-based automated Parkinson's disease motor assessment. Journal of neuroscience methods.

8. Rahimi F, Duval C, Jog M, Bee C, South A, Jog M, et al. 2011 Capturing whole-body mobility of patients with Parkinson disease using inertial motion sensors: Expected challenges and rewards. Annual International Conference of the IEEE Engineering in Medicine and Biology Society. IEEE; 2011. p. 5833–8.

9. Bonato P, Sherrill DM, Standaert DG, Salles SS, Akay M 2004. Data mining techniques to detect motor fluctuations in Parkinson's disease. 26th Annual International Conference of the IEEE Engineering in Medicine and Biology Society.p. 4766–9.

10. Skodda S.2012 Analysis of voice and speech performance in Parkinson's disease: a promising tool for the monitoring of disease progression and differential diagnosis. Neurodegenerative Disease Management ;2(5):535–45.

11. Cancela J, Pastorino M, Arredondo MT, Pansera M, Pastor-Sanz L, Villagra F, et al 2011. Gait assessment in Parkinson's disease patients through a network of wearable accelerometers in unsupervised environments. Annual International Conference of the IEEE Engineering in Medicine and Biology Society. IEEE; 2011. p. 2233–6.

12. Pastorino M, Cancela J, Arredondo MT, Pansera M, Pastor-Sanz L, Villagra F, et al.2011 Assessment of Bradykinesia in Parkinson's disease patients through a multi-parametric system. Conference proceedings : Annual International Conference of the IEEE Engineering in Medicine and Biology Society. IEEE Engineering in Medicine and Biology Society. Conference:1810–3.

13.: Movement Disorder Society Task Force on Rating Scales for Parkinson's disease. 2003 The Unified Parkinson's Disease Rating Scale (UPDRS): Status and Recommendations. Movement Disorders Vol. 18, No. 7, 738-50

14.: M. G. Tsipouras, A. T. Tzallas, G. Rigas, P. Bougia, D. I. Fotiadis, and S. Konitsiotis,2010 "Automated Levodopa-induced dyskinesia assessment," Engineering in Medicine and Biology Society EMBC 2010 Annual International Conference of the IEEE, vol. 2010, pp. 2411–14

15.: G. Rigas, A. T. Tzallas, M. G. Tsipouras, P. Bougia, E. E. Tripoliti, D. Baga, D. I. Fotiadis, S. G. Tsouli, and S. Konitsiotis, "Assessment of Tremor Activity in the Parkinson's Disease Using a Set of Wearable Sensors," Information Technology in Biomedicine, IEEE Transactions on, vol. 16, no. 3, pp. 478–87

Al and Ga Nanoparticles Array Deposited on Silicon by UTAM for Sensors Applications

W M Salih[1], A J Mohammed[2], H A Kadhum[3], N A Hussain[4] and M A Hassan[5]

Al-Mustansiriyah University, College of Science, Department of Physics (Sensors & Solar Cells Group SSCG), Baghdad, IRAQ

[1]da_lal2009@yahoo.com, [2]spiritjabir@yahoo.com, [3]haider_mu2010@yahoo.com [4]nedalali@yahoo.com and [5]marwa_alganaby@yahoo.com

Abstract. Highly order array of Al and Ga nanoparticles have been synthesized by controlling the size pores of Ultra Thin Alumina Membranes masks (UTAMs) produced by anodic alumina, the deposition of array was carried out on the silicon substrate, the roughness and the size distribution of Al and Ga nanoparticles has been investigated, also shown how to organize into order 3Dstructures. The AFM results show that the prepared films have nano dimensions, pores and have large surface area which increasing the probability of gas, chemical and /or photon interaction then they could be employed for sensing device fabrication.

1. Introduction

During the last fifteen years, metallic nanoparticles have attracted the attention of many researchers due to the double possibility of localizing light in nanometric dimension and getting high intensity local fields (excitation of localized Plasmon Resonances LPRs).Their use as plasmonic sensors has found many nanotechnology applications in biosenseing nanocircuitry, spectroscopy, photovoltaic devices and microscopy[1,2]. Typical plasmonic sensor tools are made of Silver and Gold .Nanoparticles made of these metals show LPRs in the VIS-NIR range. Because of the enormous impact of these tools in biosensing applications and also because many biological molecules and matter respond in the UV part of the spectrum, it is necessary to investigate and design plasmonic tools able to work at shorter wavelengths[3,4]. Aluminum and Gallium have recently shown to be two possible candidates to cover this spectral rang. The purpose of this contribution is to show a numerical comparison through the DDA method (Discrete Dipole Approximation) between these metals and the conventional Gold and Silver when nanoparticles made of these material are located on dielectric substrate .The hemispherical particle shape will be taken as a basic geometrical model to develop this comparison, in part stimulated by the experiments performed [5,6].

Nanometer -sized particle are one of the main building blocks of nano devices. In the application of nano -particles as in the field of optical and electronic devices, it is usually desirable to fabricate ordered array of nanoparticles on substrates to improve the controllability and performance of the nano devices. The UTAM nano-pattering has been recognized as a suitable method for fabricating ordered nano-particle array on substrates with controllable structural parameters and tunable properties [7].

2. Experimental Work

Template of Ultra Thin Alumina Membranes (UTAM) first prepared by anodization, it was made under a constant cell potential condition (voltage= 40 volt) in oxalic acid solution, after that several chemical procedures was carried out to obtain UTAM masks [8,9,10], Then Aluminum and Gallium nanoparticles produce by physical vapor deposition using an electron beam technique (EB), small pieces of (Al) and (Ga) have impurity about 99.999, put in graphite crucible after that put the substrate of Si with UTAM mask in the holder and centered above the crucible, then the deposition chamber was evacuated until reach 2×10^{-6} mbar ,next the deposition procedure was carried out by apply current until deposition rate become (0.1 Å/sec), after the deposition process was finished the samples were ready for synthesizing and characterizing measurements. Atomic Force Microscopic measurement, these measurements consist of several analysis use Gwyddion and imager 4.62 programs

3. Results and Discussions

The atomic force microscope produces images for surfaces topography at very high magnification and it is even possible to observe the atomic structure of crystal. The aim of this study with the atomic force microscope is to obtain information about the surface formation, by analyzing the AFM images using the software "Gwyddion". Line profiles and imager 4.62 were employed to estimate the Roughness, Length, height distribution, three dimension average height and average diameter of the Al and Ga nanoparticles.

The AFM images were shown in Figure (1) indicate that the particles cover the surface cavities, leading to decrease in the surface roughness; this may be due to two reasons. Firstly, using the UTAM masks [11]. Secondly, the lower rate of deposition [12].

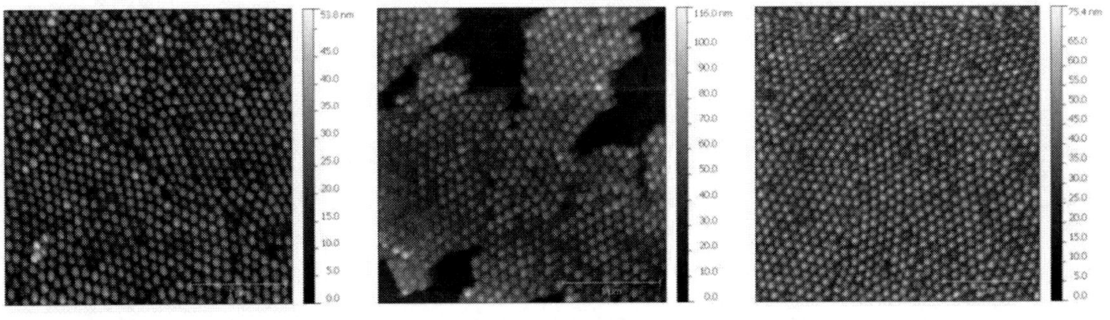

a b c

Figure1: AFM images plane view of Al nanoparticles. a)Al nanoparticles at deposition rate of 0.1(A°/sec). b) Al nanoparticles at deposition rate of 0.2(A°/sec). c) Ga nanoparticles at deposition rate of 0.1(A°/sec).

Journal of Physics: Conference Series **450** (2013) 012056 doi:10.1088/1742-6596/450/1/012056

The size value distribution of Al and Ga nanoparticles are illustrate in Figure. (2). While the length and 3D topography of Al and Ga nanoparticles shown in Figure (3)

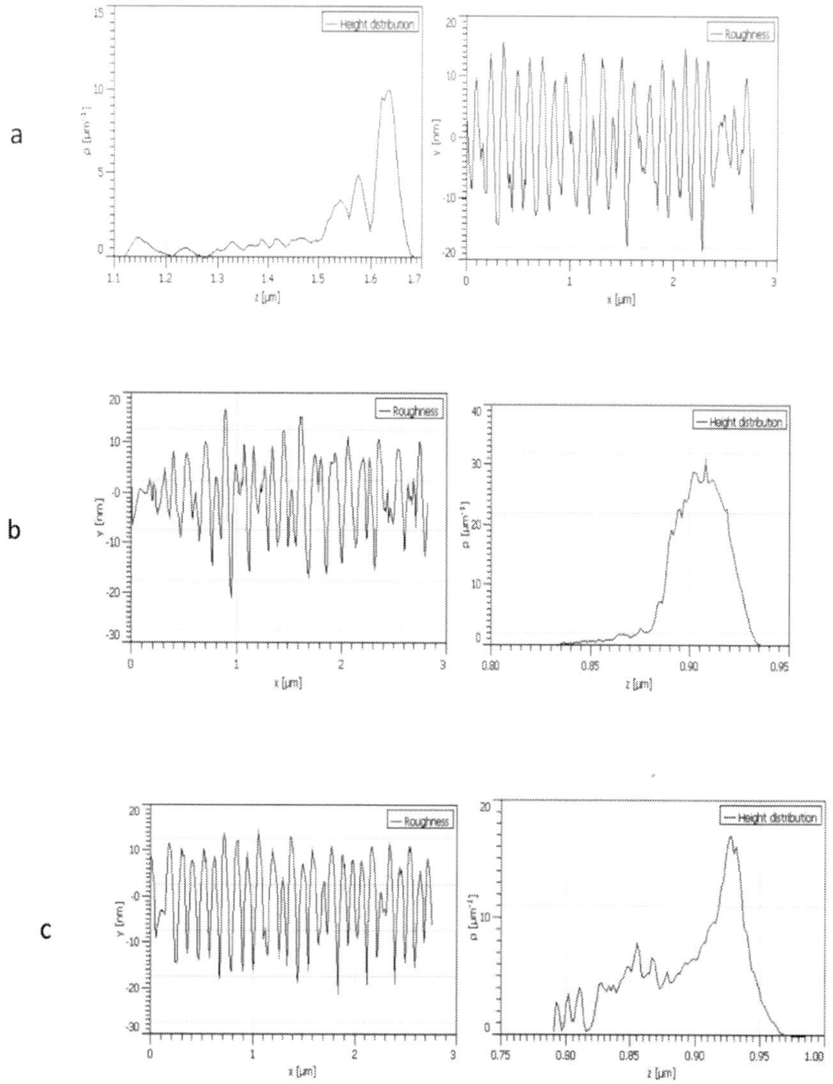

Figure 2: On the left column the roughness of Al and Ga nanoparticles.ON the right column high distribution a)Al nanoparticles at deposition rate of 0.1(A°/sec). b) Al nanoparticles at deposition rate of 0.2(A°/sec). c) Ga nanoparticles at deposition rate of 0.1(A°/sec).

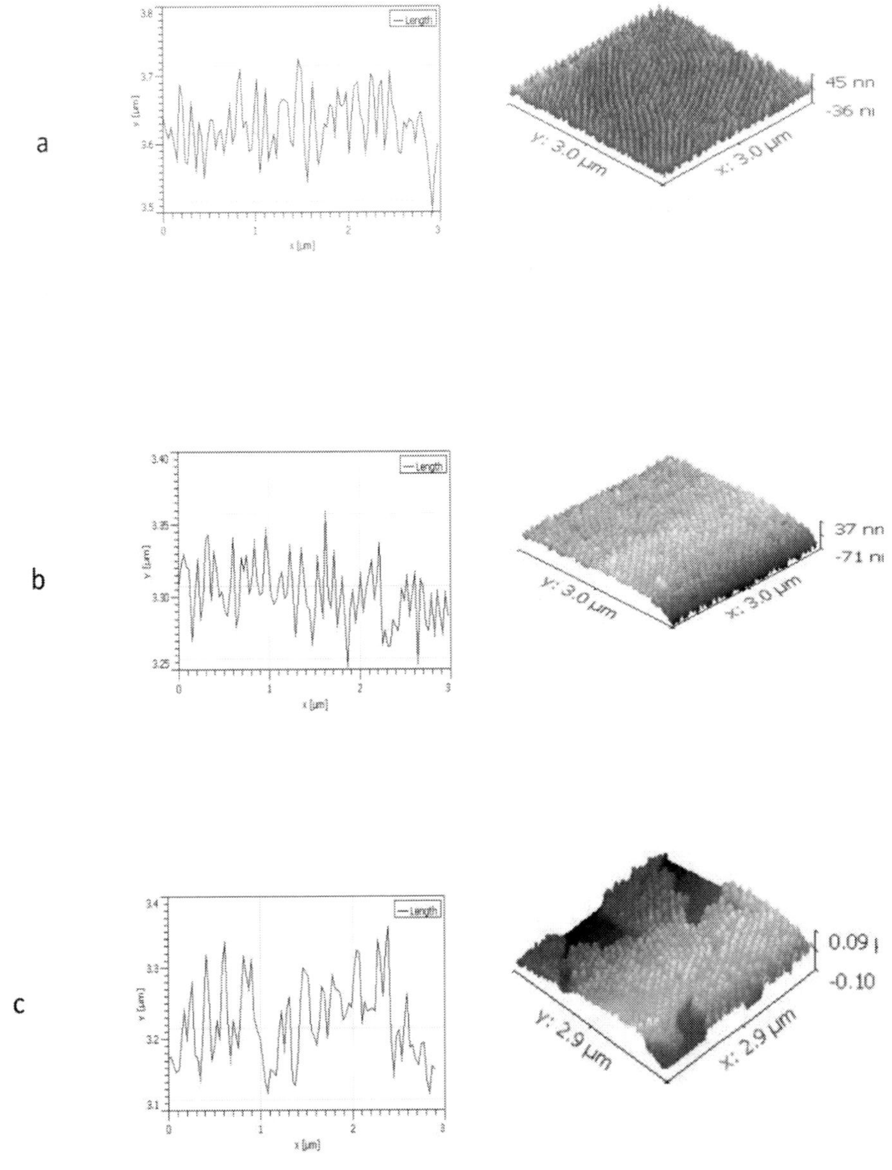

Figure 3: Length and 3D of Al and Ga nanoparticles. a)Al nanoparticles at deposition rate of 0.1(A°/sec). b) Al nanoparticles at deposition rate of 0.2(A°/sec). c) Ga nanoparticles at deposition rate of 0.1(A°/sec).

Figure (4) shows the histogram of distribution density, the results show the particle density across the film, since large size particulates have the higher angular distribution, we have observed that a huge number of Al and Ga nanoparticles (right column). Left column shows the granularity cumulation distribution.

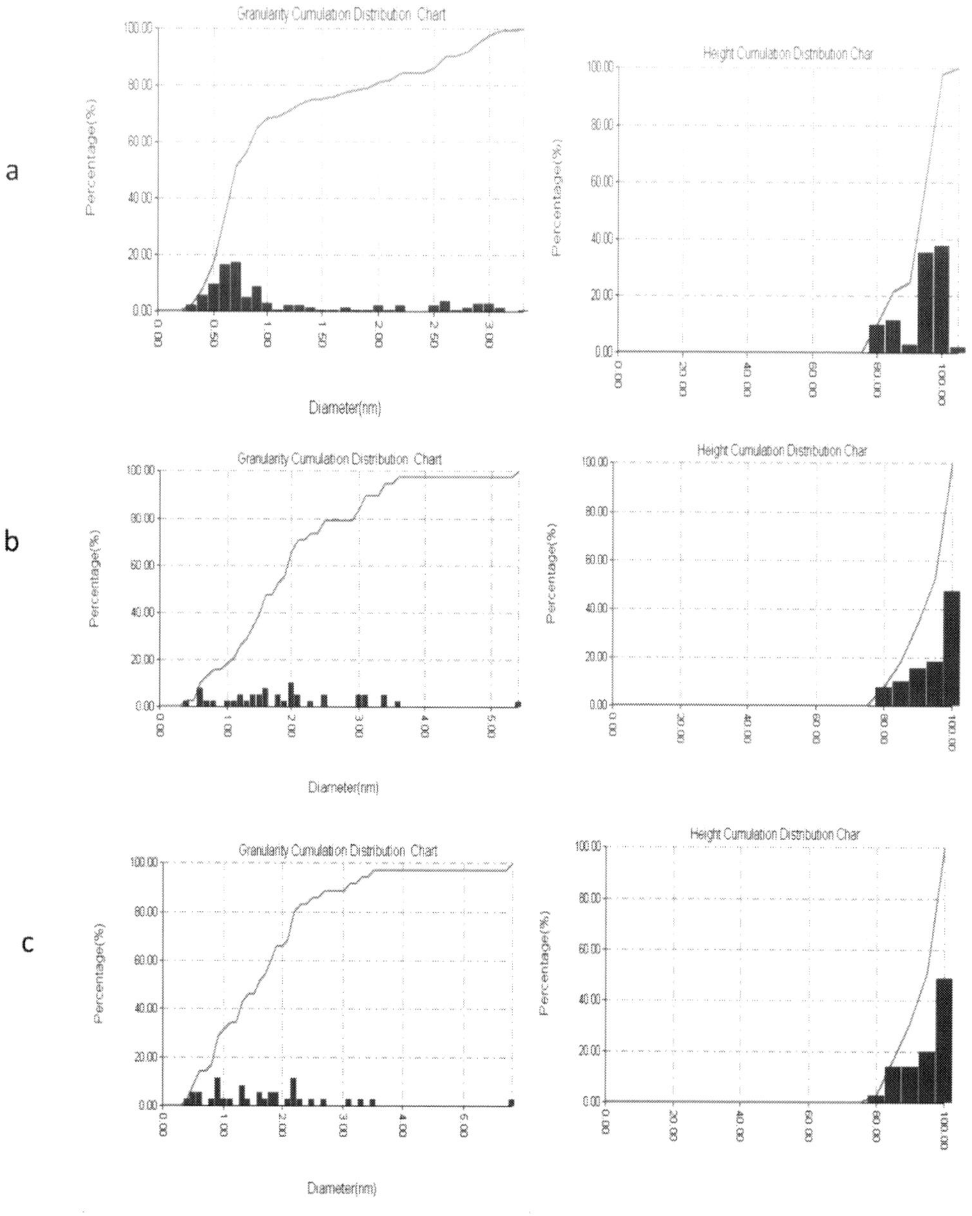

Figure 4: Histogram of the height and diameter is the function values distribution density. a) Al nanoparticles at deposition rate of 0.1(A°/sec). b) Al nanoparticles at deposition rate of 0.2(A°/sec). c) Ga nanoparticles at deposition rate of 0.1(A°/sec

Table (1) illustrates the value of height particle for Al and Ga nanoparticle respectively. The table show the average value of heights are (92.648 and 91.878 nm) for Al nanoparticles at deposition rates (0.1 and 0.2 Å/sec) respectively and (92.739 nm) for Ga nanoparticles at deposition rate (0.1 Å/sec).

Table 1: Height Cumulation Distribution a) Al nanoparticles at deposition rate of 0.1(Å/sec). b) Al nanoparticles at deposition rate of 0.2(Å/sec). c) Ga nanoparticles at deposition rate of 0.1(Å/sec).

a

Avg. Height:92.648 nm	**<=10% Height:0 nm**
<=50% Height:90.0000 nm	**<=90% Height:95.0000 nm**

Height(nm)<	Volume (%)	Cumulation(%)	Height(nm)<	Volume (%)	Cumulation(%)	Height(nm)<	Volume (%)	Cumulation(%)
80.0000	10.14	10.14	90.0000	2.90	24.64	100.000	37.68	97.83
85.0000	11.59	21.74	95.0000	35.51	60.14	105.000	2.17	100.00

b

Avg. Height:91.878 nm	**<=10% Height:80.0000 nm**
<=50% Height:90.0000 nm	**<=90% Height:95.0000 nm**

Height(nm)<	Volume (%)	Cumulation(%)	Height(nm)<	Volume (%)	Cumulation(%)	Height(nm)<	Volume (%)	Cumulation(%)
80.000	7.89	7.89	90.000	15.79	34.21	100.0000	47.37	100.00
85.000	10.53	18.42	95.000	18.42	52.63			

c

Avg. Height:92.739 nm	**<=10% Height:80.0000 nm**
<=50% Height:90.0000 nm	**<=90% Height:95.0000 nm**

Height(nm)<	Volume (%)	Cumulation(%)	Height(nm)<	Volume (%)	Cumulation(%)	Height(nm)<	Volume (%)	Cumulation(%)
80.000	2.86	2.86	90.	14.29	31.43	100.00	48.57	100.00
85.000	14.29	17.14	95.000	20.00	51.43			

4. Conclusion

From our results, we can conclude that the Al and Ga nano particles deposited on the silicon substrate to form nano structural thin films by using a simple physical deposition method. The AFM morphological studies show that the prepared films have large surface area, increasing the probability of interaction with the film materials (as chemical, gas or photon).

References:

[1] 1. Kreibig, U. and Volmer, M. *Optical Properties of Mtal Clusters* (Springer-Verlag,Heidelberg, 19

[2] T.L.Temple and.M.Bagnall,journal of Applied Physics 109,084343 (2011).

[3] Romain Quidant, Mark Kreuzer, Nature Nanotechnology, **5** (2010) 762.

[4] H.A. Atwater, A. Polman, Nat. Materials **9** (2010), 205.

[5] T. Kalkbrenner, U. Hakanson, A. Schädle, S. Burger, C. Henkel, V. Sandoghdar, Phys. Rev.L (2005) 200801.

[6] K. Lance Kelly, Eduardo Coronado, Lin Lin Zhao, and George C. Schatz, Journal Of Physical Che B **107** (2003) 668.

[7] Mustafa H Chowdhury, Krishanu Ray, Stephen K Gray, James Pond, and Joseph R Lakowicz, Ana Chemistry **81** (2009) 1397-403.

[8] P.C. Wu, T. Kim, A.S. Brown, M. Losurdo, G. Bruno and H.O. Everitt, Appl. Phys. Lett. **90** 103119.

Non-invasive measurement of cholesterol in human blood by impedance technique: an investigation by 3D finite element field modelling

Ekaterina Aristovich and Sanowar Khan

City University London, School of Engineering and Mathematical Sciences, Northampton Square, London EC1V 0HB, UK

E-mail: Ekaterina.Aristovich.1@city.ac.uk

Abstract. This paper concerns detection of particle concentration (e.g. cholesterol) in conductive media (e.g. human blood) by impedance technique. The technique is based on changes in the impedance measurement across a given conducting medium due to changes in the particle concentration. The impedance is calculated by calculating the current through the conducting media produced by electric field distribution between two electrodes. This is done by modelling and computation of 3D electric fields between the electrodes for known voltages applied between them using the well-known finite element method (FEM). The complexity of such FE models is attributed to particle distribution, their geometric and material parameters, and their shape and size which can be of many orders of magnitude smaller than the overall problem domain under investigation. This paper overcomes this problem by adopting an effective particle coagulation (aggregation) strategy in FE modelling without significantly affecting the accuracy of field computation.

1. Introduction

The impedance technique has been widely used in many areas (industry, chemical production, pharmacy, medicine, biosciences) and it has been proven to be reliable for imaging (EIT, ERT) [1-3] and quantitative analyses (BMI, fuel purity, impedance spectroscopy) [4-8]. Modelling of non-conductive particles in conductive solution has been studied for several decades. There are in existence analytical formulas and methods for modelling separate particles, particles with effective conductivity [4, 5] as part of various sub-modelling approaches. The strategy adopted in this paper is instead of taking an effective conductivity in a volume of interest, it is possible to introduce an effective aggregation scheme by which small particles are aggregated together to reduce scaling factor without significantly decreasing the FE solution accuracy (especially for high aspect ratio of particle/media conductivities) the problem.

2. Particles suspended in conductive media

The mathematical modelling methodologies developed in this work have been applied to non-invasive detection of cholesterol particles in human blood by impedance technique. This points to the development of a simple, non-invasive method for measuring total blood cholesterol in which the conductivity distribution of objects (e.g. cholesterol particles carried by lipoproteins) within a test volume (e.g. blood plasma) is measured in the form of electrical impedance. Human blood is a complex medium which is composed of cells of different sizes and shapes (red blood cells, white blood cells, platelets, etc.) distributed in a conductive aqueous solution (blood plasma). About 55% of the blood fluid is composed of blood plasma (mostly water – 91.5%) and the remaining 45% consists of various blood components (various blood cells and other inclusions) [9].This means from the electric field and conductivity point of view blood is essentially a non-homogenous and anisotropic medium. Specifically, this also means that the electrical properties of the whole blood (e.g. electrical conductivity, σ and the dielectric permittivity, ε) are very different from those of blood plasma, which contains no cells [10, 11].

Cholesterol is almost insoluble in water and it is transported in the blood stream by lipoproteins that are water-dispersible and carry cholesterol and triglycerides internally. It is important to note that a total cholesterol of 5 mmol/l corresponds to about 10^{12} lipoprotein particles (5-30 nm in size) in each mm^3 of blood. At the same time cholesterol particles have the highest contrast ratio in terms of electrical properties among other components in a given blood volume. This clearly shows a major challenge in taking into account individual particles for modelling purposes which would introduce enormous complexity and dramatically increase modelling times to an unacceptable level.

3. Mathematical models and its finite element realisation

3.1. Field modelling and impedance calculation

For a given potential distribution φ on the electrodes (Figure 1) the electric field distribution in the 3D problem domain Ω (x, y, z) between the electrodes is given by the following Laplace's equation:

$$\nabla^2\varphi = 0 \text{ in } \Omega \tag{1}$$

Under appropriate boundary conditions shown in Figure 1 the above equation is solved by finite element method (FEM) [12] in terms of electric potential φ for given conductivity (σ) and permittivity (ε) distribution in the problem Ω. It is assumed that the material properties are linear, piece-wise homogeneous and isotropic. Following the solution of equation (1), field intensity and flux density vectors **E**=–grad$V$ (V is the potential difference) and **D**=ε**E** can be calculated. From these the current I=**J**A (A is the cross-sectional area) is calculated using current density **J**= σ**E**. Finally, the impedance Z is calculated from the simple relationship between current (I) and voltage (V) between the electrodes:

$$Z = \frac{V}{I} \tag{2}$$

With frequencies lower than 100 kHz most living tissues are assumed being electrolytic conductors. At higher frequencies the dielectric properties of bio-tissue may dominate. The higher the frequency the closer tissue properties come to those of water. For the modelling studies the above field equation was solved using the commercial software package COMSOL [13].

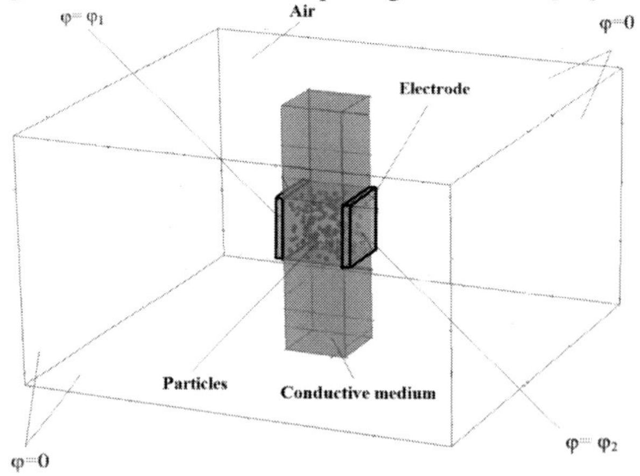

Figure 1. Basic 3D finite element (FE) model of particles in a conductive medium.

3.2. Finite element (FE) models

In order to tackle the problem defined above, full 3D FE modelling of current and potential distribution is required as the current is not confined to the measurement plane between the electrodes and changes in conductivity off this plane also contribute to calculated potential data. However, in order to establish confidence on some of the initial modelling results a number of 2D FE models were also set up. The basic 3D model used for simulation studies is shown in Figure 1 which shows the 'active' conductive medium between the electrodes (e.g. blood plasma, etc.) and beyond to account for any fringe field effects. In the absence of significant fringe field effects it can be assumed that only particles suspended between the electrodes contribute to the change in impedance between the electrodes. It also shows the boundary conditions under which equation (1) by FEM. Some of the initial 3D models were tested by comparing their results with those from corresponding 2D models with limited number of particles for which analytical results exist. This not only, to a certain extent establishes confidence on the on the 3D models but also justifies their use for subsequent simulation studies for realistic material and geometric properties.

3.3 Introduction of particles

For a given total cholesterol value it is not difficult to calculate the total volume of cholesterol particles V_c that would occupy a given volume of blood V_b in between the electrodes in Figure 1. The radius of each of the cholesterol particles r_c can then be calculated from the total number of cholesterol particles N_c contained in the volume V_c that need to be used for modelling purposes. However, as mentioned above it is not realistic and computationally efficient to do so in FE models because of the sheer number of particles that it represents. The alternative approach is to establish by modelling experiments the maximum number of cholesterol particles above which no significant change in the modelling results can be seen. This strategy of particle aggregation (coagulation) simplifies FE models (Figure 2) without significantly affecting the results of FE modelling. This can be clearly seen from Figure 3. In Figure 2 the same total volume of cholesterol particles V_c is represented by two different total numbers of particles, which are randomly distributed.

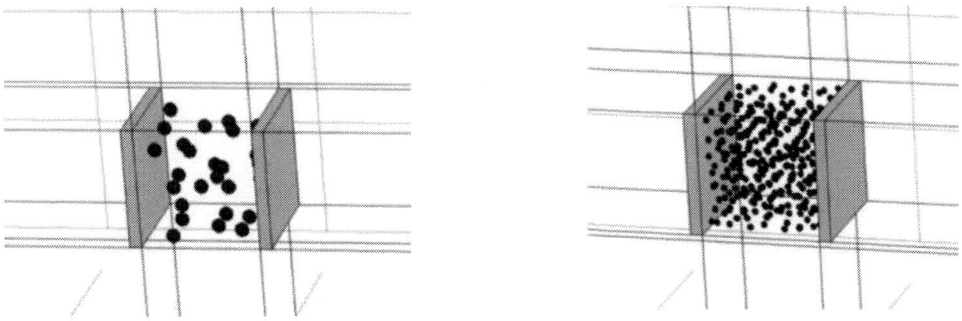

Figure 2. Approximation of a given total volume of particles by different particle numbers.

Figure 3 essentially shows that a given total volume of particles can be represented by variable number of particles and a threshold value of this number can be achieved (N_c=500) by FE modelling beyond which any further increase in the number of particles will not significantly affect the modelling results. This aggregation approach significantly reduces modelling complexity and save considerably computation time.

It should be noted that for each of the particle numbers used in Figure 3, modelling results were obtained for 10 different random distributions of these particles to quantitatively ascertain the effects of particle distribution on modelling results. Hence the upper and lower bounds of currents I for each of the particle numbers used for FE modelling. As can be seen the effect of the distribution of particles on modelling results decreases as the total number of particles increases.

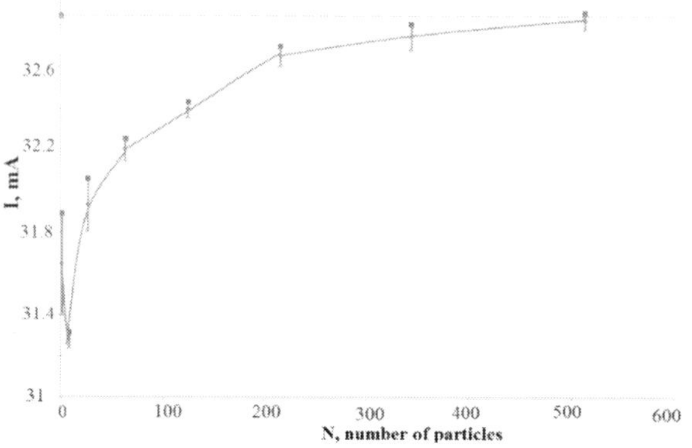

Figure 3. Number of particles used in FE models against modelling results (e.g. total current I between the electrodes, V_c=const.

4. Comparison of simplistic 3D modelling results with corresponding analytical results

Because of the complexity of 3D FE models involving a large number of particles in a conductive medium some of the initial modelling studies were carried out for simplistic conductivity distributions in the problem domain for which analytical solution and 2D modelling results exist.

4.1. Parallel cylinders and spherical particles aligned horizontally and vertically

It is clear that for FE modelling a given arrangement of spherical particles in 3D can be simulated in 2D by considering the particles are 'cylinders' extended infinitely in the z direction. A 3D equivalent of this 2D model would be cylinders of finite z-direction dimension in a 3D problem domain represented by the first set of models in Figure 4. The second set of models in Figure 4 shows similar arrangements for particles arranged vertically. So a 2D model with particles arranged vertically or horizontally equivalents to a corresponding 3D model with cylinders (which model the particles in 2D) arranged in the same way. The current values obtained from modelling of these scenarios are given in Table 1 below. As expected, a vertical arrangement of cylinders between the electrodes offers maximum 'resistance' to current paths.

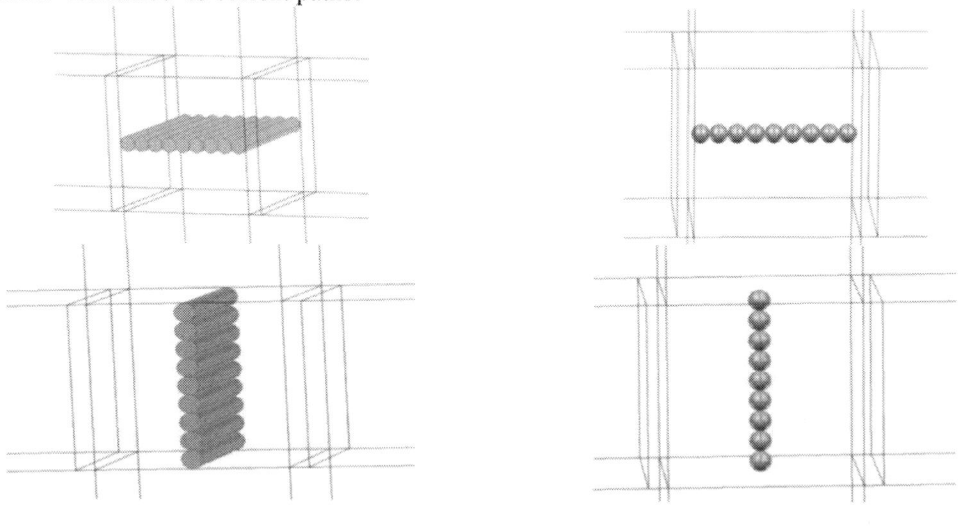

Figure 4. 3D equivalents of 2D models of particles (left) and their realistic 3D models (right).

Table 1. FE modelling results (calculated currents in mA) for cases shown in Figure 4

	3D equivalents calculated from 2D models	Cylinders in 3D models	Spherical particles in 3D models
Horizontally arranged	30.7	29.6	32.0
Vertically arranged	23.0	22.8	32.4

4.2. Extreme cases

Finally, two extreme cases were modelled in 3D: one with no suspended particles (Figure 5) and the other with one 'large' spherical particle equivalent in volume to the total volume of smaller particles suspended at the centre of the problem domain between the two electrodes (Figure 6). For the first case, it is expected to see the electric field and current density uniformly distributed in the volume between the electrodes, which can be seen in Figure 5. The theoretical value of this uniform current density (J=976.8 A/m^2) agrees very well with that obtained from modelling (graph in Figure 5).

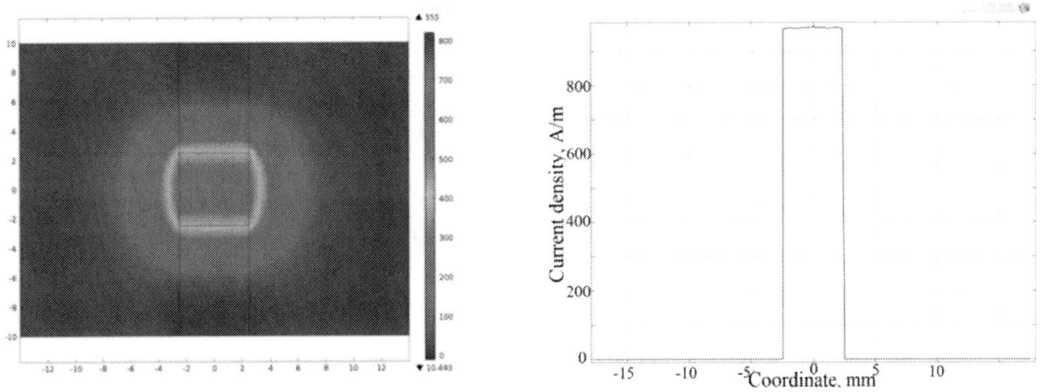

Figure 5. Electric field and current density distributions in the problem domain with no suspended particles.

Equivalent electric field and current density distributions for the second extreme case can bee seen in Figure 6 below.

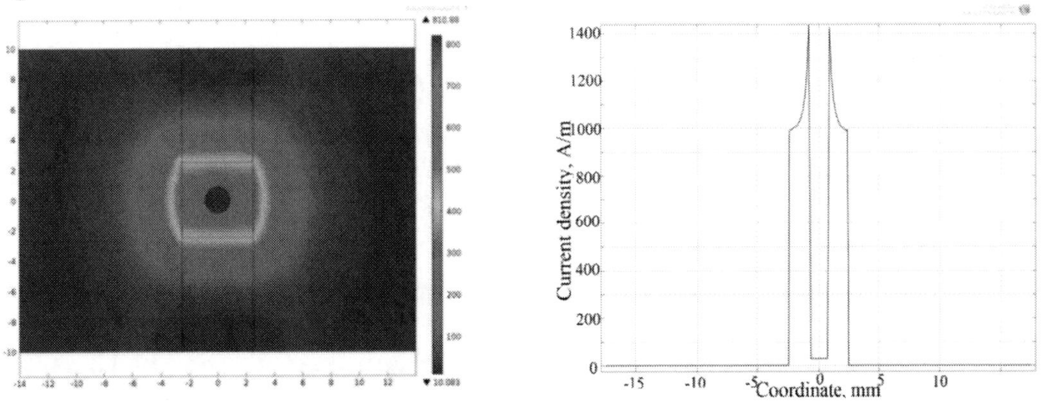

Figure 6. Electric field and current density distributions in the problem domain with one 'large' suspended particle placed at the centre of the problem domain between the electrodes.

The analytical current density values in the particle (J_{sp}=20 A/m^2) and in the surrounding conducting medium (J=1425.6 A/m^2) are in good agreement with those obtained from 3D FE modelling (graph in Figure 6).

5. Conclusions

Methodologies have been developed for FE modelling and computation of electric field distribution in conductive media containing suspended particles of various sizes, shapes and material properties. This has been used to model cholesterol particles in human blood which has so far shown encouraging results for detection of total cholesterol in the blood by impedance technique. This should ultimately lead to a practical device for non-invasive measurement of blood cholesterol.

Further work is being carried out to include even more blood components in the 3D FE models developed so far.

References

[1] Holder D 2005 *Electrical impedance tomography: methods, history and applications* (Bristol: Institute of Physics).
[2] Yungal SH. 1996 *Electrical methods in geophysical exploration of deep sedimentary basins* (London: Chapman & Hall).
[3] Scharfetter H, Brunner P, Mayer M, Brandstätter B, and Hinghofer-Szalkay H *IEEE transactions on bio-medical engineering* (New York) **52** 975-982.
[4] Grimnes S and Martinsen ØG 2008 *Bioimpedance and bioelectricity basics* (London: Academic).
[5] He B 2004 *Modeling and Imaging of Bioelectrical Activity Principles and Applications* (New York: Kluwer Academic/Plenum Publishers) 293-299.
[6] Tura A, Sbrignadello S, Barison S, Conti S and Pacini G 2007 *Biophysical Chemistry* **129** 235-41.
[7] Shiga T, Oshima Y, Kanai H, Hirata M, Hosoda K and Nakao K 2007 *13th International Conference on Electrical Bioimpedance and the 8th Conference on Electrical Impedance Tomography: ICEBI 2007, August 29th-September 2nd 2007, Graz, Austria* (Berlin ; New York: Springer) **17** 687-690.
[8] Grossi M, Lanzoni M, Pompei A, Lazzarini R, Matteuzzi D and Ricco B 2008 *Biosensors and Bioelectronics* **23** 1616-1623.
[9] Tortora G J and Derrickson B 2009 *Principles of anatomy and physiology*. (Hoboken, N.J.: Wiley) **2** 691.
[10] Hirsch F G, Texter C E, Wood L A, Ballard W C, Horan F E and Wright I S *American Society of Hematology. Blood* (Philadelphia, PA) **5** 1017-1035.
[11] Texter C E, Hirsch F G, Wood L A, Ballard W C, Horan F E, Wright I S, Starr D and Balandin B P *American Society of Hematology. Blood* (Philadelphia, PA) **5** 1036-1048.
[12] Rao S S 2005 *The finite element method in engineering* (Oxford: Butterworth-Heinemann).
[13] Saad Y and Schultz M H 1986 *SIAM Journal on Scientific and Statistical Computing* (Philadelphia: SIAM) **7** 856-869.

AUTHOR INDEX

AUTHOR INDEX